Studies in Systems, Decision and Control

Volume 153

Series editor

Janusz Kacprzyk, Polish Academy of Sciences, Warsaw, Poland
e-mail: kacprzyk@ibspan.waw.pl

The series "Studies in Systems, Decision and Control" (SSDC) covers both new developments and advances, as well as the state of the art, in the various areas of broadly perceived systems, decision making and control- quickly, up to date and with a high quality. The intent is to cover the theory, applications, and perspectives on the state of the art and future developments relevant to systems, decision making, control, complex processes and related areas, as embedded in the fields of engineering, computer science, physics, economics, social and life sciences, as well as the paradigms and methodologies behind them. The series contains monographs, textbooks, lecture notes and edited volumes in systems, decision making and control spanning the areas of Cyber-Physical Systems, Autonomous Systems, Sensor Networks, Control Systems, Energy Systems, Automotive Systems, Biological Systems, Vehicular Networking and Connected Vehicles, Aerospace Systems, Automation, Manufacturing, Smart Grids, Nonlinear Systems, Power Systems, Robotics, Social Systems, Economic Systems and other. Of particular value to both the contributors and the readership are the short publication timeframe and the world-wide distribution and exposure which enable both a wide and rapid dissemination of research output.

More information about this series at http://www.springer.com/series/13304

Haiyan Xu · Keith W. Hipel
D. Marc Kilgour · Liping Fang

Conflict Resolution Using the Graph Model: Strategic Interactions in Competition and Cooperation

Haiyan Xu
College of Economics and Management
Nanjing University of Aeronautics and Astronautics
Nanjing, Jiangsu
China

Keith W. Hipel
Department of Systems Design Engineering
University of Waterloo
Waterloo, ON
Canada

and

Centre for International Governance Innovation
Waterloo, ON
Canada

and

Balsillie School of International Affairs
Waterloo, ON
Canada

D. Marc Kilgour
Department of Mathematics
Wilfrid Laurier University
Waterloo, ON
Canada

and

Department of Systems Design Engineering
University of Waterloo
Waterloo, ON
Canada

Liping Fang
Department of Mechanical and Industrial Engineering
Ryerson University
Toronto, ON
Canada

and

Department of Systems Design Engineering
University of Waterloo
Waterloo, ON
Canada

ISSN 2198-4182 ISSN 2198-4190 (electronic)
Studies in Systems, Decision and Control
ISBN 978-3-030-08508-7 ISBN 978-3-319-77670-5 (eBook)
https://doi.org/10.1007/978-3-319-77670-5

Softcover re-print of the Hardcover 1st edition 2018

Printed on acid-free paper

This Springer imprint is published by the registered company Springer International Publishing AG part of Springer Nature
The registered company address is: Gewerbestrasse 11, 6330 Cham, Switzerland

To Ju, Sheila, Joan and Hong

Preface

The theory and practice of key advances in the ***Graph Model for Conflict Resolution (GMCR)*** are presented for strategically investigating real-world disputes arising in any field in which conflict takes place. Since humans are inherently ***competitive***, GMCR can be utilized to ascertain what is the best a particular decision maker (DM) can achieve given the social constraints of a conflict in which the DM dynamically interacts with others in terms of moves and countermoves as he or she seeks to satisfy her goals or value system. When trying to negotiate a climate change agreement, for example, each nation may act according to its own self-interests in order to fare as well as possible in the short term, by reducing its greenhouse gas emissions as little as possible. However, a country may then attempt to find out whether it can do even better if it ***cooperates*** with other nations to reach a fair climate change deal in which each nation cuts back very significantly in its greenhouse gas emissions in order for the nations of the world to do much better in the long run and thereby avoid the extreme consequences of climate change. Accordingly, the rich range of GMCR methodologies presented in this book and elsewhere can be employed in a highly competitive situation, in which all participants are out to satisfy their own goals, to ones in which there is a high level of cooperation when it is beneficial for DMs to form coalitions.

You, our valued reader, may wish to know if this book contains information that will be useful to your understanding and capability for resolving tough disputes in your domain of interest, which may range from personal disputes within a family to international trading conflicts among corporations and nations. If you are a ***researcher*** in multiple participant decision-making who wishes to refine and expand basic GMCR methodologies or to employ the latest advances in conflict resolution for tackling complex conflicts within a domain such as stakeholder satisfaction in land use development and planning, then this book should be of high value to you. If you are a ***teacher*** in operations research, systems engineering, or an applied field of application in which conflict takes place, you may wish to use this book as a course text at the upper undergraduate or graduate levels or else as a valuable informative reference in a course. If you are a ***mentor*** of students carrying out research at the Ph.D. or Master's level, or tackling tough problems involving conflict in challenging projects, you will find this book to be highly attractive for meeting your purposes. If you are a ***student*** studying conflict resolution and would like to investigate how nations or regions can learn from their past mistakes in order to discover how to avoid similar situations ever taking place again, such as a great depression from an economical perspective or a devastating war with a rogue nation from a military viewpoint, then the contents of this text constitute essential informative conflict resolution techniques to include in your tool kit. A doctoral student may wish to expand the basic GMCR methodologies based on gaps that he finds when systematically studying conflict in fields such as energy development, environmental engineering, water resources, and legal studies. If you are a ***practitioner*** or professional like a consulting engineer, urban planner, political advisor, manager, lawyer, policy analyst, or military systems engineer, this book will be compelling for you to use in resolving challenging practical problems within your professional area of expertise. For instance, as climate change intensifies and regional wars erupt, military analysts within operations research groups in defense departments will find this book to be very useful for tackling the severe security issues involved with the mass migration of affected populations, as is occurring and intensifying right now in Europe where refugees are continually arriving in increasing numbers. If you are a professional like a computer engineer or computer scientist, you may wish to utilize the basic design for a flexible decision support system (DSS) for conflict resolution put forward in this book for programming the next generation of DSSs for employment by researchers, teachers, mentors, students, and practitioners for applying the new GMCR techniques in this book to real-life situations.

To convince you, our reader, that GMCR can be actually utilized in practice for addressing challenging ***real-world disputes***, examples are provided throughout the book to demonstrate how the various ideas can be applied. These applications clearly demonstrate why "good theory means good practice" and vice versa. Hence, in the very first chapter in the book, a highly controversial groundwater contamination dispute which occurred in the town of Elmira, Ontario, Canada, is employed to explain how the conflict can be modeled and analyzed using GMCR in order to gain a better understanding and strategic insights. This same environmental conflict along with others are utilized in the book to explain how various concepts are designed and work in practice.

The basic theoretical structure of GMCR and its expansions were purposefully designed to address conflicts which actually occur in reality. To accomplish this, the underlying axioms of GMCR were formulated to reflect the key characteristics of real-world conflict, thereby forming the solid foundations upon which the theoretical framework can be properly built and expanded. For example, in a conflict situation, DMs often think like a chess player in terms of ***moves and countermoves***. If a particular DM is contemplating moving from the current situation to a more preferred state, the DM may wish to know the consequences of this possible move.

If, for instance, a car manufacturer decides to decrease the selling price of its cars and thereby hopefully gain greater market share, will the company's competitors also decrease the cost of buying their cars and put the particular company in a worse situation? If so, the company is better off not to lower its prices. In GMCR, different ways in which people may ***behave*** under conflict can be captured mathematically by what are called solution concepts or stability definitions. Furthermore, the possible moves that a DM controls can be recorded using a graph in which the scenarios or states that could occur form the vertices (nodes) while moves that the DM can make in one step are drawn as the directed arcs connecting states. Another key feature of GMCR is that only ***relative preference information*** is required which means that you only have to know if a DM prefers one state over another or if the states are equally preferred. Hence, if someone asks you if you would like to have a cup of coffee or tea, you may respond by saying that I prefer to have coffee, thank you, or it does not matter. You would certainly not give a quantitative response by saying that for me coffee has a utility value of 6.912 while tea is worth 2.591. A key design feature of GMCR is that only relative preference information is needed, which is fairly easy to obtain in practice and mimics the way people think about their preferences.

The foregoing fashion of directly thinking about a conflict in terms of moves and countermoves coupled with relative preferences is called the ***logical form*** of the game. A person can intuitively understand how a conflict can evolve and be resolved by logically explaining what can happen using moves and countermoves as DMs attempt to do the best they can in a dispute. If, for instance, from a state all of the ways in which a DM could unilaterally improve can be sanctioned by others, then this state is said to be stable for that DM according to a certain ***type of behavior***. If it is not advantageous for any of the DMs to move, the state is a possible resolution or equilibrium if it is reached during the evolution of the dispute under study. For a specific conflict, providing a logical explanation of what can happen is highly appealing. However, the information contained in a graph keeping track of moves or preferences can be stored in a matrix for computational purposes. In fact, the logical interpretation of GMCR both in terms of modeling and stability calculations can be equivalently formulated using a ***matrix representation***, which is also called algebraic form. When programming the engine for calculating the stability results, the matrix form is much more efficient than its logical counterpart in terms of the number of required calculations. Moreover, for theoretical purposes, it is much easier to expand GMCR when the matrix form is utilized. Therefore, throughout the book, both the logical and matrix representations of GMCR are provided for all of the advancements that are presented, which makes this book truly unique.

To appreciate the uniqueness and innate capabilities of GMCR, the connections and differences of GMCR with respect to other game theory methods are discussed in the second chapter. Moreover, the relationships of GMCR to other formal decision-making techniques developed in the fields of Operations Research, Systems Engineering and elsewhere are clearly explained. If a decision-making methodology like GMCR is programmed as a DSS so it can be readily applied to actual disputes, the methodology becomes an ***operational decision technology***. In practice, one may use a toolbox of decision technologies for addressing a complex problem like urban expansion for which GMCR could be used for investigating the strategic and controversial aspects of the project.

In actuality, everything affects everything else within and among societal and physical systems of systems. For instance, the utilization of fossil fuels in society's industrial, transportation and electricity generation systems in nations around the world releases massive quantities of carbon dioxide into the atmosphere, which is one of a number of deadly greenhouse gases causing average temperatures around the globe to increase significantly over time. This, in turn, alters the earth's climate system, creates extreme weather conditions, shrinks the area and thickness of sea ice, melts glaciers, makes ocean levels rise, and increases the acidity of oceans. These and other negative consequences of climate change on the earth's natural systems can adversely impact societal systems such as agriculture, industry and the economy as a whole, as well as the stakeholders who are part of these systems. Accordingly, it is highly intuitive and informative to envision any problem from a ***system of systems perspective***. Within this vision of reality, a useful tool like GMCR can be employed to investigate the myriads of conflicts that will arise among affected parties, which for the case of climate change will surely increase in number and intensity as the climate continues to deteriorate, perhaps irreversibly.

To responsibly handle complex problems connected to climate change, the Elmira groundwater contamination problem, and other tough issues facing society, an ***integrative and adaptive approach to management and governance*** can be followed in a participatory fashion with stakeholders whose interests or values must be taken into account in policy design and decision-making. In this way, solutions to problems can be found which adhere to desirable systems characteristics like sustainability, fairness, and robustness. A flexible tool like GMCR can be employed to handle disputes that may arise for which the stakeholders value systems are always considered.

After putting decision-making into perspective in Chap. 2 and explaining the vital role that GMCR has to play, various conflict models are defined in Chap. 3. As explained in Chap. 3, what is called the ***option form*** of the game is particularly powerful as a notation for keeping track of the options or courses of actions available to each DM in a dispute and recording the possible feasible states or scenarios that could occur in the conflict. These states are then used in both the ***logical form*** and ***matrix representation*** of GMCR presented in Sects. 3.2 and 3.3, respectively. Because they reflect the underlying value system of a DM, a crucial input to a conflict model is the relative preference of the DM among the feasible states that could occur.

Subsequent to modeling a given conflict in terms of DMs, states, state transitions, and relative preferences, a stability analysis is carried out in terms of investigating moves and countermoves that could occur according to four ***solution concepts*** reflecting ***human behavior*** under conflict when determining if a state is stable or not: Nash stability, general metarationality, symmetric metarationality, and sequential stability. Depending on the type of preference information that is available, these solution concepts are appropriately defined for both the logical and matrix representations of GMCR. Hence, the next four chapters in the book provide the stability definitions for the following types of preference information:

Chapter 4: ***Simple preference*** in which a given state can be more preferred, equally preferred, or less preferred to another state by a DM.

Chapter 5: ***Unknown preference*** in which a DM does not know the preference relationship for some pairs of states. This type of preference uncertainty is uniquely defined for employment with GMCR since it does arise in practice. In the last chapter in this book, it is mentioned that fuzzy sets, grey numbers, and probabilistic approaches to preference uncertainty have also been developed for employment with GMCR.

Chapter 6: In some situations, a DM may greatly prefer one state over another such as when environmentalists greatly prefer that an industrialist does not allow his company to significantly pollute the surrounding environment by releasing untreated wastes. This is referred to as ***degree of preference*** for which the degree can be taken to any level for specified pairs of states.

Chapter 7: ***Hybrid preference*** in which unknown and degree of preference can occur as well as simple preference.

As mentioned earlier, in addition to determining how well a given DM may fare when behaving independently, one should also determine if a DM can do even better by cooperating with others. Hence, in Chap. 8 ***coalitional stabilities*** are defined for the aforementioned four types of preference situations for both the logical and matrix forms of GMCR. As an important type of follow-up analysis, the possible ***evolution of a conflict*** from a specified starting or status quo state to a particular final state is presented for both the logical and matrix representations of GMCR in Chap. 9. In practice, one may wish to know whether a desirable state, such as a win/win resolution, can actually be reached by DMs who have under their control unilateral moves that they can select to levy.

The book concludes with the presentation of a ***universal design*** of future generations of DSSs for GMCR based on an internal matrix representation structure for handling the current and future expansions of GMCR in Chap. 10. These ***future opportunities*** include the capability of having systems engineering investigations in which inverse engineering and behavioral engine specification can be fully studied. Inverse engineering or inverse GMCR means ascertaining the preferences needed by DMs for a desirable final state to be an equilibrium. The behavioral engine problem is given the input and output to determine the type of behavior exhibited by the DMs.

So, our cherished readers, we trust that you will enjoy the exciting journey through our comprehensive book. But hang on to your hats: there will be a lot more to come in the future both in terms of ***new operational methodologies*** for expanding the capabilities of GMCR and also the wealth of ***pressing conflicts*** that have to be properly addressed right now, as well as challenging conflicts that may arise in the future as the earth becomes a smaller and smaller place for all of us to live and prosper.

We warmly wish you, our readers, a most revealing and exciting journey through our book.

Bon voyage!

Nanjing, China — Haiyan Xu
Waterloo, Canada — Keith W. Hipel
Waterloo, Canada — D. Marc Kilgour
Toronto, Canada — Liping Fang
March 2018

Acknowledgements

The four authors of this book have had the distinct pleasure of working with one another for many years within a stimulating academic environment of friendship, collegiality, and innovation. The academic home in which they developed many of their ideas in conflict resolution, among themselves and with colleagues and students, was the Conflict Analysis Group based in the Department of Systems Design Engineering at the University of Waterloo. The authors have been privileged to work with many gifted scholars at their own universities in Canada and China as well as at many other research institutions around the globe. Accordingly, they would like to convey their deep appreciation to their colleagues and former students who executed research on topics directly connected to the theme of this book or that complemented or inspired research contained in this text. These contributions include co-authorship of research papers that form part of the content of this book, as well as the provision of timely guidance and encouragement.

Although the solid foundations for the leading-edge ideas of the Graph Model for Conflict Resolution (GMCR) contained in this text were cleverly designed and carefully constructed over three decades ago, this book focuses on advances in GMCR achieved during the past ten years, including some material appearing in print here for the first time. Research ideas from journals or other publications are of course properly referenced. Moreover, for material that comes directly from another publication, permission of the copyright holder has been obtained.

As an expression of our sincere gratitude, we would like to record here the names of our fellow researchers who are co-authors or inspirers of ideas contained in this book. In alphabetical order, they are as follows: Taha Alhindi, Yasir Aljefri, Mubarak Al-Mutairi, Motahareh Armin, Abul Bashar, Sean Bernath Walker, Michele Mei-Ting Bristow, David Bristow, Ye (Richard) Chen, Yu Chu, Aldo Dagnino, Mitali De, Jose R. del Monte, Jianfeng Ding, Niall M. Fraser, Bing (Ben) Fu, Kei Fukuyama, Masao Fukushima, Amanda Garcia, Bingfeng Ge, Mohammad Reza Ghanbarpour, Kiyoko Hagihara, Yoshimi Hagaihara, Luai Hamouda, Xueshan Han, Yu Han, Shawei (David) He, Yuhang Hou, Kaixan (Max) Hu, Takehiro Inohara, Ju Jiang, Yangzi Jiang, Monika Karnis, Moustafa Kassab, Young-Jae Kim, Yi (Ginger)

Ke, Rami Kinsara, Hanbin (Eric) Kuang, Jonathan R. D. Kuhn, Jason K. Levy, Kevin Wu Li, Xiang-Ming (Samuel) Li, Xuemei Li, Jing Ma, Kaveh Madani, Yasser Matbouli, Darren Bruno Gerrit Meister, K. D. W. Nandalal, Leilei Ni, Donald J. Noakes, Amer Obeidi, Norio Okada, Sevda Payganeh, Xiaoyong (John) Peng, Simone Philpot, the late James Radford, Jiwu Rao, Michael R. Rooks, Sara Anne Ross, J. Fernando Molina Ruibal, Hiroyuki Sakakibara, Bader Sabtan, Maiko Sakamoto, Peter Savich, Majid Sheikhmohammady, Maisa Mendonça Silva, Aihua Song, Nigel W. Stokes, West Suhanic, Hirokazu Tatano, Marcella Maia Bezerra de Araujo Urtiga, Junjie Wang, Lizhong (George) Wang, Muhong Wang, Qian Wang, Colin Williams, Saied Yousefi, Yi Xiao, Peng Xu, Jun Yang, Xian-Pei Yin, Jing (Crystal) Yu, Dao-Zhi Zeng, Jinlong (John) Zhang, Jinshuai Zhao, Shinan Zhao, Yuming (Arthur) Zhu, and Ziming Zhu.

The authors are extremely grateful for the assistance they received in the refinement, processing, and proofreading of the final version of book. Specifically, Shinan Zhao and Yi Xiao provided a great deal of timely and expeditious help in completing the book. The authors would also like to thank professionals at Springer for the expert service they rendered in the publication of the book.

The authors are indebted for financial support provided by funding agencies in China and Canada for their long-standing support of the research contained in this book. In particular, they would like to acknowledge funding from the National Natural Science Foundation of China (NSFC) (Grant Numbers 71071076, 71471087, and 61673209) as well as the Natural Sciences and Engineering Research Council (NSERC) of Canada through its Discovery and Strategic Grants programs. Over the years, the authors' home universities have also furnished research space and in-kind support.

In closing, the authors would like to express their special appreciation of their spouses and family members who provided the love and support that enabled them to blossom in their academic careers and dedicate their creative time to challenging research projects. We all agree that it was our supportive family and social environments that made the writing of this book possible.

Nanjing, China	Haiyan Xu
Waterloo, Canada	Keith W. Hipel
Waterloo, Canada	D. Marc Kilgour
Toronto, Canada	Liping Fang
March 2018	

Contents

About the Authors

Haiyan Xu is *Professor* with the College of Economics and Management, Nanjing University of Aeronautics and Astronautics, China. She received her B.Sc. degree from Nanjing University, China, and her Master's of Mathematics and Ph.D. degree in Systems Design Engineering from the University of Waterloo in Canada. Her current research interests in Game Theory and Optimization include the development of normal algebraic techniques for conflict resolution and methods for optimization with applications in environmental management and finance. She is *recipient* of the Excellent Professor Award from the Province of Jiangsu, China.

Keith W. Hipel is *University Professor* of Systems Design Engineering at the University of Waterloo, *Officer* of the Order of Canada, former *President* of the Academy of Science (Royal Society of Canada), *Senior Fellow* of the Centre for International Governance Innovation, *Fellow* of the Balsillie School of International Affairs, and *Coordinator* of the Conflict Analysis Group at Waterloo. He is globally renowned for his interdisciplinary research in *Systems Engineering* on the development of *conflict resolution, multiple objective decision-making*, and *time series analysis* for addressing complex *system of systems* problems lying at the confluence of society, technology, and the environment, with applications in *water resources, environmental engineering, energy*, and *sustainable development.*

D. Marc Kilgour is *Professor* of Mathematics at Wilfrid Laurier University and *Adjunct Professor* of Systems Design Engineering at the University of Waterloo. Most of his many publications provide a mathematical analysis of multiparty decision problems. He has contributed innovative applications of game theory and related methodologies to international relations, arms control, environmental management, negotiation, arbitration, voting, fair division, and coalition formation, and pioneered the application of decision support systems to strategic conflict. He was *Co-editor* of the Springer *Handbook of Group Decision and Negotiation* (2010), *President* of the Peace Science Society in 2012–2013, and *President* of the INFORMS Section on Group Decision and Negotiation in 2014–2017.

Liping Fang is *Professor* of Mechanical and Industrial Engineering, former *Chair* of his Department, and *Associate Dean*, Undergraduate Programs and Student Affairs, Faculty of Engineering and Architectural Science at Ryerson University and *Adjunct Professor* of Systems Design Engineering at the University of Waterloo. His research interests in Industrial and Systems Engineering include conflict resolution, agent-based modeling and simulation, e-services, environmental and water resources management, and decision support systems. He is *Fellow* of the Canadian Academy of Engineering (FCAE), Institute of Electrical and Electronics Engineers (FIEEE), Engineering Institute of Canada (FEIC), and Canadian Society for Mechanical Engineering (FCSME).

Acronyms

CCA	Council of Canadian Academies
CDO	Canadian Opposition
CIUM	Coalition Improvement or Uncertain Move
CWAM	Cooperative Water Allocation Model
DBMS	Database Management System
DGMS	Dialog Generation and Management System
DM	Decision Maker
DSS	Decision Support System
GCGMR	General Coalitional General Metarationality
GCGS	General Coalitional Graph Model Stability
GCNash	General Coalitional Nash Stability
GCSEQ	General Coalitional Sequential Stability
GCSMR	General Coalitional Symmetric Metarationality
GDU	Garrison Diversion Unit
GGMR	General General Metarationality
GGS	General Graph Model Stability
GHG	Greenhouse Gases
GMCR	Graph Model for Conflict Resolution
GMCR II	GMCR (Graph Model for Conflict Resolution) II
GMR	General Metarationality
GNash	General Nash Stability
GS	Graph Model Stability
GSEQ	General Sequential Stability
GSMR	General Symmetric Metarationality
GWP	Global Water Partnership
IG	Integrated Graph
INBO	International Network of Basin Organizations
INFOR	Information Systems and Operational Research
MBMS	Model-base Management System
MCDA	Multiple Criteria Decision Analysis

MRCR	Matrix Representation for Conflict Resolution
MRSC	Matrix Representation of Solution Concepts
MRSCU	Matrix Representation of Solution Concepts with Preference Uncertainty
MSUI	Mild or Strong Unilateral Improvement
MSUIUM	Mild or Strong Unilateral Improvement or Uncertain Move
OR	Operations Research
SCGS	Strong Coalitional Graph Model Stability
SCGMR	Strong Coalitional General Metarationality
SCSEQ	Strong Coalitional Sequential Stability
SCSMR	Strong Coalitional Symmetric Metarationality
SEQ	Sequential Stability
SGMR	Strong General Metarationality
SGS	Strong Graph Model Stability
SMR	Symmetric Metarationality
SoS	System of Systems
SSEQ	Strong Sequential Stability
SSMR	Strong Symmetric Metarationality
UI	Unilateral Improvement
UIUM	Unilateral Improvement or Uncertain Move
UM	Unilateral Move
USS	United States Support
UUM	Unilateral Uncertain Move
WCGMR	Weak Coalitional General Metarationality
WCGS	Weak Coalitional Graph Model Stability
WCSEQ	Weak Coalitional Sequential Stability
WCSMR	Weak Coalitional Symmetric Metarationality
WGMR	Weak General Metarationality
WGS	Weak Graph Model Stability
WSEQ	Weak Sequential Stability
WSMR	Weak Symmetric Metarationality
WWI	World War I
WWII	World War II

List of Figures

List of Tables

Chapter 1
Conflict Resolution in Practice

1.1 The Pervasiveness of Conflict

Conflict occurs virtually everywhere in society. A powerful methodology called the Graph Model for Conflict Resolution (GMCR) is put forward in this book for addressing tough conflict situations. Hence, this book should prove to be especially valuable for people who must deal with social conflict now.

The goal of this chapter is to stress the importance of being able to address conflict in society and to emphasize that the GMCR methodology constitutes a powerful set of tools which possesses the innate capabilities of being able to formally investigate tough conflict situations. Therefore, this well-designed methodology should be of wide interest to practitioners, teachers, researchers, mentors and students in any field in which conflict takes place, which is virtually every area of human endeavour. To clearly demonstrate that GMCR can be conveniently utilized for modeling and analyzing real-world conflict, the procedure for formally studying an actual groundwater contamination dispute that occurred in the town of Elmira, Ontario, Canada, is presented in Sect. 1.2.

In the second last section of this chapter, Sect. 1.3, a road map is provided in terms of a flow chart and table to explain how one can navigate through the book in order to ultimately understand how to address actual disputes ranging from simple to highly complex situations. Illustrative real-world applications are employed throughout the book to demonstrate how valuable GMCR tools can be applied in practice. Although modeling and calculations can be done by hand for simple disputes, Decision Support Systems for implementing many GMCR ideas are available now for employment by practitioners, teachers, researchers, mentors and students, as pointed out in Sect. 10.2.

H. Xu et al., *Conflict Resolution Using the Graph Model: Strategic Interactions in Competition and Cooperation*, Studies in Systems, Decision and Control 153,
https://doi.org/10.1007/978-3-319-77670-5_1

1.1.1 Pressing Conflicts Facing Society

Conflict inevitably arises whenever human beings interact with one another. For example, family members may have differences of opinion regarding the choice of a restaurant at which to have dinner together. An employee may disagree with his or her employer over how much her annual salary increase should be. Besides individuals, conflict also occurs within and among organizations. In the highly competitive business world, companies aggressively promote their products to increase their market share in a given sector both within a particular nation and globally. Consider, for instance, the intense competition for a greater share of car sales around the world occurring among major players like General Motors, Toyota, Volkswagen, Honda, BMW, Renault and many other large international producers of automobiles. Political parties vie with one another to attract supporters by proposing different policies ranging from infrastructure renewal to tax reductions. Internationally, ongoing negotiations aim to find appropriate and effective measures for combating greenhouse gas emissions and thereby mitigating climate change. Environmentalists, government agencies and citizens may be in dispute with a chemical company on how to clean up a "brownfield" caused by the release of chemical wastes which polluted an underground aquifer. Air pollution in major cities in which intense industrialization is taking place coupled with a massive increase in the number of private vehicles, puts citizens in direct conflict with government and industry. Regional wars in the Middle East and terrorist attacks in Europe, the United States and China are illustrations of lethal kinds of ongoing conflict. Indeed conflict appears to be an inherent characteristic of human behavior as well as both private and public societal organizations created by people.

Due to the ubiquity of conflicts, which range from personal disagreements to international military campaigns, and which occur within practically every field of study including trade, law, engineering and health care, there is great demand for formal methodologies to assist decision makers (DMs) facing controversies, and to discover and attain stable resolutions. As mentioned by Hipel et al. (2011), the aforementioned "social" conflicts possess certain key inherent characteristics including the presence of:

- two or more DMs, such as individuals, organizations, and nations, participating in the dispute;
- different options or actions under the control of each of the DMs;
- a separate value system for each DM which has multiple dimensions or objectives, many of which are in direct conflict with the values of other participants in a conflict;
- relative preferences for a particular DM that reflect his or her value system with respect to the conflict being investigated;
- possible moves and countermoves controlled by a given DM which can be brought into play at any time or not at all;

- strategic moves and countermoves among the DMs that could dynamically take place at any time, in any order, as the conflict evolves from a starting point to an intermediate or final outcome;
- a range of ways people may behave under conflict situations such as an insightful DM who can perceive many potential moves and countermoves into the future before deciding what to do, much like a clever chess player planning his or her next move;
- participants who may act completely independently of one another or sometimes in cooperation with others by forming a coalition in which members of the coalition benefit from their cooperative actions;
- high uncertainty caused by, for instance, unknown relative preferences and many DMs taking part in a conflict;
- psychological factors like attitudes, emotions and misunderstandings.

When designing a large structure like a building or dam, one must take into account the local geophysical characteristics, such as the soil properties and possible seismic occurrences, as well meteorological conditions including temperature, wind and precipitation. In an analogous fashion, when designing a model for systematically studying social conflict, one must consider the aforesaid list of key characteristics. The Graph Model for Conflict Resolution (Kilgour et al. 1987, Fang et al. 1993) and its associated extensions (Hipel et al. 2011) were purposefully designed to handle all of these and other traits of real-world conflict.

1.1.2 Objectives of This Book

The overall purpose of this book is to present in a highly informative and user-friendly way the latest ideas of the Graph Model for Conflict Resolution (GMCR) for systematically investigating actual conflict occurring in the real-world so that a reader of this book will be in a position to readily study disputes that are of direct interest to her or him. Consider the aforementioned metaphor of designing a structure. In addition to the physical environment, a given building must be designed to provide a range of purposes or services such as retail space for stores located at ground level and offices for different kinds of organizations that will occupy higher floors. Similarly, conflict models must be structured to not only take into account the key characteristics of conflict but also to furnish desired uses or purposes. In particular, the GMCR has incorporated into its basic design the capability to (Hipel et al. 2011):

- summarize in an organized way complex and often confusing information about a conflict into a clear model structure. This process puts the conflict into perspective by focusing on the essentials of the problem within a simple, yet revealing, model framework;
- function when information or data are scarce or when one is drowning in an over-abundance of information;

- enhance the understanding of the dispute being investigated via this type of systems thinking;
- facilitate meaningful communication using the structured "graph model language" among stakeholders and interested parties;
- predict the strategic consequences of making potential choices under conflict in order to select the best possible decision given the social and strategic constraints existing in light of what others may do to advance their own positions. This will reduce making misinformed decisions having potentially highly negative and perhaps irreversible impacts;
- be cognizant of the strategic implications of other characteristics of the conflict under study such as coalition formation, preference uncertainty and psychological factors; and,
- make informed decisions based on the foregoing sensible modeling and analyses which may lead to a win/win resolution in which all disputants benefit.

1.1.3 Audience

Conflict arises in all areas of human endeavour. Therefore, the contents of this timely book on conflict resolution should be useful to people working within many disciplines or areas of study which include:

- Agriculture,
- Aquaculture,
- Business,
- Climate Science,
- Economics,
- Engineering,
- Environmental Science and Engineering,
- Food Systems,
- Law,
- Logistics and Supply Chain Management,
- Military Science,
- Political Science,
- Service Industry,
- Sociology, and
- Trade.

Within each of the above and other disciplines, the book should prove to be highly useful to

- Practitioners,
- Teachers,
- Researchers,
- Mentors, and
- Students.

The clear explanation of concepts, coupled with illustrative examples and available Decision Support Systems for implementation purposes, make the book especially attractive for adoption by all of the above-mentioned users. Consultants, for example, working in engineering or international business, can apply the GMCR methodology now to challenging problems in order to obtain sage strategic advice for their clients. Because the book includes some of the latest ideas from GMCR, it should prove to be highly useful for researchers. Problems furnished at the end of each chapter, combined with other benefits, make the book ideal for employment in a course dealing with conflict resolution. This book contains the most recent approaches to the theory and practice of conflict resolution currently available as a single informative document.

1.2 Investigating Conflict

1.2.1 Key Ideas

When a person is confronted by a conflict problem that must somehow be resolved, he or she must decide what to do. To reach a decision, the individual naturally thinks in a way that captures the key characteristics of the conflict and simplifies it to a level in which it can be more readily understood. A better comprehension of what is taking place permits the person to imagine the consequences of his actions when he makes different decisions. This type of "what-if" analytical thinking should assist in leading to a more sound and beneficial decision not only for this individual but hopefully all parties involved in this dispute.

The objective of a formal conflict model is to mimic the way a person thinks under conflict in order to aid the individual in making a more informed decision. To achieve this, the mathematical design and capabilities of the model must reflect the key aspects of a conflict problem occurring in the real-world. Recall that in Sect. 1.1.1, a list of key inherent properties or characteristics of actual conflict is provided. Subsequently, in Sect. 1.1.2 a list of features is given that was purposefully incorporated into the mathematical design of GMCR in order to make it realistic, meaningful, operational and informative. In fact, GMCR is specifically constructed to mirror the way in which a human being perceives and reacts to conflict in reality. In turn, this means that GMCR should be helpful to a person for enhancing his perception and comprehension of the conflict and thereby providing guidance on how to wisely interact with others in the best way possible within the social constraints on how others may respond. GMCR can also act as a "book-keeping" technique by keeping a systematic record of what is taking place and furnishing insightful resolutions that the individual may have inadvertently overlooked because of stress, excitement or carelessness.

What are the main components in a conflict that one must incorporate into a model of it? For a start, there obviously must be at least two participants or decision makers (DMs) who are in dispute over some issue. In the conflict under investigation, each

DM must be in control of options or courses of action he or she has at his disposal. Additionally, because each DM has his own value system which guides his behavior in the social world, a given DM has his own particular preferences over the scenarios or states that could occur. It is these differing value systems or preferences among DMs over states that provide the basic fuel for igniting and driving conflict.

A given DM in a conflict or an interested third party studying the dispute must decide what actions to take in order to fare as well as possible according to his value system or preferences. Prior to reaching a decision about what to do, the person will probably think like a chess player in terms of the potential consequences of how others may react when a particular course of action is chosen. If the conflict is at a particular state and the DM can improve on his own to reach a better position by unilaterally changing his option selections that he directly controls, he will certainly be tempted to do so. However, if one or more of the other DMs can block his improvement by invoking actions of their own that put the DM in a less preferred situation, the DM may be better off to maintain his current position and not move. In other words, the present situation is stable for that DM. It is this kind of "what-if" thinking that is systematically investigated at what is called the stability analysis stage of a formal conflict study. If, for instance, a given state is stable for all DMs according to a certain type of potential human behavior under conflict, it is referred to as a resolution or equilibrium. The purposes of stability analyses include determining what resolutions are most likely to occur and can DMs do even better by forming coalitions. A key advantage of carrying out a conflict study using a formal approach like GMCR is that one can execute a range of what are called follow-up analyses such as how could the conflict evolve from a status quo state to an eventual final equilibrium and what are the strategic impacts of various kinds of sensitivity analyses. If, for example, the preferences of a key DM were slightly different could this result in a much better resolution taking place that may be a win/win equilibrium for all parties.

Figure 1.1 provides a flowchart of the main steps involved when carrying out a conflict study. As can be seen, the modeling stage is followed by an analysis stage which includes stability and follow-up analyses. The theoretical and practical aspects on how this is actually done are outlined in the next three subsections and presented in detail in later chapters in this book. Additionally, the ways in which GMCR can be utilized in practice are described in Sect. 1.2.5. To clarify the explanations given in this section and make it more convincing, an actual groundwater contamination dispute is utilized.

1.2.2 Modeling

1.2.2.1 Groundwater Contamination Dispute

Brownfields constitute a very serious type of pollution problem and are associated with intense conflict facing all industrialized countries. A brownfield is land which has been polluted by previous or current industrial activities and frequently involves

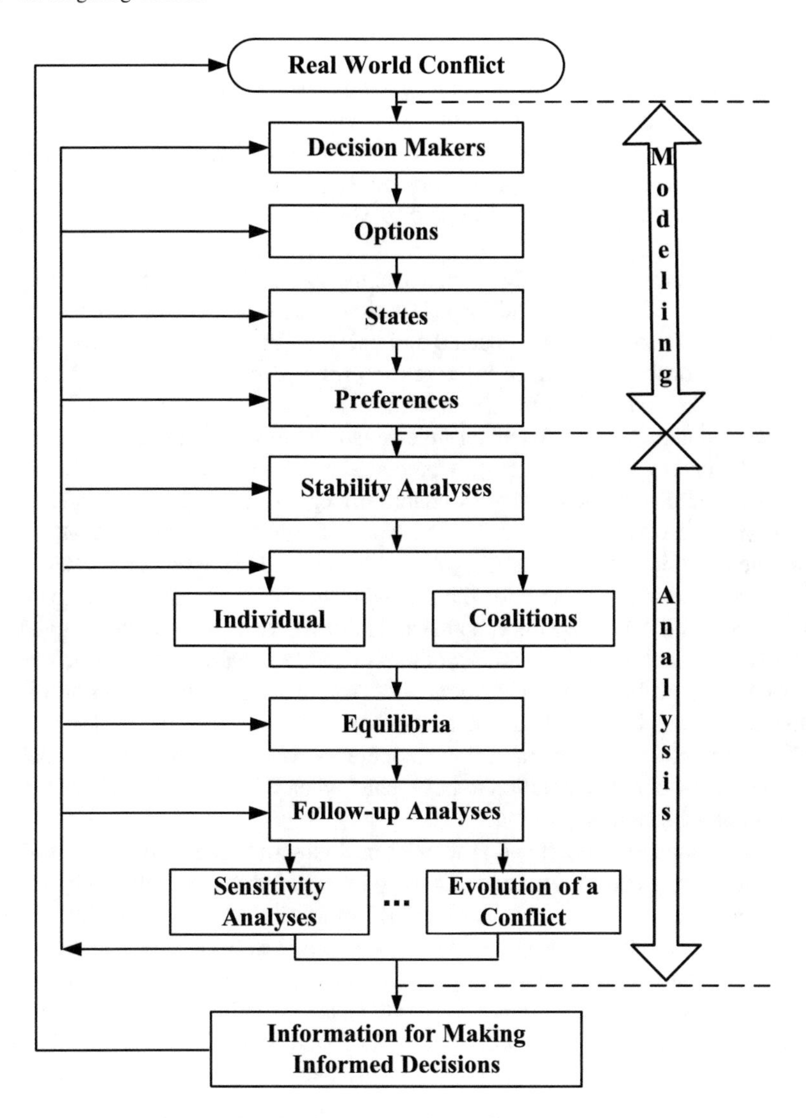

Fig. 1.1 Main steps for applying GMCR to a specific conflict

the contamination of underground aquifers which serve as water supplies for communities (USEPA 1997). Hence, brownfields often pose a high risk to adversely affecting human health as well as other life forms. The terminology brownfield arose in direct contrast to the label greenfield on which lush green vegetation grows since the land is not polluted. A range of conflicts have arisen over the years regarding the rehabilitation, sale and redevelopment of brownfield properties in former industrial areas usually located in or near cities in industrialized nations. Because brownfield redevelopment projects have social, economic and environmental impacts

(Greenberg and Lewis 2000, McCarthy 2002, De Sousa 2003), the associated disputes tend to be highly controversial although good government policy can help to lessen their occurrence and severity (Bernath Walker et al. 2010, Hipel and Bernath Walker 2012). Moreover, both land and water can also be severely polluted by agricultural activities which include the applications of fertilizers and insecticides on a large scale basis and intense irrigation practices. This in turn can create conflict among stakeholders.

As an example of a real-world conflict over the pollution of an underground aquifer by industry, an actual case study from Canada is considered. The three Canadian authors of this book actually modeled and analyzed this highly publicized dispute while it was ongoing based on information provided by an interested third party who was a domain expert with respect to the history and evolution of this extremely controversial dispute which received widespread national and international attention. In fact, the investigation team closely predicted what actually occurred just after its study was completed and provided valuable strategic insights into why and how the conflict was resolved. In Sect. 1.2.5, the types of situations in which GMCR may be applied and how it can be used by consultants for advising clients are described. The procedure explained now for the groundwater pollution dispute is close to the way a consultant would use the valuable and informative decision technology. Additionally, because this kind of water pollution dispute is representative of a class of environmental problems occurring or about to take place in many nations around the globe, it provides insightful guidance as to how they may be better understood and resolved. Finally, it demonstrates how cooperation via the formation of a coalition can cause a reasonable resolution to be reached when independent behavior by each DM cannot achieve this.

Canada's industrial heartland is located in Southern Ontario for which three of the Great Lakes - Huron, Erie and Ontario - form its border with the United States and Canada's largest city, Toronto, is located, as indicated on the map in Fig. 1.2. The Regional Municipality of Waterloo, which is situated just over 100 km to the west of Toronto and marked by the darkened area in Fig. 1.2, contains fertile farmland and a spectrum of industries. Elmira is a relatively small agricultural town of about 7500 people, situated about 15 km north of the city of Waterloo, which is one of Canada's most renowned high technology centers. Although Elmira has some industries, it is best known for its famous maple syrup festival, which takes places at the end of the winter season every year. Because it is located about 100 km away from any of the Great Lakes, most communities in the Region of Waterloo, including Elmira and the city of Waterloo, obtain their fresh water from underlying aquifers. When carrying out chemical tests in 1989, the Ontario Ministry of the Environment (MoE) discovered that the aquifer supplying water to Elmira was contaminated by N-nitroso demethylamine (NDMA). Due to the fact that NDMA is a carcinogen which can cause cancer to form in humans, the citizens of Elmira were shocked and outraged to find out that their drinking water was poisonous. Not surprisingly, this great risk to human health and the environment made headline news across Canada and internationally over a relatively long period of time.

Since NDMA is a by product of its production processes, Uniroyal Chemical Ltd. (UR) was suspected as being the source of the pollution found in the aquifer. In

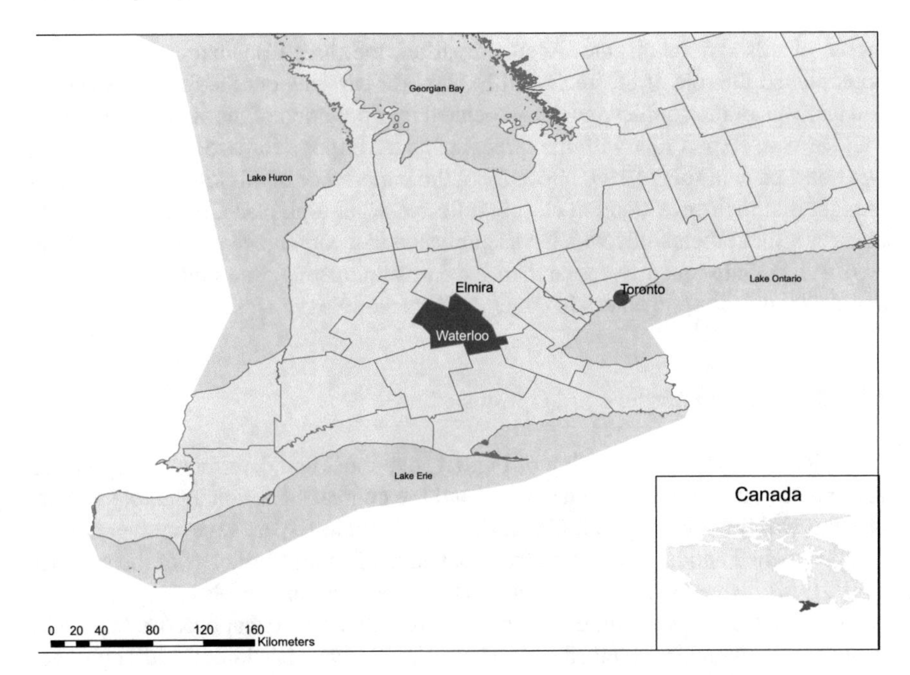

Fig. 1.2 Location of the Elmira groundwater contamination dispute in Southern Ontario, Canada

accordance with Ontario's environmental laws, what is called a Control Order was issued by MoE which requested UR to take relatively expensive measures to cleanse the aquifer, remove buried metal drums containing chemicals that were leaking, and properly treat future wastes released into a local stream. As permitted under the environmental law, UR immediately appealed the Control Order.

Because the Regional Municipality of Waterloo and the Township of Woolwich, which falls within the Waterloo Region and is where Elmira is located, viewed the pollution problem in the same way, they are considered in this conflict as a single unit called Local Government (LG). In fact, LG spent a considerable amount of financial resources on legal costs to determine its possible role in resolving this conflict as well as on engineering studies. Quite naturally, the elected officials representing LG wanted its citizens to be protected from consuming contaminated water and it sided with MoE which is overall in charge of addressing environmental problems in the Province of Ontario. Costly bore holes had to be drilled and testing carried out to ascertain how the plume of pollution from the pollution source had spread underground via the groundwater.

A formal conflict study was executed when the conflict over UR appealing the Control Order was ongoing during the summer of 1991. Hence, the investigation was done in the way an interested party or consultant would tackle a current dispute. A local environmentalist by the name of Dr. Murray Height who lived in Elmira at that time acted as the Domain Expert to provide information to calibrate a conflict model of this serious dispute. As noted earlier, the three Canadian authors of this book played the role of the Technical Experts for carrying out the formal modeling and analysis of the conflict and the associated interpretation of the strategic findings. The Domain Expert met with the three Technical Experts for two meetings lasting two hours each in July of 1991. Because of the importance of this kind of conflict and the fact that their investigation closely reflected what took place, the authors wrote an article about their study which was published in a conference proceedings (Hipel et al. 1993). Subsequently, researchers used this interesting case study in the testing and development of advances in the GMCR methodology.

1.2.2.2 Decision Makers and Options

As indicated in the upper portion of Fig. 1.1, one must first determine the key DMs participating in the dispute under study and the courses of action available to each of them with respect to the conflict. After the Domain Expert gave an overview of what happened and is currently taking place in the Elmira conflict as of the summer of 1991, the Technical Experts first asked him who were the people or organizations having real decision-making power in the conflict. He stated that because this was a major environmental problem, the MoE had to be involved. In addition, although there was also a smaller chemical company located in Elmira, UR was clearly the prime suspect in contaminating the aquifer. From the start, the Domain Expert thought that the two levels of government consisting of the Regional Municipality of Waterloo at the higher level and Township of Woolwich at the lower level should be thought of as a single DM which he called Local Government, since they were working together to solve the problem and thereby protect the health of their joint citizens. Some discussion arose over whether or not local environmental groups should be considered as additional DMs in the dispute. Since they held no direct power connected to the negotiations taking place among the aforementioned three parties, the Domain Expert thought that they should not be entertained as DMs in the study. There were, of course, interested parties and stakeholders in what eventually happens who could try to influence others, but they hold no specific actions that they could levy in the actual negotiations. Hence, to keep the study as accurate and simple as possible, they were not incorporated into the model. The left hand side of Table 1.1 lists the three DMs participating in the Elmira negotiations during the summer of 1991.

Comments about Decision Makers:

A range of labels have been used in the literature as an alternative to the words Decision Maker (DM). These names include player, actor, stakeholder and participant. The advantage of employing the phrase DM is that the label directly implies that only someone or organization having real decision-making power is considered in the conflict model. Moreover, a specific DM may stand for a single person, an organization or even a nation. In practice, the authors have found that GMCR works well for any combination of different types of DMs.

Table 1.1 Decision makers and options in the Elmira conflict

Decision makers	Option names	Explanations
MoE (Ontario Ministry of the Environment)	1. Modify	Modify the initial Control Order
UR (Uniroyal)	2. Delay	Delay or lengthen negotiations by "dragging its feet"
	3. Accept	Accept the current Control Order
	4. Abandon	Abandon its Elmira plant by closing it down
LG (Local government)	5. Insist	Insist that the original Control Order be implemented

Table 1.2 Feasible states for the Elmira conflict

MoE									
1. Modify	N	Y	N	Y	N	Y	N	Y	–
UR									
2. Delay	Y	Y	N	N	Y	Y	N	N	–
3. Accept	N	N	Y	Y	N	N	Y	Y	–
4. Abandon	N	N	N	N	N	N	N	N	Y
LG									
5. Insist	N	N	N	N	Y	Y	Y	Y	–
State number	s_1	s_2	s_3	s_4	s_5	s_6	s_7	s_8	s_9

After agreeing upon the DMs who should be included in the conflict model, the Technical Experts asked the Domain Expert what are the specific options, courses of action, or powers, each DM has in the Elmira dispute. About one hour of discussion was required to come up with the list of options for which a summary label for each option is given in the middle column of Table 1.1 with an explanation provided in the right column. Notice that MoE has the single option of modifying the Control Order, UR possesses three options and LG has the single option of insisting that the original Control Order be implemented by UR.

An option can be either selected or not chosen by a given DM. Table 1.2 lists the feasible states or scenarios that could occur in the Elmira dispute after the infeasible ones are removed. In this table, each column stands for a state and for convenience of explanation each state or column is assigned a state number as indicated at the bottom. To interpret what each column means, consider state s_1 given as the left column, where a Y opposite a specific option means "yes" the option is taken whereas N indicates "no" the option is not selected. Hence, for state s_1, notice that MoE is not modifying the Control Order by not choosing the option numbered as 1, UR is delaying the negotiations by selecting the option numbered as 2 in its list of three options but not options 3 and 4, and LG is not insisting that the original Control Order is followed by not choosing option 5. This scenario is actually the situation that existed as the status quo when the Elmira conflict was studied as a current dispute in July, 1991.

Because each option can be selected or not, there are $2^5 = 32$ mathematically possible states that could occur. However, states that cannot take place in the real-world can be removed. Therefore, the three Technical Experts asked the Domain Expert if there are option selections that could not occur, such as those which are mutually exclusive. He thought that for UR options 2, 3 and 4 are mutually exclusive and at most one of those three options can be chosen at the same time. Moreover, UR would select at least one option. Notice that this is the case for the nine feasible states given as columns in Table 1.2. Finally, if UR decides to abandon its chemical plant in Elmira, the conflict is essentially over no matter what other options are taken as indicated by the dashes for state s_9 where each dash ("–") means N or Y. The process followed to obtain s_9 is referred to as state combination since the four dashes mean that column s_9 actually contains $2^4 = 16$ states, which are in essence the same and therefore can be interpreted as one state.

Comments about Options:

In practice, one can usually obtain the DMs and options in a conflict fairly expeditiously. Additionally, infeasible situations can be easily identified. Keep in mind that a user, or in this case the Domain Expert, does not have to write down the final list of feasible states, since this can easily be done using a user-friendly computer package referred to as a Decision Support System (DSS) (see Sect. 2.3.3 and Chap. 10). For the Elmira conflict, the size of the conflict was substantially reduced from thirty-two to only nine states, which is usually the case for most conflicts. Theoretically, the GMCR methodology can handle any finite number of DMs and options. Even if the conflict has a large number of feasible states, it can easily be modeled and analyzed using a DSS since the Domain Expert only has to supply the DMs, options, infeasible situations and relative preference information for each DM with respect to the feasible states. This clever way of recording a conflict is referred to as Option Form and was proposed by Nigel Howard (1971).

1.2.2.3 Relative Preferences

In the natural world, the laws of physics dictate how physical objects move and interact with one another, as they move in space and time. In the social world, value systems and related preferences drive human behavior and motivate human beings to take actions. Differences in value systems and preferences among people involved with a specific issue are what causes conflict and associated moves and countermoves as they all jockey for better positions. Accordingly, the most crucial input for modeling a conflict is the preference for each DM with respect to the feasible states or outcomes that could occur.

In a social situation, a friend may ask you if you would like to have a cup of coffee or tea. You may reply that you prefer to drink coffee over tea or it doesn't matter. When you only have to indicate whether coffee is more preferred, equally preferred or less preferred than tea, this is referred to as relative preference information. If you were to say to a friend that coffee is worth 9.623 to me and tea has a utility value of 2.529, your friend may respond that he does not have time after all to have a refreshment with you, since your response is so weird. Hence, in social situations, including many kinds of social conflicts ranging from sincere cooperation to warfare, one cannot usually obtain cardinal numbers in a meaningful way to represent preference in disputes.

An inherent design feature of GMCR is that only relative preferences are needed for each of the DMs among the feasible states. Notice from Table 1.2 that for the case of the Elmira conflict, there are only nine feasible states. Therefore, one can easily rank the nine states from most to least preferred by simply rearranging the ordering of the states until a sensible ranking is determined for a specific DM. This is what the Technical Experts did in consultation with the Domain Expert for the case of the Elmira dispute. During the second hour of their first meeting, a relative ranking of states for each DM was determined. Table 1.3 displays the ordering or ranking of feasible states from most preferred on the left to least preferred on the right from MoE's viewpoint. Notice that eight states containing an N opposite the fourth option of abandon are preferred over the situation, state 9, in which UR abandons its plant, since MoE is part of the Ontario Government which does not want to see jobs lost in Ontario due to a plant closing. Also, as attested by the Y opposite the third option in which UR accepts the current Control Order, the four states on the left in Table 1.3 are preferred over those for which UR does not accept it. As demonstrated by this discussion, a DM naturally thinks in terms of preference statements like the foregoing two examples. Table 1.4 shows the preference statements for MoE listed from most important at the top to least important at the bottom. An algorithm is available under the assumption of what is called transitive preferences (see comments on relative preferences given below) for taking these preference statements and ordering the states as given in Table 1.3. By comparing the hierarchical preference statements to the ranking of states in Table 1.3, one can appreciate how they work. Since the statements involve options, this procedure is referred to as Option Prioritization (Hipel et al. 1997, Fang et al. 2003a), which constitutes a refinement of the preference

Table 1.3 Ranking of states for MoE in the Elmira conflict from most to least preferred

MoE									
1. Modify	N	N	Y	Y	N	N	Y	Y	–
UR									
2. Delay	N	N	N	N	Y	Y	Y	Y	–
3. Accept	Y	Y	Y	Y	N	N	N	N	–
4. Abandon	N	N	N	N	N	N	N	N	Y
LG									
5. Insist	Y	N	N	Y	Y	N	N	Y	–
MoE's preference	s_7	s_3	s_4	s_8	s_5	s_1	s_2	s_6	s_9

Table 1.4 Option prioritization for MoE in the Elmira conflict

Preference statements
MoE most prefers that Uniroyal not abandon its Elmira plant. Hence, it does not want to see UR select option 4
Next, MoE would like Uniroyal to accept the current Control Order by choosing option 3
MoE then prefers that Uniroyal not delay the appeal process by UR not taking option 2
MoE would not like to modify the Control Order by selecting option 1 which it controls
MoE prefers that Local Government insists that the original Control Order be applied by taking option 5, if and only if (iff) it does not modify the Control Order (does not take option 1)

Table 1.5 Ranking of states for the three decision makers in the Elmira conflict

Decision maker	Ranking of states from most to least preferred								
MoE	s_7	s_3	s_4	s_8	s_5	s_1	s_2	s_6	s_9
UR	s_1	s_4	s_8	s_5	s_9	s_3	s_7	s_2	s_6
LG	s_7	s_3	s_5	s_1	s_8	s_6	s_4	s_2	s_9

tree approach first put forward by Fraser and Hipel (1988). Option Prioritization is embedded in three existing DSSs, which are discussed in Sect. 10.1.2.

Besides the ordering of states according to preference for MoE as ranked in Table 1.3, the Domain and Technical Experts separately ranked the states for UR and LG. Table 1.5 displays the ordering of the nine states for the three DMs in the Elmira conflict, where the ranking of states for each DM is from most preferred on the left to least preferred on the right. Of course, the ordering of states for MoE in Table 1.3, where option form is used, is the same as the ranking in Table 1.5 where only the state numbers are given. For an interpretation of what a state number means in actuality in Table 1.5 or elsewhere, one can refer to the option form presented in Table 1.2 or Table 1.3.

Comments on Relative Preferences:

GMCR was purposefully designed to handle a rich range of preference types that can occur in reality. To accomplish this, the deep foundations of GMCR rest upon the bedrock of a very basic type of preference situation on which to build a very solid theoretical structure. The most fundamental type of preference arises when one compares two objects or ideas according to one's preference. In the illustration given earlier, a friend may ask you if you prefer to have coffee or tea to drink to which you may respond that you prefer to have coffee, thank you. When a cardinal number, such as a utility value, is not used to stand for preference, the preference information is said to be relative or qualitative. Simple relative preference refers to the situation where coffee is either more preferred, equally preferred or less preferred to tea. Although one could dig deeper to ascertain how underlying values affect preference, this is not necessary since usually it is fairly easy to directly obtain relative preference information. For instance, if you prefer coffee to tea, it would be difficult even for you to explain how this preference is dictated by or connected to your underlying system of values. However, it should be noted that what is called multiple criteria decision analysis (MCDA) has been used to compare states in a conflict according to criteria (see, for instance, Ke et al. (2012a, b) and Silva et al. (2017a, b)).

Now consider situations that can arise when comparing three objects, ideas or states according to preference, such as three beverages from which your friend or host allows you to choose. Suppose you must select either coffee, tea or coca cola to drink. You may prefer coffee to tea because coffee smells better, tea to coke since tea is healthier. If you also prefer coffee to coke since coffee is hot and the coke is cold, then your preferences are said to be transitive as indicated in the diagram at the top of Fig. 1.3. However, suppose that you actually prefer coke to coffee because coke is cold. Then, you have the situation in the lower part of Fig. 1.3 which is called intransitive preference. GMCR is designed to take care of both transitive and intransitive preference information since both can arise in the real-world.

The ordering or ranking of states according to preference of a given DM in Tables 1.3 and 1.5 satisfy the assumption of transitivity. Hence, whenever states or objects are ordered according to preference, for which ties or equal preference can also be present, the ranking is automatically said to be transitive. Because a ranking of states or objects according to cardinal numbers contains a ranking for which a higher number means more preferred, cardinal numbers also satisfy the definition of transitivity. Hence, GMCR can handle transitive preference information even if it is expressed as cardinal numbers.

Sometimes, relative preferences may contain more information than just simple preference in what is called degree or strength of preference. For example, you may

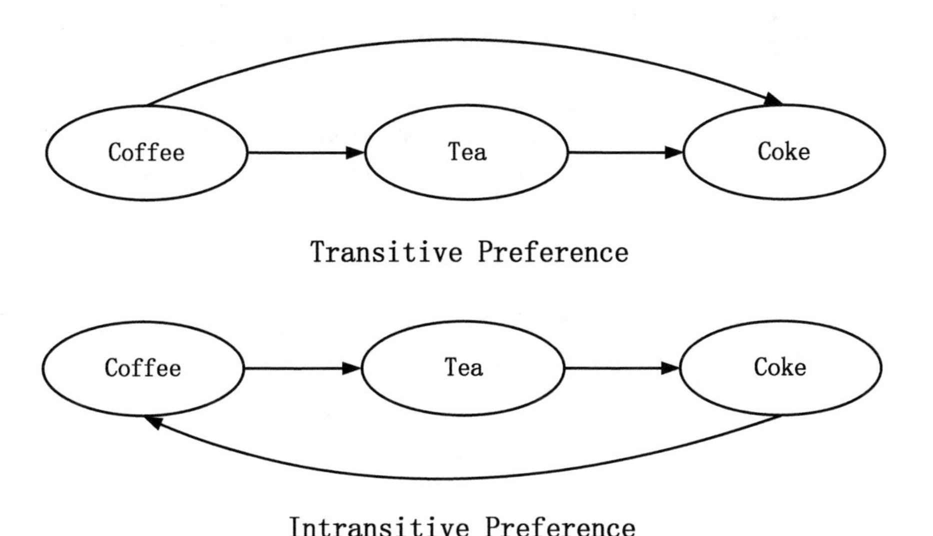

Fig. 1.3 An example of transitive and intransitive preferences

greatly prefer to have tea over coffee because tea is much healthier. An environmental agency may greatly prefer that a company clean up the pollution that it has caused. Once again, GMCR can also accommodate this kind of important preference which is the subject of Chap. 6 in this book.

Because preferences for DMs are often the most difficult information to obtain when calibrating a conflict model, uncertain preferences are an important topic in conflict resolution. A variety of formal approaches have been developed for modeling uncertain preferences. In some cases, part of the preference information for a DM may be simply "unknown", which is the focus of Chap. 5 in this book. As pointed out in Sect. 10.3.1, uncertain preferences can also be modeled using fuzzy sets, grey numbers and probability. Finally, different types of preference information can be combined in what is called hybrid preference. In Chap. 7, degrees of preference and unknown preference are combined for simultaneous employment within the GMCR methodology when modeling and analyzing a conflict when this is required.

A point to emphasize is that preferences for a specific DM depend not only what that DM can do on his or her own, but also on what other DMs choose to do. For instance, as indicated in Table 1.4, the top three preference statements for MoE depend on actions taken by UR and not itself. Moreover, by modeling the preference of each DM according to the way that DM evaluates reality makes one better appreciate the values, preferences and concerns of all parties involved in a dispute.

1.2.3 Stability Analysis

In a conflict situation, people think strategically like a chess player. Before making a specific move at any point in the game, the chess player thinks of the moves and countermoves that could take place after making his or her move. If the player believes that he will ultimately end up in a worse position and possibly lose the game, he will consider other moves and the consequences thereof before deciding what to do.

In a conflict study, the stability analysis stage is similar to a chess game. One now knows the DMs and their options, the feasible states which can occur, and the relative preferences of each DM. The preferences for a DM provide the motivation for moves that bring about possible improvement and those which can be used for sanctioning possible moves by competing DMs. If the game is at a particular state, the question arises as to whether or not it is advantageous for a DM to move unilaterally to a more preferred state or remain where he is. For instance, if the DM can improve on his own but the other DMs can move in a way that puts the DM in a less preferred position, he is better off to stay where he is. If this is the case, the state is said to be stable for that particular DM according to the type of human behavior under conflict that is being considered. If it is individually stable for all of the DMs, the state constitutes a possible resolution or equilibrium.

The characteristics of actual conflict and the dynamic behavior that can take place in a dispute are listed in Sect. 1.1.1, while the way in which GMCR has been designed to capture these features of actual conflict is mentioned in Sect. 1.1.2. Because people may behave differently under conflict, a range of solution concepts, describing potential behavior, have been defined. A specific solution concept, which is also called a stability definition, is a mathematical definition using ideas from set theory, logic and graph theory, which specifies how moves and countermoves take place to create a stable state. As explained in Chaps. 4–7, these logical definitions of solution concepts can also be written in what is called matrix form for different kinds of preference situations.

Table 1.6 lists the main solution concepts that are employed in Chaps. 4–8 in this book while additional stability definitions are mentioned in Sect. 10.3.1. The column on the left in this table provides the technical names of the solution concepts, the associated acronyms and the original references. The mathematical definitions of these solution concepts within the paradigm of GMCR and for different preference structures are given in Chaps. 4–7 for both the logical and matrix forms. The four columns on the right in Table 1.6 furnish qualitative characterizations of the solution concepts according to the four criteria of foresight, knowledge of preferences, disimprovement and strategic risk. More specifically, foresight reflects the number of moves and countermoves that a DM can envision when deciding on the stability of the state under consideration. Knowledge of preferences refers to the preferences that are required when carrying out a stability analysis. Therefore, when calculating stability for the top three solution concepts in Table 1.6, one only has to know the preferences of the focal DM and not those of the sanctioning DMs. The criterion labelled disimprovement reflects the tendency of a DM to put himself in a less preferred state to block unilateral improvements by an opponent. The threat of the particular DM ending up in a worse situation as a consequence of unilateral moves by the sanctioning DMs is sufficient to cause stability even if it is not advantageous for them. On the other hand, for sequential stability, the sanctioning DMs will only levy sanctions which are unilateral improvements for themselves. The criterion in the right column of Table 1.6, strategic risk, refers to the attitude of a DM towards strategic risk.

When carrying out a stability analysis for a given dispute, one analyzes every state for stability from each DM's viewpoint for all of the solution concepts listed in Table 1.6. For the case of simple preferences, the overall stability findings for the Elmira groundwater contamination dispute are provided later in Table 4.8, along with examples on how to calculate stability by hand.

The stability analysis discussed thus far assumes that DMs are behaving independently. This permits a DM to ascertain what is the best that he or she can hope to achieve by acting noncooperatively. The next step in an analysis is to determine if one can fare even better by cooperating with others by forming a coalition. To answer this question, one can employ the coalitional or cooperative versions of the solution concepts listed in Table 1.6 which are defined in Chap. 8 for different kinds of preference structures.

As mentioned in Sect. 1.2.2, the Elmira groundwater contamination dispute was modeled and analyzed using GMCR by the three Canadian authors of this book who fulfilled the role of the Technical Experts for GMCR and a Domain Expert who

Table 1.6 Solution concepts describing human behavior under conflict (based on Table 1 in Hipel et al. (1997))

Solution concepts	Stability descriptions	Foresight	Knowledge of preferences	Disimprovement	Strategic risk
Nash Stability (R) (Nash 1950, 1951)	Focal DM cannot move unilaterally to a preferred state	Low	Own	Never	Ignores risk
General metarationality (GMR) (Howard 1971)	All focal DM's unilateral improvements are sanctioned by subsequent unilateral moves by others	Medium	Own	By opponents	Avoids risk: conservative
Symmetric metarationality (SMR) (Howard 1971)	All focal DM's unilateral improvements are sanctioned, even after response by the focal DM	Medium	Own	By opponents	Avoids risk: conservative
Sequential stability (SEQ) (Fraser and Hipel 1979, 1984)	All focal DM's unilateral improvements are sanctioned by subsequent unilateral improvements by others	Medium	All	Never	Takes some risks: satisfies

was extremely familiar with this serious pollution problem. Recall that in ongoing negotiations during the summer and early fall of 1991, UR (Uniroyal Chemicals Inc.) was appealing a Control Order issued by the MoE (Ontario Ministry of the Environment) to cleanse the aquifer which it had polluted. LG (Local Government) has the option of insisting that the original Control Order be implemented. The three Technical Experts met with the Domain Expert on two occasions to study this conflict. In the first two-hour session, they determined the DMs and their options (see Table 1.1), decided upon the feasible states (Table 1.2) and the relative preferences of

Table 1.7 Evolution of the Elmira conflict from the status quo to a transitional noncooperative equilibrium and to a final cooperative coalition equilibrium

DMs	Status quo		Noncooperative equilibrium		Cooperative equilibrium
MoE					
1. Modify	N		N	⟶	Y
UR					
2. Delay	Y		Y	⟶	N
3. Accept	N		N	⟶	Y
4. Abandon	N		N		N
LG					
5. Insist	N	⟶	Y		Y
State	s_1		s_5		s_8

each of the DMs over the feasible states (Tables 1.3 and 1.5). After the first meeting, the three Technical Experts carried out a thorough stability analysis for the conflict model they constructed based on guidance from the Domain Expert. At the second two-hour meeting which took place about two weeks later, the three Technical Experts showed their key strategic findings to the Domain Expert in order to get feedback and possibly make further refinements.

Table 1.7 provides an overview of the strategic insight gleaned from the stability calculations. As can be seen on the left, the status quo state during July of 1991 is state s_1 in which MoE has not modified its Control Order (an N is placed opposite its option called Modify), UR has selected its option to delay the negotiation (as indicated by the Y opposite option number 2, Delay, which it controls), and LG is not strongly insisting that the original Control Order be implemented (an N written opposite option number 5). LG thus caused the conflict to move from state s_1 to state s_5. This movement is referred to as a unilateral move by LG since only LG changes its option selections while the other DMs' option choices remain fixed. Moreover, as can be seen in Table 1.5, LG prefers state s_5 to s_1 since s_5 is further to the left in the ranking of states for LG. Because s_5 is more preferred to s_1 by LG, this move is referred to as a Unilateral Improvement (UI) for LG. In fact, because a directed graph is used to systematically keep track of the unilateral moves by each DM in one step, the overall methodology is called Graph Model for Conflict Resolution (GMCR). A directed graph can be drawn separately for each DM or the separate directed graphs can be combined into a single overall integrated directed graph if the conflict is not too large. Figure 1.4 displays the integrated graph model for the three DMs in the Elmira dispute, which is also shown as Fig. 3.5 in Sect. 3.2.3 and Fig. 4.8 in Sect. 4.5. In this figure, each node stands for one of the nine feasible states given in Tables 1.2 and 1.3. An arc joining two states represents a move in one step for the DM written on the arc. Hence, notice in the left position of the graph there is an arc which joins state s_1 and state s_5 with LG written on it to show the

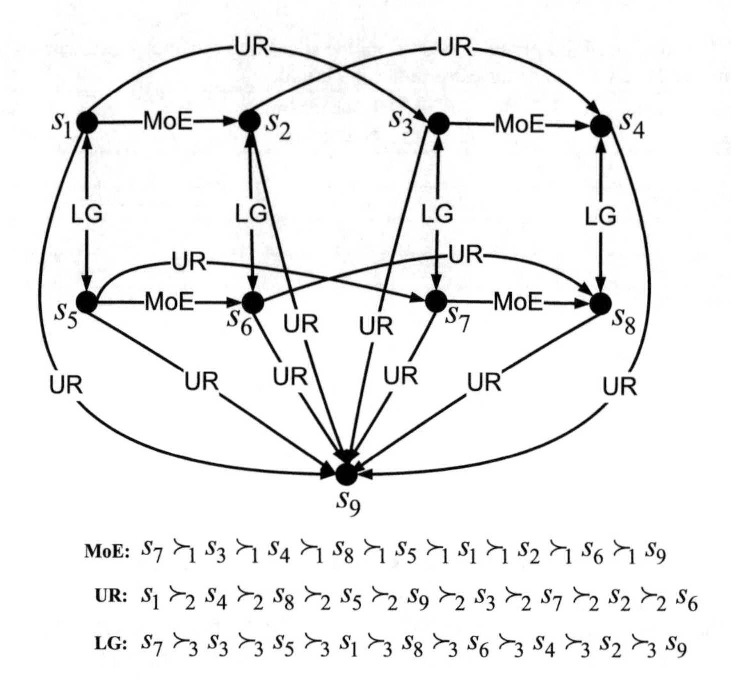

Fig. 1.4 Integrated graph of the Elmira conflict

situation between the left two states in Table 1.7. The double arrowheads indicate that movement can take place in either direction. In Table 1.7, only the movement from s_1 to s_5 is indicated in the evolution of the dispute. As illustrated in Fig. 1.4, movement in one direction, which means the move is irreversible, is depicted using a single arrowhead. Therefore, the arc controlled by UR going from s_1 to s_9 only has an arrowhead pointing towards s_9 to indicate that closing down its chemical plant in Elmira is essentially an irreversible move. Notice that below the integrated graph in Fig. 1.4, the ranking of states from most to least preferred is provided (where $s_k >_i s_e$ means state s_k is more preferred to state s_e by DM i). This is because a complete graph model consists of the feasible states (which can be formed by option selections of the DMs in the dispute), movements in one step for each DM, and relative preferences for each DM. Keep in mind that given the DMs and options the resulting states and the movements in one step for each DM among states can be determined using a DSS and do not have to be supplied by the Domain Expert.

Returning to Table 1.7, the three Technical Experts pointed out to the Domain Expert in their second meeting that LG can unilaterally cause the conflict to change from sate s_1 to s_5 as a result of its UI. However, once the conflict arrives at state s_5, it is not advantageous for any DM to unilaterally move away from it. As explained in detail in Sect. 4.5, this state is individually stable for all three DMs and hence is a noncooperative equilibrium. The conflict will remain at s_5 unless DMs somehow cooperate and act together as a coalition. Nonetheless, notice from the graph model in Fig. 1.4 as well as Table 1.5, both MoE and UR prefer state s_8 to s_5. Therefore, if they cooperate and move together as shown for the two states on the right in Table 1.7, MoE and UR have a joint UI from state s_8 to s_5. Specifically, MoE can modify the Control Order by changing option 1 from not selecting it in state s_5 to choosing it as indicated by the Y written for state s_8 opposite option 1 as well as the arrow joining the N and the Y for option 1 at the top right of Table 1.7. The arrows joining state s_5 and s_8 opposite options 2 and 3 indicate that UR simultaneously stops delaying the negotiations and accepts the modified Control Order.

The modeling and analysis of the Elmira negotiation during the summer of 1991 as a current ongoing dispute at that time reflect well what actually occurred. On October 7, 1991, MoE and UR suddenly announced that they had reached an agreement. MoE modified the Control Order and UR accepted the modification and thereby ceased delaying the negotiations. Clearly, both parties recognized that forming a coalition for this joint improvement to take place was mutually beneficial, as shown in the joint UI from s_5 to s_8 in Table 1.7. The dramatic announcement of an agreement was a surprise to LG, which was angry over not being included in the secret agreement. This is because LG had devoted significant effort, time and money to the negotiation process as well as the events and activities that took place from the time the carcinogen was first discovered in Elmira's water supply to October of 1991.

Comments about the History of the Elmira Contamination:

To supply the town of Elmira with fresh uncontaminated water, a 15 km pipeline was built from the City of Waterloo to Elmira. Even though only the wells in Elmira located close to the UR chemical plant were contaminated, all of the wells in the town were closed and, hence, all of Elmira's water supply is from Waterloo, which continues to be the case up until the present time. To stop the underground contaminated water plume from spreading and also for treating the polluted water, it is being pumped to the surface, treated using a fairly simple process of exposing the water to ultraviolet light and then releasing it into the nearby creek. Besides cleaning the polluted part of the aquifer, all ongoing NDMA effluents are being treated.

As of early 2012, a private company called Chemtura was cleansing the water for MoE by operating:

- a shallow aquifer containment and treatment system on the Uniroyal property to prevent discharge of contaminated water from reaching the adjacent creek;
- a municipal aquifer property boundary containment and treatment system along the property boundary to stop the most contaminated water from leaving the property; and
- an off-site containment and treatment system consisting of wells at four locations to prevent further movement of the plume. One of the locations is the former municipal supply wells at the south end of town.

This type of continuing treatment of contaminated water has already been in operation for a quarter of a century and could continue for decades, if not centuries. In fact, the Elmira dispute is expanding since residents and environmentalists are, as of 2017, concerned about pollutants from the chemical plant deposited in the bed of the stream flowing through the site and elsewhere. Philpot et al. (2017) completed a strategic study of this particular dispute using GMCR.

Elmira exemplifies the importance of appropriately treating effluent wastes when a chemical process is first brought into operation and not storing it in unsafe facilities for later treatment or discharging it directly into the environment. Once pollution penetrates an aquifer, it could persist for thousands of years and continue to spread outward from the source. As pointed out at the start of Sect. 1.2.2, brownfields, land polluted by industrial and agricultural activities, will be a source of great controversy around the globe for a long time to come. Accordingly, developing countries would be wise to not follow the path of older industrial countries in Europe and North America by knowingly polluting the landscape under the banner of economic growth according to the assumption that the pollution can be cleansed later using the profits of industrialization. This pathway leads to astronomical costs to just contain the pollution and perhaps never being able to completely eliminate it and the associated huge risks to human health. A connected problem is how to handle chemicals of emerging concern: new synthetic chemicals developed by humans for which the health and environmental impacts are unknown. Sound scientific practices and effective conflict resolution procedures will clearly play major roles. Other types of challenging conflicts to which GMCR has already been applied are mentioned in Sects. 1.2.5 and 2.4.

1.2.4 Follow-Up Analyses

As illustrated using the Elmira groundwater contamination dispute, individual and coalitional stability analyses can be carried out to obtain strategic insights as an aid for better understanding of the conflict and what actions can be taken to resolve the dispute. Procedures for extending the GMCR techniques beyond those outlined in the previous sections form part of what is referred to as Follow-up Analyses. Usually, the characteristics of a given conflict being investigated suggest what type of Follow-up Analyses to pursue. Two types of Follow-up Analysis, Evolution of a Conflict and Sensitivity Analysis, are described in this section. In Sect. 10.3, ongoing and planned expansions of the GMCR methodology are discussed. Many of these developments could be classified under Follow-up Analyses.

1.2.4.1 Evolution of a Conflict

As mentioned in Sect. 1.2.3, in a stability analysis, each state is analyzed for stability from each DM's viewpoint according to a range of solution concepts describing potential human behavior under conflict such as those listed in Table 1.6. This can be carried out for both individual competitive behavior and cooperative behavior for which coalitions can form. Possible coalition formation is considered since one may wish to discover if DMs can do even better via cooperation and thereby reach a win/win resolution at least for the coalition members. Recall that this is the case for the groundwater contamination dispute analyzed in Sect. 1.2.3.

In many conflict situations one may wish to ascertain if a noncooperative or cooperative resolution can be reached from a starting state which is often the current situation, or status quo, for a conflict being formally studied. Therefore, finding possible paths which can be followed to possibly reach a particular state is also referred to as Status Quo Analysis. Because of the usefulness and importance of this topic, it is the focus of Chap. 9 in this book.

Table 1.7 for the Elmira conflict was included in Sect. 1.2.3 in order to provide an explanation of what happened in this groundwater contamination dispute as part of the overall stability analysis discussion. In fact it is an informative example of how a conflict can evolve over time from a starting status quo state to a final resolution. As explained in Sect. 1.2.3, by selecting its option to insist that the initial Control Order be implemented, LG causes the conflict to move from state s_1 to s_5, which was calculated to be a noncooperative equilibrium. Via cooperation, MoE and UR can make the conflict go from state s_5 to state s_8 which is the final cooperative resolution to the dispute. Recall that state s_8 is preferred to state s_5 by both MoE and UR and, therefore, this coordinated movement constitutes a joint unilateral improvement.

1.2.4.2 Sensitivity Analyses

In a sensitivity analysis, one wishes to determine how meaningful changes in certain model parameters can influence the stability findings. As just pointed out, the types of sensitivity analyses to pursue are dictated by the particular circumstances and properties of the conflict being investigated. The kinds of sensitivity analyses to be considered include:

- preference changes,
- uncertain preference,
- option modification or expansion,
- additional option to produce a desirable result,
- bringing other DMs into the game,
- consideration of other kinds of human behavior (solution concepts),
- misunderstandings (called hypergames), and
- entertainment of other modes to bargaining and negotiation.

As an example of one type of sensitivity analysis, consider preference changes. In a particular conflict, the final resolution may depend heavily upon the preferences of a key DM. Hence, one could alter part of the ranking of an important DM, such as a company in a pollution dispute, to see if a better result can be obtained. If, for instance, the equilibria do not essentially change as a result of a subsequent analysis using the new preference information, then one can conclude that the findings are robust to its alteration. When preferences are uncertain (see Chap. 5 for unknown preference), contain strength, degree or level (see Chap. 6), or are a combination of these two preference characteristics (see Chap. 7 for hybrid preferences), one could carry out analyses to determine how the strategic findings change.

1.2.5 Application Approaches

The Elmira groundwater contamination dispute described in Sect. 1.2.2 was a serious environmental conflict which took place in Southern Ontario, Canada. This controversial environmental conflict is utilized for explaining how one can formally model, analyze and carry out follow-up analyses of an actual dispute in Sects. 1.2.2–1.2.4, respectively, using a highly flexible methodology called the Graph Model for Conflict Resolution (GMCR) which is the focus of this book. Moreover, this dispute is employed for illustration purposes in many sections in later chapters in this textbook.

Because the Elmira conflict was formally investigated using GMCR when it was ongoing, it constitutes an insightful case study on how one can analyze a current dispute. As listed below, in addition to an ongoing dispute, one may also wish to study historical and hypothetical controversies for reasons mentioned:

- *Current*
  - Need to make wise and informed decisions.
- *Historical*
  - Appreciate why a Pareto-inferior solution took place.
  - By understanding the past, avoid making similar mistakes in the future.
- *Hypothetical*
  - Study strategic interactions in a generic conflict.
  - Test new solution concepts and other theoretical developments.

An historian, for example, may wish to ascertain why a Pareto-inferior outcome took place when a better outcome could have occurred, perhaps via better communication among the competitors. An outcome is called Pareto-inferior when there exists another state which is not less preferred by any DM and more preferred by at least one DM. A desirable outcome which is not Pareto-inferior can sometimes be reached for a set of DMs when they have meaningful communication. As the maxim goes, "by learning from one's past one can make wiser decisions in the future."

Quite often, a very simple conflict can be used to reflect the key characteristics of a general class of disputes. For instance, Prisoner's Dilemma is a generic conflict used to represent situations in which one must decide if one should act independently according to one's short-term self-interest or cooperate with another party to fare better in the longer term. The Sustainable Development Game first proposed by Hipel (2001) mirrors a basic situation in which an environmental agency is in dispute with a developer whose project could possibly harm the environment. This generic conflict is employed for explaining ideas later in this book in Sect. 3.1. Sometimes it is informative to employ the format of what is called a 2×2 game which is explained in Sect. 3.1. Finally, as noted above, hypothetical conflicts, as well as smaller actual disputes, provide a convenient mechanism for checking and refining new theoretical expansions to GMCR such as those presented in Chaps. 3–9 and other advancements mentioned in Sect. 10.3.

The most convincing way for demonstrating the efficacy of a decision methodology is via real-world applications ranging from the simple to highly complicated situations. The Elmira dispute constitutes an insightful application of GMCR to a current conflict which provides useful strategic insights. Table 1.8 lists a wide variety of real-world conflicts to which GMCR has been applied in many different fields. Table 1.9 provides a listing of the applications used in this book for explaining many key ideas in GMCR and demonstrating how these concepts can be applied in practice. In fact, because GMCR realistically captures the key features or characteristics of a conflict in its basic design, the GMCR methodology is independent of the field or discipline of study and can be utilized for investigating virtually any types of "social" conflict. GMCR is clearly a good "systems" approach to investigating conflict as explained later in Sect. 2.3.2.

As emphasized in Sect. 1.1.1, as well as in the next chapter in Sects. 2.3 and 2.4, there are a plethora of important conflicts to address in order to obtain meaningful strategic insights for making informed decisions. For example, one of the greatest dangers now facing society on a global scale is climate change and associated controversies on how to fairly and effectively cut back on greenhouse gas releases, as outlined in Sect. 2.4. In his book on "The Future: Six Drivers of Global Change", Gore (2013) describes six emergent revolutionary changes on a global basis, all of which involve intense conflict. Certainly, in an over-crowded world fraught with increasing disagreements, conflict will always be present and, hence, there is a great need for access to realistic conflict methodologies like GMCR.

Hipel et al. (2001b) describe three main types of situations in which GMCR can be used as listed below.

- *Analysis and simulation tool for a DM in a conflict, or a DM's agent.* Moves and countermoves following a DM's possible actions can be analyzed, and the potential consequences of certain actions assessed, in order to improve the DM's position. Preparations and assessments can be carried out at different times as the conflict unfolds.
- *Communication and analysis tool in mediation.* GMCR can be utilized by a mediator to assess the possible consequences of DMs' various preferences, without confirmation by the DMs as to which ones correctly describe their preference. Options that are beneficial, detrimental, or irrelevant to all parties can be ascertained by using this process.
- *Analysis tool for a third-party analyst.* Based on the observed outcome and evolution of a conflict, an analyst can estimate the DMs' preferences. How the structure of the conflict influenced the DMs' behavior and better ways to structure a future conflict can also be studied.

Table 1.8 Application areas

Area	Explanation	References
Aquaculture	Fish farming in the oceans, lakes, rivers, and small ponds causes controversy because of the fear of the spread of disease to wild fish stocks and pollution	Noakes et al. (2003, 2005), Hamouda et al. (2004, 2005)
Brownfields	The polluted land and aquifers by former and existing industry are called brownfields. This is a massive problem in all industrialized and developing countries. In the Elmira conflict (see Sects. 1.2.2, 1.2.3, 1.2.4 and 4.5), the waste from a chemical factory caused a serious conflict in Canada because the aquifer was polluted by a carcinogen (cancer causing agent)	Hu et al. (2009), Bernath Walker et al. (2010), Hipel et al. (2010), Yousefi et al. (2010c, 2011), Hipel and Bernath Walker (2012), Philpot et al. (2017)
Construction management	Conflicts arise in construction among the owner of a project, the construction company building the project for the owner and labor over delays, shoddy workmanship, unexpected costs, strikes and other reasons. Public-private partnerships for building, operating and maintaining infrastructure can be controversial	Kassab et al. (2006, 2010, 2011), Yousefi et al. (2010a, b)
Energy	Energy conflicts can arise for many reasons including environmental concerns such as greenhouse gas releases, pollution of groundwater by fracking, and potential radiation leaks from nuclear plants. The routing of electrical transmission lines as well as oil pipelines through different political jurisdictions can create serious conflicts	Armin et al. (2012), Matbouli et al. (2015), Xiao et al. (2015), Garcia et al. (2016), O'Brien and Hipel (2016)
First Nations (Aboriginal People)	In many countries, the honoring of aboriginal rights can create disputes when these rights are abused or perceived to be ignored in problems including hydroelectric power development, resource ownership and fishing right. In Canada, aboriginal people are referred to as being First Nations	Ma et al. (2005), Obeidi et al. (2006)

Table 1.8 (continued)

Area	Explanation	References
Military and peace support	Von Clausewitz said that warfare is the extension of politics by other means. Negotiations can avoid war such as the avoidance of all-out nuclear war between the Soviet Union and USA in the Cuban Missile Crisis of 1962. Wars over water in the Middle East have been avoided in the past by Third Party Mediation. Ongoing skrmishes between Azerbaijan and Armenia have erupted over the control of Armenian enclave of Nagorno-Karabakh in Azerbaijan which is territory now controlled by Armenia since almost all of the people living there are Christian Armenians. Peace support in areas such as Bosnia attempts to stop fighting among different cultural groups in a region	Fraser et al. (1990), Kilgour et al. (1998), Hipel (2011), Hipel et al. (2014)
Softwood lumber (international trade)	The conflict over the export of softwood lumber from Canada to the United States over what American lumber companies claim to be subsidized prices is an example of an ongoing international trade dispute	Hipel et al. (1990, 2001b)
Sustainable development	The conflict over maintaining a stable environment in the face of societal development activities such as industrialization and agriculture arises within and among countries around the globe	Levy et al. (1995), Hipel and Obeidi (2005), Ghanbarpour and Hipel (2009)
Water exports and diversions	The potential exportation of water in bulk quantities from Canada has caused intense conflict to arise in Canada. One form of water exportation is diverting water such as in China where it is being diverted from the south to the north	Obeidi et al. (2002), Hipel and Obeidi (2005), Obeidi and Hipel (2005)

Table 1.8 (continued)

Area	Explanation	References
Water resources management	The utilization, management and control of water among different users within and across political jurisdictions have caused serious disputes to arise around the globe. Because water is such a key resource to society, the area of water resources management and governance is highly developed (Hipel et al. (2008), see Sect. 2.4.2 in this book). Wolf (2002) claims that water is a mechanism for cooperation and peace rather than war	Hipel et al. (1999, 2001a, 2015, 2016), Gopalakrishnan et al. (2005), Nandala and Hipel (2007), Ma et al. (2011, 2013), Madani and Hipel (2011), Chu et al. (2015), Philpot et al. (2016), Garcia et al. (2017)

Under the first category, a consultant may advise a company on how to interact with its competitors to obtain a greater market share of a given product. At a company retreat, the company Vice Presidents could use GMCR in "role playing" by pretending that a given Vice President represents a specific company. By putting itself in the "shoes" of its competitors, the organization can better comprehend the consequences of any actions it may take.

Under the second classification, a mediator, for instance, could employ GMCR in negotiations between labor and management over designing a new contract. Between sessions, the mediator could use GMCR to determine how only small concessions and minimum shifts in preferences by each side may result in a mutually beneficial agreement. These valuable insights could be utilized by the mediator during the face-to-face negotiations between the parties, or via behind-the-scenes communication, to reach a final agreement.

Under the third category, an interested third-party may wish to know how conflicts may be resolved. For example, even though China is not a direct combatant in the ongoing conflict in the Middle East, it is very interested in knowing what could take place since this could affect its trade opportunity with countries in this region. Although Canada is not directly involved in disputes over the control of islands located in the South China Sea, it does want to have an understanding of the security and economic consequences of those disputes over time.

In the Elmira dispute studied in Sects. 1.2.2–1.2.4, the three Canadian authors of this book acted as Technical Experts for analyzing this conflict for which background information was provided by a Domain Expert. Because the Domain Expert was not taking part in the actual negotiations, but was rather an interested third party who was an environmentalist and property owner in Elmira, this conflict falls under the third category in the above list. In fact, "consultants" could be used in all three categories given in this list. Although, a consultant would mainly be hired by an organization to

Table 1.9 Descriptions of the cases used in this book

Cases	Description	Locations
Sustainable development conflict	Generic dispute between the developer of a project and an environmental agency who would like to see the Developer to act responsibly to minimize environment damages	Sections 3.1, 4.2, 5.2, 5.3, 6.3, and 6.5
Elmira groundwater contamination dispute (**Elmira**)	Negotiations among a chemical company which polluted the underground water supply aquifer of the town of Elmira located in Southern Ontario, Canada, the Ontario Ministry of the Environment, and the Local Government over cleaning the aquifer	Sections 1.2, 4.5, 9.2, and 9.3
Lake Gisborne bulk water export conflict (**Gisborne**)	Dispute among a company, government and environmentalists over the proposed bulk export of water from Lake Gisborne located in the Canadian Atlantic province of Newfoundland and Labrador	Sections 5.4, 7.5, 8.10, and 9.3
Garrison diversion unit (**GDU**) conflict	Dispute among the proponents of constructing a large scale irrigation project in the American State of North Dakota, the Canadian Government representing the interests of Canadian whose waters would be polluted downstream, and the International Joint Commission (a neutral body formed by the Boundary Waters Treaty of 1909 between Canada and the US to investigate disputes over water)	Sections 6.6 and 9.3

investigate a present conflict, they could be used to carry out tasks for a client related to current, historical and hypothetical disputes mentioned in the first list given in this subsection.

Below, a possible procedure is outlined on how a consultant in conflict resolution may perform his or her duties for a client. In fact, this approach is close to that used in Sects. 1.2.2–1.2.4 for formally studying the Elmira groundwater contamination dispute. Clearly, there are many advantages to using GMCR in practice as is summarized in the next subsection.

- First Meeting
  - Morning: In highly interactive discussions the client provides the consultant with information about the decision makers and options in the conflict under study.
  - Afternoon: In-depth discussions take place to determine the preference statements for each decision maker.
- Analytical Calculations

  The consultant uses the model input data elicited from the client to calculate analytical results and discover strategic insights employing the DSS GMCR (Graph Model for Conflict Resolution) II. A preliminary report is written.
- Second Meeting

  Consultant and client use the initial report as a basis to confirm insights and discover new ones. Strategic advice for decision-making under conflict is proposed.
- Final Report

  Consultant prepares the final report and sends it to the client.

1.2.6 Benefits

The Graph Model for Conflict Resolution (GMCR) possesses a solid mathematical design which makes it ideally suited for systematically investigating real-world conflict. As pointed out in Sect. 1.1.1, actual conflict contains key inherent characteristics such as the way decision makers (DMs) think and behave in terms of moves and countermoves as they attempt to do the best they can both independently and perhaps in cooperation with others. As emphasized in Sect. 1.1.2, these important features of reality are purposely embedded into the axiomatic foundations of GMCR in order to provide the underlying capability of GMCR to reflect reality via a range of powerful functions like forecasting possible resolutions based upon different types of behavior in diverse conflict situations. When implemented as a user-friendly program as summarized in Sect. 2.3.3 and described in more detail in Chap. 10, the GMCR methodology can be readily utilized by both practitioners and researchers to formally study conflict ranging from the simple to highly complicated ones in virtually any field where conflict arises. Moreover, when conflict studies are carried out within an overall systems thinking approach to governance, the strategic insights and guidance obtained can be fully appreciated, as explained in Sect. 2.4. The rich range of applications of GMCR across many different fields mentioned in Sect. 1.2.5 and listed in Table 1.8 confirms the ingrained capability of GMCR to address tough conflicts occurring in the real-world. Finally, the realistic foundational design of GMCR means that the basic methodology can be easily expanded to handle all of the advancements presented in detail in this book as well as many others mentioned in Sect. 10.3.

Because of the aforesaid and other reasons, it is not surprising that the utilization of GMCR can provide a wide variety of benefits to users. These benefits, which overlap with those already listed in Sect. 1.1.2 and elsewhere in the book, include:

- Provides a systematic structure of a conflict.
- Can be conveniently applied in practice.
- Handles any finite number of decision makers and options.
- Can take into account a rich range of preference structures.
- Keeps track of all feasible movements by DMs among states.
- Can model both reversible and irreversible moves, as well as common moves explained in Sect. 3.2.
- Describes the ways in which people or organizations may behave under conflict.
- Generates states or possible scenarios.
- Forecasts compromise resolutions or equilibria.
- Determines most likely conflict resolutions.
- Performs extensive follow-up analyses, including tracing the evolution of a conflict and sensitivity analyses.
- Points out where more useful information is needed.
- Can be easily expanded to take into account new developments such as unknown preference (see Chap. 5), degrees or strength of preferences (Chap. 6), and hybrid preference which constitutes a combination of unknown and degrees of preference (Chap. 7).
- Can be made available as a decision support system for convenient application purposes.
- Permits easy and convenient communication.
- Allows systematic book-keeping.
- Provides valuable strategic insights.
- Suggests paths for optimal decision-making for a given decision maker.
- Can be easily expanded to handle a rich range of conflict characteristics or situations such as those given in this book and Chap. 10.

1.3 Journeys Through the Book

You, the reader of our book, are about to embark on a fascinating journey through the realm of informative conflict resolution using flexible methodologies for addressing

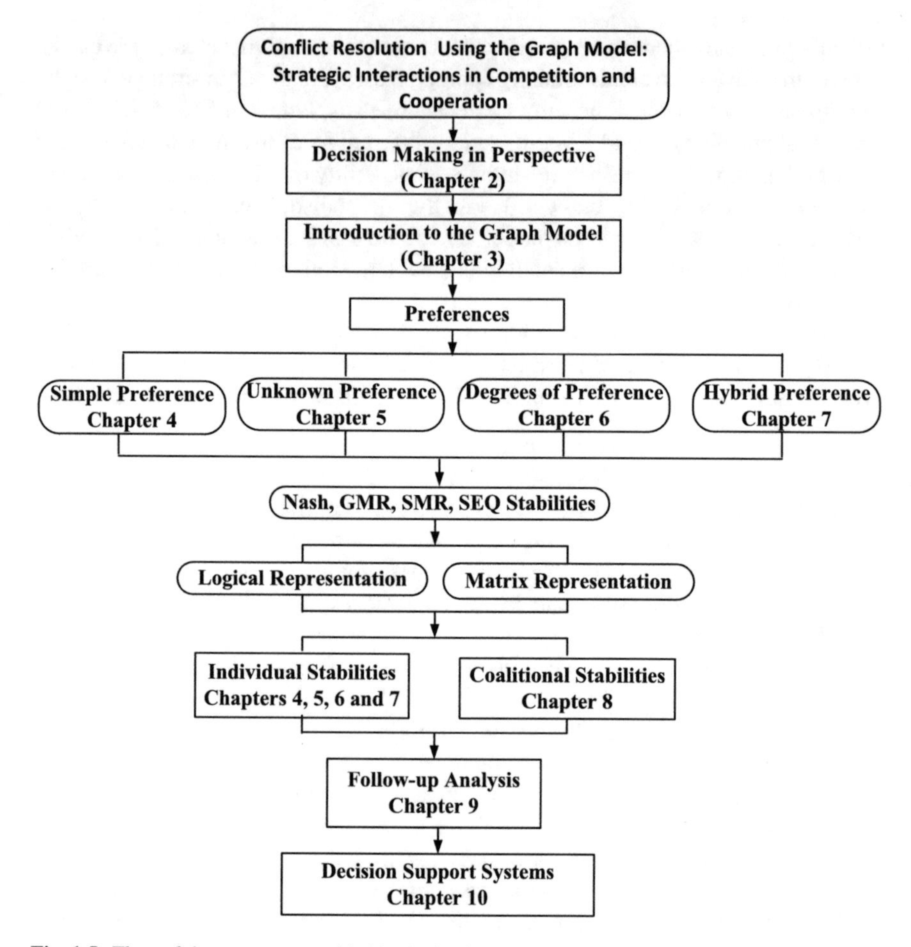

Fig. 1.5 Flow of the contents provided in the book

challenging disputes that can arise in the real-world. As is the case for many interesting ventures, there are a number of possible routes that can be followed through this kingdom. Figure 1.5 provides a road map on how you can navigate through our book, depending upon your available time and particular interests. Additionally, Table 1.10 furnishes a brief description of the contents of each chapter along with the applications that are used to illustrate how key ideas in the chapters can be utilized in practice. An overview of each of the disputes is provided in Table 1.9. With this travel information in hand, let us discuss a number of pathways that you can consider exploring.

A popular English expression is the "proof of the pudding is in the eating". In other words, try applying the Graph Model for Conflict Resolution (GMCR) to an actual dispute of direct interest to you in order to convince yourself that it is useful for systematically investigating a conflict of your choice and thereby become inspired

Table 1.10 Key contents in the book

Chapter	Title	Contents	Conflict examples
Chapter 1	Conflict resolution in practice	Objectives of the book. Demonstrate how GMCR can be applied to an actual conflict	Elmira groundwater contamination dispute
Chapter 2	Decision-making in perspective	An overview of conflict resolution and formal decision-making techniques	
Chapter 3	Conflict models in graph form	Three kinds of games are introduced: normal form, option form and the graph model	1. Sustainable development conflict; 2. Elmira groundwater contamination dispute
Chapter 4	Stability definitions: simple preference	Logical and matrix representations of the four basic stabilities are described	1. Sustainable development conflict; 2. Elmira groundwater contamination dispute
Chapter 5	Stability definitions: unknown preference	Unknown preference and associated logical and matrix representations of the four stabilities are described	1. Extended sustainable development conflict; 2. Gisborne bulk water export conflict
Chapter 6	Stability definitions: degrees of preference	Multiple degrees of preferences and related logical and matrix representations of the four stabilities are presented	1. Three-degree sustainable development conflict; 2. Garrison Diversion Unit conflict
Chapter 7	Stability definitions: hybrid preference	Hybrid preferences and associated logical and matrix representations of the four stabilities are put forward	Extended Gisborne conflict
Chapter 8	Coalitional stabilities	Logical and matrix representations of four coalitional stabilities for various preference structures are developed	Gisborne conflict
Chapter 9	Follow-up analysis: conflict evolution	The conflict evolutionary paths from any status quo to an equilibrium for various preference structures are described	1. Elmira conflict; 2. Gisborne conflict; 3. Garrison Diversion Unit conflict
Chapter 10	Design of a decision support system for conflict resolution	Existing decision support systems (DSSs), a universal design for a graph model DSS and future opportunities in conflict resolution are described	

to ultimately end up reading most of the book. Accordingly, if you would like to be able to begin by modeling and analyzing an existing conflict as soon as possible, the shortest path is as follows. As is illustrated in Sect. 1.2 for the Elmira groundwater contamination conflict, an informative way to record a conflict is to employ option form which provides a meaningful way or "language" to discuss it with others. By reaching an improved understanding of what is taking place in the conflict, better strategic insights can be obtained as an aid for enhancing decision-making. Whatever the case, if you want to apply GMCR as expeditiously as possible first read about option form described in Sect. 3.1.2 followed by the graph model in Sect. 3.2. As can be seen in the middle part of Fig. 1.5, four approaches to modeling preferences are available in this book with others mentioned in Sect. 10.3. For now, stay with the simple preference described in Sect. 3.2.4 and in Chap. 4 when employed at the analysis stage in which moves and countermoves are considered to determine stability for a particular state from a given decision maker's (DM's) point of view as well as the associated overall equilibria. Of the four key stability definitions, consisting of Nash, general metarational (GMR), symmetric metarational (SMR) and sequential stability (SEQ) given in Sects. 4.2.1 and 4.2.3 for the two-DM and n-DM (two or more DM) cases, respectively, concentrate on employing Nash and SEQ in your application. The concept of SEQ stability is realistic to employ in practice since a DM will not do something that hurts himself or herself. Moreover, only look at the logical representation of a conflict for now since a direct interpretation explains how people think in terms of moves and countermoves in a conflict situation. You can find out how stability calculations are actually carried out by reading Sect. 4.5 for the case of the Elmira dispute. If your conflict is not too large, you can first do your calculations by hand to fully appreciate why this way of formally investigating conflict reflects what takes place in reality. If you plan to analyze larger conflicts, thereby making hand calculations too tedious, investigate many conflicts, or check your manual calculations, you may wish to employ an existing decision support system (DSS) (see Sect. 10.1.2) or write your own program (see Sect. 10.2 for a universal design and for specifics about designing the input, engine and output for a DSS).

For a fairly comprehensive tour through the book depicted in Fig. 1.5, again read Sect. 3.1.2 on option form and Sect. 3.2 on the graph model. Next, pass through Chap. 4 dealing with simple preference only and stick to the sections in Chap. 4 as well as Chap. 8 (Coalitional Stabilities) and Chap. 9 (Follow-up Analysis: Conflict Evolution) that are concerned with the logical form of systematically studying a dispute.

As can be seen in the central part of Fig. 1.5, there are four key types of preference structures for consideration in each of these chapters as well as Chaps. 8 and 9. If, for instance, you suspect that uncertain preferences are of import in the types of conflict you wish to study, you could include Chap. 5 regarding Unknown Preference as part of the pathway you wish to follow. If levels or strength of preference are of relevance, you may wish to incorporate Chap. 6 into your travel plans. For example, an environmentalist may greatly prefer that a planned industrial complex not pollute the surrounding environment. This kind of "emotion" is reflected in the preference structure described in Chap. 6 as well as relevant parts of Chaps. 8 and 9. Finally, if you may deal with both unknown and levels of preferences in disputes that you plan to address, you can read Chap. 7 on Hybrid Preferences which simultaneously takes into account both of these preferences within a single framework. Once again, pertinent sections from Chaps. 8 and 9 can be utilized in combination with hybrid preferences.

The logical form of a conflict, expressed in terms of moves and countermoves, is a truly intuitive and natural way to think about conflict. On the other hand, the matrix formulation of GMCR provides a solid foundation upon which to theoretically expand the basic paradigm of GMCR in many exciting and useful directions as described in Sect. 10.3. As the saying goes, "good theory means good practice." Accordingly, the universal design of a DSS for allowing you to effectively study any kind of conflict from the small to large and from the simple to complex depends heavily upon this clever concept of the matrix form. This is especially true for the engine design of a DSS in which massive calculations must be executed for large conflicts. Therefore, besides the logical form, in order to be able to understand in great depth the GMCR paradigm for both theoretical and practical purposes, you may wish to also include in your trip itinerary the matrix representation of GMCR given in Sect. 3.3 as well as relevant sections of all of the remaining chapters of the book. Depending on the kinds of preferences you wish to study, you can include all or some of Chaps. 4–7 in your travels through the GMCR countryside.

As can been seen in Table 1.9 and in the right column of Table 1.10, the illustrative applications utilized in this book are environmental. These applications were selected because of their particular conflict characteristics for simply explaining how to conveniently apply ideas in this book rather than their field of application. In fact, the GMCR methodology constitutes a truly general systems approach to conflict resolution (see Sects. 2.3.2 and 2.4.1) which is independent of the domain of application. As listed in Table 1.8 on application areas in which GMCR has been utilized in the past, it has been successfully applied to conflicts in numerous domains ranging from peaceful international trade to violent military confrontations. Hence, the roadmap drawn in Fig. 1.5 can be used in the domain of application which is of direct concern or importance to you. As mentioned in Sect. 1.1.3, the ideas in our book should be of great relevance to practitioners, researchers, teachers and students working in any particular field of study in which conflict may arise. Because conflict seems to almost always arise when two or more humans are gathered together and interacting with one another, our basic human nature over our conflicting value systems makes conflict inevitable in virtually any field of study involving people. Moveover, it arises among plants and animals as they are all competing for resources.

This book has been purposefully designed for use by both practitioners and researchers. The authors and their colleagues and students have been engaged for many decades in developing sound theoretical methodologies for tackling real conflict situations arising in the real-world. Problems are provided at the end of each chapter for those who would like practice in using the ideas described in these chapters. The book is designed for use by a consultant who may be part of an operations research or systems engineering team in a large company or a person who owns his own law or engineering consulting firm in which he or she must deal with conflict. As explained in Sect. 2.4, one may utilize these clearly powerful sets of techniques for addressing tough disputes within an overall system of systems perspective and in an integrative and adaptive fashion. Often the set of GMCR tools may be utilized in combination with other game theory methods (Sect. 2.2) and other formal techniques from Operations Research (Sect. 2.3.1) and Systems Engineering (Sect. 2.3.2). When implemented as a flexible DSS (see Sect. 2.3.3 and Chap. 10), the GMCR methodology can be readily applied to tough conflict problems of direct concern to you.

1.4 Problems

1.4.1 As a responsible citizen in the nation in which you reside, briefly discuss a main challenge facing your country. Describe the key stakeholders or decision makers involved in the issue. How do you think this problem can be resolved?

1.4.2 At the international level, which issue facing humanity around the globe is of most concern to you? Briefly outline this problem, discuss the conflict underlying it, mention the key decision makers and suggest how the dispute can be fairly and appropriately resolved.

1.4.3 From your own perspective and personal experience in dealing with conflict, explain why you think using a formal method to investigate conflict could be beneficial to you.

1.4.4 Discuss how you think certain types of conflict can enhance society and how other kinds of conflict can harm it.

1.4.5 In your field of study, employment or interest, make a list of the range of conflicts which commonly arise. How do you think they can be solved?

1.4.6 When you are personally involved in a conflict, why do you think it is important to know the preferences or values of the other people with whom you are interacting?

1.4.7 Why is it important to think like a chess player in terms of moves and countermoves, when investigating a conflict situation? Describe a dispute in the past in which you were involved and how this idea of thinking about the consequences of decisions that you could make naturally led you to think like a chess player when deciding upon what course of action to take.

1.4.8 In Sect. 1.2.2.3, it is pointed out that in a social conflict often it is only possible to obtain relative preference information for each involved decision maker. Do you agree with this statement? Describe a situation for which you think it may be possible and desirable to have cardinal preference information such as utility values or dollars.

1.4.9 Find a dispute that is of direct interest to you on the internet or in the newspaper. Using the option form given in Sect. 1.2.2, write down the decision makers (DMs), each DM's options or courses of action and the feasible states that could occur. By utilizing the idea of preference statements given in Table 1.4 for the Elmira conflict, record what you think are the relative preferences for each of the key DMs.

1.4.10 As explained in Sect. 1.2.5, one can apply a conflict resolution methodology such as the Graph Model for Conflict Resolution to a current, historical or generic conflict. An important type of generic or underlying conflict is called Prisoner's Dilemma. Explain the basic idea underlying this general kind of conflict and why it is of such great importance.

1.4.11 In Sect. 1.2.5, three general situations in which the Graph Model for Conflict Resolution may be used are explained. Describe a specific real-world example of each of these three situations.

1.4.12 Table 1.8 provides a list of application areas, in which the Graph Model for Conflict Resolution (GMCR) has been successfully applied to specific real-world conflicts. Select one reference for a conflict formally studied using GMCR which is of direct interest to you. Briefly describe the conflict and then explain the strategic insights that were gained into resolving this conflict using GMCR.

References

Armin, M., Hipel, K. W., & De, M. (2012). The Ontario nuclear power dispute: A strategic analysis. *Environmental Systems Research*, *1*(11), 1–16. https://doi.org/10.1186/2193-2697-1-11.

Bernath Walker, S., Boutilier, T., & Hipel, K. W. (2010). Systems management study of a private brownfield renovation. *Journal of Urban Planning and Development*, *136*(3), 249–260. https://doi.org/10.1061/(asce)0733-9488(2010)136:3(249).

Chu, Y., Hipel, K. W., Fang, L., & Wang, H. (2015). Systems methodology for resolving water conflicts: The Zhanghe river water conflict in China. *International Journal of Water Resources Development*, *31*(1), 106–119. https://doi.org/10.1080/07900627.2014.933096.

De Sousa, C. A. (2003). Turning brownfields into green space in the City of Toronto. *Landscape and Urban Planning*, *62*(4), 181–198. https://doi.org/10.1016/s0169-2046(02)00149-4.

Fang, L., Hipel, K. W., & Kilgour, D. M. (1993). *Interactive decision making: The graph model for conflict resolution*. New York: Wiley. https://doi.org/10.2307/2583940.

Fang, L., Hipel, K. W., Kilgour, D. M., & Peng, X. (2003). A decision support system for interactive decision making, part 1: Model formulation. *IEEE Transactions on Systems, Man and Cybernetics, Part C: Applications and Reviews*, *33*(1), 42–55. https://doi.org/10.1109/tsmcc.2003.809361.

Fraser, N. M., & Hipel, K. W. (1979). Solving complex conflicts. *IEEE Transactions on Systems, Man, and Cybernetics*, *9*(12), 805–816. https://doi.org/10.1109/tsmc.1979.4310131.

Fraser, N. M., & Hipel, K. W. (1984). *Conflict analysis: Models and resolutions*. New York: North-Holland. https://doi.org/10.2307/2582031.

Fraser, N. M. & Hipel, K. W. (1988). Decision support systems for conflict analysis. In M. Singh, D. Salassa & K. Hindi (Eds.), *Proceedings of the IMACS/IFOR First International Colloquium on Managerial Decision Support Systems and Knowledge-Based Systems* (pp. 13–21), Manchester, UK. Amsterdam: North-Holland.

Fraser, N. M., Hipel, K. W., Jaworsky, J., & Zuljan, R. (1990). A conflict analysis of the Armenian Azerbaijani dispute. *The Journal of Conflict Resolution, 34*(4), 652–677. https://doi.org/10.1177/0022002790034004004.

Garcia, A., Obeidi, A., & Hipel, K. W. (2016). Two methodological perspectives on the Energy East Pipeline conflict. *Energy Policy, 91*, 397–409. https://doi.org/10.1016/j.enpol.2016.01.033.

Garcia, A., Hipel, K. W., & Obeidi, A. (2017). Water pricing conflict in British Columbia. *Hydrological Research Letters, 11*(4), 194–200.

Ghanbarpour, M. R., & Hipel, K. W. (2009). Sustainable development conflict over freeway construction. *Environment, Development and Sustainability, 11*(2), 241–253. https://doi.org/10.1007/s10668-007-9107-2.

Gopalakrishnan, C., Levy, J., Li, K. W., & Hipel, K. W. (2005). Water allocation among multiple stakeholders: Conflict analysis of the Waiahole Water Project, Hawaii. *Water Resources Development, 21*(2), 283–295. https://doi.org/10.1080/07900620500108494.

Gore, A. (2013). *The future: Six drivers of global change*. New York: Random House.

Greenberg, M., & Lewis, M. J. (2000). Brownfields redevelopment, preferences and public involvement: A case study of an ethnically mixed neighbourhood. *Urban Studies, 37*(13), 2501–2514. https://doi.org/10.1080/00420980020080661.

Hamouda, L., Hipel, K. W., & Kilgour, D. M. (2004). Shellfish conflict in Baynes sound: A strategic perspective. *Environmental Management, 34*(4), 474–486. https://doi.org/10.1007/s00267-004-0227-2.

Hamouda, L., Hipel, K. W., Kilgour, D. M., Noakes, D. J., Fang, L., & McDaniels, T. (2005). The salmon aquaculture conflict in British Columbia: A graph model analysis. *Ocean and Coastal Management, 48*(7–8), 571–587. https://doi.org/10.1016/j.ocecoaman.2005.02.001.

Hipel, K. W. (2001). Conflict resolution. In M. K. Tolba (Ed.), *Our fragile world: Challenges and opportunities for sustainable development (Forerunner to the Encyclopedia of Life Support Systems)* (Vol. 1, pp. 935–952). Oxford: Eolss.

Hipel, K. W. (2011). A systems engineering approach to conflict resolution in command and control. *The International C2 Journal, 5*(1), 1–56.

Hipel, K. W., & Obeidi, A. (2005). Trade versus the environment: Strategic settlement from a systems engineering perspective. *Systems Engineering, 8*(3), 211–233. https://doi.org/10.1002/sys.20031.

Hipel, K. W., & Bernath Walker, S. (2012). Brownfield redevelopment. In R. Craig (Ed.), *Ecosystem management and sustainability* (pp. 44–48). Barrington, MA, USA: Berkshire Publishing. https://doi.org/10.1177/0891242416656049.

Hipel, K. W., Fang, L., & Kilgour, D. M. (1990). A formal analysis of the Canada-U.S. softwood lumber dispute. *European Journal of Operational Research, 46*(2), 235–246. https://doi.org/10.1016/0377-2217(90)90134-w.

Hipel, K. W., Fang, L., Kilgour, D. M., & Haight, M. (1993). Environmental conflict resolution using the graph model. *Proceedings of the 1993 IEEE International Conference on Systems, Man, and Cybernetics* (Vol. 1, pp. 153–158). https://doi.org/10.1109/ICSMC.1993.384737.

Hipel, K. W., Kilgour, D. M., Fang, L., & Peng, X. (1997). The decision support dystem GMCR in environmental conflict management. *Applied Mathematics and Computation, 83*(2–3), 117–152. https://doi.org/10.1016/s0096-3003(96)00170-1.

Hipel, K. W., Kilgour, D. M., Fang, L., & Peng, X. (1999). The decision support system GMCR II in negotiations over groundwater contamination. *Proceedings of the 1999 IEEE International Conference on Systems, Man, and Cybernetics* (Vol. 5, pp. 942–948). Tokyo, Japan: IEEE. https://doi.org/10.1109/ICSMC.1999.815681.

Hipel, K. W., Kilgour, D. M., Fang, L., & Peng, X. (2001a). Applying the decision support system GMCR II to negotiation over water. In U. Shamir, (Ed.) *Negotiation over water* (pp. 50–70). The International Hydrological Programme, Technical Document in Hydrology No. 53, United Nations Educational, Science and Cultural Organization (UNESCO), Paris, France.

Hipel, K. W., Kilgour, D. M., Fang, L., & Peng, X. (2001b). Strategic decision support for the services industry. *IEEE Transactions on Engineering Management*, *48*(3), 358–369. https://doi.org/10.1109/17.946535.

Hipel, K. W., Obeidi, A., Fang, L., & Kilgour, D. M. (2008). Adaptive systems thinking in integrated water resources management with insights into conflicts over water exports. *INFOR*, *46*(1), 51–69. https://doi.org/10.3138/infor.46.1.51.

Hipel, K. W., Hegazy, T., & Yousefi, S. (2010). Combined strategic and tactical negotiation methodology for resolving complex brownfield conflicts. *Pesquisa Operacional*, *30*(2), 281–304. https://doi.org/10.1590/s0101-74382010000200003.

Hipel, K. W., Kilgour, D. M., & Fang, L. (2011). The graph model for conflict resolution. In J. J. Cochran, L. A. Cox, P. Keskinocak, J. P. Kharoufeh, & J. C. Smith (Eds.), *Wiley encyclopedia of operations research and management science* (Vol. 3 of 8, pp. 2099–2111). New York: Wiley. https://doi.org/10.1002/9780470400531.eorms0882.

Hipel, K. W., Kilgour, D. M., & Kinsara, R. A. (2014). Strategic investigations of water conflicts in the Middle East. *Group Decision and Negotiation*, *23*(3), 355–376. https://doi.org/10.1007/s10726-012-9325-3.

Hipel, K. W., Fang, L., Cullmann, J., & Bristow, M. (Eds.). (2015). *Conflict resolution in water resources and environmental management*. Heidelberg: Springer. https://doi.org/10.1007/978-3-319-14215-9.

Hipel, K. W., Sakamoto, M., & Hagihara, Y. (2016). Third party intevention in conflict resolution: Dispute between Bangladesh and India over control of the Ganges River. In K. Hagihara & C. Asahi (Eds.), *Coping with regional vulberability: Preventing and mitigating damages from environmental disasters* (pp. 329–355). Tokyo, Japan: Springer. https://doi.org/10.1007/978-4-431-55169-0_17.

Howard, N. (1971). *Paradoxes of rationality: Theory of metagames and political behavior*. Cambridge: MIT Press. https://doi.org/10.2307/1266876.

Hu, K., Hipel, K. W., & Fang, L. (2009). A conflict model for the international hazardous waste disposal dispute. *Journal of Hazardous Materials*, *172*(1), 138–146. https://doi.org/10.1016/j.jhazmat.2009.06.153.

Kassab, M., Hipel, K. W., & Hegazy, T. (2006). Conflict resolution in construction disputes using the graph model. *Journal of Construction Engineering and Management*, *132*(10), 1043–1052. https://doi.org/10.1061/(asce)0733-9364(2006)132:10(1043).

Kassab, M., Hegazy, T., & Hipel, K. W. (2010). Computerized decision support system for construction conflict resolution under uncertainty. *Journal of Construction Engineering and Management*, *136*(12), 1249–1257. https://doi.org/10.1061/(ASCE)CO.1943-7862.0000239.

Kassab, M., Hipel, K. W., & Hegazy, T. (2011). Multi-criteria decision analysis for infrastructure privatization using conflict resolution. *Structure and Infrastructure Engineering- Maintenance, Management and Life-Cycle Design and Performance*, *11*(9), 661–671. https://doi.org/10.1080/15732470802677649.

Ke, Y., Fu, B., De, M., & Hipel, K. W. (2012a). A hierarchical multiple criteria model for eliciting relative preferences in conflict situations. *Journal of Systems Science and Systems Engineering*, *21*(1), 56–76. https://doi.org/10.1007/s11518-012-5187-0.

Ke, Y., Li, K. W., & Hipel, K. W. (2012b). An integrated multiple criteria preference ranking approach to the Canadian west coast port congestion problem. *Expert Systems with Applications*, *39*(10), 9181–9190. https://doi.org/10.1016/j.eswa.2012.02.086.

Kilgour, D. M., Hipel, K. W., & Fang, L. (1987). The graph model for conflicts. *Automatica*, *23*, 41–55. https://doi.org/10.1016/0005-1098(87)90117-8.

Kilgour, D. M., Fang, L., Last, D., Hipel, K. W., & Peng, X. (1998). Peace support, GMCR II, and Bosnia. In A. Woodcock & D. Davis (Eds.), *Analysis for peace operations* (pp. 268–282). Clementsport, Nova Scotia: Canadian Peacekeeping Press.

Levy, J. K., Hipel, K. W., & Kilgour, D. M. (1995). Holistic approach to sustainable development: The graph model for conflict resolution. *Information and Systems Engineering*, *1*, 159–177.

Ma, J., Hipel, K. W., & De, M. (2005). Strategic analysis of the James Bay hydroelectric dispute in Canada. *Canadian Journal of Civil Engineering*, *32*, 868–880. https://doi.org/10.1139/l05-028.
Ma, J., Hipel, K. W., & De, M. (2011). Devils Lake emergency outlet diversion conflict. *Journal of Environmental Management*, *92*(2), 437–447. https://doi.org/10.1016/j.jenvman.2010.08.027.
Ma, J., Hipel, K. W., & McLachlan, S. M. (2013). Cross-border conflict resolution: Sediment contamination dispute in Lake Roosevelt. *Canadian Water Resources Journal*, *38*(1), 73–82. https://doi.org/10.1080/07011784.2013.773773.
Madani, K., & Hipel, K. W. (2011). Non-cooperative stability definitions for strategic analysis of generic water resources conflicts. *Water Resources Management*, *25*(8), 1949–1977. https://doi.org/10.1007/s11269-011-9783-4.
Matbouli, Y., Hipel, K. W., & Kilgour, D. M. (2015). Strategic analysis of the great Canadian hydroelectric power conflict. *Energy Strategy Reviews*, *4*, 43–51. https://doi.org/10.1016/j.esr.2014.08.002.
McCarthy, L. (2002). The brownfield dual land-use policy challenge: Reducing barriers to private redevelopment while connecting reuse to broader community goals. *Land Use Policy*, *19*(4), 287–296. https://doi.org/10.1016/s0264-8377(02)00023-6.
Nandalal, K. W. D., & Hipel, K. W. (2007). Strategic decision support for resolving conflict over water sharing among countries along the Syr Darya River in the Aral Sea Basin. *Journal of Water Resources Planning and Management*, *133*(4), 289–299. https://doi.org/10.1061/(asce)0733-9496(2007)133:4(289).
Nash, J. F. (1950). Equilibrium points in *n*-person games. *Proceedings of the National Academy of Sciences of the United States of America*, *36*(1), 48–49. https://doi.org/10.1073/pnas.36.1.48.
Nash, J. F. (1951). Noncooperative games. *Annals of Mathematics*, *54*(2), 286–295. https://doi.org/10.1515/9781400884087-008.
Noakes, D. J., Fang, L., Hipel, K. W., & Kilgour, D. M. (2003). An examination of the salmon aquaculture conflict in British Columbia using the graph model for conflict resolution. *Fisheries Management and Ecology*, *10*, 123–137. https://doi.org/10.1046/j.1365-2400.2003.00336.x.
Noakes, D. J., Fang, L., Hipel, K. W., & Kilgour, D. M. (2005). The Pacific Salmon Treaty: A century of debate and an uncertain future. *Group Decision and Negotiation*, *14*(6), 501–522. https://doi.org/10.1007/s10726-005-9005-7.
Obeidi, A., & Hipel, K. W. (2005). Strategic and dilemma analyses of a water export conflict. *INFOR*, *43*(3), 247–270. https://doi.org/10.1080/03155986.2005.11732727.
Obeidi, A., Hipel, K. W., & Kilgour, D. M. (2002). Canadian bulk water exports: Analyzing the Sun Belt conflict using the graph model for conflict resolution. *Knowledge, Technology, and Policy*, *14*(4), 145–163. https://doi.org/10.1007/s12130-002-1020-2.
Obeidi, A., Hipel, K. W., & Kilgour, D. M. (2006). Turbulence in Miramichi Bay: The Burnt Church conflict over native fishing rights. *Journal of the American Water Resources Association*, *42*(12), 1629–1645. https://doi.org/10.1111/j.1752-1688.2006.tb06025.x.
O'Brien, N. L., & Hipel, K. W. (2016). Strategic analysis of the New Brunswick, Canada fracking controversy. *Energy Economics*, *55*, 69–78. https://doi.org/10.1016/j.eneco.2015.12.024.
Philpot, S., Hipel, K. W., & Johnson, P. A. (2016). Strategic analysis of a water rights conflict in the south western United States. *Journal of Environmental Management, 180*, 247–256. https://doi.org/10.1016/j.jenvman.2016.05.027.
Philpot, S. L., Johnson, P. A., & Hipel, K. W. (2017). Analysis of a brownfield management conflict in Canada. *Hydrological Research Letters*, *11*(3), 141–148. https://doi.org/10.3178/hrl.11.141.
Silva, M. M., Hipel, K. W., Kilgour, D. M., & Costa, A. P. C. S. (2017a). Urban planning in Recife, Brazil: Evidence from a conflict analysis on the New Recife project. *Journal of Urban Planning and Development*, *143*(3), 05017007–1–05017007–11. https://doi.org/10.1061/(asce)up.1943-5444.0000391.
Silva, M. M., Kilgour, D. M., Hipel, K. W., & Costa, A. P. C. S. (2017b). Probabilistic composition of preferences in the graph model with application to the New Recife project. *Journal of Legal Affairs and Dispute Resolution in Engineering and Construction*, *9*(3), 05017004. https://doi.org/10.1061/(asce)la.1943-4170.0000235.

USEPA. (1997). Brownfields economic redevelopment initiative. Technical report, Solid Waste and Emergency Response. Washington, DC: United States Environmental Protection Agency.

Wolf, A. T. (Ed.). (2002). *Conflict prevention and resolution in water systems*. Cheltenham, UK: Edward Elgar.

Xiao, Y., Hipel, K. W., & Fang, L. (2015). Strategic investigation of the Jackpine mine expansion dispute in the Alberta oil sands. *International Journal of Decision Support System Technology*, *7*(1), 50–62. https://doi.org/10.4018/ijdsst.2015010104.

Yousefi, S., Hipel, K. W., & Hegazy, T. (2010a). Attitude-based negotiation methodology for the management of construction disputes. *Journal of Management in Engineering*, *26*, 114–122. https://doi.org/10.1061/(asce)me.1943-5479.0000013.

Yousefi, S., Hipel, K. W., & Hegazy, T. (2010b). Attitude-based strategic negotiation for conflict management in construction projects. *Project Management Journal*, *41*(4), 99–107. https://doi.org/10.1002/pmj.20193.

Yousefi, S., Hipel, K. W., & Hegazy, T. (2010c). Considering attitudes in strategic negotiation over brownfield disputes. *ASCE Journal of Legal Affairs and Dispute Resolution in Engineering and Construction*, *2*(4), 1–10. https://doi.org/10.1061/(asce)la.1943-4170.0000034.

Yousefi, S., Hipel, K. W., & Hegazy, T. (2011). Optimum compromise among environmental dispute issues using attitude-based negotiation. *Canadian Journal of Civil Engineering*, *38*(2), 184–190. https://doi.org/10.1139/l10-125.

Chapter 2
Decision-Making in Perspective

2.1 Overview

The goal of this chapter is to put the field of conflict resolution and associated domain of game theory into perspective so that a reader will be able to fully appreciate the inherent value of utilizing Graph Model for Conflict Resolution (GMCR) as a highly informative and operational means for studying actual social conflict that does take place in reality. Within Sect. 1.2 of the previous chapter, some of the key ideas underlying the formal investigation of conflict are discussed. Within this book, the detailed explanation of how these concepts are defined and operationalized using real-world examples are presented in subsequent chapters. In Sect. 2.2, the evolution and development of a rich variety of game theory methods is put into perspective to highlight the important and central role that GMCR plays for sensibly and flexibly modeling and analyzing social conflict. An insightful classification of game theory methods permits one to understand situations in which various approaches can be employed, with GMCR being the key methodology for formally studying societal conflict occurring in many different disciplines such as engineering, law and military science.

Two key fields in which a wide range of formal decision-making tools have been developed since the late 1930s and 1940s are Operations Research (OR) and Systems Engineering, respectively. The history and development of decision technologies within these two dynamic fields of study are presented in Sects. 2.3.1 and 2.3.2 for OR and Systems Engineering, respectively. To permit a given decision-making approach to be applied to practical problems, a user-friendly set of programs and associated databases should be made available in what are called Decision Support Systems (DSSs) outlined in Sect. 2.3.3 and described in detail for GMCR in Chap. 10. As explained in Sect. 2.4.1, a System of Systems (SoS) interpretation of reality provides a solid foundation on which an informative decision tool like GMCR can be utilized, since DMs or agents compete and cooperate with one another within and among systems. As discussed in Sect. 2.4.2, responsible governance can be achieved and

H. Xu et al., *Conflict Resolution Using the Graph Model: Strategic Interactions in Competition and Cooperation*, Studies in Systems, Decision and Control 153,
https://doi.org/10.1007/978-3-319-77670-5_2

key systems values like robustness, sustainability and fairness can be met, when this is carried out using integrative and adaptive management concepts.

2.2 Game Theory Methods: Classifications

As explained in the next subsection, because conflict is so ubiquitous in society, researchers and practitioners have developed a rich range of conflict analysis approaches for investigating many different kinds of situations. In fact, to put conflict analysis methods into perspective, an insightful way for classifying conflict analysis or game theory techniques is put forward in Sect. 2.2.2. By knowing the key characteristics of a specific real-world dispute being studied, one can select an appropriate game theory method, or set of techniques, that has the structural capabilities for modeling and analyzing it.

2.2.1 The Evolution of Game Theory Methods

As vividly described in Sect. 1.1, conflict is an inherent characteristic of human nature which dictates how individuals, groups of people, organizations and nations interact with one another. For instance, warfare is recorded in the history of early civilizations that existed in Mesopotamia, China and India, up until the present time when nasty regional wars have been taking place in the Middle East for many decades. International companies in the automobile industry, electronics, information technology and many other fields are currently fiercely competing on a global basis to capture larger market shares in each of these areas. On a personal level, people might still disagree over how work should be fairly allocated when completing a given task as mundane as cleaning the rooms in a house.

The fact that conflict is so prevalent attracted the attention of scholars in many fields of study. Because conflict involves people, researchers in the social sciences were among the first to explain and categorize conflict in fields such as sociology, law and economics. Attempts to formalize the study of conflict by developing mathematical models of disputes are more recent. The areas in which formal mathematical models of conflict have been developed are often collectively referred to as game theory. Early work in game theory can be traced back to the year 1654 when the French mathematicians Pierre de Fermat and Blaise Pascal studied a specific kind of parlor game and established the foundations for the theory of probability. Nonetheless, it was the groundbreaking work of von Neumann (1928) and, particularly, von Neumann and Morgenstern (1944, 1953) that decisively brought game theory into the modern era as a distinct domain of mathematical enquiry having a rich range of conflict problems to tackle. Moreover, since conflict arises in virtually every field of human endeavor, contributions to game theory have been made by experts working in many areas. As is explained in Sect. 2.3, mathematicians, scientists and engineers working in the fields of Operations Research and Systems Engineering have designed a wide range of formal decision-making methods including various game theory techniques. Operations Research was started by the British military just prior to the start of World War II (WWII) in Europe while Systems Engineering was largely initiated shortly afterwards. In addition, since the cessation of hostilities after WWII, there has been a great proliferation of research in game theory. As a matter of fact, it is difficult to keep track of what has been accomplished in game theory across a large number of fields and how to put the many contributions to game theory into perspective.

2.2.2 Classifying Formal Game Theory Techniques

To permit researchers and practitioners to wisely select the most appropriate game theory tools to utilize for addressing different kinds of conflict problems and for deciding upon where there are needs for refining and extending game theory tools as well as designing new methods, one must be able to classify game theory techniques in meaningful ways. In general, one requires useful criteria for categorizing game theory methods. Some of these criteria include types of preference information, number of decision makers, number of options or strategies, size of the conflict, kinds of human behavior under conflict, types of available information, kinds of uncertainty, and level of cooperation which can range from highly noncooperative competition to increasing levels of cooperation ending at a universal coalition (Fang et al. 1993, Sect. 1.4). By being aware of the criteria under which game theory techniques are categorized as well as the key characteristics of an actual conflict being investigated, one can choose an appropriate set of game theory tools that possess the theoretical capability, as expressed by the criteria, to model the main characteristics of the dispute. In other words, one makes a one-on-one linkage between model criteria and problem characteristics to select the appropriate set of tools. Additionally, one can discover a gap in the literature which indicates where more research is needed if tools are not currently available to address certain problem characteristics.

As suggested by Hipel and Fang (2005), an especially informative way to classify formal game theory methods is according to type of preference. Figure 2.1 displays a genealogy of game theory methods for categorizing game theory techniques with respect to relative and cardinal preferences. As mentioned in Sect. 1.2.2, when a person asks a friend whether she would like to have coffee or tea to drink, the companion would probably respond that she would prefer to have a cup of coffee. If having coffee or tea is equally preferred, the person may respond that it does not matter which drink you serve me. This is what is called a relative preference and it constitutes a nonquantitative way of expressing a preference. On the other hand, if utility values conveyed as real numbers or benefits given as dollars are used to indicate preference, these are called cardinal preferences. It does not make sense to state that a cup of coffee is worth 6.2 utility units to a person and tea 2.8 in place of simply stating that she prefers to drink coffee. However, when the profits made by a

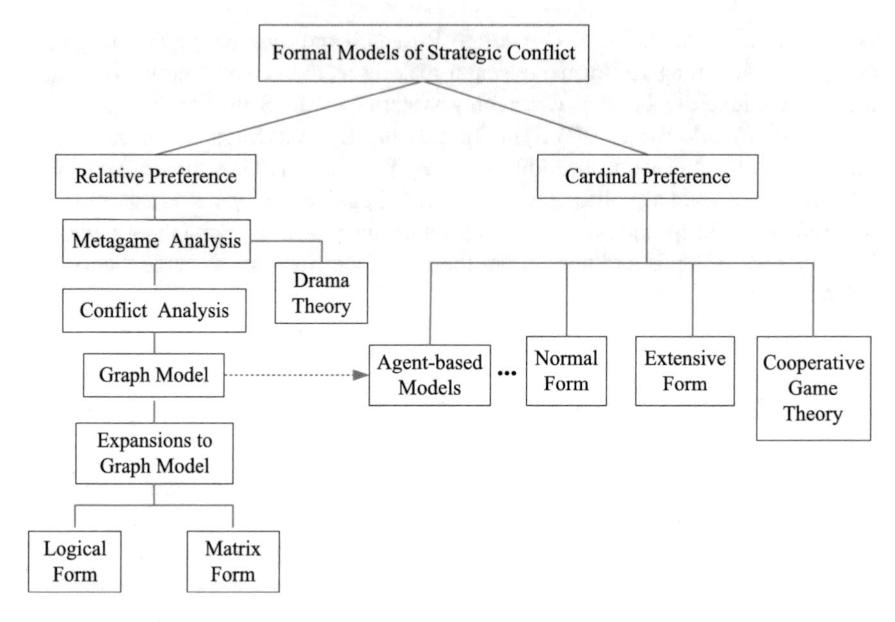

Fig. 2.1 Genealogy of formal multiple participant decision-making models

company in producing two different products are 100 and 35 dollars then it may be meaningful to use cardinal numbers given as dollars to represent the preference of the company to manufacture more of the product which brings in higher profits. Even though the qualitative methods listed in the left branch in Fig. 2.1 only depend upon relative preference information, these techniques, like the ones in the right branch, constitute formal mathematical game theory techniques. In fact, as explained in upcoming chapters in the book the kinds of mathematical concepts used to build the qualitative game theory methods come from set theory, logic, graph theory and matrix algebra - the mathematics for expressing relationships.

The focus of this book is the Graph Model for Conflict Resolution (GMCR), which is given under relative preferences in the left branch in Fig. 2.1. These game theory methods listed in the left branch are especially useful for modeling and analyzing real-world social conflicts such as environmental disputes, trading conflicts among nations and an argument between neighbors over where the fence between their properties should be located. When reading from the top to the bottom of the left branch, the earliest method created under this category is the pioneering technique of metagame analysis which was developed by Howard (1971). Subsequently, Fraser and Hipel (1979, 1984) expanded the scope of metagame analysis through the development of a methodology called conflict analysis which was further significantly enhanced by Kilgour et al. (1987) and Fang et al. (1993) in the construction of the comprehensive approach labelled as GMCR. As indicated at the bottom of the left branch, GMCR has been further improved by the design of a matrix form of this approach, which is utilized throughout this book as well as by many other enhancements presented in Chaps. 3–9. Moreover, GMCR has been significantly broadened in scope by the addition of many other developments presented in Chaps. 4–9 plus those mentioned in Sect. 10.3. Over the years, summary papers describing the capabilities of the GMCR methodology as well as opportunities for future development have been written (see, for instance, Hipel et al. (2003) and Kilgour and Hipel (2005, 2010)). The original logical form for logically explaining stability calculations in terms of moves and countermoves has been further improved by the design of a matrix form of this approach used extensively in this book and is especially important in carrying out stability calculations in the engine of a DSS for GMCR described in Chap. 10. Additionally, as depicted in the left branch, metagame analysis was expanded by Howard et al. (1992), Howard (1999) and Bryant (2003, 2015) through their development of a procedure called drama theory for nonquantitatively modeling the dynamic aspects of conflict based on the metaphor of a drama.

In their book entitled "Theory of Games and Economic Behavior", von Neumann and Morgenstern (1944, 1953) mainly deal with quantitative game theory methods, which are calibrated using cardinal preferences and often referred to as classical game theory methods. Three popular kinds of techniques from classical game theory are normal form, extensive form, and cooperative game theory which are listed in the right branch drawn in Fig. 2.1. In classical normal form, one assumes that two or more DMs interact one time only. The normal form of the game is defined in Sect. 3.1 in this book under the assumption of having relative preferences, rather than cardinal. A convenient way to display the normal form is to use a matrix in which the row player or DM controls the strategies represented by the rows while the column player is in charge of the column strategies. Each cell in the matrix represents a possible scenario or state. Within the extensive form, multiple interactions among DMs are depicted using a tree-like structure that keeps track of all possible evolutions of the game.

Cooperative methods are used to examine the interaction of DMs who must cooperatively decide how to fairly divide a "pie" or some resource in an equitable fashion. These methods are often employed to analyze coalition formation, voting problems or optimal resource allocation procedures. The Cooperative Water Allocation Model (CWAM), for example, constitutes a large-scale optimization model based on ideas from cooperative game theory, economics and hydrology to fairly allocate water among competing users in a river basin (Wang et al. 2003, 2007, 2008a, b, Hipel et al. 2013b). Based on a systems approach, CWAM considers not only the physical systems consisting of hydrological and environmental factors but also the societal system. Moreover, CWAM has been expanded to handle demand-side management to promote water use efficiency (Xiao et al. 2016). Furthermore, CWAM has been successfully applied to fair water allocation problems in the South Saskatchewan River Basin located in the Canadian Province of Alberta (Wang et al. 2008a, b, Hipel et al. 2013b) as well as the Aral Sea region (Wang et al. 2007).

As shown in the central part of Fig. 2.1, another approach in which cardinal preferences are assumed is agent-based modeling. In this procedure, the actions and interactions of agents are simulated in order to assess their impacts on the overall system. Hence, this method can be employed to test whether or not a given policy will function according to expectations in practice. For example, one may wish to determine if a cap and trade method will significantly reduce the amount of greenhouse gases released by society into the atmosphere. Agent-based modeling can be interpreted as a bottom-up approach to performance assessment since it determines if individual decision-making units, often referred to as autonomous agents, interact in a way that causes the policy to meet its goals. As pointed out by Hipel and Fang (2005), researchers in agent-based modeling often directly import concepts from classical game theory for utilization in their formal analyses of rules or protocol governing the high level behavior of interacting agents as is done by Rosenschein and Zlotkin (1994). Therefore, Hipel and Fang (2005) recommended that solution concepts describing possible moves and countermoves among DMs within the GMCR paradigm be utilized in agent-based modeling for policy assessment. This was accomplished for the first time by Bristow et al. (2014) when they examined the responsible utilization of common pool resources such as water and the atmosphere in order to avoid a Tragedy of the Commons (Hardin 1968, Ostrom et al. 1994, 1999) in which a common resource is destroyed via entirely competitive rather than cooperative behavior. A dotted line showing the connection of GMCR to agent-based modeling is drawn in the central part of Fig. 2.1.

A number of books and papers have been written in which different approaches to game theory have been described and compared. For instance, Hipel (2009a, b) and Kilgour and Eden (2010) have produced edited books in which experts from many fields have written papers on a range of conflict analysis methodologies in group decision and negotiation for application in many different areas. The chapters in these handbooks largely concentrate on methodologies listed in the left branch of Fig. 2.1 but techniques coming under the right branch and elsewhere are also presented. Hipel and Bernath Walker (2011) and Hipel et al. (2016) provide an overview of the employment of conflict analysis methods in environmental management that span both branches in Fig. 2.1.

2.3 Formal Decision-Making Techniques

As discussed in Sect. 2.2 and depicted in Fig. 2.1, game theory is comprised of a rich range of mathematically-based techniques for formally investigating conflict. Fortunately, a wide variety of formal decision-making tools have been developed for investigating many different kinds of decision situations. Because decision-making arises in many areas of human endeavor, from engineering design to international trade, approaches to tackling decision problems have been put forward by researchers and practitioners from many different disciplines. Two specific disciplines, or fields, in which many different types of formal or mathematically-founded techniques have been developed starting about the time of World War II (WWII) are Operations Research and Systems Engineering. Accordingly, a brief history of these two disciplines along with an overview of the types of tools that have been developed within them are put forward in the next two subsections, respectively. The reason why it is important to be aware of the existence of a valuable range of formal tools is because when addressing tough systems problems, such as those arising in energy use and climate change, usually a number of specific tools can be selected for assisting in realistically addressing a particular problem. Moreover, due to their inherent mathematical design, most of these methods are readily available as decision support systems containing comprehensive computer programs and databases for permitting them to be conveniently applied to practical problems, as outlined in Sect. 2.3.3 and explained in more detail in Chap. 10, including issues related to governance discussed in Sect. 2.4.

2.3.1 Operations Research

Operations Research (OR) or Operational Research, as the British call OR, constitutes a systematic approach for scientifically solving real-world problems. The term "scientific" is used in the definition because only formal techniques that actually work in practice for enhancing decision-making are utilized, often in the face of sparse information and high uncertainty (Kimball and Morse 1951, Hipel 1981, Ravindran et al. 1987). OR was originally conceived by the British as a response to a potential military threat just before the outbreak of WWII in Europe. As described by Lardner (1979), the British military was concerned about how to defend the United Kingdom against potential air attacks from Germany since German bombers could reach the UK in a very short period of time. In fact, by the mid-1930s Germany was the dominant economic and military power in continental Europe and it was acting very aggressively against its neighbors in response to the unfair treatment that it thought it received under the Treaty of Versailles signed in 1919 just after World War I (WWI). By 1935, radar was recognized by the British as a viable means of detecting enemy aircraft before they reached the British Isles. Accordingly, the British established a system of radar bases in the southern and eastern parts of England. When the British tested their system of radar bases against mock air attacks from their own air squadrons launched from air bases in France, the system failed to work. There were, for instance, poor communication among radar stations and a lack of systematic defensive strategies from fighter aircraft that took off from Royal Air Force facilities in England to disrupt or stop the attack. As a consequence, in July 1938, research into the operational aspects of radar systems was initiated. The effectiveness of OR was confirmed by the successful air exercises carried out during the summer of 1939.

The first major employment of OR in WWII actually saved the UK from defeat by Germany. In particular, at the outbreak of WWII on September 1, 1939, the OR Section was attached to the Headquarters of the Royal Air Force Fighter Command. On May 10th, 1940, Winston Churchill replaced Neville Chamberlain as Prime Minister of the UK. On the same day, the Germans launched Fall Gelb (Operation Yellow) which led to the rout of the French army and British Expeditionary Forces by the Wehrmacht (refer to Bennett and Dando (1977, 1979) for a history of the Battle of France and to Fraser and Hipel (1984), Sect. 4.2, for a conflict analysis of the strategic surprise used in Fall Gelb when the main German forces unexpectedly attacked through the Ardennes). Subsequently, from July 10th, 1940 to September 15th, 1940, the German Luftwaffe attempted to defeat the UK by aerial bombardment during the Battle of Britain. Largely because of OR, Germany was not successful in defeating the UK and in reality suffered massive losses of military aircraft and personnel. OR scientists can be accredited with saving the Royal Air Force from being obliterated during the Battle of France so it could survive to be victorious in the Battle of Britain. A relatively small OR study demonstrated that based on current losses and replacement rates at that time, the Germans would have destroyed the entire Royal Air Force within two weeks. A graphical presentation of these findings on May 15th, 1940, convinced Churchill not only to stop sending more fighter squadrons to France but also to withdraw all of the British air squadrons which were in France at that time.

In addition to the air force, the other UK armed services also employed OR teams for solving specific large-scale military problems. A well-known naval illustration is how allied shipping losses as a result of attacks by German submarines in the North Atlantic against ships transporting supplies and personnel to the UK from Canada and the United States were reduced by increasing the size of escorted convoys. After the United States entered the war on December 7, 1941 as a direct result of the unexpected Japanese aerial attack on Pearl Harbor by planes launched from aircraft carriers, the American armed forces used OR in its military decision-making.

Besides their OR teams, both the British and American armed forces utilized the talents of gifted mathematicians, scientists and engineers to break encoded messages sent by the Germans and Japanese, respectively. In particular, at Bletchley Park located 80 km northwest of London, the location of the UK's Government Code and Cypher School, mathematicians like Alan Turing and William "Bill" Tutte helped to break the German Enigma and Lorenz ciphers, respectively. What was called Ultra intelligence at Bletchley Park may have shortened the war by as much as two to four years (Aldrich 2010, Briggs 2011, Grey 2012). In the Pacific arena of the war, personnel at the United States Navy's Combat Intelligence Unit were able to decipher encoded messages sent by the Japanese Navy throughout the war and thereby knew about Japanese naval maneuvers before they took place (Winton 1993, Benson 1997). Because of the deciphering of a message giving the flight plans of Admiral Isoroku Yamamoto, Commander-in-Chief of the Combined Japanese Fleet, American pilots in P-38 fighters killed Admiral Isoroku Yamamoto on April 18, 1943 by shooting down the plane carrying him as it was about to land at Bougainville in the Solomon Islands.

As explained by Fang et al. (1993, Sect. 1.4) and many other authors, OR is both an art and a craft. The art is composed of a general approach to solving complicated operational problems, whereas the craft component consists of a wide range of mathematical methods for furnishing reasonable findings when properly applied to specific problems. Methods that are commonly considered to be part of OR include optimization techniques such as linear, nonlinear and integer programming; probabilistic techniques like Markov Chains, queuing theory and certain kinds of

Table 2.1 Classification of decision-making models

		Objectives: One	Objectives: Two or More
Decision Makers	**One**	**Most OR Models**	**Multiple Criteria Decision Analysis**
	Two or More	**Team Theory**	**Game Theory**

time-series models; and some game theory approaches. As is also the situation for the game theory methods mentioned in Sect. 2.2.2 and summarized in Fig. 2.1, there is a range of criteria that could be employed for categorizing OR techniques. In Table 2.1, OR techniques are classified with respect to two criteria: number of decision makers (DMs) and number of objectives. As indicated, many OR methods or models represent the perspective of one DM having a single objective. For instance, linear programming can be employed as an optimization tool by a company to minimize its costs expressed as a linear algebraic objective function which is minimized within a feasible region constrained by linear algebraic inequalities. Multiple criteria decision analysis (MCDA) methods (see, for example, MacCrimmon (1973), Keeney and Raiffa (1976), Saaty (1980), Hwang and Yoon (1981), Goicoechea et al. (1982), Vincke (1992), Roy (1996), Rajabi et al. (1998), Hobbs and Meier (2000), Belton and Stewart (2002), Chen et al. (2008), Chen et al. (2011), Hipel et al. (2009a), Kuang et al. (2015)) are purposefully designed for discovering the set of more preferred alternative solutions to a problem when the discrete alternatives are evaluated against criteria ranging from cost (a quantitative criterion) to aesthetics (a nonquantitative or qualitative criterion). The evaluations of the criteria for each alternative are indications of the achievements of objectives or preferences of the DM. Because many decisions in most fields ultimately involve making a discrete choice for a given DM, such as deciding upon the specific type of car to purchase, MCDA techniques have been applied to a diverse range of fields spanning from water resources (Hipel 1992) to energy problems (Hobbs and Meier 2000). This important set of tools is given as an example of a decision model containing one DM having two or more objectives as listed in the top right cell in Table 2.1.

As indicated in the bottom left cell in Table 2.1, team theory is an example of a technique having two or more DMs but only one objective since each team participating in a sporting event has the single goal or objective of winning. In a card game such as poker, each player possesses the single objective of winning the most money.

The focus of this book is the general decision-making situation in which there are multiple DMs, each of whom can have more than one objective. As indicated in the bottom right cell in Table 2.1, the game theory methods outlined in Sect. 2.2.2 and

Fig. 2.1 fall within this category. An example of a flexible game theory method is GMCR which constitutes the theme of this book and is contained in the left branch of Fig. 2.1. As explained in Sect. 2.2.2 and by authors such as Hipel (2009a, b) and Kilgour and Eden (2010), a wide variety of game theory methods have been developed over the years for tackling different kinds of multiple participant-multiple objective decision-making situations (Hipel et al. 1993). In fact, this is the category of OR for which there is great demand for the development of decision techniques but where OR researchers have devoted the least effort. Accordingly, a key goal of this book is to significantly extend the field of conflict resolution such that researchers and practitioners will possess more comprehensive tools for effectively addressing complex problems arising in multiple participant-multiple objective decision making.

The terminologies of normative and descriptive methods are often utilized for characterizing OR methods. A normative technique stipulates what a DM should do in order to reach a well-defined objective. For instance, the Cooperative Water Allocation Model (CWAM) mentioned in Sect. 2.2.2, which is formulated as an overall nonlinear programming model, can be optimized to specify how water can be fairly allocated among competitors in a river basin. Since fairness ideas from cooperative game theory are contained in CWAM, this model falls under the right branch in Fig. 2.1. Alternatively, a descriptive model captures the main characteristics of a problem in order to describe their relationships and a range of consequences that could occur. For example, conflict analysis techniques contained in the left branch of Fig. 2.1 can mainly be interpreted as being descriptive because they describe a variety of possible compromise resolutions as well as the various social interactions that can cause these equilibria to take place. Nevertheless, a conflict analysis method like GMCR can also be thought of as containing a normative component. This is because the findings of a GMCR investigation can be used to furnish a DM with a better understanding of the conflict under study and strategic advice on how to interact with his or her competitors in order to reach his most preferred equilibrium within the social constraints of the conflict. When a specific equilibrium is recommended for resolving a conflict, along with a particular path for reaching it, GMCR can be interpreted as being used in a normative fashion. Finally, to make both the descriptive and normative aspects of a conflict analysis study readily available, it must be implemented as a DSS, as outlined in Sect. 2.3.3 and explained in more detail in Chap. 10.

As pointed out earlier, OR was conceived and originally developed by the armed forces just prior to and throughout WWII to tackle urgent operational military problems as they arose or were anticipated. During the first few decades after the war, OR researchers and practitioners focused on designing highly mathematical and quantitative methods that are useful for addressing well-defined problems especially at the tactical level of decision-making. For instance, within an industrial organization, OR teams regularly employ mathematical programming techniques for solving difficult technical problems in resource allocation at the tactical level. Nonetheless, although some advances have been made more recently in developing formal techniques for utilization at the strategic level of decision-making, where the information base is often qualitative in nature, much work remains to be accomplished. Within and among most organizations, strategic decision-making almost always involves

Table 2.2 Two levels of decision-making

Tactical level	Stratagic level	References
Tactical	Strategic	Radford (1988, 1989), Rosenhead (1989)
Regular problem	Messes	Ackoff (1981)
Technical	Practical	Ravetz (1971)
Tame	Wicked	Rittel and Webber (1973)
Hard systems	Soft systems	Checkland (1981)
High ground	Swamp	Schon (1987)
Components	System or system of systems	Hipel et al. (2009b)

multiple DMs, each of whom has multiple goals. Accordingly, a key goal of this book is to assist in meeting this current need for extending the domain of OR.

Researchers and practitioners commonly refer to two major levels of decision-making: tactical and strategic. Moreover, many authors highlight the need for constructing more procedures for addressing less structured problems occurring at the strategic level. A range of labels that have been coined for describing these lower and higher levels are provided in the first and second columns of Table 2.2, respectively, along with references in the third column.

OR is the most widely known field for producing formal decision-making methods. Many of the problems studied using OR tend to be large-scale and highly complicated. Because of this, when investigating a specific problem often OR practitioners and researchers have backgrounds in many different disciplines and work as a team when addressing the various aspects of the overall problem using many different techniques. The team must obtain reasonable solutions in a scientific and expedient manner. Stated otherwise, the team must efficiently solve complicated well-structured problems in order to meet specified objectives. Due to the great success of OR for systematically solving tough problems, after WWII OR Societies were formed in many industrialized societies and associated OR journals were founded. For example, the world's oldest OR society was started in the UK as the Operational Research Club in April, 1948, which later became the OR Society in 1953. Since 1950, this society has been publishing the Journal of the Operational Research Society. In the USA, the Institute for Operations Research and the Management Sciences (INFORMS) publishes many journals for which the flagship journals are Operations Research and Management Science. The Group Decision and Negotiation Section of INFORMS produces its own journal entitled Group Decision and Negotiation. In Canada, the Canadian Operational Research Society publishes the journal called Information Systems and Operational Research (INFOR).

Outside of the military sciences, one of the first fields to take an OR and systems approach to problem solving was water resources. Hence, for instance, many applications and developments in OR can be found in journals such as Water Resources Research (published by the American Geophysical Union), Journal of Water Resources Planning and Management (American Society of Civil

Engineers) and the Canadian Water Resources Journal (sponsored by the Canadian Water Resources Association). Other disciplines in which OR is widely utilized and expanded include transportation, urban planning, systems design engineering, systems analysis (Miser and Quade 1985, 1988), management sciences, systems thinking (Checkland 1981), industrial engineering and business. Together the foregoing disciplines are often referred to as the "Systems Sciences". The comprehensive encyclopedia on systems and control edited by Singh (1987) contains definitions and explanations of decision-making techniques from the systems sciences, artificial intelligence, and elsewhere. A discipline or field that utilizes ideas from OR but goes well beyond that is Systems Engineering which is now described.

2.3.2 Systems Engineering

The key underlying philosophy of Systems Engineering is to tackle problems from a holistic or overall viewpoint. One must first see the entire "forest" before trying to solve a problem involving a specific "tree" which is, of course, a subset of the forest. This concept of envisioning a complete system connected to a problem, which is composed of interconnected components synergistically serving the overarching goals of the system is natural and very pleasing to the mind. In Japan, one can contemplate for hours while viewing a rock garden which consists of various sets of rock formations situated at satisfying but perhaps surprising locations in a sea of sand marked with intersecting flowing patterns. This is the way people like to view reality: artistically, systems thinking is Eastern in derivation but technically more Western. One of the first physical systems drawn in one of the most creative and insightful phases of all human history - the Renaissance - was the Hydrological Cycle depicted by the great Leonardo Da Vinci.

Among other authors, Hipel et al. (2009a) provided a comparison of OR and Systems Engineering. Because OR attempts to be scientific, it is founded upon reductionist concepts. Hence, OR attempts to understand a phenomenon by comprehending its components and their relationships. Since these relationships and interconnections are often complex, an OR approach may not capture the entire picture and the emergent behavior which can arise as a result of complexity. Rather, the system behavior is determined by precise cause-and-effect relationships (see, for instance, Ackoff 1962 and Keys 1991). Therefore, OR techniques are quantitative in nature and most applicable to well-defined problems at the tactical level shown on the left in Table 2.2.

Compared to OR, Systems Engineering is more qualitative and less analytical and is designed for tackling unstructured and complex problems (Sage 1992, Warfield 2006, Haimes 2016). As explained by Hipel et al. (2009a), Systems Engineering focuses on:

- quantitative and qualitative methods,
- strategic and tactical levels of decision-making,
- integration of technology, institutional perspective and value judgment,

- entire system including the components and their synergistic connections,
- holistic viewpoint,
- unstructured and complex problems, and
- single and multiple decision makers.

As pointed out in Sect. 2.3.1, the terminology of OR was coined by the British military who carried out "research" into the "operational" aspects of radar systems in 1938, since having a reliable defensive system against potential German bombing raids of the United Kingdom was of great concern to the British. The label Systems Engineering was first utilized in the Bell Telephone Laboratories in the 1940s (Schlager 1956) and this flexible approach to creative problem solving and design was quick to be adopted by many other organizations including NASA (National Aeronautics and Space Administration) in the United States of America and industry. The field of Systems Engineering continues to be developed at an expanding rate by both practitioners and researchers. Leading research papers on Systems Engineering can be found in journals such as the IEEE Transactions on Systems, Man, and Cybernetics: Systems; IEEE Systems Journal; as well as the journal Systems Engineering which is published by the International Council on Systems Engineering (INCOSE). In fact, INCOSE regularly releases reports on the latest advances in Systems Engineering. Departments of Systems Engineering exist in many universities situated in many nations around the globe. Systems Engineering groups exist in most large industrial organizations and many departments of Defence. Most professionals working in Systems Engineering and OR are fully aware of the developments in both of these fields and do not hesitate to utilize any relevant available methods from either area for addressing tough problems.

A classic book on Systems Engineering was written by one of its greatest pioneers, the late Andrew P. Sage, in 1992 (Sage 1992). Because Systems Engineering is such a dynamic and exciting field, an encompassing and universally adopted definition of Systems Engineering is difficult to find. In his highly innovative and informative approach to risk assessment, Haimes (2016) carries out risk studies within a Systems Engineering and multiple objective decision-making framework. A definition provided by K.W. Hipel is "Systems Engineering is an integrative and multidisciplinary approach to creative problem solving which takes into account stakeholders' value systems and satisfies important societal, environmental, economic and other criteria in order to enhance the decision-making process when designing, implementing, operating and maintaining a system or system of systems to meet societal needs in a fair, ethical and sustainable manner throughout the system's life cycle" (Hipel et al. 2007, 2009b).

Because thinking in terms of systems for problem solving is so widely accepted, publications regarding the systematic solving of challenging problems appear in journals in many disciplines. Basic systems-type methodologies that are closely related to Systems Engineering, and often thought of as being part of it, include control theory (Clarke et al. 1998), complex adaptive systems (Lansing 2009) and chaos theory (Thiétart and Forgues 1995).

Hipel et al. (2007) discussed the future of Systems Engineering in terms of application domains and research methods. As noted in the abstract of their paper "The methods [Systems Engineering] must be refined and expanded to meet the changing needs of the 21st century: from a system to a system-of-system; from a disciplinary outlook to a multidisciplinary outlook; from a mass production to a mass customization focus; from a steady state to a real-time perspective; and from an optimal to an adaptive approach." Accordingly, the important concepts of system of systems and adaptive management are explained in Sects. 2.4.1 and 2.4.2, respectively, with respect to their relevance to conflict resolution.

2.3.3 Decision Support Systems

A rich range of formal decision-making techniques have been developed in the fields of OR and Systems Engineering, as explained in Sects. 2.3.1 and 2.3.2, respectively. Moreover, a wide variety of game theory techniques are available for application purposes as pointed out in Sect. 2.2.2 and summarized in Fig. 2.1. The focus of this book is the Graph Model for Conflict Resolution (GMCR) for which many useful and powerful techniques are presented in detail as summarized in Sect. 1.3 and listed in the Table of Contents. Moreover, many extensions of GMCR are currently underway while others are planned, as discussed in Sects. 10.3.1 and 10.3.2, respectively.

To permit practitioners and researchers to conveniently apply mathematically-based techniques to physical-based, societal-founded, or combined systems problems, Decision Support Systems (DSSs) are needed. In this way, a user can focus on the insights gained from a rigorous investigation rather than on spending a significant amount of time programming an approach or a set of techniques, for solving the problem. Previously, DSSs were simply referenced to as "user-friendly" programs. As reflected in its title, the goal of a DSS is to aid or support decision-making by making known methodologies and associated data sets immediately available to a user, analyst or DM for applying to a problem of interest to him or her.

As emphasized by Hipel et al. (2008a), a formal model constitutes a representation of a system having a clearly defined mathematical structure. A properly designed model captures the key characteristics of the system or part of the system being studied to allow the system to be better understood so that informed decisions can be made regarding it. The mathematical analysis of a realistic model of a system can be highly effective in investigating the properties of the system and forecasting or simulating system behavior. When carrying out sensitivity analyses, the impacts of meaningful changes to one or more model parameters can be determined by comparing strategic findings before and after sensitivity analyses. Accordingly, one can obtain answers to "what-if" questions about the system. As noted above, a formal model, or collection of models, can be employed to rigorously examine physical, societal or hybrid systems.

A DSS is an easy-to-use computer package containing modeling and analytical capabilities, for one or more formal mathematical techniques. The DSS allows

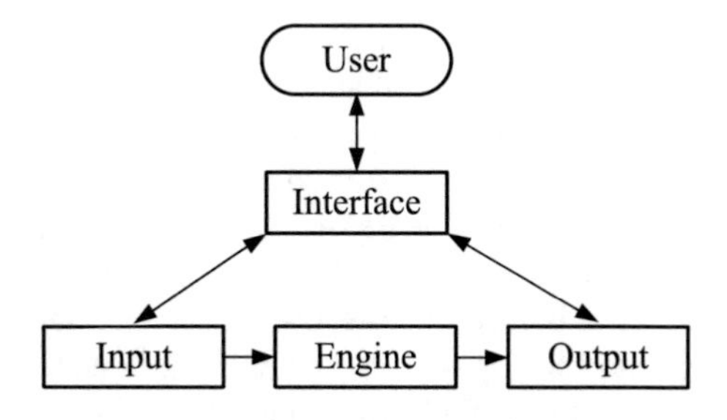

Fig. 2.2 Model-based decision support system for conflict resolution

practitioners and researchers to expeditiously create, revise, refine and analyze a model to support decisions. DSS technologies form one of the most important areas in the field of Information Technology (IT) which encompasses the development and application of computer software and hardware. In his landmark book on Decision Support Systems Engineering, Sage (1991) describes the main components of a DSS as being the Model-base Management System (MBMS) and Database Management System (DBMS) which are connected to a user via a Dialog Generation and Management System (DGMS). In fact, because of the great import of DSS engineering, in general, and in the field of conflict resolution, in particular, Chap. 10 is entirely devoted to the design of a DSS for applying to real-world conflict situations. A general discussion on DSSs is provided in Sect. 10.1.1 along with Fig. 10.1 depicting Sage's general design of DSS (Sage 1991). The rest of Chap. 10 focuses on DSSs for conflict resolution.

Figure 2.2 displays a simplified version of a model-based DSS for conflict resolution when using GMCR or another similar conflict model. The key input information required from the user via the interface (DGMS) is the decision makers (DMs), each DM's options or courses of actions and each DM's relative preferences among the feasible states or scenarios that could occur, as outlined in Sect. 1.2.2. What is particularly advantageous about GMCR is that a minimum amount of information is required by the user to build or calibrate a conflict model. The "grunt" work can be done by the DSS based on this information. The engine is used to carry out the stability calculations for each state from each DM's viewpoint according to a range of solution concepts describing potential human behaviors under conflict as discussed in Sect. 1.2.3. When a state is stable for all DMs according to a particular solution concept, it forms an equilibrium or potential resolution to the dispute under study. The engine can also determine outcomes for situations where DMs may cooperate with one another by forming coalitions. Follow-up analyses, such as the determination of the potential evolution of a conflict over time and various kinds of sensitivity analyses, are also carried out by the engine. As explained in Sect. 10.2.3, a highly efficient engine can be designed and constructed based on the matrix formulation of GMCR. Finally, the output from the DSS in Fig. 2.2 contains important strategic information calculated by the engine such as the potential resolutions and which resolutions can be reached from the current status quo situation.

A DSS furnishes a mechanism by which practitioners, researchers and society, in general, can take full advantage of advances in research in a given field of interest. From a researcher's perspective, a DSS is the means by which he or she can more directly contribute to the enhancements of society. Therefore, the final chapter of this book is entirely devoted to DSSs in conflict resolution. A person, who is trying to learn how moves and countermoves can take place in a conflict as DMs interact with one another, is encouraged to do some calculations by hand in order to fully understand the process and appreciate why GMCR is such a realistic decision technology. Keeping this in mind, a DSS is absolutely essential for GMCR to be widely adopted for helping to resolve conflicts ranging from the simple to the complex. Besides describing existing DSSs for GMCR in Sect. 10.1.2, a universal design for a DSS for GMCR is provided in Sect. 10.2. In this way, companies, government organizations, research teams and others can readily construct their own DSSs if the existing DSSs do not possess all of the capabilities that they require. Moreover, as pointed out in Sect. 10.3, new GMCR developments can be easily added to a properly built DSS.

2.4 Conflict Resolution in Responsible Governance

As explained in Sects. 2.2 and 2.3, a rich range of formal game theory and systems science tools, respectively, have been developed for providing advice to enhance the decision-making process. As pointed out in Sects. 2.3.1 and 2.3.2, many of these techniques were designed within the fields of Operations Research and Systems Engineering, respectively. Moreover, a variety of approaches to decision-making, such as value-focused thinking (Keeney 1992) and concentrating on the interests of the stakeholder (Fisher and Ury 1981, Fisher et al. 1991), have been proposed. Finally, general procedures for improving decision-making have been put forward in fields such as business administration, law, political science and sociology. The aforementioned and other procedures for making decisions can be employed within the general governance procedure outlined in this section.

Because humans live in a highly interconnected world in which the actions of one group of stakeholders can directly affect others, including the natural environment, a truly innovative systems thinking approach to governing society in a highly realistic and fair fashion is required. Accordingly, in the next section a System of Systems framework within which governance systems can be based is proposed. In Sect. 2.4.2, an integrative and adaptive paradigm for governance is described in which the value systems of the key interest groups are taken into account in a participatory way in order to reach desirable systems objectives such as resiliency, sustainability and fairness. The flowchart in Fig. 2.3 summarizes this realistic approach to responsible governance.

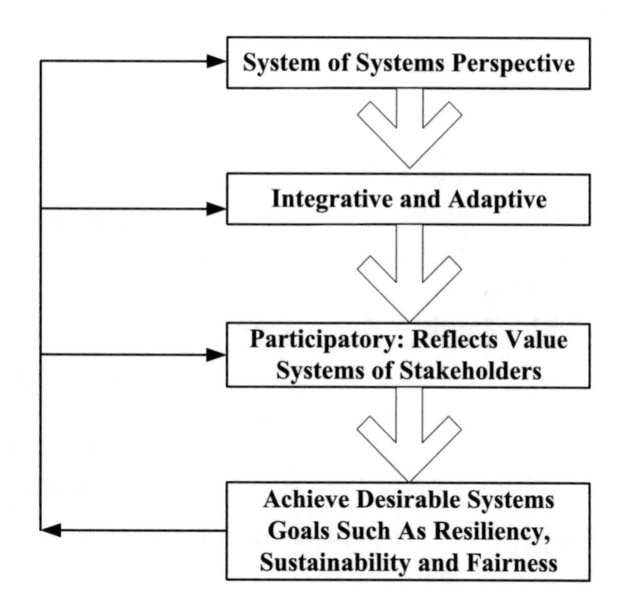

Fig. 2.3 Systems thinking in responsible governance

2.4.1 System of Systems

Multiple decision makers or participants, having their own objectives or value systems, inhabit the main sets of systems, existing on planet Earth (Hipel and Fang 2005). As depicted in Fig. 2.4, these key systems containing multiple stakeholders can be categorized into four main kinds of systems: environmental, societal, intelligent and integrated systems. Because each of these four groups contains many systems, it is referred to as a System of Systems (SoS). Illustrations of environmental systems are the atmospheric, geological, hydrological, zoological, botanical, and ecological systems. Examples of societal systems include agricultural, industrial, economic, political, governmental, infrastructure and urban systems. Within societal systems, creative people and organizations design, build and maintain intelligent systems, such as robotic, mechatronic and automated production systems for satisfying human demands and requirements. Integrated systems, such as individuals and software agents bidding for products over the internet using eBay, are formed by a combination of societal and intelligent systems. A modern commercial aircraft like a Boeing B787 Dreamliner or an Airbus A380, the world's largest passenger airliner, is another example of an integrated system since these planes can be flown automatically using specially designed intelligent systems or under the control of a pilot.

By referring to Fig. 2.5, one can envision how societal SoS and environmental SoS are interconnected (Hipel et al. 2009b). Notice on the left and right sides in this figure that the societal and environmental SoSs, respectively, consist of many systems. For instance, as can be seen on the right, the environmental SoS is composed of complex interrelated atmospheric, water, land and biological systems. As

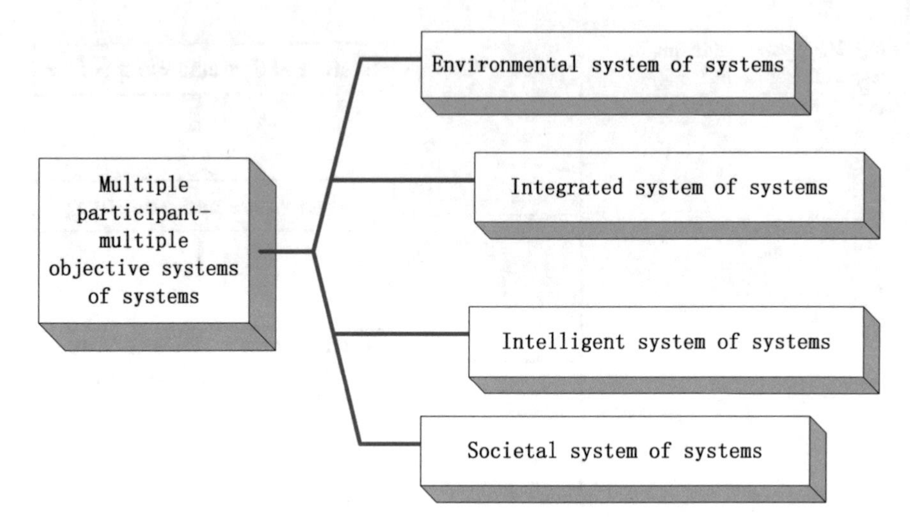

Fig. 2.4 Kinds of multiple participant-multiple objective systems of systems

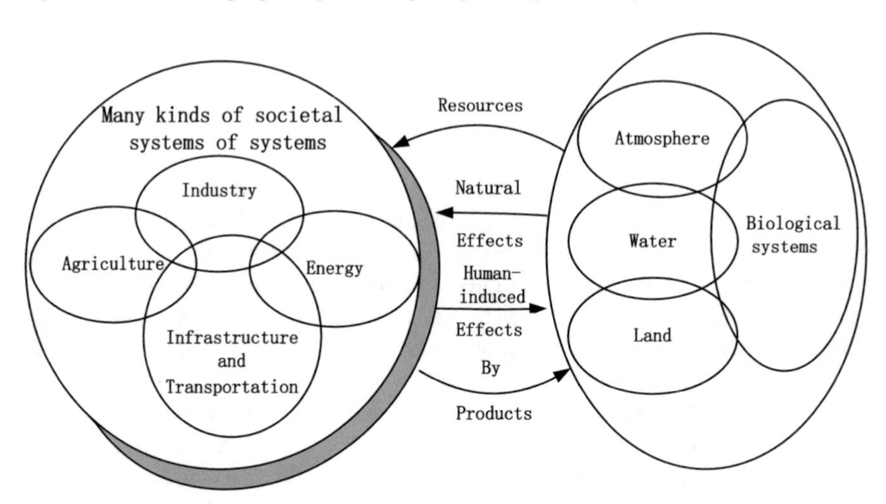

Fig. 2.5 Societal and environmental systems of systems

indicated by the arrow at the top linking these two sets of systems, societal SoS extract resources from the environmental SoS in order to function. For instance, the steel industry depends upon iron ore and energy sources from the environment to be able to operate. Unfortunately, by products from the range of activities occurring within the societal SoS are released into the environmental SoS. For example, carbon dioxide and other air pollutants are emitted from the smokestacks of steel plants which contribute to global warming and climate change while other pollutants are released into nearby bodies of water, thereby degrading water quality. As illustrated by the arrows in the middle, humans can affect both environmental and societal SoS while the natural world has a direct influence over societal SoS. Due to the large-scale release of greenhouse gases by many societal systems and associated land-use changes, humans are causing climate change which in turn degrades societal systems such as agriculture.

The concept of an SoS was developed to capture situations in which systems cooperate and interact with one another under certain circumstances but can also act as independent systems on other occasions. In their research, Sage and Biemer (2007) define an SoS as "a large-scale, complex system, involving a combination of technologies, humans, and organizations, and consisting of components that are systems themselves, achieving a unique end-state by providing synergistic capability from its component systems". Research into the development of the idea of an SoS include contributions by Maier (1998), Sage and Cuppan (2001), Hipel and Fang (2005), Hipel et al. (2009b) and Jamshidi (2009). Based on earlier research by authors such as Maier (1998) and Sage and Cuppan (2001), the authors Sage and Biemer (2007) maintain that an SoS possesses a majority of the following characteristics: (1) operational independence of the individual systems; (2) managerial independence of the separate systems; (3) geographical distribution of the individual systems; (4) emergent behavior in which the SoS performs functions not possible by any of the individual systems on their own; (5) evolutionary development created by continuous interoperability relationships among systems; (6) self-organization; and (7) adaptation. By definition, an SoS constitutes a complex system, in which each individual system is autonomous because it may be evolving independently from other systems in the SoS.

Over the years, research has been carried out to model complex systems which includes contributions in fields like complex adaptive systems (Lansing 2009), chaos theory (Thiétart and Forgues 1995) and cybernetics (Wiener 1948). Contributions have been made on the design of SoS architecture for addressing a class of problems (see, for instance, Ge et al. (2013, 2014b)) including the employment of GMCR within this type of design (Ge et al. 2014a). An SoS approach has also been utilized for addressing quality control strategies for a complex product (Liu and Hipel 2012), risk management of extreme events (Bristow et al. 2012), global food security (Hipel et al. 2010), and water resources management (Hipel et al. 2011, 2013a). In order to model the decision making and physical systems characteristics of systems, one requires a broad range of flexible tools for utilization in these two realms, as explained by Hipel et al. (2008a, b). Sections 2.1 to 2.3 deal with formal decision-making tools with a focus on methods for employment in multiple participant-multiple objective decision-making. In fact, as mentioned in Sect. 1.1.2, the purpose of this text is to present the latest ideas in conflict resolution in order to address the strategic aspects of pressing real-world problems such as widespread pollution of the natural environment, climate change, international trade and regional wars.

As an explanation of how one would employ both decision-making and physical systems tools, consider the groundwater pollution problem described in Sect. 1.2. In this real-life dispute, by products from a chemical plant located in Elmira, Ontario, Canada, polluted the aquifer underlying this town with a carcinogen. The local government hired engineering consultants to determine the seriousness of the pollution and how far the plume containing the pollutant had spread. Bore holes were drilled at many locations to obtain samples of the groundwater and physical systems models expressed as differential equations were used to model the flow of the pollutant. In addition, methods for cleansing the aquifer and preventing further pollution were investigated. Within the societal aspects of this problem, the negotiations that took place in 1991 among the local government, company and the Ministry of the Environment in Ontario are formally modeled and analyzed in Sect. 4.5 to obtain strategic insights into how this problem could be resolved. As of late 2017, one quarter of a century after the carcinogen was discovered, the Elmira aquifer is continuing to be cleansed by pumping water from the underground aquifer and treating it before releasing it into a nearby stream, the company is treating its pollutants at the Elmira plant before discharging them, and the town of Elmira is receiving all of its water via a pipeline from the city of Waterloo situated 15 km to the south. This is clearly a very serious and expensive situation which would be much worse if a strategic settlement to rectify the situation had not been reached.

The Elmira groundwater contamination is an actual illustration of what is called a brownfield: land and groundwater which are contaminated by industrial activities. In reality, this is a widespread phenomenon in North America, Europe and the developing world. There are more than one-half million brownfields in the USA, 360,000 in Germany and 33,000 in Canada (NRTEE 2003, Hipel and BernathWalker 2011). The rapid industrialization of Asian nations is creating brownfields on a massive scale. Even though brownfields could be largely prevented by investing heavily in pollution control equipment during industrialization, nations prefer to first obtain wealth via industrialization in the quickest and cheapest way possible and to clean up the brownfields later. This, of course, is much more expensive in the long term and some of the damage may be irreversible. Enhanced decision-making methodologies such as the game theory techniques presented in this book may assist in making more informed decisions since they directly take into account both the long and short term values of the various decision makers.

Today, the nations of the world are knowingly dumping massive amounts of greenhouse gases (GHG) into the "atmospheric commons" even though they are aware that this short-term behavior is causing temperatures to significantly rise over time with increasingly devastating consequences. In other words, the countries are playing the self-interest version of Prisoner's Dilemma, rather than the long-sighted version in which the nations cooperate and thereby greatly reduce greenhouse gas releases starting now so they will all benefit more in the long run. The main uncertainty that currently remains is whether the climate system has already reached the point of no-return in which global warming will be irreversible no matter what society does. An important fact to mention is that the physical systems and other related solutions to this global warming problem are already known. In particular, nuclear power plants, renewable energy, energy efficiency and life-style changes are well recognized solutions which can be expeditiously implemented to greatly reduce the use of fossil fuels. Coal fired-plants for generating electricity can be quickly converted to gas-fired plants to reduce carbon dioxide emissions by 50% in the short term and subsequently close down in the medium term when they can be replaced by cleaner energy sources. Where society has failed up to the present time is to solve the societal and strategic aspects of climate change, especially with respect to reaching meaningful negotiated agreements at the international, national, provincial and local levels to significantly cut back on greenhouse gas emissions.

Within Canada, Magna International Inc., a large car parts manufacturing corporation based in Ontario, commissioned the Council of Canadian Academies (CCA) to put together an Expert Panel to investigate the latest evidence regarding energy use and climate change and what can be done to reduce GHG emissions. Subsequently, one year later, the Expert Panel on Energy Use and Climate Change, consisting of an interdisciplinary group of eight people and co-chaired by Keith W. Hipel and Paul R. Portney, completed its report entitled "Technology and Policy Options for a Low Emission Energy System in Canada", which was officially released by the CCA on October 27, 2015 (CCA 2015). The three key findings of the Expert Panel based on the existing evidence are:

- Technology options are available now to move to a low-emission energy system. In fact, Canada could reduce its GHG emissions from 60% to 90% at manageable cost.
- The electricity supply system should be decarbonized by replacing coal and gas-fired electricity generation plants by non-GHG energy generation systems such as renewable, nuclear and hydro.
- Economy-wide policies are needed to meet stringent GHG cutback requirements, but flexible in the sense that different measures could be taken to accomplish this. For instance, the Province of British Columbia already has a carbon tax which has worked extremely well whereas Ontario and Quebec have elected to use a cap and trade system.

Within Canada, negotiations involving the Federal Government, ten Canadian Provinces and three Territories are needed to come up with a harmonized set of policies and associated physical systems solutions to significantly reduce GHG releases. Moreover, since an agreement among the nations of the world reached in Paris on December 11, 2015 has no mandatory or legal requirements for countries to greatly reduce their GHG emissions, intense negotiations will still have to take place to accomplish this. Fortunately, the countries did state that they would not like temperature increases to go beyond 1.5 °C, which in turn means drastic GHG emission cutbacks are necessary. All of this indicates that comprehensive negotiation tools could prove to be extremely useful for reaching binding and lasting agreements. However, the election of American President Donald Trump on November 8, 2016, meant that the United States may weaken or withdraw its commitments to reducing its GHG emissions put forward in Paris. In fact, in June 2017, the US announced its intention to withdraw from the Paris Accord, for which the earliest effective date of US withdrawal is November 2020. The game theoretic procedures presented in this book are designed to assist with reaching agreements among stakeholders having conflicting objectives in order to resolve tough SoS problems such as climate change, as exemplified by Bernath Walker and Hipel (2017) and He et al. (2017).

Conflict resolution tools mentioned in Sect. 2.2, economics, and ideas from the social sciences can be employed within a responsible governance system to solve climate change and many other tough decision problems facing society. A governance system includes policies, agreements, laws, regulations, compliance methods, monitoring, institutions and management. The key aspects needed in governance systems are discussed next and also noted in Fig. 2.3.

2.4.2 Integrative and Adaptive Management

Because water resources engineers and managers must deal with both the physical aspects of water management as well as the societal components, they are global leaders on how to design effective governance systems. Moreover, they were one of the first groups of professionals outside of the military to adopt, enhance and expand operations research and systems engineering tools for solving challenging SoS problems. In addition, they have developed a range of decision support systems (DSSs) for implementing both the physical and societal systems parts of SoS problem solving to allow these tools to be utilized by both practitioners and researchers (Hipel et al. 2008a, b).

Integrative and adaptive management constitute two key interrelated concepts needed for achieving effective water governance within an SoS structure. As explained in "A Handbook for Integrated Water Resources Management in Basins" published in 2009 by the Global Water Partnership (GWP) and International Network of Basin Organizations (INBO) (GWP and INBO 2009), "The integrated water resources management approach helps to manage and develop water resources in a sustainable and balanced way, taking account of social, economic and environmental interests. It recognizes the many different and competing interest groups, the sectors that use and abuse water, and the needs of the environment." This systems thinking approach to water resources management directly states that many interconnected factors must be taken into account and that the value systems of competing stakeholders and environmental requirements must be entertained. Additionally, because of the physical reality that a watershed controls the flow of water, water management practices must be implemented at the basin level in consonance with policy, laws and regulations at the local, regional, provincial or state, national and international levels. For instance, the International Joint Commission (IJC) (IJC 2009) recommends an integrated approach to river basin management for basins that are intersected by the Canada-US border. The IJC is an independent binational body having investigative, regulatory and adjudicative roles for implementing the 1909 Boundary Waters Treaty between Canada and the USA for impartially addressing water and environmental problems between the two nations.

In addition to being integrative, water and other types of governance must be adaptive to be capable of handling the largely unpredictable behavior of environmental and societal SoSs caused by their intrinsic complexity, uncertainty and interconnectedness. As a result of this unpredictability, one does not know in advance how well a given policy is going to function for appropriately tackling both anticipated and unexpected events. Within the concept of passive adaptive management, one utilizes new knowledge garnered by monitoring and experience to iteratively refine plans to improve existing management approaches and related decision-making. When practicing active adaptive management, strategies are purposefully changed in order to scientifically test new hypotheses, and thereby learn by experimentation to determine the best management strategy. Therefore, adaptive management is popularly referred to as "learning by doing". The clever idea of adaptive management was originally put forward by Holling (1978) and other scientists as a consequence of investigating resiliency theories of ecological systems, and since that time a rich body of literature has amassed such as contributions by Walters (1986), Gunderson and Holling (2002), Noble (2004), NRC (2004), AWRA (2006, 2009), Gunderson and Light (2006), and Williams and Jackson (2007).

In their research, Hipel et al. (2008b, 2009b) recommend that adaptive integrative management be conducted within a SoS framework in combination with the employment of formal decision-making tools, from operations research, systems engineering and other systems science fields (see Sect. 2.3), as well as ideas from the social sciences and humanities. Additionally, any policy or agreement should reflect the value systems of the stakeholders it serves and contain a dispute resolution mechanism that guides disputants in a positive direction towards a win/win resolution. This can be strengthened in practice by adopting a participatory approach to involving stakeholders in the practice of responsible governance and making the various decision technologies available as DSSs.

Throughout this book, case studies involving actual disputes in a range of different fields are employed to demonstrate how the various related conflict resolution methods can be conveniently applied in practice to better understand problems and gain strategic insights for improving decision-making. However, the reader is encouraged to keep in mind that conflict resolution is usually employed in conjunction with other societal and physical systems tools within an integrative and adaptive approach to governance embedded in an SoS framework.

2.5 Important Ideas

The decisions that you make determine your destiny. Therefore, you want to make the most informed possible decision to help you reach your goals, keeping in mind that in decisions involving interactions with others, you must understand their values and the effects they can have upon you depending on which actions they select. In a given conflict situation, you may wish to consider joining a coalition to ascertain if you can do even better via meaningful cooperation with others to achieve a win/win resolution.

The Graph Model for Conflict Resolution (GMCR) has a key role to play in tackling tough decision problems involving multiple participants each of whom possesses his or her own values, objectives or preferences. Among an array of available game theoretical methods described in Sect. 2.2 and portrayed in Fig. 2.1, GMCR is especially effective in addressing actual complex conflicts in almost any field ranging from very simple ones to large disputes. In fact, GMCR constitutes a unique and relatively new formal decision-making method which nicely complements the rich range of decision tools developed in fields such as Operations Research (Sect. 2.3.1), Systems Engineering (Sect. 2.3.2) and elsewhere. When implemented as a Decision Support System (Sect. 2.3.3 and Chap. 10), GMCR constitutes a truly powerful decision technology for formally studying real-world conflicts. When utilized within a system of systems engineering perspective (Sect. 2.4.1) to integrative and adaptive management in a participatory fashion (Sect. 2.4.2), GMCR can be utilized as an important and complementary methodology for effectively carrying out responsible governance.

2.6 Problems

2.6.1 In Sect. 2.2.2 and Fig. 2.1, game theory methods are classified and put into perspective. In the right branch in Fig. 2.1, agent-based models are listed as a set of techniques. Find a journal paper or textbook on agent-based modeling and briefly explain the basic idea underlying it and how it is used in practice. Outline how it could be used in policy design and analysis.

2.6.2 A game theoretical method listed in the right branch in Fig. 2.1 is extensive form. By referring to a journal paper or book, outline how extensive form works. According to Chap. 4 in the book by Fang et al. (1993) qualitatively explain how the Graph Model for Conflict Resolution and extensive form are connected. What is a key drawback of extensive form?

2.6.3 Drama Theory is listed in the left branch of Fig. 2.1. Locate a journal paper or book dealing with Drama Theory to use as a basis for successfully outlining how Drama Theory works.

2.6.4 The conflict analysis approach of Fraser and Hipel (1979, 1984) is part of the left branch in Fig. 2.1. Describe three improvements to Conflict Analysis which are embedded in the Graph Model for Conflict Resolution. Be sure to supply references to the literature to support the enhancements that you mention.

2.6.5 As noted in Sect. 2.3.1, Operations Research (OR) was founded by the British military just prior to the outbreak of WWII in Europe. Many classified documents regarding the development and employment of OR during the war were released 50 years after the war ended. Select an interesting paper describing the successful utilization of OR during the war and outline the points in the paper that you personally found to be of interest to you.

2.6.6 How did the American fleet in the Pacific Ocean during WWII use a "soft" systems approach to encryption to communicate among their ships, especially their lethal aircraft carriers. (Hint: Find a reference involving the famous "code talkers".)

2.6.7 One of the most famous code breakers working at Bletchley Park northwest of London during WWII to break the German codes was Dr. William Tutte. What great feat did he accomplish and at which university did he help found a famous Faculty of Mathematics after WWII?

2.6.8 Locate a paper in the IEEE Transactions on Systems, Man and Cybernetics: Systems which deals with system of systems engineering. Outline the main contributions in the paper and explain why a system of systems engineering approach was useful and insightful in dealing with the problems addressed in the paper.

2.6.9 The International Council on Systems Engineering (INCOSE) is a strong proponent of the system of system (SoS) engineering approach and publishes the journal called Systems Engineering. Select an SoS engineering report published by INCOSE or else a journal article dealing with this topic. Summarize the key contributions of the article and explain why an SoS engineering approach was highly effective for solving the problem studied in the article.

2.6.10 Why is the Graph Model for Conflict Resolution an indispensable approach for using within a system of systems engineering approach to problem solving and creative decision-making?

2.6.11 Select a reference on decision support systems (DSSs) that is of high interest to you. Based on this reference, outline the basic design of a DSS and why DSSs are essential for employment in real-world decision-making.

2.6.12 Most of the key ideas in integrative and adaptive management addressed in Sect. 2.4.2 came from the field of water resources. Choose a key journal paper or book from the water resources literature which deals with integrated and adaptive water resources management in a participatory fashion. Outline the key contents of this paper and explain why conflict resolution has a key role to play in this insightful approach to management.

References

Ackoff, R. L. (1962). *Scientific method: Optimizing applied research decisions*. New York: Wiley. https://doi.org/10.1038/sj.bdj.4808633.

Ackoff, R. L. (1981). The art and science of mess management. *Interfaces, 11*(1), 20–26. https://doi.org/10.1287/inte.11.1.20.

Aldrich, R. J. (2010). *GCHQ: The uncensored story of Britain's most secret intelligence agency*. London: HarperPress.

AWRA. (2006). Special issue on adaptive management of water resources. *Water Resources IMPACT, 8*(3), American Water Resources Association (AWRA).

AWRA. (2009). Special issue on application of adaptive management of water resources. *Water Resources IMPACT, 11*(3), American Water Resources Association (AWRA).

Belton, V., & Stewart, T. (2002). *Multiple criteria decision analysis: An integrated approach*. Norwell: Kluwer.

Bennett, P. G., & Dando, M. (1977). Fall Gelb and other games: A hypergame perspective of the fall of France, 1940. *Journal of the Conflict Research Society, 1*(2), 1–33.

Bennett, P. G., & Dando, M. (1979). Complex strategic analysis: A hypergame study of the fall of France. *Journal of the Operational Research Society, 30*(1), 23–32. https://doi.org/10.2307/3009663.

Benson, R. L. (1997). *A history of US communications intelligence during World War II: Policy and administration*. Washington, D.C.: Center for Cryptologic History, National Security Agency. https://doi.org/10.2307/120221.

Bernath Walker, S., & Hipel, K. W. (2017). Strategy, complexity and cooperation: The Sino-American climate regime. *Group Decision and Negotiation, 26*(5), 997–1027. https://doi.org/10.1007/s10726-017-9528-8.

Briggs, A. (2011). *Secret days: Codebeaking in Bletchley Park*. Barnsley, UK: Frontline Books.

Bristow, M., Fang, L., & Hipel, K. W. (2012). System of systems engineering and risk management of extreme events: Concepts and case study. *Risk Analysis, 32*(11), 1935–1955. https://doi.org/10.1111/j.1539-6924.2012.01867.x.

Bristow, M., Fang, L., & Hipel, K. W. (2014). From values to ordinal preferences for strategic governance. *IEEE Transactions on System, Man, and Cybernetics: Systems, 44*(10), 1364–1383. https://doi.org/10.1109/tsmc.2014.2308154.

Bryant, J. (2003). *The six dilemmas of collaboration: Inter-organisational relationships as drama*. Chichester, UK: Wiley.

Bryant, J. (2015). *Acting strategically using drama theory*. Boca Raton, FL: CRC Press. https://doi.org/10.1201/b18912.

CCA. (2015). *Technology and policy options for a low-emission energy system in Canada*. Council of Canadian Academies (CCA), Ottawa, Canada: The Expert Panel on Energy Use and Climate Change.

Checkland, P. (1981). *Systems thinking, systems practice*. Chichester, UK: Wiley. https://doi.org/10.1016/0016-3287(82)90032-5.

Chen, Y., Li, K. W., Kilgour, D. M., & Hipel, K. W. (2008). A case-based distance model for multiple criteria ABC analysis. *Computers & Operations Research, 35*(3), 776–796. https://doi.org/10.1016/j.cor.2006.03.024.

Chen, Y., Kilgour, D. M., & Hipel, K. W. (2011). An extreme-distance approach to multiple criteria ranking. *Mathematical and Computer Modelling, 53*(5), 646–658. https://doi.org/10.1016/j.mcm.2010.001.

Clarke, F. H., Ledyaev, Y. S., Stern, R. J., & Wolenski, P. R. (1998). *Nonsmooth analysis and control theory*. New York: Springer.

Fang, L., Hipel, K. W., & Kilgour, D. M. (1993). *Interactive decision making: The graph model for conflict resolution*. New York: Wiley. https://doi.org/10.2307/2583940.

Fisher, R., & Ury, W. (1981). *Getting to yes: Negotiating agreement without giving in*. New York: Penguin Books.

Fisher, R., Ury, W., & Patton, B. (1991). *Getting to yes: Negotiating agreement without giving in* (2nd ed.). New York: Penguin Books.

Fraser, N. M., & Hipel, K. W. (1979). Solving complex conflicts. *IEEE Transactions on Systems, Man, and Cybernetics, 9*(12), 805–816. https://doi.org/10.1109/tsmc.1979.4310131.

Fraser, N. M., & Hipel, K. W. (1984). *Conflict analysis: Models and resolutions*. New York: North-Holland. https://doi.org/10.2307/2582031.

Ge, B., Hipel, K. W., Yang, K., & Chen, Y. (2013). A data-centric capability-focused approach for system-of-systems architecture modeling and analysis. *Systems Engineering, 16*(3), 363–377. https://doi.org/10.1002/sys.21253.

Ge, B., Hipel, K. W., Fang, L., Yang, K., & Chen, Y. (2014a). An interactive portfolio decision analysis approach for system-of-systems architecting using the graph model for conflict resolution. *IEEE Transactions on Systems, Man, and Cybernetics: Systems, 44*(10), 1328–1346. https://doi.org/10.1109/tsmc.2014.2309321.

Ge, B., Hipel, K. W., Yang, K., & Chen, Y. (2014b). A novel executable modeling approach for system-of-system architecture. *IEEE Systems Journal, 8*(1), 4–13. https://doi.org/10.1109/jsyst.2013.2270573.

Goicoechea, A., Hansen, D., & Duckstein, L. (1982). *Multiobjective decision analysis with engineering and business applications*. New York: Wiley. https://doi.org/10.2307/2581355.

Grey, C. (2012). *Decoding organization: Bletchley Park, codebreaking and organization studies*. Cambridge, UK: Cambridge University Press. https://doi.org/10.1017/cbo9780511794186.

Gunderson, L. H., & Holling, C. S. (2002). *Panarchy: Understanding transformations in human and natural systems*. Washington, D.C.: Island Press.

Gunderson, L., & Light, S. S. (2006). Adaptive management and adaptive governance in the Everglades ecosystem. *Policy Sciences, 39*(4), 323–334. https://doi.org/10.1007/s11077-006-9027-2.

GWP and INBO. (2009). *A handbook for integrated water resources management in basins*. Stockholm, Sweden and Paris, France: Global Water Partnership and International Network of Basin Organizations.

Haimes, Y. Y. (2016). *Risk modeling, assessment, and management* (3rd ed.). New York: Wiley.

Hardin, G. (1968). The tragedy of the commons. *Science, 162*(5364), 1243–1248. https://doi.org/10.1080/19390450903037302.

He, S., Kilgour, D. M., & Hipel, K. W. (2017). A general hierarchical graph model for conflict resolution with application to greenhouse gas emission disputes between USA and China. *European Journal of Operational Research, 257*(3), 919–932. https://doi.org/10.1016/j.ejor.2016.08.014.

Hipel, K. W. (1981). Operational research techniques in river basin management. *Canadian Water Resources Journal, 6*(4), 205–226. https://doi.org/10.4296/cwrj0604205.

Hipel, K. W. (Ed.). (1992). *Multiple objective decision making in water resources*. Bethesda, MD: American Water Resources Association. https://doi.org/10.1111/j.1752-1688.1992.tb03150.x.

Hipel, K. W. (Ed.). (2009a). *Conflict resolution* (Vol. 1). Oxford, UK: Eolss Publishers.

Hipel, K. W. (Ed.). (2009b). *Conflict resolution* (Vol. 2). Oxford, UK: Eolss Publishers.

Hipel, K. W., & Fang, L. (2005). Multiple participant decision making in societal and technological systems. In T. Arai, S. Yamamoto, & K. Makino (Eds.), *Systems and human science—For safety, security, and dependability* (pp. 3–31). Amsterdam, The Netherlands: Elsevier. https://doi.org/10.1016/b978-044451813-2/50003-8.

Hipel, K. W., & Bernath Walker, S. (2011). Conflict analysis in environmental management. *Environmetrics, 22*(3), 279–293. https://doi.org/10.1002/env.1048.

Hipel, K. W., Radford, K. J., & Fang, L. (1993). Multiple participant multiple criteria decision making. *IEEE Transactions on Systems, Man, and Cybernetics, 23*(4), 1184–1189. https://doi.org/10.1109/21.247900.

Hipel, K. W., Kilgour, D. M., Fang, L., & Li, W. (2003). Resolution of water conflicts between Canada and the United States. In K. Nandalal & S. Simonovic (Eds.), *State-of-the-art report on systems analysis methods for resolution of conflicts in water resources management* (pp. 62–75). Paris, France: United Nations Educational, Scientific and Cultural Organization (UNESCO).

Hipel, K. W., Jamshidi, M. M., Tien, J. M., & White III, C. (2007). The future of systems, man and cybernetics: Application domains and research methods. *IEEE Transactions on Systems, Man, and Cybernetics, Part C, Applications and Reviews, 37*(5), 726–743. https://doi.org/10.1109/tsmcc.2007.900671.

Hipel, K. W., Fang, L., & Kilgour, D. M. (2008a). Decision support systems in water resources and environmental management. *Journal of Hydrologic Engineering, 13*(9), 761–770. https://doi.org/10.1061/(asce)1084-0699(2008)13:9(761).

Hipel, K. W., Obeidi, A., Fang, L., & Kilgour, D. M. (2008b). Adaptive systems thinking in integrated water resources management with insights into conflicts over water exports. *INFOR, 46*(1), 51–69. https://doi.org/10.3138/infor.46.1.51.

Hipel, K. W., Kilgour, D. M., Rajabi, S., & Chen, Y. (2009a). Operations research and refinement of courses of action. In A. P. Sage & W. B. Rouse (Eds.), *Handbook of systems engineering and management* (pp. 1171–1222). New York: Wiley.

Hipel, K. W., Obeidi, A., Fang, L., & Kilgour, D. M. (2009b). Sustainable environmental management from a system of systems perspective. In M. Jamshidi (Ed.), *System of systems engineering: Innovations for the 21st century* (pp. 443–481). New York: Wiley. https://doi.org/10.1002/9780470403501.ch18.

Hipel, K., W., Fang, L., & Heng, M. (2010). System of systems approach to policy development for global food security. *Journal of Systems Science and Systems Engineering, 19*(1), 1–21. https://doi.org/10.1007/s11518-010-5122-1.

Hipel, K. W., Kilgour, D. M., & Fang, L. (2011). Systems methodologies in vitae systems of systems. *Journal of Natural Disaster Science, 32*(2), 63–77. https://doi.org/10.2328/jnds.32.63.

Hipel, K. W., Fang, L., Ouarda, T. B. M. J., & Bristow, M. (2013a). An introduction to the special issue on tackling challenging water resources problems in Canada: A systems approach. *Canadian Water Resources Journal, 38*(1), 3–11. https://doi.org/10.1080/07011784.2013.773643.

Hipel, K. W., Fang, L., & Wang, L. (2013b). Fair water resources allocation with application to the South Saskatchewan River basin. *Canadian Water Resources Journal, 38*(1), 47–60. https://doi.org/10.1080/07011784.2013.773767.

Hipel, K. W., Fang, L., & Xiao, Y. (2016). Conflict resolution. In V. Singh (Ed.), *Handbook of applied hydrology*. New York: McGraw-Hill. https://doi.org/10.1007/978-3-319-31217-0_10.

Hobbs, B. F., & Meier, P. (2000). *Energy decisions and the environment: A guide to the use of multicriteria methods*. Dordrecht, The Netherlands: Kluwer.

Holling, C. S. (Ed.). (1978). *Adaptive environmental assessment and management*. New York: Wiley. https://doi.org/10.1007/s00267-008-9187-2.

Howard, N. (1971). *Paradoxes of rationality: Theory of metagames and political behavior*. Cambridge, MA: MIT Press. https://doi.org/10.2307/1266876.

Howard, N. (1999). *Confrontation analysis: How to win operations other than war*. Washington D.C.: Command and Control Research Program (CCRP), Office of the Assistant Secretary of Defense, Department of Defense.

Howard, N., Bennett, P. G., Bryant, J. W., & Bradley, M. (1992). Manifesto for a theory of drama and irrational choice. *Journal of the Operational Research Society, 44*, 99–103. https://doi.org/10.2307/2584447.

Hwang, C., & Yoon, K. (1981). *Multiple attribute decision making: Methods and applications*. Berlin, Germany: Springer.

IJC. (2009). *The International Watersheds Initiative: Implementing a New Paradigm for Transboundary Basins: Third Report to Governments on the International Watersheds Initiative*. Ottawa, Canada and Washington, D.C., USA: International Joint Commission.

Jamshidi, M. (2009). *System of systems engineering: Innovations for the twenty-first century*. New York: Wiley.

Keeney, R. L. (1992). *Value focused thinking: A path to creative decision making*. Cambridge, MA: Harvard University Press.

Keeney, R. L., & Raiffa, H. (1976). *Decision analysis with multiple conflicting objectives*. New York: Wiley.

Keys, P. (1991). *Operational research and systems*. New York: Springer. https://doi.org/10.1007/978-1-4899-0667-0.

Kilgour, D. M., & Hipel, K. W. (2005). The graph model for conflict resolution: Past, present, and future. *Group Decision and Negotiation, 14*(6), 441–460. https://doi.org/10.1007/s10726-005-9002-x.

Kilgour, D. M., & Eden, C. (Ed.). (2010). *Handbook of group decision and negotiation*. Dordrecht, The Netherlands: Springer. https://doi.org/10.1007/978-90-481-9097-3.

Kilgour, D. M., & Hipel, K. W. (2010). Conflict analysis methods: The graph model for conflict resolution. In D. M. Kilgour & C. Eden (Eds.), *Handbook of group decision and negotiation* (pp. 203–222). Dordrecht, The Netherlands: Springer.

Kilgour, D. M., Hipel, K. W., & Fang, L. (1987). The graph model for conflicts. *Automatica, 23*, 41–55. https://doi.org/10.1016/0005-1098(87)90117-8.

Kimball, G. E., & Morse, P. M. (1951). *Methods of operations research*. New York: Wiley. https://doi.org/10.2307/3006426.

Kuang, H., Kilgour, D. M., & Hipel, K. W. (2015). Grey-based PROMETHEE II with application to evaluation of source water protection strategies. *Information Sciences, 294*, 376–389. https://doi.org/10.1016/j.ins.2014.09.035.

Lansing, J. S. (2009). Complex adaptive systems. *Annual Review of Anthropology, 32*(4), 183–204.

Lardner, H. (1979). The origins of operational research. In *Operational Research 78, Proceedings of the Eighth IFORS International Conference on Operational Research* (pp. 3–12), Toronto, Canada. Amsterdam: North-Holland.

Liu, Y., & Hipel, K. W. (2012). A hierarchical decision model to select quality control strategies for a complex product. *IEEE Transactions on Systems, Man, and Cybernetics Part A: Systems and Humans, 42*(4), 814–826. https://doi.org/10.1109/tsmca.2012.2183363.

MacCrimmon, K. R. (1973). An overview of multiple objective decision making. In J. Cochrane & M. Zeleny (Eds.), *Multiple criteria decision making* (pp. 18–44). Columbia, SC: University of South Carolina Press.

Maier, M. W. (1998). Architecting principles for systems of systems. *Systems Engineering, 1*(4), 267–284. https://doi.org/10.1002/(sici)1520-6858(1998)1:4<267::aid-sys3>3.0.co;2-d.

Miser, H. J., & Quade, E. S. (Eds.). (1985). *Handbook of systems analysis: Overview of uses, procedures, applications, and practice*. New York: North-Holland. https://doi.org/10.2307/2582574.

Miser, H. J., & Quade, E. S. (Eds.). (1988). *Handbook of systems analysis: Craft issues and procedural choices*. New York: North-Holland. https://doi.org/10.2307/2583791.

Noble, B. (2004). Applying adaptive environmental management. In B. Mitchell (Ed.), *Resources and environmental management in Canada: Addressing conflict and change* (pp. 442–466). Toronto: Oxford University Press.

NRC. (2004). *Adaptive management for water resources project planning*. Washington, D.C.: The National Academies Press, National Research Council (NRC).

NRTEE. (2003). *Cleaning up the past, building the future: A national Brownfield redevelopment strategy for Canada*. Ottawa, Canada: National Round Table on the Environment and the Economy (NRTEE).

Ostrom, E., Gardner, R., & Walker, J. (1994). *Rules, games, and common-pool resources*. Ann Arbor, MI: University of Michigan Press. https://doi.org/10.3998/mpub.9739.

Ostrom, E., Burger, J., Field, C. B., Norgaard, R. B., & Policansky, D. (1999). Revisiting the commons: Local lessons, global challenges. *Science, 284*(5412), 278–282. https://doi.org/10.1126/science.284.5412.278.

Radford, K. J. (1988). *Strategic and tactical decisions*. New York: Springer. https://doi.org/10.1007/978-1-4613-8815-9.

Radford, K. J. (1989). *Individual and small group decisions*. New York: Springer. https://doi.org/10.1007/978-1-4757-2068-6.

Rajabi, S., Kilgour, D. M., & Hipel, K. W. (1998). Modeling action-interdependence in multiple criteria decision making. *European Journal of Operational Research, 110*(3), 490–508. https://doi.org/10.1016/s0377-2217(97)00318-4.

Ravetz, J. R. (1971). *Scientific knowledge and its social problems*. Oxford, UK: Oxford University Press. https://doi.org/10.2307/2218013.
Ravindran, A., Capelo, J. J. S., & Bonome, D. T. P. (1987). *Operations research: Principles and practice*. New York: Wiley.
Rittel, H. W., & Webber, M. M. (1973). Dilemmas in a general theory of planning. *Policy Sciences, 4*(2), 155–169. https://doi.org/10.1007/bf01405730.
Rosenhead, J. (1989). *Rational analysis for a problematic world: Structuring methods for complexity, uncertainty and conflict*. Chichester, UK: Wiley.
Rosenschein, J. S., & Zlotkin, G. (1994). *Rules of encounter: Designing conventions for automated negotiation among computers*. Cambridge, MA: MIT Press.
Roy, B. (1996). *Multicriteria methodology for decision aiding*. Dordrecht, The Netherlands: Kluwer. https://doi.org/10.1007/978-1-4757-2500-1.
Saaty, T. L. (1980). *The analytic hierarchy process*. New York: McGraw-Hill. https://doi.org/10.1002/0470011815.b2a4a002.
Sage, A. P. (1991). *Decision support systems engineering*. New York: Wiley.
Sage, A. P. (1992). *Systems engineering*. New York: Wiley. https://doi.org/10.1201/9781420010855.ch4.
Sage, A. P., & Cuppan, C. D. (2001). On the systems engineering and management of systems of systems and federations of systems. *Information, Knowledge, Systems Management, 2*(4), 325–345.
Sage, A. P., & Biemer, S. M. (2007). Processes for system family architecting, design, and integration. *IEEE Systems Journal, 1*(1), 5–16. https://doi.org/10.1109/11196.2007.900240.
Schlager, K. J. (1956). Systems engineering—Key to modern development. *IRE Transactions on Engineering Management, 3*(3), 64–66. https://doi.org/10.1109/iret-em.1956.5007383.
Schon, D. A. (1987). *Educating the reflective practitioner: Toward a new design for teaching and learning in the professions*. San Francisco, CA: Jossey-Bass.
Singh, M. G. (1987). *Systems and control encyclopedia: Theory, technology, applications*. Oxford, U.K.: Pergamon Press.
Thiétart, R. A., & Forgues, B. (1995). Chaos theory and organization. *Organization Science, 6*(1), 19–31. https://doi.org/10.1287/orsc.6.1.19.
Vincke, P. (1992). *Multicriteria decision-aid*. New York: Wiley. https://doi.org/10.2307/2584205.
Von Neumann, J. (1928). Die Zerlegung eines Intervalles in abzählbar viele kongruente Teilmengen. *Fundamenta Mathematicae, 1*(11), 230–238. https://doi.org/10.4064/fm-11-1-230-238.
Von Neumann, J., & Morgenstern, O. (1944). *Theory of games and economic behavior*. Princeton, NJ: Princeton University Press. https://doi.org/10.2307/2550081.
Von Neumann, J., & Morgenstern, O. (1953). *Theory of games and economic behavior* (3rd ed.). Princeton, NJ: Princeton University Press.
Walters, C. (1986). *Adaptive management of renewable resources*. New York: McGraw Hill.
Wang, L., Fang, L., & Hipel, K. W. (2003). Water resources allocation: A cooperative game theoretic approach. *Journal of Environmental Informatics, 2*(2), 11–22. https://doi.org/10.3808/jei.200300019.
Wang, L., Fang, L., & Hipel, K. W. (2007). Mathematical programming approaches for modeling water rights allocation. *Journal of Water Resources Planning and Management, 133*(1), 50–59. https://doi.org/10.1061/(asce)0733-9496(2007)133:1(50).
Wang, L., Fang, L., & Hipel, K. W. (2008a). Basin-wide cooperative water resources allocation. *European Journal of Operational Research, 190*(3), 798–817. https://doi.org/10.1016/j.ejor.2007.06.045.
Wang, L., Fang, L., & Hipel, K. W. (2008b). Integrated hydrologic-economic modeling of coalitions of stakeholders for water allocation in the South Saskatchewan River basin. *Journal of Hydrologic Engineering, 13*(9), 781–792. https://doi.org/10.1061/(asce)1084-0699(2008)13:9(781).
Warfield, J. N. (2006). *An introduction to systems science*. Singapore: World Scientific. https://doi.org/10.1142/9789812774040.

Wiener, N. (1948). *Cybernetics or control and communication in the animal and the machine*. New York: Wiley. https://doi.org/10.2307/2226579.

Williams, J. W., & Jackson, S. T. (2007). Novel climates, no-analog communities, and ecological surprises. *Frontiers in Ecology and the Environment, 5*(9), 475–482. https://doi.org/10.1890/1540-9295(2007)5[475:ncncae]2.0.co;2.

Winton, J. (1993). *Ultra in the Pacific: How breaking Japanese codes and cyphers affected naval operations against Japan 1941–45*. Annapolis, MD: Naval Institute Press.

Xiao, Y., Hipel, K. W., & Fang, L. (2016). Incorporating water demand management into a cooperative water allocation framework. *Water Resources Management, 30*(9), 2997–3012. https://doi.org/10.1007/s11269-016-1322-x.

Chapter 3
Conflict Models in Graph Form

As depicted in Fig. 2.1 in Sect. 2.2.2, many models are available for describing strategic conflicts. For example, in the left branch of Fig. 2.1, metagame analysis employs option form (Howard 1971) for recording a conflict, while in the right branch, normal form is often written using a tabular or matrix format for the case of two decision makers (DMs). For the Graph Model for Conflict Resolution (GMCR) listed at the bottom of the left branch in Fig. 2.1, the movements in one step by a given DM are captured within a directed graph for that DM. As mentioned in Sect. 2.2.2, the models for the approaches given on the left in Fig. 2.1 only require relative preference information for each DM, while those in the right branch need cardinal preferences.

As explained in Sect. 1.2.2 and portrayed in Fig. 1.1, the key ingredients in any conflict model are the DMs, states or scenarios that could take place, and the preferences of each DM. The main purpose of this chapter is to define in detail these main modeling components with respect to GMCR. Because smaller conflicts are often conveniently recorded using what is called normal form, this type of abstract game model is described in Sect. 3.1.1. A very flexible format for writing down small, medium, and large conflict is option form which is defined in Sect. 3.1.2. Moreover, the exact linkages of the normal and option forms to the graph model are explained in this chapter. A simple conflict over sustainable development is employed to show how these three forms are used in practice and the connections among them. Additionally, a small conflict written in graph form is used in Sect. 3.2 to illustrate a situation which cannot be captured by either the normal or option form. Finally, the graph model is developed in a new direction, called matrix representation of the graph model, which is given in Sect. 3.3 and constitutes an equivalent way to represent the graph model. In fact, the matrix representation for GMCR is utilized throughout Chaps. 3–9 for addressing a range of situations (Xu et al. 2007, 2009a, b, c, 2010a, b, c, d, 2011, 2013, 2014, Bernath Walker et al. 2013, Hou et al. 2015).

H. Xu et al., *Conflict Resolution Using the Graph Model: Strategic Interactions in Competition and Cooperation*, Studies in Systems, Decision and Control 153,
https://doi.org/10.1007/978-3-319-77670-5_3

3.1 Normal Form and Option Form

3.1.1 Normal Form

In game theory, normal form is a way of describing a game using a list of strategies for each DM, together with preference information. Its formal definition is as follows.

Definition 3.1 (*Game in Normal Form*) A game G in normal form is usually written as a triplet $G = \langle N, \{T_i\}_{i \in N}, \{u_i\}_{i \in N} \rangle$, where

- $N = \{1, 2, \cdots, n\}$ is a nonempty set of DMs;
- for each DM $i \in N$, T_i is the nonempty strategy set of DM i;
- for each DM $i \in N$, $u_i : T_1 \times T_2 \times \cdots \times T_n \rightarrow \mathbb{R}$ is the utility of DM i.

In the above definition, let t_{ik} be a specific strategy for DM i, where $t_{ik} \in T_i$ and $m_i = |T_i|$ denote the number of strategies for DM i in T_i. Then $t = (t_{1a}, t_{2b}, \cdots, t_{ir}, \cdots, t_{nw})$ is called a strategy profile, where $t_{1a} \in T_1$, $t_{2b} \in T_2, \cdots, t_{ir} \in T_i, \cdots,$ and $t_{nw} \in T_n$. The symbol "$\times$" indicates the Cartesian product for which $T_1 \times T_2 \times \cdots \times T_n$ represents the set of all strategy profiles. Each element or strategy profile in this Cartesian product set is formed by selecting one element from each T_j and all possible combinations of these selections are used to create the total set of strategy profiles. When comparing the normal form with the graph form defined later, a strategy profile is also called a state so the state set $S = T = T_1 \times T_2 \times \cdots \times T_n$. The "$u_i$" denotes the von Newmann–Morgenstern (1953) utility function for DM i. For a given DM, a utility function maps each state to a real number for which a higher number means more preferred. In many disputes, a DM is interested in whether a state is more preferred to another state but not by how much. Therefore, one often employs $s >_i q$ to express that DM i prefers state s to state q.

To calculate the stability of a state for a given DM according to the different types of stability definitions presented in Chap. 4, one must define the set of movements in one step controlled separately by each DM in the conflict. Hence, one can expand the definition of a game in normal form by explicitly defining movement among states as is done in Sect. 3.2. When using $s \in S$ to represent a strategy as is done in the next definition, let $s = (s_{1a}, s_{2b}, \cdots, s_{ir}, \cdots, s_{nw})$ which indicates the strategy that each DM controls to form state s. Equivalently, this means that $s \in T_1 \times T_2 \times \cdots \times T_n$.

Definition 3.2 (*Unilateral Move in Normal Form*) For a game in normal form, the set of states to which DM $i \in N$ can unilaterally cause the game to move from state $s \in S$ is defined as:

$$
\begin{aligned}
R_i(s) = \{q \in S : q_{il} \neq s_{il} \text{ for some } 1 \leq l \leq m_i \text{ and} \\
q_{jk} = s_{jk} \text{ for any } j \in N \setminus \{i\} \text{ and } 1 \leq k \leq m_j\},
\end{aligned}
$$

where $s_{il}, q_{il} \in T_i$, $s_{jk}, q_{jk} \in T_j$, and $\setminus$ refers to "set subtraction".

In words, this definition means that for a state q to be a unilateral move by DM i from state s (i.e. $q \in R_i(s)$), the strategy for DM i in state q is different from that in state s (i.e. $q_{il} \neq s_{il}$) and the strategies of the other DMs (i.e. $N \setminus \{i\}$) in state q remain the same as in state s ($q_{jk} = s_{jk}$ for any $j \in N \setminus \{i\}$). When ascertaining stability, as explained later in Chap. 4, one also must determine unilateral improvements by a DM as now defined.

Definition 3.3 (*Unilateral Improvement in Normal Form*) For a game in normal form, the set of unilateral improvements from state $s \in S$ for DM $i \in N$ is defined as:

$$R_i^+(s) = \{q \in R_i(s) : u_i(q) > u_i(s)\},$$

where u_i is DM i's utility.

A clear way to write down the normal form for a two-DM game is to use what is called 'matrix' form, as displayed in Table 3.1. As can be seen, DM 1, on the left, controls the two strategies $T_1 = \{t_{11}, t_{12}\}$ depicted as rows while DM 2 is in charge of the two strategies $T_2 = \{t_{21}, t_{22}\}$ given as columns. Each of the four cells in the matrix is a state s_k for which DM 1 and DM 2 have selected a strategy and contains the utility values of the state for DM 1 and DM 2. As can be seen at the bottom of Table 3.1, a state can be written as a situation in which each DM selects a strategy. Hence, state $s_3 = (t_{12}, t_{21})$ and this state appears as the bottom left cell in Table 3.1 for which the utility values for DMs 1 and 2 are $u_1(s_3)$ and $u_2(s_3)$, respectively.

As specified in Definition 3.2, when a given DM unilaterally causes the conflict to move from one state to another, the strategies of the other DMs remain the same. Referring to Table 3.1, notice that if DM 2 remains fixed at strategy t_{21} in the left column of the matrix, then DM 1 can cause the conflict to move from state s_1 to s_3 by changing his or her strategy from t_{11} to t_{12}. Similarly, DM 1 can make the game proceed from state s_3 to s_1 by changing his strategy from t_{12} in $s_3 = (t_{12}, t_{21})$ to the strategy t_{11} to form state $s_1 = (t_{11}, t_{21})$. Moreover, if state s_1 is more preferred by DM 1 to state s_3 (i.e. $u_1(s_1) > u_1(s_3)$), then the unilateral move from s_3 to s_1 is also

Table 3.1 2 × 2 game in normal form

		DM 2	
		t_{21}	t_{22}
DM 1	t_{11}	s_1: $u_1(s_1), u_2(s_1)$	s_2: $u_1(s_2), u_2(s_2)$
	t_{12}	s_3: $u_1(s_3), u_2(s_3)$	s_4: $u_1(s_4), u_2(s_4)$

$s_1 = (t_{11}, t_{21})$, $s_2 = (t_{11}, t_{22})$, $s_3 = (t_{12}, t_{21})$, $s_4 = (t_{12}, t_{22})$

Table 3.2 Sustainable development game in normal form

		DM 2	
		SD	NSD
DM 1	P	s_1 10, 5	s_2 6, 2
	NP	s_3 8, 7	s_4 1, 4

s_1 = (P, SD), s_2 = (P, NSD), s_3 = (NP, SD), s_4 = (NP, NSD)

a unilateral improvement according to Definition 3.3. Therefore, $s_1 \in R_1(s_3)$, $s_3 \in R_1(s_1)$, and $s_1 \in R_1^+(s_3)$. Finally, when examining unilateral moves by DM 2, one must fix the strategy on row, of DM 1. For instance, if DM 1 remains at strategy t_{11} on the first row in Table 3.1, then DM 2 can unilaterally cause the conflict to move from state s_1 to s_2 and back again. Therefore, $s_2 \in R_2(s_1)$ and $s_1 \in R_2(s_2)$. If DM 2 prefers state s_1 more than s_2 (i.e. $u_2(s_1) > u_2(s_2)$), then $s_1 \in R_2^+(s_2)$.

Example 3.1 (*Sustainable Development Conflict in Normal Form*) A specific illustration of the general 2 × 2 game displayed in Table 3.1 is the sustainable development game shown in Table 3.2. This environmental dispute was proposed by Hipel (2001) to model a basic conflict which could arise between an environmental agency (DM 1) and a developer (DM 2).

Therefore, the set of DMs is given by

$$N = \{DM\ 1,\ DM\ 2\}.$$

DM 1 can be either proactive (P) in encouraging responsible behavior by the developer with respect to environmental issues or not proactive (NP). As can be seen in Table 3.2, DM 1, on the left, controls its two strategies depicted as rows, where the strategy set for DM 1 is

$$T_1 = \{proactive\ (P),\ not\ proactive\ (NP)\} = \{P,\ NP\}.$$

DM 2, the developer of the project under consideration, can practice sustainable development (SD) or not adhere to sustainable development (NSD), which are displayed as columns in Table 3.2. Thus, DM 2 controls the strategy set

$$T_2 = \{sustainable\ development\ (SD),\ not\ sustainable\ development\ (NSD)\}$$
$$= \{SD,\ NSD\}.$$

When each DM selects a strategy, a state is created. Each of the four cells in Table 3.2 is a state for which DM 1 and DM 2 have selected a strategy. For instance, the cell labeled as state s_2 in Table 3.2 is the situation for which DM 1 selects strategy P and DM 2 chooses strategy NSD to produce state $s_2 = (P, NSD)$. Accordingly, the set of states in the sustainable development conflict is

$$T = T_1 \times T_2 = \{P, NP\} \times \{SD, NSD\} = \{(P, SD), (P, NSD), (NP, SD), (NP, NSD)\}.$$

If $s_1 = (P, SD)$, $s_2 = (P, NSD)$, $s_3 = (NP, SD)$, and $s_4 = (NP, NSD)$, then

$$T = \{s_1, s_2, s_3, s_4\}.$$

The two numbers written in each cell of the matrix in Table 3.2 represent the preference or utility of DM 1 and DM 2, respectively, where a high number means more preferred. Specifically, the utility values of DM 1 are

$$u_1(s_1) = 10, u_1(s_2) = 6, u_1(s_3) = 8, \text{ and } u_1(s_4) = 1;$$

while the utility values of DM 2 are

$$u_2(s_1) = 5, u_2(s_2) = 2, u_2(s_3) = 7, \text{ and } u_2(s_4) = 4.$$

The utility values for each of the two DMs can be used to order the states from most to least preferred such that:

$$s_1 >_1 s_3 >_1 s_2 >_1 s_4 \text{ for DM 1; and}$$

$$s_3 >_2 s_1 >_2 s_4 >_2 s_2 \text{ for DM 2.}$$

Note that the most preferred state for DM 1 (the environmental agency) is state s_1 for which DM 1 is proactive (P) and DM 2 is practicing sustainable development (SD). The least preferable state for DM 2 is state s_2 for which DM 2 has a preference value of 2. As explained for the example in Table 3.1, to determine the unilateral moves for DM 1, one first fixes DM 2 at a specified column. Hence, in Table 3.2, suppose that DM 2 has chosen strategy SD in the left column. Then DM 1 can cause the game to move from state s_1 to s_3 by changing his strategy selection from P to NP. Therefore, $R_1(s_1) = \{s_3\}$ and since DM 1 can change his strategy from NP to P in the same column, $R_1(s_3) = \{s_1\}$. Likewise, when DM 2 selects strategy NSD, DM 1 controls the movement in the second column and, therefore, $R_1(s_2) = \{s_4\}$ and $R_1(s_4) = \{s_2\}$.

As was also explained above for Table 3.1, to ascertain the states to which DM 2 can unilaterally cause to the game to move, DM 1's strategy must remain the same as either the first or the second row in Table 3.2. Accordingly, DM 2's unilateral moves are $R_2(s_1) = \{s_2\}$ and $R_2(s_2) = \{s_1\}$ when DM 1 remains at strategy P on the first row, and $R_2(s_3) = \{s_4\}$ and $R_2(s_4) = \{s_3\}$ when DM 1 is fixed at strategy NP in the second row.

Similarly, the unilateral improvements of DM 1 from the four states, respectively, are

$$R_1^+(s_1) = \emptyset, R_1^+(s_2) = \emptyset, R_1^+(s_3) = \{s_1\}, \text{ and } R_1^+(s_4) = \{s_2\};$$

while the unilateral improvements of DM 2 from the four states are

$$R_2^+(s_1) = \emptyset, R_2^+(s_2) = \{s_1\}, R_2^+(s_3) = \emptyset, \text{ and } R_2^+(s_4) = \{s_3\}.$$

3.1.2 Option Form

In a strategic conflict, a DM usually controls various courses of actions which are referred to as options. Let n be the number of DMs and O_i denote the option set of DM i, where o_{ij} is DM i's jth option. Then, the set of all options in a conflict model is $O = \bigcup_{i \in N} O_i$ in which index i indicates which DM controls the options. It may also be expressed as $O = \{O_1, O_2, \cdots, O_i, \cdots, O_n\}$, where the number of options in O_i is h_i. When a given DM decides which of his or her options to select or not a specific strategy is formed.

Definition 3.4 (*Strategy in Option Form*) Let O_i denote the option set of DM i for $i \in N$ for which $o_{ij} \in O_i$. A strategy for DM i is a mapping $g : O_i \rightarrow \{0, 1\}$, such that for $j = 1, 2, \cdots, h_i$

$$g(o_{ij}) = \begin{cases} 1 & \text{if DM } i \text{ selects option } o_{ij}, \\ 0 & \text{otherwise,} \end{cases}$$

where o_{ij} is DM i's jth option.

One can assign $g(o_{ij})$ a value of 1 to indicate that DM i will select option o_{ij}. Similarly, $g(o_{ij}) = 0$ means that DM i will not choose this option. A state is formed when each DM has selected a specific strategy. In other words, for each option the DM controlling the option has decided whether or not he or she will choose it. The formal definition for a state is as follows.

Definition 3.5 (*State in Option Form*) Let $O = \bigcup_{i \in N} O_i$ be the set of all options in a conflict for $o_{ij} \in O_i, i = 1, 2, \cdots, n$. A state is a mapping $f : O \rightarrow \{0, 1\}$, such that for $i = 1, 2, \cdots, n$,

$$f(o_{ij}) = \begin{cases} 1 & \text{if DM } i \text{ selects option } o_{ij}, \\ 0 & \text{otherwise.} \end{cases} \tag{3.1}$$

Let h denote the total number of options available to the DMs. A state can be treated as an h-dimensional column vector consisting of having an element of 0 or 1.

Therefore f_s is used to express the h-dimensional column vector to denote state s. Hence, f_s may be written as $[(g^s(O_1))^T, \cdots, (g^s(O_i))^T, \cdots, (g^s(O_n))^T]^T$ in which $g^s(O_i)$ denotes DM i's strategy corresponding to state s for $i = 1, 2, \cdots, n$ and is an h_i-dimensional column vector whose elements are

$$g^s(o_{ij}) = \begin{cases} 1 & \text{if DM } i \text{ selects option } o_{ij}, \\ 0 & \text{otherwise,} \end{cases}$$

A concise way to represent the set of all possible states in a conflict is to use the concept of a power set written as $\{0, 1\}^O$, where O is the set of all options, each of which can be not chosen or selected as indicated by 0 or 1, respectively. Therefore, the set of all mathematically possible states in a conflict model is $\{0, 1\}^O$. In mathematics, given a set O, the power set of O, written as 2^O, is the set of all subsets of O. Then, the power set of O contains $2^{|O|} = 2^h$ elements. Every state s can also be equivalently expressed as a subset of O, for which the mapping f is defined by Eq. 3.1. Although $2^{|O|} = 2^h$ states are mathematically possible, only a part of them are feasible in practice due to various option constraints, as explained in Sect. 3.2.2. The symbol S is used to designate the set of feasible states.

The option form is especially useful for practical applications because it can readily handle conflicts having any finite numbers of DMs, each of whom controls a finite number of option or courses of action. Consequently, as is done throughout this book, often option form is employed for writing down a conflict as part of the GMCR methodology. Because the number of states is typically much larger than the number of options in a conflict, when option form is employed in practice, the user only has to supply the relatively short list of options, for which the states can be automatically generated using a computer program. The option form is formally defined as follows.

Definition 3.6 (*Game in Option Form*) A game G in option form is usually written as $G = \langle N, \{O_i\}_{i \in N}, S, \{\succ_i, \sim_i\}_{i \in N}\rangle$, where

- $N = \{1, 2, \cdots, n\}$ is a nonempty set of DMs;
- for each DM $i \in N$, O_i is the nonempty option set of DM i;
- $S = \{s_1, s_2, \cdots, s_m\}$ is a nonempty set of feasible states;
- for each DM $i \in N$, $\{\succ_i, \sim_i\}$ represents i's preference where $s_k \succ_i s_t$ means that DM i prefers state s_k to state s_t while $s_k \sim_i s_t$ indicates that DM i has equal preference for these two states or is indifferent between them.

Note that the precise mathematical properties of $\{\succ_i, \sim_i\}$ are given in Sect. 3.2.4.

Similarly, one can expand the definition of a game in option form by explicitly defining unilateral moves and unilateral improvements among states as is done in Sect. 3.1.1. Let $h_i = |O_i|$ denote the cardinality of DM i's option set O_i and f_s stand for the mapping from options in the set O to state s.

Definition 3.7 (*Unilateral Moves in Option Form*) For a game in option form, the set of unilateral moves of DM $i \in N$ from state $s \in S$ is defined as:

$$R_i(s) = \{q \in S : g^q(o_{il}) \neq g^s(o_{il}) \text{ for some } o_{il} \in O_i \text{ and}$$

$$g^q(o_{jk}) = g^s(o_{jk}) \text{ for any } j \in N \setminus \{i\}\},$$

where $1 \leq l \leq h_i$ and $1 \leq k \leq h_j$.

One can define unilateral improvements by a DM for the option form based on Definition 3.7.

Definition 3.8 (*Unilateral Improvement in Option Form*) For a game in option form, the set of unilateral improvements from state $s \in S$ for DM $i \in N$ is defined as:

$$R_i^+(s) = \{q \in R_i(s) \text{ and } q \succ_i s\}.$$

Example 3.2 (*Sustainable Development Conflict in Option Form*) The sustainable development game for Example 3.1 has two DMs:

$$N = \{DM\ 1, DM\ 2\}.$$

DM 1 has the single option

$$O_1 = \{o_{11}\} = \{proactive\ (P)\} \text{ while}$$

DM 2 controls the option

$$O_2 = \{o_{21}\} = \{sustainable\ development\ (SD)\}.$$

When each DM decides upon which of his options to select or not, a state is formed. Table 3.3 presents the option form of the sustainable development game introduced in Example 3.1. The left column in this table lists each of the two DMs followed by the option which it controls. The four columns of Ys or Ns given on the right in Table 3.3 constitute the set of the four feasible states in this dispute where

$$S = \{s_1, s_2, s_3, s_4\}.$$

Table 3.3 Sustainable development game in option form

DM 1: Environmental agency				
1. Proactive (P)	Y	Y	N	N
DM 2: Developer				
2. Sustainable development (SD)	Y	N	Y	N
States	s_1	s_2	s_3	s_4

Preferences $s_1 \succ_1 s_3 \succ_1 s_2 \succ_1 s_4$ for DM 1 and $s_3 \succ_2 s_1 \succ_2 s_4 \succ_2 s_2$ for DM 2

Rather than use 0 or 1 to indicate whether or not an option is taken as is done in Definitions 3.4 and 3.5, a Y or N is utilized, respectively, since these letter symbols are easier to interpret. Specifically, a "Y" indicates that an option is selected by the DM controlling it while an "N" means that the option is not chosen. Therefore, DM 1's two strategies are being proactive (Y) or not (N) and DM 2's two strategies are practicing sustainable development (Y) or not (N). DM 1's two strategies can also be expressed by $g(o_{11}) = 1$ and $g(o_{11}) = 0$ according to Definition 3.4. Similarly, $g(o_{21}) = 1$ and $g(o_{21}) = 0$ represent DM 2's two strategies. A state is any combination of Y's and N's opposite all of these DMs' options. Hence, a state is formed after each DM selects a strategy, so there are four states in the sustainable development game, which are written

$$s_1 = \begin{pmatrix} Y \\ Y \end{pmatrix} \text{ or } s_1 = (Y\ Y)^T,$$

$$s_2 = \begin{pmatrix} Y \\ N \end{pmatrix} \text{ or } s_2 = (Y\ N)^T,$$

$$s_3 = \begin{pmatrix} N \\ Y \end{pmatrix} \text{ or } s_3 = (N\ Y)^T,$$

and

$$s_4 = \begin{pmatrix} N \\ N \end{pmatrix} \text{ or } s_4 = (N\ N)^T,$$

where $(Y\ N)^T$, for example, denotes DM 1 will select the proactive option and DM 2 will not choose sustainable development. At the bottom of Table 3.3, the states are ranked or ordered by preference for each of the two DMs from most preferred on the left to least preferred on the right. Because the states are ordered according to preference, this type of preference is referred to as ordinal preference (see Sect. 3.2.4). Moreover, since there are no equally preferred states for each of the DMs, the preferences are said to be strict ordinal.

In the literature, the symbols $\succ_i$ and $\sim_i$ are often not used when it is known that the states are ranked from most to least preferred for a given DM. Accordingly, when employing option form, the ordering of states for each DM is as shown below:

$$\begin{pmatrix} Y & N & Y & N \\ Y & Y & N & N \\ s_1 & s_3 & s_2 & s_4 \end{pmatrix} \text{ and } \begin{pmatrix} N & Y & N & Y \\ Y & Y & N & N \\ s_3 & s_1 & s_4 & s_2 \end{pmatrix}.$$

Ordering of states for DM 1 Ranking of states for DM 2

To determine the unilateral move or moves for DM 1, one first fixes DM 2's strategy. Hence, in Table 3.3, suppose that DM 2 has chosen strategy SD as indicated by the Y located opposite SD and directly above s_1. Then, DM 1 can unilaterally cause the game to move from state s_1 to s_3 by changing his strategy selection from Y to

N. Therefore, since $g^{s_3}(o_{11}) \neq g^{s_1}(o_{11})$ and $g^{s_3}(o_{21}) = g^{s_1}(o_{21})$, then $R_1(s_1) = \{s_3\}$ according to Definition 3.7. Because DM 1 can also change his strategy from N to Y, $R_1(s_3) = \{s_1\}$. Likewise, when DM 2 selects strategy N, DM 1 controls the movement from s_2 to s_4 or from s_4 to s_2. Accordingly, $R_1(s_2) = \{s_4\}$ and $R_1(s_4) = \{s_2\}$. Similarly, to ascertain the states to which DM 2 can unilaterally cause the game to move, DM 1's strategy must be fixed. Hence, DM 2's unilateral moves are $R_2(s_1) = \{s_2\}$ and $R_2(s_2) = \{s_1\}$ when DM 1 remains at strategy $g^s(o_{11}) = Y$ and $R_2(s_3) = \{s_4\}$ and $R_2(s_4) = \{s_3\}$ when DM 1 is fixed at strategy $g^s(o_{11}) = N$.

In a similar fashion, the unilateral improvements of DM 1 from the four states, respectively, are

$$R_1^+(s_1) = \emptyset \text{ and } R_1^+(s_3) = \{s_1\} \text{ because } s_1 \succ_1 s_3,$$

$$\text{as well as } R_1^+(s_2) = \emptyset \text{ and } R_1^+(s_4) = \{s_2\} \text{ since } s_2 \succ_1 s_4,$$

while the unilateral improvements of DM 2 from the four states are

$$R_2^+(s_1) = \emptyset \text{ and } R_2^+(s_2) = \{s_1\} \text{ since } s_1 \succ_2 s_2,$$

$$\text{as well as } R_2^+(s_3) = \emptyset \text{ and } R_2^+(s_4) = \{s_3\} \text{ because } s_3 \succ_2 s_4.$$

3.2 Graph Model

The normal form of Sect. 3.1.1 provides a means for easily determining the states in a game, especially for the situation in which there are only two DMs. The option form defined in Sect. 3.1.2 can be conveniently utilized for ascertaining the states for both simple and complex games. In particular, when recording a conflict, as is done in Table 3.3 for the sustainable development conflict, one simply writes in the left column of the table the name of each of the DMs followed by all of options that the DM controls. The set of feasible states can then be written by hand on the right side of the table using the Y-N notation, as is done in Table 3.3. For a conflict having a relatively large number of DMs and options, a computer program can be used to generate the mathematically possible set of states, from which any infeasible states can be easily removed, as explained later in Sect. 3.2.2. Whatever the case, one ends up with the set of feasible states over which each DM has her or his own relative preferences. The definition of the graph model starts with the assumption that the set of feasible states are already known, and, for example, may have been generated using option form.

Definition 3.9 (*Graph Model*) A *graph model* is a structure

$$G = \langle N, S, \{A_i, \succeq_i, i \in N\} \rangle,$$

where

- N is a nonempty, finite set, called the set of DMs.
- S is a nonempty, finite set, called the set of feasible states.
- For each DM i, $A_i \subseteq S \times S$ is DM i's set of oriented arcs, which contains the movements in one step controlled by DM i.
- Precise mathematical properties of the preference relation for DM i, $\succeq_i$, is presented in Sect. 3.2.4

Note that $G_i = (S, A_i)$ is DM i's directed graph in which S denotes the vertex set and each oriented arc in $A_i \subseteq S \times S$ indicates that DM i can make a one-step unilateral move from the initial state of the arc to its terminal state. For simple models, sometimes it is informative to combine all of the DMs' directed graphs, $\{G_i : i \in N\}$, along with their preferences, to create what is called an integrated graph model.

Definition 3.10 The *integrated graph* of a graph model G is the structure $IG = \langle S, \{A_i, i \in N\}\rangle$.

In the above definition, an integrated graph IG, with vertex set S and arc set $A = \{A_i : i \in N\}$, contains all of the DM's individual graph $\{G_i : i \in N\}$. The arcs in A that are associated with DM i are considered to be labeled by i, or colored with color i. Thus, G may have multiple copies of an arc, but each copy has a different color. A two-DM conflict model is used to illustrate the components comprising a graph model.

Example 3.3 (*Sustainable Development Conflict in Graph Form*) The sustainable development game is first presented in Example 3.1. The graph model of this conflict is shown in Fig. 3.1, where the directed graph and relative preferences for DMs 1 and 2 are displayed on the left and right sides, respectively. For each DM, a given arc represents the unilateral movement, in one step, under that DM's control. Hence, for instance, DM 1 controls movement from state s_1 to state s_3, and back again, as

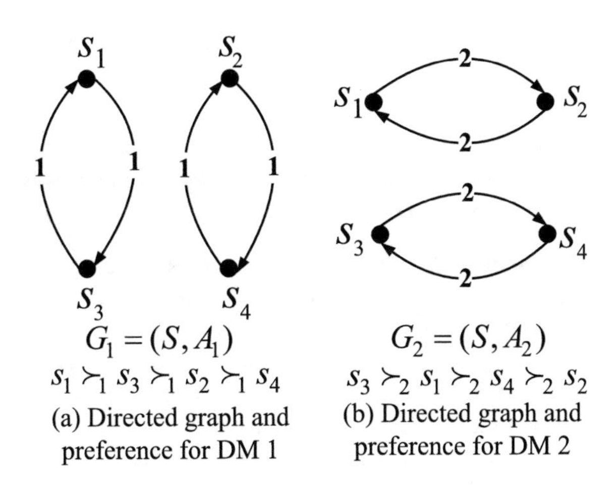

Fig. 3.1 Graph model for the sustainable development conflict

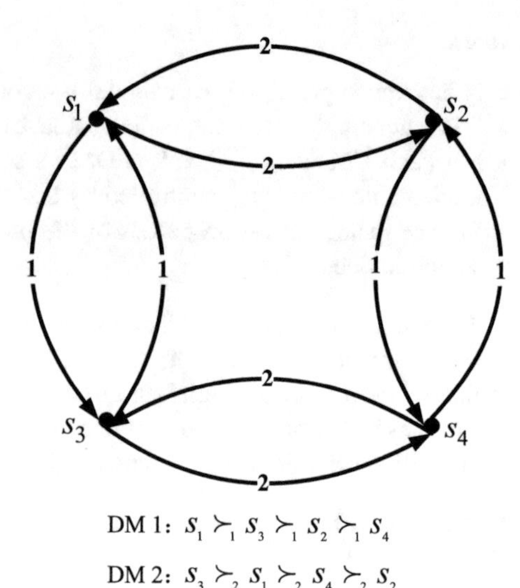

Fig. 3.2 Integrated graph model for the sustainable development conflict

indicated by the two arcs on the left in Fig. 3.1a, for which an arrow indicates the direction of movement between the two states.

Specifically, the graph model of the sustainable development conflict presents

- the DM set, $N = \{1, 2\}$;
- the state set, $S = \{s_1, s_2, s_3, s_4\}$;
- the directed graphs for the two DMs, $G_1 = (S, A_1)$ and $G_2 = (S, A_2)$ depicted in Fig. 3.1, where

$$A_1 = \{(s_1, s_3), (s_3, s_1), (s_2, s_4), (s_4, s_2)\} \text{ and } A_2 = \{(s_1, s_2), (s_2, s_1), (s_3, s_4), (s_4, s_3)\}.$$

- the preference information for the two DMs, which consists of

$$s_1 \succ_1 s_3 \succ_1 s_2 \succ_1 s_4 \text{ and } s_3 \succ_2 s_1 \succ_2 s_4 \succ_2 s_2;$$

The integrated graph IG, in combination with preference information, is called the integrated graph model. Figure 3.2 displays the integrated graph model for the sustainable development conflict.

In summary, a graph model contains the DMs, feasible states, the movements controlled by each DM which can be drawn as the set of separate directed graphs for the DMs or as a single integrated graph, and preference information. Although the normal form and option form can also represent the sustainable development game (see Sect. 3.1), they have a number of drawbacks. Particularly, movements among states in the normal and option formats are automatically restricted by their special structures. Figure 3.3 shows an example of a graph model that cannot be represented

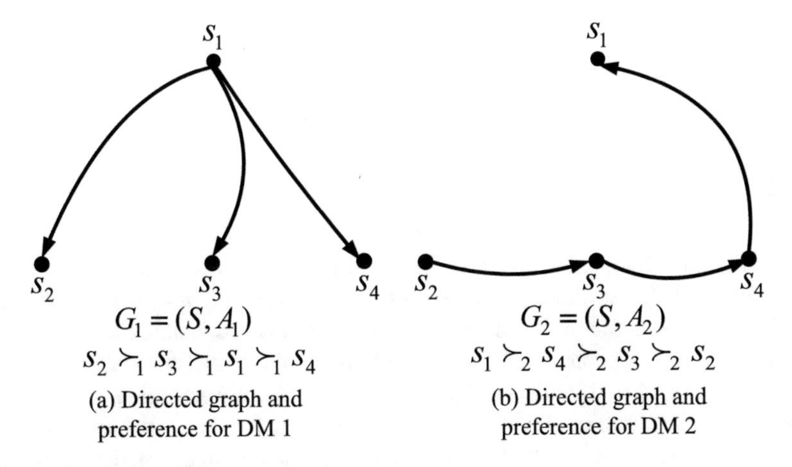

Fig. 3.3 A graph model

using the normal and option forms. Consider, for instance, Fig. 3.1a which is derived from the normal and option forms in Tables 3.2 and 3.3, respectively. Notice that in Fig. 3.1a, DM 1 cannot move from s_1 to s_2 and s_1 to s_4, as he or she can in Fig. 3.3a. Therefore, the states in Fig. 3.3 are derived in a way that permits this more general type of movement. In addition to permitting more flexible types of movement, the graph model possesses other advantages as discussed in Sects. 1.2 and 3.2.3 and elsewhere in the book.

3.2.1 Decision Makers

A strategic conflict is a situation in which two or more DMs with different objectives interact with one another. A DM may be an individual or a group, such as an industrial or governmental organization. For example, in a conflict in which family members are arguing over where to spend their next vacation, each DM is a person. In a trading conflict among car manufacturers which are trying to increase their shares of the automobile market, each DM represents a large company having many directors, shareholders, and employees. In previous research, a DM is also referred to as a player, actor, stakeholder, or participant. The term decision maker is used in this book, because it can stand for individuals or groups of people who can make decisions that affect a given conflict.

In the sustainable development conflict in Example 3.1, the environmental agency is a DM consisting of many people whose role is to protect the local environment from potential harm caused by the activities of a developer. Because the objective of the developer, which could be a large company, is to maximize profits, this DM's goal is in conflict with the aim of the environmental agency.

Any subset H of DMs in the set N is called a coalition. If $|H| > 0$, then the coalition H is nonempty. If $|H| > 1$, then the coalition H is nontrivial.

3.2.2 States

As mentioned before, states can be defined using normal form (Sect. 3.1.1) or option form (Sect. 3.1.2). Additionally, as shown by the types of movement in Fig. 3.3, which cannot be captured by the normal and option forms, states can be specified by other means. Nonetheless, option form is particularly useful for defining states in a rich range of real-world conflicts that can be readily investigated within the paradigm of the graph model. According to Definition 3.5, because each option can be either chosen or not, a conflict with h options has 2^h mathematically possible states. However, only a portion of them may be feasible in practice due to various option constraints. Additionally, each state can also be represented by a column indicating which options are selected (denoted by "Y") or not (indicated by "N"). As an example of a real-world dispute, consider the groundwater contamination conflict first mentioned in Sect. 1.2.2. In this dispute, Uniroyal Chemicals Ltd. (UR) polluted the aquifer underlying the town of Elmira located in Southern Ontario, Canada, from which the town previously obtained its water supplies. After the discovery of the pollutant, which is a carcinogen, the Ministry of the Environment (MoE) for the Province of Ontario, issued a Control Order in which it requested UR to treat its liquid discharges and cleanse the aquifer. The model in option form shown in Table 3.4 is for the negotiations that took place among the three DMs when UR appealed the Control Order which is a right it can exercise according to provincial law. Table 3.4, which is also given in Chap. 1 as Table 1.1, provides an explanation of the options controlled by the DMs. The feasible states for the negotiation are presented in Table 3.5.

Because each state can be either taken or not, a conflict having a total of five options as in Table 3.5 contains $2^5 = 32$ mathematically possible states. However, only the feasible states that could take place in reality are listed in Table 3.5 as columns. For convenience of explanation, each column or state is assigned a state number.

Table 3.4 Options for the Elmira model

MoE (Ministry of the Environment)
1. **Modify** the Control Order to make it more acceptable to UR
UR (Uniroyal Chemicals Ltd.)
2. **Delay** the appeal process
3. **Accept** the current Control Order
4. **Abandon** its Elmira operation
LG (Local Government)
5. **Insist** that the original Control Order be applied

Table 3.5 Feasible states for the Elmira model

MoE									
1. Modify	N	Y	N	Y	N	Y	N	Y	–
UR									
2. Delay	Y	Y	N	N	Y	Y	N	N	N
3. Accept	N	N	Y	Y	N	N	Y	Y	N
4. Abandon	N	N	N	N	N	N	N	N	Y
LG									
5. Insist	N	N	N	N	Y	Y	Y	Y	–
State number	s_1	s_2	s_3	s_4	s_5	s_6	s_7	s_8	s_9

For instance, state s_5 is the scenario in which MoE is not modifying the Control Order, UR is delaying the negotiations but is not accepting the current Control Order and is not abandoning its factory in Elmira, and LG is insisting that the original Control Order be accepted by UR. As can be appreciated, UR's three options are mutually exclusive and, hence, UR cannot select two or more options at the same time. Therefore, any state in which UR chooses more than one of its three options is removed from the conflict since it is infeasible. Moreover, because UR is expected to do something, UR will choose at least one of its options. Finally, if UR abandons its plant, it does not matter what other options are selected by the other two DMs. Therefore, the set of resulting states are essentially the same and are represented as the single state s_9 in which a dash "–" indicates either Y or N.

3.2.3 State Transitions

One advantage of the graph model is its innate capability to systematically keep track of state transitions. State transition is the process by which a conflict model moves from one state to another. If a DM can cause a state transition on his or her own, then this transition is called a unilateral move (UM) for that DM. Let $R_i(s)$ denote DM i's reachable list from state s by UMs. This set contains all states to which DM i can move from state s in one step, and, hence, $R_i(s) = \{q \in S : (s, q) \in A_i\}$, where S is the set of feasible states and A_i is the set of arcs connecting two states which are controlled by DM i. For instance, in the sustainable development conflict shown in Fig. 3.2, Environment agency (DM 1) can move to state s_3 and Developer (DM 2) can reach state s_2 by one step from state s_1. Therefore, $R_1(s_1) = s_3$ and $R_2(s_1) = s_2$. Allowable state transitions constitute an important modeling component, as they determine the arc structure of a graph model and reflect the dynamic aspects of conflict in terms of potential moves and countermoves DMs can interactively take as they attempt to reach their goals.

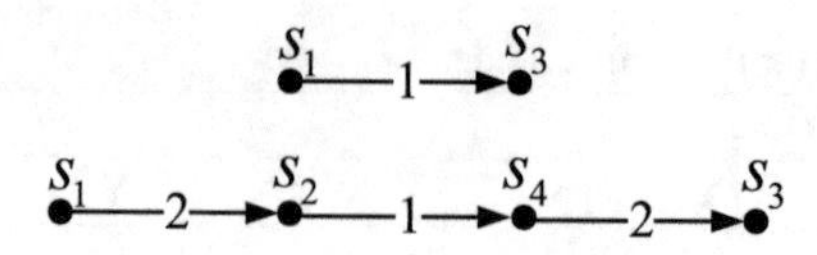

Fig. 3.4 Movements from state s_1 to state s_3 for the sustainable development conflict

In models based on option form, it is assumed that a DM has a UM from one state to another if and only if the two states differ in one or more options selected by that DM. In a graph model, moves controlled by each DM can be intuitively understood and seen as moves within each DM's directed graph or in combination within an integrated graph. The evolution of a conflict can be viewed as starting from a status quo (initial state) and then passing from one state to another, according to moves and countermoves controlled by individual DMs, until it eventually stops at some state such as an equilibrium or a compromise resolution. This interesting research topic about status quo analysis is discussed in Chap. 9. For example, all possible moves from state s_1 to state s_3 for the sustainable development conflict obtained from the integrated graph in Fig. 3.2 are depicted in Fig. 3.4. Note that an important restriction of a graph model is that no DM can move twice in succession along any path.

In the alternative normal form shown in Table 3.2, DM 1 may change the current position of the sustainable development conflict by changing the row but not the column, and DM 2 may change the column but not the row. For example, DM 1 can move from (P, SD) to (NP, SD), but not to (P, NSD) or (NP, NSD).

In option form, a DM can unilaterally cause the conflict to move from one state to another by changing his option choices when the other DM does not alter his option selections. For example, DM 1 can move unilaterally from $(\mathrm{Y\ Y})^T$ to $(\mathrm{N\ Y})^T$, but not to $(\mathrm{Y\ N})^T$ or $(\mathrm{N\ N})^T$.

Figure 3.5 shows the integrated graph for the Elmira model. By examining this figure or using an appropriate algorithm from Chap. 9, one can see that the following six possible paths connect state s_1 to s_8 where the letters on an arc indicate the DM controlling the movement between the two associated states for that arc.

$$s_1 \longrightarrow s_5 \longrightarrow s_6 \longrightarrow s_8,$$

$$s_1 \longrightarrow s_5 \longrightarrow s_7 \longrightarrow s_8,$$

$$s_1 \longrightarrow s_3 \longrightarrow s_4 \longrightarrow s_8,$$

$$s_1 \longrightarrow s_2 \longrightarrow s_4 \longrightarrow s_8,$$

$$s_1 \longrightarrow s_3 \longrightarrow s_7 \longrightarrow s_8,$$

$$s_1 \longrightarrow s_2 \longrightarrow s_6 \longrightarrow s_8.$$

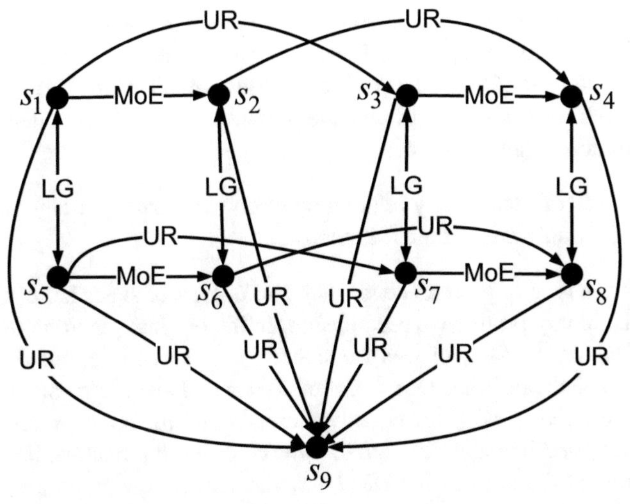

MoE: $s_7 \succ_1 s_3 \succ_1 s_4 \succ_1 s_8 \succ_1 s_5 \succ_1 s_1 \succ_1 s_2 \succ_1 s_6 \succ_1 s_9$

UR: $s_1 \succ_2 s_4 \succ_2 s_8 \succ_2 s_5 \succ_2 s_9 \succ_2 s_3 \succ_2 s_7 \succ_2 s_2 \succ_2 s_6$

LG: $s_7 \succ_3 s_3 \succ_3 s_5 \succ_3 s_1 \succ_3 s_8 \succ_3 s_6 \succ_3 s_4 \succ_3 s_2 \succ_3 s_9$

Fig. 3.5 Integrated graph for the Elmira model

If the name of each DM in an integrated graph is replaced by assigning a distinct color to any arc controlled by that DM, this produces what is called a colored graph (Xu et al. 2009b, 2013) (see Sect. 3.3.1).

3.2.4 *Preferences*

Obviously, preference information plays an important role in decision analysis. Each DM has preferences among the possible states that can arise. One way to express preferences is to use real numbers. For example, one object may have a monetary value of \$5 and another \$10. However, \$5 and \$10 may be worth much more to a poor person than a rich one. Therefore, the concept of utility theory was proposed to reflect the worth or utility of an object. More specifically, cardinal utility refers to a measurement scale for utility, often expressed as utils, that permits one to quantitatively compare the utility of objects. For the case of conflict resolution, utility values would reflect the preferences of a person or DM among the feasible states where a higher number means more preferred. The graph model requires only relative preference information for each DM, but can of course use cardinal information; moreover, it can handle both intransitive and transitive preferences. The formal definition for transitive preference is given below.

Definition 3.11 Let R denote any relation between two states. For any $k, s, q \in S$, if $k\ R\ s$ and $s\ R\ q$ imply $k\ R\ q$, then R is transitive.

The most basic type of preference is when two objects or states are compared in what is called a binary preference relationship. In the original graph model, simple preference (Fang et al. 1993) of DM i is coded by a pair of relations $\{\sim_i, \succ_i\}$ on S, where $s \succ_i q$ indicates that DM i prefers s to q and $s \sim_i q$ means that DM i is indifferent between s and q (or equally prefers s and q). Strict preference $\succ$ is transitive in many graph models, though in some cases it is intransitive. It is assumed that the preference relations of each DM $i \in N$ have the following properties:

(i) $\sim_i$ is reflexive and symmetric (i.e., $\forall s, q \in S, s \sim_i s$, and if $s \sim_i q$, then $q \sim_i s$);
(ii) $\succ_i$ is asymmetric (i.e., $s \succ_i q$ and $q \succ_i s$ cannot occur simultaneously);
(iii) $\{\sim_i, \succ_i\}$ is strongly complete.

Property (iii) implies that, for any $s, t \in S$, exactly one of the following statements is true: $s \succ_i t$, $t \succ_i s$, or $s \sim_i t$. The conventions that $s \succeq_i q$ is equivalent to either $s \succ_i q$ or $s \sim_i q$, and that $s \prec_i q$ is equivalent to $q \succ_i s$, are convenient.

If the definition for transitive preferences given in Definition 3.11 does not hold then the preferences are said to be intransitive. In Sect. 1.2.2 and Fig. 1.3, an example is provided for when a person, say DM i, compares three beverages according to preference. For the case of transitivity: Coffee $\succ_i$ Tea and Tea $\succ_i$ Coke implies Coffee $\succ_i$ Coke; For the case of intransitivity, the above relationship does not hold. Coffee $\succ_i$ Tea and Tea $\succ_i$ Coke but Coke $\succ_i$ Coffee. For the graph model, transitivity of preferences is not required, and all results hold whether preferences are transitive or intransitive.

The state set S can be divided into subsets based on preference relative to a fixed state $s \in S$. These subsets are essential components in stability analysis. The descriptions of these subsets for simple preference are presented as follows:

- $\Phi_i^+(s) = \{q : q \succ_i s\}$ denotes states preferred to state s by DM i;
- $\Phi_i^=(s) = \{q : q \sim_i s\}$ denotes states equally preferred to state s by DM i;
- $\Phi_i^-(s) = \{q : s \succ_i q\}$ denotes states less preferred than state s for DM i.

3.2.5 *Directed Graph*

A graph is a pair (V, E) of sets satisfying $E \subseteq V \times V$. A directed graph $G = (V, A, \psi)$, which is also called a digraph (Dieste 1997), is a set of vertices (nodes) V and a set of oriented edges (arcs) A with $\psi : A \rightarrow V \times V$. If $a \in A$ satisfies

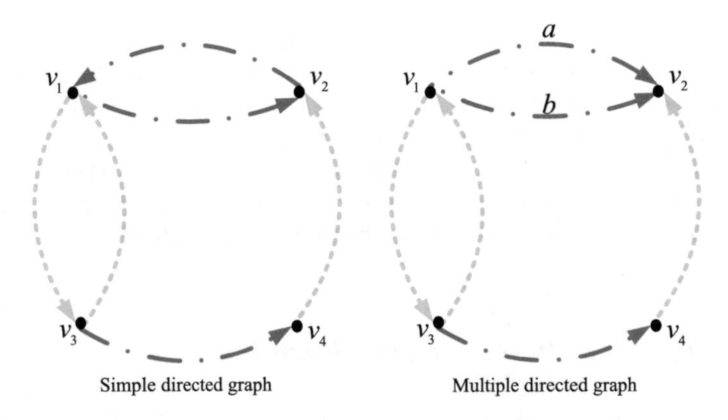

Fig. 3.6 Directed graphs

$\psi(a) = (u, v)$, then we say that a has initial vertex u and terminal vertex v. A multidigraph is a digraph containing multiple edges, i.e., it may contain $a, b \in A$ such that $a \neq b$ and $\psi(a) = \psi(b)$, in which case a and b are said to be multiple arcs. A digraph with no multiple edges is called a simple digraph (Dieste 1997). Figure 3.6 depicts a simple directed graph and a multidigraph. For the multidigraph, a and b are multiple edges, i.e., $\psi(a) = \psi(b) = (v_1, v_2)$. If there exists $a \in A$ such that $\psi(a) = (u, v)$, then u is said to be *adjacent* to v and (u, v) is said to be incident from u and incident to v. Hence, (u, v) is called in-incident to v and out-incident to u. When G is drawn, it is common to represent the direction of an edge with an arrowhead. One generally assumes loop-free graphs; i.e., for any $a \in A$, if $\psi(a) = (u, v)$, then $u \neq v$.

3.3 Matrix Representation of a Graph Model

It is well-known that matrices can efficiently describe adjacency of vertices, and incidence of arcs and vertices in a graph, thereby permitting tracking of paths between any two vertices (Godsil and Royle 2001). Matrices possess various algebraic properties, which can be exploited to develop improved algorithms for solving a variety of problems in a graph. As such, extensive research has been conducted to design effective algorithms and efficient search procedures by exploring relationships between matrices and paths (Gondran and Minoux 1979, Shiny and Pujari 1998, Hoffman and Schiebe 2001). Because a graph model consists of several interrelated graphs, it is natural to use well-known results of Algebraic Graph Theory to help to analyze it. The adjacency matrix can be applied to represent some directed graphs. However, if a graph model contains multiple arcs between the same two states controlled by different DMs, the adjacency matrix would be unable to track all aspects of conflict evolution from the status quo state. It is well known that the incidence matrix can represent multidigraphs if all edges are labeled (Godsil and Royle 2001).

It is common to combine all DMs' graphs, $\{G_i : i \in N\}$, into an *integrated* graph G with vertex set S and arc set $A = \cup\{A_i : i \in N\}$. The arcs in A that are associated with DM i are considered to be labeled by i, or colored with color i. Thus G may have multiple copies of an arc, but each copy is a different color. A unique edge-labeling rule for colored multidigraphs is proposed in the next subsection.

3.3.1 Definitions from Algebraic Graph Theory

Definition 3.12 For a digraph $G = (V, A, \psi)$, edge $a \in A$ and edge $b \in A$ **are consecutive (in the order** ab**)** iff $\psi(a) = (u, v)$ and $\psi(b) = (v, s)$, where $u, v, s \in V$.

Definition 3.13 For a digraph $G = (V, A, \psi)$, the **line digraph** $L(G) = (A, LA)$ of G is a simple digraph with vertex set A and edge set $LA = \{d = (a, b) \in A \times A : a$ and b are consecutive (in the order ab)$\}$.

An example is given in Fig. 3.7 with the directed graph and the line graph underneath it.

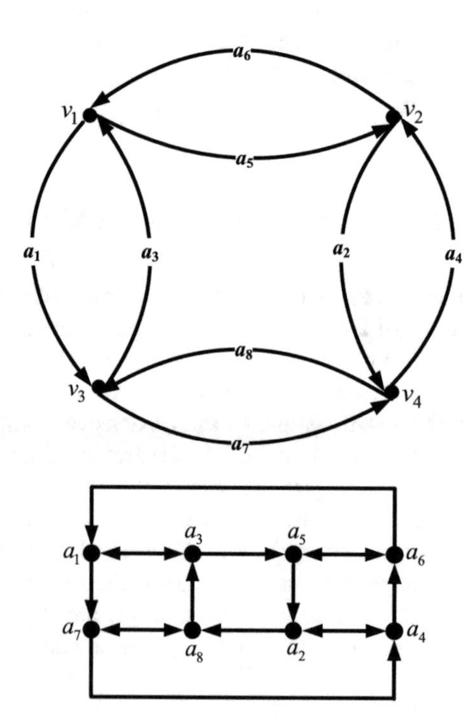

Fig. 3.7 A directed graph and its line graph

Definition 3.14 For a digraph $G = (V, A, \psi)$, a **path** from vertex $u \in V$ to vertex $s \in V$ is a sequence of vertices in G starting with u and ending with s, such that consecutive vertices are adjacent.

Note that in this book a path may contain the same vertex more than once (Buckley and Harary 1990). The length of a path is the number of edges therein.

Definition 3.15 For two $m \times m$ matrices M and Q, the **Hadamard product** for the two matrices is the $m \times m$ matrix $H = M \circ Q$ with (s, q) entry

$$H(s, q) = M(s, q) \cdot Q(s, q).$$

$$\text{If } M = \begin{pmatrix} m_{11} & m_{12} & m_{13} & m_{14} \\ m_{21} & m_{22} & m_{23} & m_{24} \\ m_{31} & m_{32} & m_{33} & m_{34} \\ m_{41} & m_{42} & m_{43} & m_{44} \end{pmatrix} \text{ and } Q = \begin{pmatrix} q_{11} & q_{12} & q_{13} & q_{14} \\ q_{21} & q_{22} & q_{23} & q_{24} \\ q_{31} & q_{32} & q_{33} & q_{34} \\ q_{41} & q_{42} & q_{43} & q_{44} \end{pmatrix},$$

$$\text{then } H = \begin{pmatrix} m_{11} \cdot q_{11} & m_{12} \cdot q_{12} & m_{13} \cdot q_{13} & m_{14} \cdot q_{14} \\ m_{21} \cdot q_{21} & m_{22} \cdot q_{22} & m_{23} \cdot q_{23} & m_{24} \cdot q_{24} \\ m_{31} \cdot q_{31} & m_{32} \cdot q_{32} & m_{33} \cdot q_{33} & m_{34} \cdot q_{34} \\ m_{41} \cdot q_{41} & m_{42} \cdot q_{42} & m_{43} \cdot q_{43} & m_{44} \cdot q_{44} \end{pmatrix}.$$

Let "$\vee$" denote the disjunction operator ("or") on two matrices. Assuming that H and G are two $m \times m$ matrices, the disjunction operation on matrices H and G is defined by:

Definition 3.16 For two $m \times m$ matrices H and G, **disjunction matrix** of H and G is the $m \times m$ matrix $M = H \vee G$ with (u, v) entry

$$M(u, v) = \begin{cases} 1 \text{ if } H(u, v) + G(u, v) \neq 0, \\ 0 \text{ otherwise.} \end{cases}$$

$$\text{If } H = \begin{pmatrix} 1 & 0 & 0 & 1 \\ 0 & 0 & 1 & 1 \\ 1 & 1 & 0 & 1 \\ 0 & 0 & 0 & 1 \end{pmatrix} \text{ and } G = \begin{pmatrix} 1 & 0 & 0 & 0 \\ 1 & 0 & 0 & 1 \\ 1 & 0 & 0 & 1 \\ 1 & 0 & 0 & 1 \end{pmatrix},$$

$$\text{then } M = H \vee G = \begin{pmatrix} 1 & 0 & 0 & 1 \\ 1 & 0 & 1 & 1 \\ 1 & 1 & 0 & 1 \\ 1 & 0 & 0 & 1 \end{pmatrix}.$$

Definition 3.17 The **sign function**, $sign(\cdot)$, maps an $m \times m$ matrix with (u, v) entry $M(u, v)$ to the $m \times m$ matrix

$$sign[M(u,v)] = \begin{cases} 1 & M(u,v) > 0, \\ 0 & M(u,v) = 0, \\ -1 & M(u,v) < 0. \end{cases}$$

If $M = \begin{pmatrix} 1.8 & 0 & -9.7 & 1 \\ 0 & -11.3 & 0 & 117.9 \\ -1.4 & 12.3 & 0 & 89.5 \\ 0 & -77.9 & 0 & 96.5 \end{pmatrix}$, then $sign(M) = \begin{pmatrix} 1 & 0 & -1 & 1 \\ 0 & -1 & 0 & 1 \\ -1 & 1 & 0 & 1 \\ 0 & -1 & 0 & 1 \end{pmatrix}$.

Important matrices associated with a digraph include the adjacency matrix and the incidence matrix (Godsil and Royle 2001). Let $m = |V|$ denote the number of vertices and $l = |A|$ be the number of edges of the directed graph G. Then,

Definition 3.18 For a multidigraph $G = (V, A, \psi)$, the **adjacency matrix** is the $m \times m$ matrix J with (u, v) entry

$$J(u,v) = \begin{cases} 1 & \text{if } (u,v) \in A, \\ 0 & \text{otherwise,} \end{cases}$$

where $u, v \in V$.

For the directed graph shown in Fig. 3.7, the adjacency matrix is expressed as

$$J = \begin{pmatrix} 0 & 1 & 1 & 0 \\ 1 & 0 & 0 & 1 \\ 1 & 0 & 0 & 1 \\ 0 & 1 & 1 & 0 \end{pmatrix}.$$

The adjacency matrix is extended to an edge consecutive matrix in the next definition.

Definition 3.19 For a multidigraph $G = (V, A, \psi)$, the edge consecutive matrix is the $l \times l$ matrix LJ with (a, b) entry

$$LJ(a,b) = \begin{cases} 1 & \text{if edges } a \text{ and } b \text{ are consecutive in order } ab \text{ in the graph } G, \\ 0 & \text{otherwise,} \end{cases} \tag{3.2}$$

where $a, b \in A$.

By definitions of the adjacency matrix and the line graph, the edge consecutive matrix is the adjacency matrix of the line graph of G. The directed graph shown in Fig. 3.7 is used as an example to construct the edge consecutive matrix as follows.

$$LJ = \begin{pmatrix} 0 & 0 & 1 & 0 & 0 & 0 & 1 & 0 \\ 0 & 0 & 0 & 1 & 0 & 0 & 0 & 1 \\ 1 & 0 & 0 & 0 & 1 & 0 & 0 & 0 \\ 0 & 1 & 0 & 0 & 0 & 1 & 0 & 0 \\ 0 & 1 & 0 & 0 & 0 & 1 & 0 & 0 \\ 1 & 0 & 0 & 0 & 1 & 0 & 0 & 0 \\ 0 & 0 & 0 & 1 & 0 & 0 & 0 & 1 \\ 0 & 0 & 1 & 0 & 0 & 0 & 1 & 0 \end{pmatrix}. \tag{3.3}$$

Definition 3.20 For a multidigraph $G = (V, A, \psi)$, the **incidence matrix** is the $m \times l$ matrix B with (v, a) entry

$$B(v, a) = \begin{cases} -1 & \text{if } a = (v, x) \text{ for some } x \in V, \\ 1 & \text{if } a = (x, v) \text{ for some } x \in V, \\ 0 & \text{otherwise,} \end{cases}$$

where $v \in V$ and $a \in A$.

According to the signed entries, the incidence matrix can be separated into the in-incidence matrix and the out-incidence matrix.

Definition 3.21 For a multidigraph $G = (V, A, \psi)$, the **in-incidence matrix** B_{in} and the **out-incidence matrix** B_{out} are the $m \times l$ matrices with (v, a) entries

$$B_{in}(v, a) = \begin{cases} 1 \text{ if } a = (x, v) \text{ for some } x \in V, \\ 0 \text{ otherwise,} \end{cases}$$

and

$$B_{out}(v, a) = \begin{cases} 1 \text{ if } a = (v, x) \text{ for some } x \in V, \\ 0 \text{ otherwise,} \end{cases}$$

where $v \in V$ and $a \in A$.

It is obvious that $B_{in} = (B + abs(B))/2$ and $B_{out} = (abs(B) - B)/2$, where $abs(B)$ denotes the matrix in which each entry equals the absolute value of the corresponding entry of B. For the directed graph shown in Fig. 3.7, the incidence matrix, the in-incidence matrix, and the out-incidence matrix are respectively expressed by

$$B = \begin{pmatrix} -1 & 0 & 1 & 0 & -1 & 1 & 0 & 0 \\ 0 & -1 & 0 & 1 & 1 & -1 & 0 & 0 \\ 1 & 0 & -1 & 0 & 0 & 0 & -1 & 1 \\ 0 & 1 & 0 & -1 & 0 & 0 & 1 & -1 \end{pmatrix},$$

$$B_{in} = \begin{pmatrix} 0 & 0 & 1 & 0 & 0 & 1 & 0 & 0 \\ 0 & 0 & 0 & 1 & 1 & 0 & 0 & 0 \\ 1 & 0 & 0 & 0 & 0 & 0 & 0 & 1 \\ 0 & 1 & 0 & 0 & 0 & 0 & 1 & 0 \end{pmatrix}, \text{ and } B_{out} = \begin{pmatrix} 1 & 0 & 0 & 0 & 1 & 0 & 0 & 0 \\ 0 & 1 & 0 & 0 & 0 & 1 & 0 & 0 \\ 0 & 0 & 1 & 0 & 0 & 0 & 1 & 0 \\ 0 & 0 & 0 & 1 & 0 & 0 & 0 & 1 \end{pmatrix}.$$

One finds that the incidence matrix depends on the label of each edge in a directed graph. To effectively analyze the graph model for conflict resolution using matrix representation, a unique rule of priority to label colored arcs is introduced in the next subsection.

Definition 3.22 A **colored multidigraph** (V, A, N, ψ, c) is a multidigraph (V, A, ψ) and a set of colors N, and a function $c : A \rightarrow N$ such that $c(a) \in N$ is the color of $a \in A$, provided that multiple edges of (V, A, ψ) are assigned different colors, i.e., if $a \neq b$, but $\psi(a) = \psi(b)$, then $c(a) \neq c(b)$.

If $a \in A$ such that $\psi(a) = (u, v)$ and $c(a) = i$ for $i \in N$, then a can be written as $a = d_i(u, v)$. The line digraph of $G = (V, A, N, \psi, c)$, $L(G)$, is a simple digraph and each vertex in $L(G)$ corresponds to an edge in the multidigraph G. Hence, coloring edges in G is equivalent to assigning colors to vertices in $L(G)$.

Definition 3.23 For a colored multidigraph (V, A, N, ψ, c), an **edge colored path** is a path in the multidigraph (V, A, ψ) in which each constituent edge has different colors.

If any two consecutive edges are restricted to having different colors in the edge consecutive matrix, this matrix is called the edge colored consecutive matrix. Its formal definition is as follows.

Definition 3.24 For a colored multidigraph $G = (V, A, N, \psi, c)$, the **edge colored consecutive matrix** LJ_c is the $l \times l$ matrix with (a, b) entry

$$LJ_c(a, b) = \begin{cases} 1 & \text{if edges } a \text{ and } b \text{ are consecutive in order } ab \\ & \text{and have different colors in the graph } G, \\ 0 & \text{otherwise.} \end{cases} \tag{3.4}$$

From algebraic graph theory (Godsil and Royle 2001), the following Lemma 3.1 that describes the relation between the adjacency matrix and incidence matrix can easily follow.

Lemma 3.1 *For a colored multidigraph $G = (V, A, N, \psi, c)$, the adjacency matrix J is expressed as*

$$J = sign[(B_{out}) \cdot (B_{in})^T]. \tag{3.5}$$

The following lemma that establishes the relation between the incidence matrix and the edge consecutive matrix is obtained based on Definition 3.21, on the in-incidence and out-incidence matrices B_{in} and B_{out}, and Definition 3.19, on the matrix LJ.

Lemma 3.2 *For a colored multidigraph $G = (V, A, N, \psi, c)$, B_{in} and B_{out} are the in-incidence matrix and out-incidence matrix of the graph G, respectively. Then, the edge consecutive matrix LJ satisfies $LJ = (B_{in})^T \cdot (B_{out})$.*

$$s \bullet \xrightarrow{a_k} \bullet_q \xrightarrow{a_h} \bullet u$$

Fig. 3.8 a_k and a_h are consecutive in order $a_k a_h$

Proof Let $M = (B_{in})^T \cdot (B_{out})$. Any (k, h) entry of matrix M can be expressed as $M(k, h) = e_k^T \cdot M \cdot e_h = [(B_{in}) \cdot e_k]^T \cdot [(B_{out}) \cdot e_h]$, where e_k^T denotes the transpose of the kth standard basis vector of the l-dimensional Euclidean space.

Therefore, $M(k, h) \neq 0$ iff $B_{in}(q, a_k) \cdot B_{out}(q, a_h) \neq 0$ for some $q \in S$ such that $\psi(a_k) = (s, q)$ and $\psi(a_h) = (q, u)$ for $s, u \in S$. This implies that $M(a_k, a_h) \neq 0$ iff a_k and a_h are consecutive from a_k to a_h (See Fig. 3.8).

Hence, based on Definition 3.19, $M(a_k, a_h) \neq 0$ iff $LJ(a_k, a_h) \neq 0$. Since M and LJ are 0–1 matrices, then, $LJ = (B_{in})^T \cdot (B_{out})$ follows. □

3.3.2 A Rule of Priority to Label Colored Arcs

A colored multidigraph may contain several arcs with the same initial and terminal vertices, but each arc in this case must be assigned a different color (Xu et al. 2009b, 2013). To work with the set of all arcs, they must be carefully labeled. Assume that all colors and nodes are pre-numbered. Therefore, the vertex set V and the color set N in $G = (V, A, N, \psi, c)$ are numbered as $V = \{1, 2, \cdots, m\}$ and $N = \{1, 2, \cdots, n\}$, respectively. Let c_i denote the cardinality of arc set assigned color i, i.e., $c_i = |A_i|$, where $A_i = \{x \in A : c(x) = i\}$ for each $i \in N$.

To label the arcs in a colored multidigraph $G = (V, A, N, \psi, c)$, set $\varepsilon_0 = 0$ and $\varepsilon_i = \sum_{j=1}^{i} c_j$ for $i \in N$, and note that $l = \varepsilon_n = \sum_{i=1}^{n} c_i$ is the cardinality of A in G. The arcs, $a_1, a_2, \ldots, a_l$, will be labeled according to the color order; within each color, according to the sequence of initial nodes; and within each color and initial node, according to the sequence of terminal nodes. The ordering, referred to as the *Rule of Priority*, has the following properties:

1. If $\varepsilon_{i-1} < k \leq \varepsilon_i$, then $c(a_k) = i$, i.e., a_k has color i;
2. For $k < h$, if a_k and a_h both have color i for some $i \in N$, and if $\psi(a_k) = (v_x, v_y)$ and $\psi(a_h) = (v_z, v_w)$, then $x \leq z$ and, if $x = z$, then $y < w$.

If all arcs in a graph model have been labeled according to the Rule of Priority, then the index of an arc uniquely determines the DM controlling it. Therefore, $A_i = \{a_{\varepsilon_{i-1}+1}, \ldots, a_{\varepsilon_i}\}$, where A_i denotes the set of arcs with color i.

Recall that c_i denotes the cardinality of the arc set in color i and let E_{c_i} denote a $c_i \times c_i$ matrix with each entry being set to 1 for $i = 1, 2, \cdots, n$. Then, D is defined as the following block diagonal matrix

$$D = \begin{pmatrix} E_{c_1} & 0 & \cdots & 0 \\ 0 & E_{c_2} & \cdots & 0 \\ \vdots & \vdots & \ddots & \vdots \\ 0 & 0 & \cdots & E_{c_n} \end{pmatrix}. \tag{3.6}$$

It is obvious that this matrix D encodes the color scheme in the graph G, where the dimension of each diagonal block E_{c_i} depends on the number of edges in color i. More specifically, recall that $\varepsilon_i = \sum_{j=1}^{i} c_j$ for $1 \leq i \leq n$. According to the Rule of Priority for labeling edges, for any $a_k \in A$ and $\varepsilon_{i-1} < k \leq \varepsilon_i$, the edge a_k has color i. Hence, for any $a_k, a_h \in A$, if there exists $1 \leq i \leq n$ such that $k, h \in (\varepsilon_{i-1}, \varepsilon_i]$, then edges a_k and a_h have the same color i, and $D(k, h) = 1$. Also, $D(k, h) = 0$ iff edges a_k and a_h have different colors.

This matrix captures the adjacency relation between pairs of consecutive edges without considering the color(s) of the consecutive edges. Another conversion function is thus presented next to transform the original problem of searching edge-colored paths in a colored multidigraph to the standard problem of finding paths in a simple digraph without color constraints. The conversion function can now be obtained in matrix form by the following lemma.

Lemma 3.3 *For a colored multidigraph $G = (V, A, N, \psi, c)$, let E_l be the $l \times l$ matrix with each entry equal to 1. Then the edge colored consecutive matrix LJ_c satisfies $LJ_c = LJ \circ (E_l - D)$, where "$\circ$" denotes the Hadamard product.*

Proof Let $LJ(k, h)$ and $(E_l - D)(k, h)$ denote the (k, h) entries of matrices LJ and $E_l - D$, respectively. Then, $LJ(k, h) \cdot (E_l - D)(k, h) \neq 0$ iff $LJ(k, h) \neq 0$ and $D(k, h) = 0$. Based on the definitions of matrices LJ and D, $LJ(k, h) \neq 0$ iff edges a_k and a_h are consecutive in order $a_k a_h$. $D(k, h) = 0$ iff edges a_k and a_h have different colors. Obviously, based on the definition of matrix LJ_c, $LJ_c = LJ \circ (E_l - D)$. □

Lemmas 3.2 and 3.3 together present a conversion function $F(B)$ such that

$$F(B) = [(B_{in})^T \cdot B_{out}] \circ (E_l - D), \tag{3.7}$$

where $B_{in} = (B + abs(B))/2$ and $B_{out} = (abs(B) - B)/2$. Therefore, $F(B)$ transforms a problem of searching colored paths in an edge colored digraph to a standard problem of finding paths in a simple digraph with no color constraints. Note that the incident relations between vertices and edges of a graph can uniquely characterize the graph. Therefore, the incidence matrix is treated as the original graph and used for computer implementation.

Example 3.4 (*Rule of Priority and Edge Colored Consecutive Matrix.*) A sustainable development game to model a conflict between an environmental agency and a developer was considered by Hipel (2001) and Li et al. (2004). The conflict is

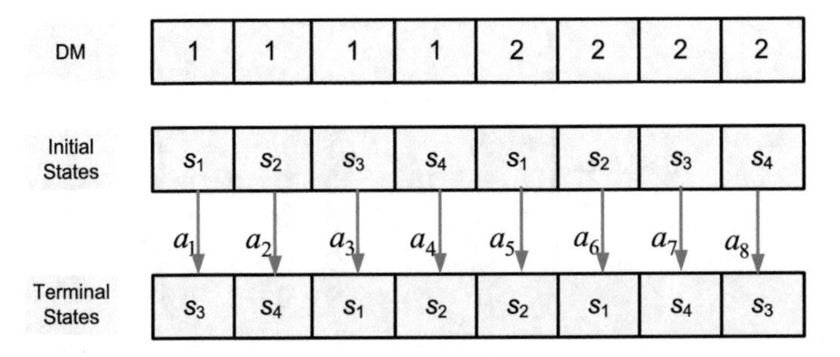

Fig. 3.9 The labels of edges

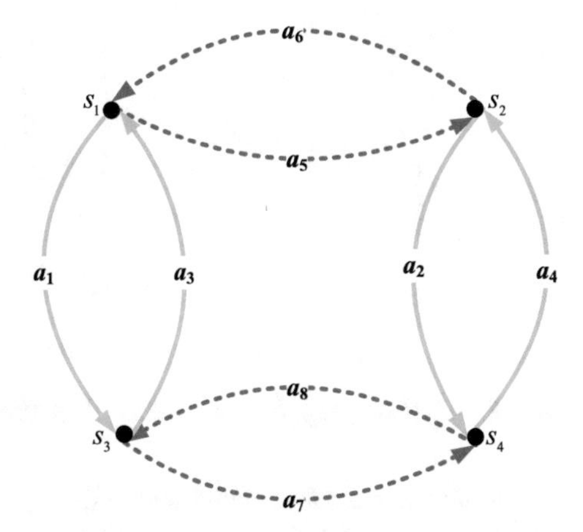

Fig. 3.10 Labeled graph model for the sustainable development conflict

modeled by two DMs: an environmental agency (DM 1) and a developer (DM 2). The graph model $G = (S, A, c)$ for the sustainable development conflict is depicted in Fig. 3.2, where vertices designate states and arcs represent movement between states. The number on a given arc indicates which DM controls the movement while the arrowhead shows the direction of movement. According to the Rule of Priority, label the edges of the graph model $G = (S, A, c)$ and calculate its edge colored consecutive matrix.

Assume that the DM set $N = \{1, 2\}$ and state set $S = \{s_1, s_2, s_3, s_4\}$. The cardinalities of the arc sets A_1 and A_2 are 4, respectively. Then, according to the Rule of Priority, the oriented arcs are numbered as in Fig. 3.9. The sustainable development game is expressed as the labeled graph model presented in Fig. 3.10 in which the full curves and dotted curves denote DM 1 and DM 2, respectively. Specifically, $a_1 = (s_1, s_3)$ and $c(a_1) = 1$; $a_2 = (s_2, s_4)$ and $c(a_2) = 1$; $a_3 = (s_3, s_1)$ and $c(a_3) = 1$; $a_4 = (s_4, s_2)$ and $c(a_4) = 1$; $a_5 = (s_1, s_2)$ and $c(a_5) = 2$; $a_6 = (s_2, s_1)$ and $c(a_6) = 2$; $a_7 = (s_3, s_4)$ and $c(a_7) = 2$; and $a_8 = (s_4, s_3)$ and $c(a_8) = 2$. Therefore, $A_1 = \{a_1, a_2, a_3, a_4\}$ and $A_2 = \{a_5, a_6, a_7, a_8\}$. Since,

$$E_l - D = \begin{pmatrix} 0 & 0 & 0 & 0 & 1 & 1 & 1 & 1 \\ 0 & 0 & 0 & 0 & 1 & 1 & 1 & 1 \\ 0 & 0 & 0 & 0 & 1 & 1 & 1 & 1 \\ 0 & 0 & 0 & 0 & 1 & 1 & 1 & 1 \\ 1 & 1 & 1 & 1 & 0 & 0 & 0 & 0 \\ 1 & 1 & 1 & 1 & 0 & 0 & 0 & 0 \\ 1 & 1 & 1 & 1 & 0 & 0 & 0 & 0 \\ 1 & 1 & 1 & 1 & 0 & 0 & 0 & 0 \end{pmatrix},$$

and LJ given in Eq. 3.3, one can obtain

$$LJ_c = LJ \circ (E_l - D) = \begin{pmatrix} 0 & 0 & 0 & 0 & 0 & 0 & 1 & 0 \\ 0 & 0 & 0 & 0 & 0 & 0 & 0 & 1 \\ 0 & 0 & 0 & 0 & 1 & 0 & 0 & 0 \\ 0 & 0 & 0 & 0 & 0 & 1 & 0 & 0 \\ 0 & 1 & 0 & 0 & 0 & 0 & 0 & 0 \\ 1 & 0 & 0 & 0 & 0 & 0 & 0 & 0 \\ 0 & 0 & 0 & 1 & 0 & 0 & 0 & 0 \\ 0 & 0 & 1 & 0 & 0 & 0 & 0 & 0 \end{pmatrix}.$$

3.3.3 Adjacency Matrix and Reachable List

Important matrices associated with a digraph include the adjacency matrix and the incidence matrix (Godsil and Royle 2001). Let $m = |V|$ denote the number of vertices and $l = |A|$ be the number of edges of the directed graph G. Then,

Definition 3.25 For a graph model $G = (S, A)$, DM i's **adjacency matrix** is the $m \times m$ matrix J_i with (s, q) entry

$$J_i(s, q) = \begin{cases} 1 & \text{if } (s, q) \in A_i, \\ 0 & \text{otherwise,} \end{cases}$$

where $s, q \in S$.

Let $i \in N$ and $s \in S$. $R_i(s)$ denotes DM $i's$ reachable list from a state s, containing all states to which DM i can move from state s in one step. $R_i(s)$ represents DM $i's$ unilateral moves (UMs). If $R_i(s)$ is written as a 0–1 row vector, then DM i's adjacency matrix J_i and reachable list from state s have the relation

$$R_i(s) = e_s^T \cdot J_i,$$

where e_s^T denotes the transpose of the sth standard basis vector of the m-dimensional Euclidean space.

For the graph model of the sustainable development game presented in Fig. 3.2, the adjacency matrices for DM 1 and DM 2 are

$$J_1 = \begin{pmatrix} 0 & 0 & 1 & 0 \\ 0 & 0 & 0 & 1 \\ 1 & 0 & 0 & 0 \\ 0 & 1 & 0 & 0 \end{pmatrix} \text{ and } J_2 = \begin{pmatrix} 0 & 1 & 0 & 0 \\ 1 & 0 & 0 & 0 \\ 0 & 0 & 0 & 1 \\ 0 & 0 & 1 & 0 \end{pmatrix}.$$

3.3.4 Preference Matrices

Preference information plays an important role in a graph model. A set of preference matrices can represent preference relations between any two states with different requirements. Two $m \times m$ preference matrices for DM i in the graph model with simple preference are defined as follows:

$$P_i^+(s,q) = \begin{cases} 1 & \text{if } q \succ_i s, \\ 0 & \text{otherwise,} \end{cases} \tag{3.8}$$

and

$$P_i^{-,=}(s,q) = \begin{cases} 1 & \text{if } s \succ_i q \text{ or } (s \sim_i q \text{ and } s \neq q), \\ 0 & \text{otherwise.} \end{cases} \tag{3.9}$$

The preference matrix P_i^+ may be used to represent more preferred relations and the preference matrix $P_i^{-,=}$ can represent less preferred or equally preferred relations between any two states. Since simple preference structure is complete, matrices P_i^+ and $P_i^{-,=}$ have the relation $P_i^+ = E - I - P_i^{-,=}$, where E is the $m \times m$ matrix with each entry equal to 1 and I the $m \times m$ identity matrix. For example, the sustainable development model provides the preference information for two DMs as follows:

$$s_1 \succ_1 s_3 \succ_1 s_2 \succ_1 s_4 \text{ and } s_3 \succ_2 s_1 \succ_2 s_4 \succ_2 s_2.$$

Therefore, two DMs' preference matrices are expressed by

$$P_1^+ = \begin{pmatrix} 0 & 0 & 0 & 0 \\ 1 & 0 & 1 & 0 \\ 1 & 0 & 0 & 0 \\ 1 & 1 & 1 & 0 \end{pmatrix}, P_1^{-,=} = \begin{pmatrix} 0 & 1 & 1 & 1 \\ 0 & 0 & 0 & 1 \\ 0 & 1 & 0 & 1 \\ 0 & 0 & 0 & 0 \end{pmatrix},$$

$$P_2^+ = \begin{pmatrix} 0 & 0 & 1 & 0 \\ 1 & 0 & 1 & 1 \\ 0 & 0 & 0 & 0 \\ 1 & 0 & 1 & 0 \end{pmatrix}, \text{ and } P_2^{-,=} = \begin{pmatrix} 0 & 1 & 0 & 1 \\ 0 & 0 & 0 & 0 \\ 1 & 1 & 0 & 1 \\ 0 & 1 & 0 & 0 \end{pmatrix}.$$

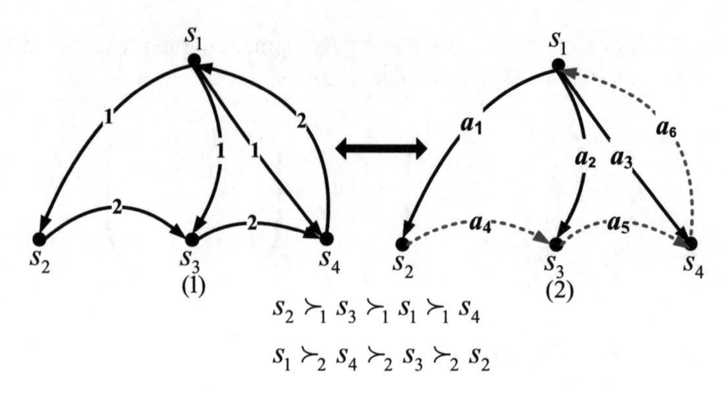

Fig. 3.11 Different representations of a graph model

3.3.5 *Incidence Matrix and Graph Model*

The incidence matrix based on the Rule of Priority with preference matrices can completely represent a graph model. Figure 3.11 depicts a graph model and its edge labeled graph. Although no DM is explicitly shown in the labeled graph, the index number of an arc uniquely determines the DM who controls it when all arcs have been numbered according to the Rule of Priority. Specifically, based on the number of arcs in i's graph G_i for $i = 1, 2$, $|A_1| = 3$, and $|A_2| = 3$, arcs a_1 to a_3 are controlled by DM 1, arcs a_4 to a_6 by DM 2.

The incidence matrix of the labeled graph

$$B = \begin{pmatrix} -1 & -1 & -1 & 0 & 0 & 1 \\ 1 & 0 & 0 & -1 & 0 & 0 \\ 0 & 1 & 0 & 1 & -1 & 0 \\ 0 & 0 & 1 & 0 & 1 & -1 \end{pmatrix},$$

and the preference matrices base on DMs' preference information

$$P_1^+ = \begin{pmatrix} 0 & 1 & 1 & 0 \\ 0 & 0 & 0 & 0 \\ 0 & 1 & 0 & 0 \\ 1 & 1 & 1 & 0 \end{pmatrix}, \text{ and } P_2^+ = \begin{pmatrix} 0 & 0 & 0 & 0 \\ 1 & 0 & 1 & 1 \\ 1 & 0 & 0 & 1 \\ 1 & 0 & 0 & 0 \end{pmatrix},$$

can represent the graph model shown in Fig. 3.11.

3.4 Important Ideas

Conflicts arise across a wide range of scales and settings. To model a strategic conflict, the normal form, option form, graph model form and its matrix representation are introduced in this chapter. Compared with the normal and option forms to represent strategic conflicts, the graph model has several advantages, including its ability to

- handle irreversible moves,
- model common moves,
- provide a flexible framework for defining, comparing, and characterizing solution concepts, and
- be easily applied to actual conflicts.

For small or generic conflicts, the normal form of the game explained in Sect. 3.1.1 can be a convenient notation to employ, as is shown in Table 3.2 for the 2×2 sustainable development game. In practice, the option form of the game defined in Sect. 3.1.2 and illustrated in Table 3.3 for the sustainable development conflict constitutes a flexible format to use in practice for recording conflicts ranging from simple to complicated ones. In fact, option form is utilized in the vast majority of cases for defining states needed in the graph model formulation. After a graph model is converted to a labeled digraph based on the proposed Rule of Priority, it can be represented by using a set of matrices that can be utilized to analyze a graph model using algebraic graph theory. In the next chapter, stability definitions (or solution concepts) are defined logically, in terms of the underlying graphs, and formulated explicitly using matrices for the case of what is called simple preference.

3.5 Problems

3.5.1 The normal form of the game is displayed in Table 3.2 for the sustainable development game. Because each of the two DMs controls two strategies, this is called a 2×2 game. These small 2×2 games represent the simplest possible game that could occur and can be highly informative for clearly explaining the strategic interpretation of conflict situations that can arise in the real-world, such as the sustainable development game. A widely known 2×2 game is called Prisoner's Dilemma which is used to reflect the situation in which a DM must decide whether to act in his or her own interest in the short term or to cooperate with another DM, in order to reach a better result in the longer term. The 2×2 normal form of this conflict is written as given in Table 3.6.

In Prisoner's Dilemma, notice that if the two decision makers labeled as DM 1 and DM 2 cooperate with one another, they both fare reasonable well (state s_1) compared to the situation in which they do not (state s_4).

(a) By referring to a well known book or paper on 2×2 games, explain in English what conflict is taking place between the two prisoners.

Table 3.6 Prisoners Dilemma in normal form

		DM 2	
		Cooperate	Do Not Cooperate
DM 1	Cooperate	3, 3 s_1	1, 4 s_2
	Do Not Cooperate	4, 1 s_3	2, 2 s_4

Table 3.7 The game of Chicken in normal form

		DM 2	
		Do Not Swerve	Swerve
DM 1	Do Not Swerve	1, 1 s_1	4, 2 s_2
	Swerve	2, 4 s_3	3, 3 s_4

(b) Using a real-world example, explain how a situation involving labour and management could be reasonably modeled using Prisoner's Dilemma.
(c) Describe how and why the conflict over climate change could be interpreted in its simplest form using Prisoner's Dilemma.

3.5.2 For the Prisoner's Dilemma game mentioned in Problem 3.5.1:

(a) Record the option form of this conflict.
(b) Show the graph model version of this dispute.

3.5.3 In repeated Prisoner's Dilemma, the two competitors deal with each other on a regular basis over time. By referring to the literature, explain the best strategy to follow in repeated Prisoner's Dilemma.

3.5.4 The famous game of Chicken is another well known 2×2 game which can be written in normal form as shown in Table 3.7.

In this high risk confrontation, two drivers, called DM 1 and DM 2, are driving at high speed towards one another. The driver, who swerves off the road to avoid a collision in which both drivers would be killed, loses the game and is called a chicken. Notice in the game of Chicken that the worst situation is when both drivers do not swerve, which is state s_1.

(a) Explain why the preferences for each DM in Chicken make sense.
(b) The Cuban Missile Crisis of 1962 is sometimes modeled as a game of Chicken. By locating appropriate references, outline what happened in the Cuban Missile Crisis. Write down the Cuban Missile Crisis in normal form as a game of Chicken.
(c) Describe another situation involving the game of Chicken which could take place in the real-world.

3.5.5 In Problem 3.5.4, it is mentioned that the Cuban Missile Crisis is sometimes interpreted as a game of Chicken. Rather than using the game of Chicken, Fraser and Hipel (1984, Chap. 2) and also Hipel (2011) develop a much more realistic model of the Cuban Missile Crisis in option form.

(a) Show the normal form of the Cuban Missile Crisis mentioned above.
(b) Write down the option form of the Cuban Missile Crisis.
(c) Show the graph model for the Cuban Missile Crisis.

3.5.6 Sometimes misunderstandings can arise in a conflict situation, which is referred to as a hypergame as mentioned in Sect. 10.3.1. By referring to the literature, qualitatively explain what is meant by a hypergame. Explain why the Cuban Missile Crisis would be best modeled as a hypergame.

3.5.7 Write down the matrix or algebraic form of the Prisoner's Dilemma game mentioned in Problem 3.5.1.

3.5.8 Show the matrix or algebraic formulation of the game of Chicken mentioned in Problem 3.5.4.

3.5.9 For the graph model shown in Fig. 3.11,

(a) label the graph model (1) to present all processes according to the Rule of Priority;
(b) calculate its edge consecutive matrix and edge colored consecutive matrix.

3.5.10 A superpower nuclear confrontation (Fang et al. 1993) is modeled using two DMs and six options. These options determine five feasible states as listed in Table 3.8. Note that state W represents a nuclear winter. The graph model is shown in Fig. 3.12.

For the graph model shown in Fig. 3.12:

(a) label the graph model to present all processes according to the Rule of Priority;
(b) calculate its incidence matrix and edge colored consecutive matrix.

Table 3.8 Decision makers, options and feasible states for the superpower nuclear confrontation conflict

DM 1					
1. Peace (P)	Y	Y	N	N	N
2. Conventional attack (C)	N	N	Y	Y	N
3. Full nuclear attack (W)	N	N	N	N	Y
DM 2					
1. Peace (P)	Y	N	Y	N	N
2. Conventional attack (C)	N	Y	N	Y	N
3. Full nuclear attack (W)	N	N	N	N	Y
States	**PP**	**PC**	**CP**	**CC**	**W**

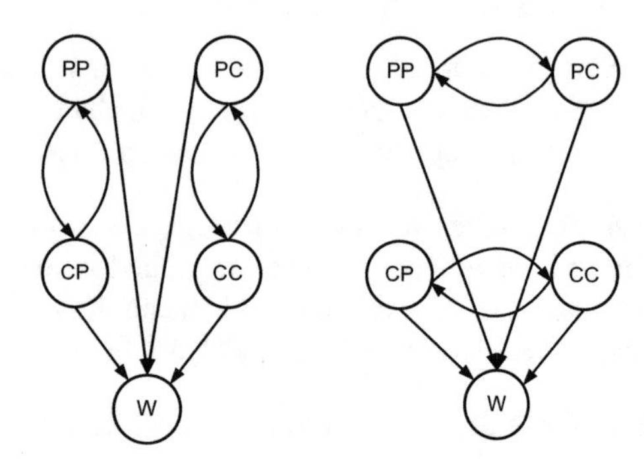

(a) Graph model for DM 1 (b) Graph model for DM 2

DM 1: $PP \succ_1 CP \succ_1 CC \succ_1 PC \succ_1 W$
DM 2: $PP \succ_2 PC \succ_2 CC \succ_2 CP \succ_2 W$

Fig. 3.12 The graph model of the superpower nuclear confrontation conflict

References

Bernath Walker, S., Hipel, K. W., & Xu, H. (2013). A matrix representation of attitudes in conflicts. *IEEE Transactions on Systems, Man, and Cybernetics: Systems*, *43*(6), 1328–1342. https://doi.org/10.1109/tsmc.2013.2260536.

Buckley, F., & Harary, F. (1990). *Distance in graphs*. Reading: Addison-Wesley.

Dieste, R. (1997). *Graph theory*. New York: Springer. https://doi.org/10.1007/978-1-4419-1153-7_402.

Fang, L., Hipel, K. W., & Kilgour, D. M. (1993). *Interactive decision making: The graph model for conflict resolution*. New York: Wiley. https://doi.org/10.2307/2583940.

Fraser, N. M., & Hipel, K. W. (1984). *Conflict analysis: Models and resolutions*. New York: North-Holland. https://doi.org/10.2307/2582031.

Godsil, C., & Royle, G. (2001). *Algebraic graph theory*. New York: Springer. https://doi.org/10.1007/978-1-4613-0163-9.

Gondran, M., & Minoux, M. (1979). *Graphs and algorithms*. New York: Wiley.

Hipel, K. W. (2001). Conflict resolution. In M. K. Tolba (Ed.), *Our fragile world: Challenges and opportunities for sustainable development (Forerunner to the Encyclopedia of Life Support Systems)* (Vol. 1, pp. 935–952). Oxford, United Kingdom: Eolss.

Hipel, K. W. (2011). A systems engineering approach to conflict resolution in command and control. *The International C2 Journal*, *5*(1), 1–56.

Hoffman, A. J., & Schiebe, B. (2001). The edge versus path incidence matrix of series-parallel graphs and greedy packing. *Discrete Applied Mathematics*, *113*, 275–284. https://doi.org/10.1016/s0166-218x(00)00294-8.

Hou, Y., Jiang, Y., & Xu, H. (2015). Option prioritization for three-level preference in the graph model for conflict resolution. In B. Kaminski, G. E. Kersten & T. Szapiro (Eds.), *Outlooks and insights on group decision and negotiation, the Proceedings of the 2015 International Conference on Group Decision and Negotiation* (pp. 269–280). Cham, Switzerland: Springer.

Howard, N. (1971). *Paradoxes of rationality: Theory of metagames and political behavior*. Cambridge: MIT Press. https://doi.org/10.2307/1266876.

Li, K. W., Hipel, K. W., Kilgour, D. M., & Fang, L. (2004). Preference uncertainty in the graph model for conflict resolution. *IEEE Transactions on Systems, Man, and Cybernetics, Part A, Systems and Humans*, *34*(4), 507–520. https://doi.org/10.1109/tsmca.2004.826282.

Shiny, A. K., & Pujari, A. K. (1998). Computation of prime implicants using matrix and paths. *Journal of Logic and Computation*, *8*(2), 135–145. https://doi.org/10.1093/logcom/8.2.135.

Von Neumann, J., & Morgenstern, O. (1953). *Theory of games and economic behavior* (3rd ed.). Princeton: Princeton University Press.

Xu, H., Kilgour, D. M., & Hipel, K. W. (2007). Matrix representation of solution concepts in graph models for two decision-makers with preference uncertainty. *Dynamics of Continuous, Discrete and Impulsive Systems*, *14*(S1), 703–707.

Xu, H., Hipel, K. W., & Kilgour, D. M. (2009a). Matrix representation of solution concepts in multiple decision maker graph models. *IEEE Transactions on Systems, Man, and Cybernetics, Part A: Systems and Humans*, *39*(1), 96–108. https://doi.org/10.1109/tsmca.2009.2007994.

Xu, H., Li, K. W., Hipel, K. W., & Kilgour, D. M. (2009b). A matrix approach to status quo analysis in the graph model for conflict resolution. *Applied Mathematics and Computation*, *212*(2), 470–480. https://doi.org/10.1016/j.amc.2009.02.051.

Xu, H., Li, K. W., Kilgour, D. M., & Hipel, K. W. (2009c). A matrix-based approach to searching colored paths in a weighted colored multidigraph. *Applied Mathematics and Computation*, *215*, 353–366. https://doi.org/10.1016/j.amc.2009.04.086.

Xu, H., Hipel, K. W., Kilgour, D. M., & Chen, Y. (2010a). Combining strength and uncertainty for preferences in the graph model for conflict resolution with multiple decision makers. *Theory and Decision*, *69*(4), 497–521. https://doi.org/10.1007/s11238-009-9134-6.

Xu, H., Kilgour, D. M., & Hipel, K. W. (2010b). An integrated algebraic approach to conflict resolution with three-level preference. *Applied Mathematics and Computation*, *216*(3), 693–707. https://doi.org/10.1016/j.amc.2010.01.054.

Xu, H., Kilgour, D. M., & Hipel, K. W. (2010c). Matrix representation and extension of coalition analysis in group decision support. *Computers and Mathematics with Applications*, *60*(5), 1164–1176. https://doi.org/10.1016/j.camwa.2010.05.040.

Xu, H., Kilgour, D. M., Hipel, K. W., & Kemkes, G. (2010d). Using matrices to link conflict evolution and resolution in a graph model. *European Journal of Operational Research*, *207*, 318–329. https://doi.org/10.1016/j.ejor.2010.03.025.

Xu, H., Kilgour, D. M., & Hipel, K. W. (2011). Matrix representation of conflict resolution in multiple-decision-maker graph models with preference uncertainty. *Group Decision and Negotiation*, *20*(6), 755–779. https://doi.org/10.1007/s10726-010-9188-4.

Xu, H., Kilgour, D. M., Hipel, K. W., & McBean, E. A. (2013). Theory and application of conflict resolution with hybrid preference in colored graphs. *Applied Mathematical Modelling*, *37*(3), 989–1003. https://doi.org/10.1016/j.apm.2012.03.009.

Xu, H., Kilgour, D. M., Hipel, K. W., & McBean, E. A. (2014). Theory and implementation of coalitional analysis in cooperative decision making. *Theory and Decision*, *76*(2), 147–171. https://doi.org/10.1007/s11238-013-9363-6.

Chapter 4
Stability Definitions: Simple Preference

Strategic conflicts, or situations in which two or more decision makers (DMs) with different objectives interact, occur often in the real-world. As discussed in Chap. 3, many models are available to represent strategic conflicts, such as the normal-form conflict model, the option-form conflict model, and the graph model. Conflict resolution has been investigated within many disciplines (Hipel 2009) including international relations, psychology, and law, as well as from mathematical and engineering perspectives (Saaty and Alexander 1989, Howard et al. 1992, Fang et al. 1993, Bennett 1995). Among the formal methodologies that handle strategic conflict, the graph model (or Graph Model for Conflict Resolution (GMCR)) (Kilgour et al. 1987, Fang et al. 1993) provides a remarkable combination of simplicity and flexibility.

The main goal of this chapter is to define stabilities in graph models with simple preference structure, based on a strict preference and an indifference relation, to be discussed in Sect. 4.1. As explained in Sect. 4.2, when determining the stability of a state for a given DM, a logical structure is employed for tracking the moves and countermoves that could take place if the DM decides to improve his or her situation. If the DM perceives that he or she will end up in a less preferred situation as a result of these potential interactions with others, the state is deemed to be stable. However, these logical representations of stabilities often require complex calculations and are difficult to code. In particular, the construction of reachable lists of a coalition having two or more DMs is a complicated process. The restriction that no DM may move twice consecutively does not constrain a coalition in the way that it limits an individual DM. For example, if there are only two DMs in a model, then a response to a unilateral improvement (UI) by one of them is necessarily a single move. But if there are more than two DMs in the model, a response to one DM's UI may consist of a sequence of many moves, provided no specific DM moves twice consecutively. The subset of DMs levying the moves under the control of group members is called a coalition. The sequence of actions by members of a coalition may constitute an action to sanction a UI by another DM or the coalition members may be moving to

H. Xu et al., *Conflict Resolution Using the Graph Model: Strategic Interactions in Competition and Cooperation*, Studies in Systems, Decision and Control 153,
https://doi.org/10.1007/978-3-319-77670-5_4

a state which is more preferred by all members of the coalition which is referred to as a coalition improvement.

The foregoing types of situations led to the development of matrix representations of a graph model and explicit matrix calculations to determine the stabilities introduced in Sect. 4.3. Because the graph model consists of several interrelated graphs, well-known results of graph theory can help to analyze a graph model. This analysis involves searching paths in a graph, subject to the important restriction that no DM can move twice in succession along any path. Therefore, a graph model must be treated as an edge colored digraph in which each arc represents a unilateral move and distinct colors refer to different DMs. An algebraic approach to searching colored paths in a colored digraph is presented in Sect. 4.3. The computational complexity of employing the matrix formulation of the graph model is investigated in Sect. 4.4. The sustainable development conflict is used throughout this chapter to illustrate how stability calculations are executed for both the logical and matrix formulations of the graph model. In Sect. 4.5, the Elmira dispute is employed to demonstrate how stability calculations are carried out using the matrix representation. Finally, part of the presentation appearing in this chapter is based upon research published earlier (Xu et al. 2007, 2009, 2010a, b, 2011, 2014).

4.1 Simple Preference

In the original form, a graph model could be calibrated using only a relative preference relation, "$\succ$ preferred", and an "equality" relation, "$\sim$ indifferent", to represent a DM's preference for one state with respect to another. The features and properties of this type of preference, called a simple preference structure, were discussed in Sect. 3.2.4. Specifically, simple preference of DM i is represented by a pair of relations $\{\succ_i, \sim_i\}$ on S, where $s \succ_i q$ indicates that DM i prefers s to q and $s \sim_i q$ means that DM i is indifferent between s and q (or equally prefers s and q). Note that, for each i, $\succ_i$ is assumed irreflexive and asymmetric, and $\sim_i$ is assumed reflexive and symmetric. Also, it is assumed that, for any $s, q \in S$, either $s \succ_i q$, $s \sim_i q$, or $q \succ_i s$. The conventions that $s \succeq_i q$ is equivalent to either $s \succ_i q$ or $s \sim_i q$, and that $s \prec_i q$ is equivalent to $q \succ_i s$, are convenient. Based on such preference information, DM i's reachable lists from a status quo state along the arcs of the directed graph, important components of stability analysis, can be defined for a graph model, as will be accomplished next.

4.1.1 Reachable Lists of a Decision Maker

Let S and N denote the state set and the DM set. The state set S can be partitioned into subsets based on preference relative to a fixed state $s \in S$. These subsets, which are essential in stability analysis, are described as follows:

- $\Phi_i^+(s) = \{q : q \succ_i s\}$, the states preferred to state s by DM i;
- $\Phi_i^=(s) = \{q : q \sim_i s\}$, the states indifferent to state s by DM i;
- $\Phi_i^-(s) = \{q : s \succ_i q\}$, the states less preferred than state s for DM i.

Let $i \in N$ and $s \in S$ be arbitrary. Denote the intersection operation by $\cap$. Recall that each arc of $A_i \subseteq S \times S$ indicates that DM i can make a unilateral move (in one step) from the initial state to the terminal state of the arc. DM i's reachable lists from state $s \in S$ for simple preference are defined as follows:

Definition 4.1 For a graph model G, A_i denotes the arcs controlled by DM i for $i \in N$. DM i's reachable lists from $s \in S$ are subsets of S as follows:

(i) $R_i(s) = \{q \in S : (s, q) \in A_i\}$ is DM i's reachable list from s by unilateral moves (UMs);
(ii) $R_i^+(s) = \{q \in S : (s, q) \in A_i \text{ and } q \succ_i s\}$ is DM i's reachable list from s by unilateral improvements (UIs);
(iii) $R_i^=(s) = \{q \in S : (s, q) \in A_i \text{ and } q \sim_i s\}$ is DM i's reachable list from s by equally preferred moves; and
(iv) $R_i^-(s) = \{q \in S : (s, q) \in A_i \text{ and } s \succ_i q\}$ is DM i's reachable list from s by unilateral disimprovements.

From the above definitions, the relationships among the subsets of S and the corresponding reachable lists from state s for DM i are depicted in Fig. 4.1. For ease of use, some additional notation is defined by $\Phi_i^{-,=}(s) = \Phi_i^-(s) \cup \Phi_i^=(s)$.

Example 4.1 A graph model with two DMs $N = \{1, 2\}$ and four feasible states $S = \{s_1, s_2, s_3, s_4\}$ is depicted in Fig. 4.2. The labels on the arcs of the graph indicate the DM who can make the move. Preference information about the states is given below the directed graph. If $s = s_1$ is selected as the status quo state, the subsets of S separated by DM i, $\Phi_i^+(s)$, $\Phi_i^=(s)$ and $\Phi_i^-(s)$, and DM i's reachable lists from s, $R_i(s)$, $R_i^+(s)$, $R_i^=(s)$ and $R_i^-(s)$ for $i \in N$, can be calculated easily.

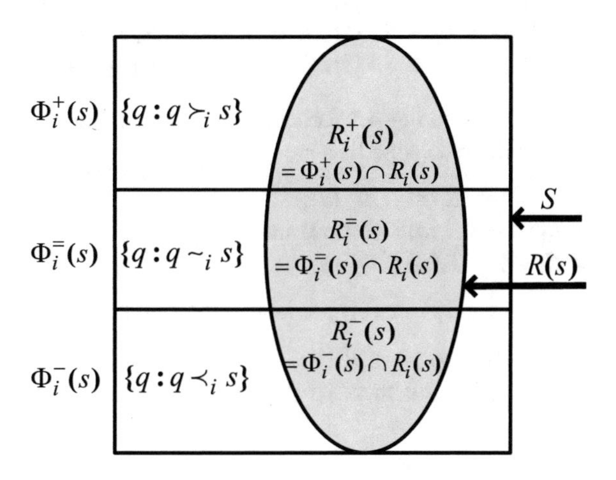

Fig. 4.1 Relations among the subsets of S and the corresponding reachable lists

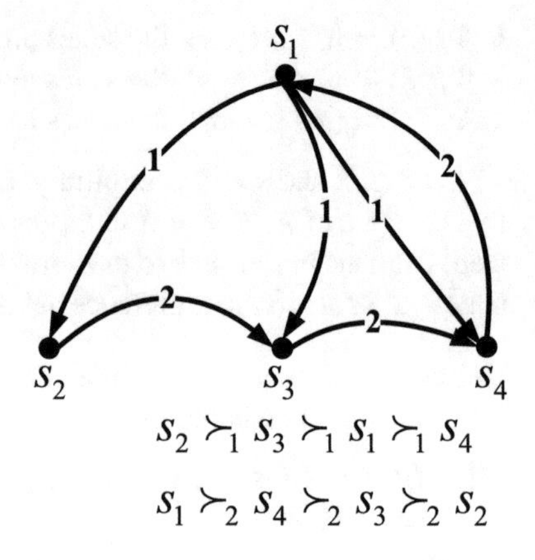

Fig. 4.2 Graph model for a two DM model

Note that the preference information

$$s_2 \succ_1 s_3 \succ_1 s_1 \succ_1 s_4 \text{ and } s_1 \succ_2 s_4 \succ_2 s_3 \succ_2 s_2$$

in Fig. 4.2 implies that the preference relations $\succ_1$ and $\succ_2$ are transitive. According to the descriptions of the subsets of S,

- $\Phi_1^+(s_1) = \{q : q \succ_1 s_1\} = \{s_2, s_3\}$ and $\Phi_2^+(s_1) = \{q : q \succ_2 s_1\} = \emptyset$;
- $\Phi_1^=(s_1) = \{q : q \sim_1 s_1\} = \emptyset$ and $\Phi_2^=(s_1) = \{q : q \sim_2 s_1\} = \emptyset$; and
- $\Phi_1^-(s_1) = \{q : s_1 \succ_1 q\} = \{s_4\}$ and $\Phi_2^-(s_1) = \{q : s_1 \succ_2 q\} = \{s_2, s_3, s_4\}$.

Clearly, DM 1's arc set is $A_1 = \{(s_1, s_2), (s_1, s_3), (s_1, s_4)\}$ and DM 2's arc set is $A_2 = \{(s_2, s_3), (s_3, s_4), (s_4, s_1)\}$. According to Definition 4.1, the DMs' reachable lists from s_1 are

- $R_1(s_1) = \{q \in S : (s_1, q) \in A_1\} = \{s_2, s_3, s_4\}$ and $R_2(s_1) = \{q \in S : (s_1, q) \in A_2\} = \emptyset$;
- $R_1^+(s_1) = \{q \in S : (s_1, q) \in A_i \text{ and } q \succ_1 s_1\} = \{s_2, s_3\}$ and $R_2^+(s_1) = \emptyset$;
- $R_1^=(s_1) = \{q \in S : (s_1, q) \in A_i \text{ and } q \sim_1 s_1\} = \emptyset$ and $R_2^=(s_1) = \emptyset$; and
- $R_1^-(s_1) = \{q \in S : (s_1, q) \in A_i \text{ and } s_1 \succ_1 q\} = \{s_4\}$ and $R_2^-(s_1) = \emptyset$.

As shown in Fig. 4.1 for Example 4.1, the relations among the subsets of the state set and the reachable lists are

- $R_i^+(s) = R_i(s) \cap \Phi_i^+(s)$;
- $R_i^=(s) = R_i(s) \cap \Phi_i^=(s)$; and
- $R_i^-(s) = R_i(s) \cap \Phi_i^-(s)$.

DM i's oriented arcs A_i are related to reachable lists as follows:

- $A_i = \{(p, q) : q \in R_i(p)\}$ is DM i's UM arc set;
- if $s \in S$ is fixed, then $A_i(s) = \{(s, q) \in A_i : q \in R_i(s)\}$ are DM i's UM arcs from s.

DM i's UM arcs, A_i, can be partitioned as follows:

- $A_i^+ = \{(p, q) \in A_i : q \succ_i p\}$ is DM i's UI arc set;
- $A_i^= = \{(p, q) \in A_i : q \sim_i p\}$ is DM i's equally preferred arc set; and
- $A_i^- = \{(p, q) \in A_i : p \succ_i q\}$ is DM i's less preferred arc set.

According to the completeness property of preference, $A_i = A_i^+ \cup A_i^= \cup A_i^-$. Now fix $s \in S$. Then

- $A_i^+(s) = \{(s, q) \in A_i : q \in R_i^+(s)\}$ are DM i's UI arcs from s;
- $A_i^=(s) = \{(s, q) \in A_i : q \in R_i^=(s)\}$ are DM i's equally preferred arcs from s; and
- $A_i^-(s) = \{(s, q) \in A_i : q \in R_i^-(s)\}$ are DM i's less preferred arcs from s.

Note that $A_i(s)$ is a subset of the arc set A_i while $R_i(s)$ is a subset of the state set S.

4.2 Logical Representation of Stability Definitions

In a graph model, a stability definition (solution concept) is a procedure for determining whether a state is stable for a decision maker (DM), and identifying a situation in which the DM would have no incentive to move away from the state unilaterally. An equilibrium of a graph model, or a possible resolution of the conflict it represents, is a state that all DMs find stable under an appropriate stability definition. Many solution concepts have been formulated to represent various decision styles and contexts. In this book, four basic solution concepts—Nash stability (Nash 1950, 1951), general metarationality (GMR) (Howard 1971), symmetric metarationality (SMR) (Howard 1971), and sequential stability (SEQ) (Fraser and Hipel 1979)—are emphasized. Recently, Li et al. (2004) extended these four solution concepts to models having preference uncertainty, which will be introduced in Chap. 5. As well, Hamouda et al. (2004, 2006) proposed new stability definitions that take strength of preference (strong or mild) into account, which will be discussed in Chap. 6.

The logical representations of Nash, GMR, SMR, and SEQ stabilities in the graph model with simple preference are given below. The four stability definitions for two-DM models are introduced first.

4.2.1 Two Decision Maker Case

Let $N = \{i, j\}$ and $s \in S$ in the following definitions.

Definition 4.2 State s is Nash stable for DM i, denoted by $s \in S_i^{Nash}$, iff $R_i^+(s) = \emptyset$.

For Nash stability (Nash 1950, 1951), DM i expects that DM j will stay at any state DM i moves to, and consequently that any state that i moves to will be final state.

Table 4.1 Nash stability of the sustainable development game with simple preferences

State	$R_i^+(s)$		Nash stability		Equilibrium
	DM 1	DM 2	DM 1	DM 2	
s_1	$\emptyset$	$\emptyset$	s	s	Eq
s_2	$\emptyset$	$\{s_1\}$	s	u	
s_3	$\{s_1\}$	$\emptyset$	u	s	
s_4	$\{s_2\}$	$\{s_3\}$	u	u	

Example 4.2 (*Nash Stability for the Sustainable Development Model*) The sustainable development game was presented in normal form, option form, and graph form in Tables 3.2 and 3.3, and Fig. 3.2, respectively. The graph model of this model is shown in Fig. 3.2 with the state set $S = \{s_1, s_2, s_3, s_4\}$ and the DM set $N = \{1, 2\}$. The letter on a given arc indicates which DM controls the movement while the arrowhead shows the direction of movement. The two DMs' preference information is presented underneath the digraph.

State s_1 is now analyzed to ascertain if it is Nash stable for DM i. From Fig. 3.2, DM 1 has a unilateral move from s_1 to s_3. However, $s_1 \succ_1 s_3$ based on the preference information, so the move by DM 1 from s_1 to s_3 is not a unilateral improvement and, therefore, state s_1 is Nash stable for DM 1 according to Definition 4.2. Next, consider the Nash stability of s_1 for DM 2. Clearly, DM 2 has a unilateral move from s_1 to s_2. Because $s_1 \succ_2 s_2$, the move by DM 2 from s_1 to s_2 is not a unilateral improvement and, therefore, state s_1 is Nash stable for DM 2. Accordingly, s_1 is an equilibrium in the sense of Nash stability. Similarly, the other three states can be assessed for Nash stability.

Nash stability results are listed in Table 4.1 in which $R_i^+(s)$ denotes DM i's UIs from $s \in S$ for $i \in N$. The letter "s" indicates that a state is Nash stable for a given DM, whereas "u" denotes that the state is Nash unstable. The letter "Eq" means that the state is an equilibrium, that is Nash stable for both DMs.

State $s \in S$ is GMR for DM i iff, whenever DM i makes any UI from s, then i's opponent can move to sanction i (that is, hurt i) in response. (A "sanction" must be an opponent's move.) The formal definition is given next.

Definition 4.3 State s is GMR stable (or, simply, GMR) for DM i, denoted by $s \in S_i^{GMR}$, iff for every $s_1 \in R_i^+(s)$ there exists at least one $s_2 \in R_j(s_1)$ with $s_2 \in \Phi_i^{-,=}(s)$ (or $s \succeq_i s_2$).

For GMR, DM i expects that its opponent j will respond by hurting i, so s is GMR stable for i iff DM j can hurt i if i takes any UI.

Example 4.3 (*GMR Stability for the Sustainable Development Model*) From Definitions 4.2 and 4.3, one can see that if $R_i^+(s) = \emptyset$, then s is Nash stable and GMR stable for DM i. Hence, for instance, s_3 is GMR stable for DM 2 for the sustainable development model. Let us assess whether s_3 is GMR for DM 1. DM 1

Table 4.2 GMR stability of the sustainable development game with simple preferences

State	$R_i^+(s)$		$R_i(s)$		GMR stability		Equilibrium
	DM 1	DM 2	DM 1	DM 2	DM 1	DM 2	
s_1	$\emptyset$	$\emptyset$	$\{s_3\}$	$\{s_2\}$	s	s	Eq
s_2	$\emptyset$	$\{s_1\}$	$\{s_4\}$	$\{s_1\}$	s	u	
s_3	$\{s_1\}$	$\emptyset$	$\{s_1\}$	$\{s_4\}$	s	s	Eq
s_4	$\{s_2\}$	$\{s_3\}$	$\{s_2\}$	$\{s_3\}$	u	u	

has a unilateral improvement from s_3 to s_1 and DM 2 has a unilateral move from s_1 to s_2. However, s_2 is less preferred than s_3 for DM 1, hence, s_3 is GMR for DM 1 according to Definition 4.3. The stabilities of other three states for the two DMs can be determined, similarly. GMR stability results are listed in Table 4.2, where, as usual, $R_i^+(s)$ denotes DM i's UIs from $s \in S$, "s" indicates GMR stable, "u" indicates GMR unstable, and "Eq" indicates a GMR equilibrium.

SMR is a similar but more restrictive stability definition compared to GMR. Under SMR, DM i expects to have a chance to counterrespond to its opponent's response to i's original move.

Definition 4.4 State s is SMR stable for DM i, denoted by $s \in S_i^{SMR}$, iff for every $s_1 \in R_i^+(s)$ there exists at least one $s_2 \in R_j(s_1)$ such that $s_2 \in \Phi_i^{-,=}(s)$ (or $s \succeq_i s_2$) and $s_3 \in \Phi_i^{-,=}(s)$ (or $s \succeq_i s_3$) for every $s_3 \in R_i(s_2)$.

Example 4.4 (*SMR Stability for the Sustainable Development Model*) By comparing Definitions 4.2–4.4, one can see that if $R_i^+(s) = \emptyset$, then s is Nash stable, GMR stable, and SMR stable for DM i. Therefore, for instance, s_3 is SMR stable for DM 2 for the sustainable development model. Let us determine whether s_3 is SMR stable for DM 1. DM 1 has a unilateral improvement from s_3 to s_1 and DM 2 has a unilateral move from s_1 to s_2, then DM 1 has only a unilateral move from s_2 to s_4. Because s_2 and s_4 are less preferred than s_3 for DM 1 and, hence, s_3 is SMR stable for DM 1 by Definition 4.4. The stabilities of other three states for the two DMs can be assessed, similarly. SMR stability results are obtained and listed in Table 4.3, where, similarly, "s" indicates SMR stable, "u" indicates SMR unstable, and "Eq" indicates a SMR equilibrium.

SEQ is similar to GMR, but includes only sanctions that are "credible". A credible action is a unilateral improvement.

Definition 4.5 State s is SEQ stable for DM i, denoted by $s \in S_i^{SEQ}$, iff for every $s_1 \in R_i^+(s)$ there exists at least one $s_2 \in R_j^+(s_1)$ with $s_2 \in \Phi_i^{-,=}(s)$ (or $s \succeq_i s_2$).

Example 4.5 (*SEQ Stability for the Sustainable Development Model*) Similar to GMR stability, if $R_i^+(s) = \emptyset$, then s is SEQ stable for DM i. Therefore, for the

Table 4.3 SMR stability of the sustainable development game with simple preferences

State	$R_i^+(s)$		$R_i(s)$		SMR stability		Equilibrium
	DM 1	DM 2	DM 1	DM 2	DM 1	DM 2	
s_1	∅	∅	$\{s_3\}$	$\{s_2\}$	s	s	Eq
s_2	∅	$\{s_1\}$	$\{s_4\}$	$\{s_1\}$	s	u	
s_3	$\{s_1\}$	∅	$\{s_1\}$	$\{s_4\}$	s	s	Eq
s_4	$\{s_2\}$	$\{s_3\}$	$\{s_2\}$	$\{s_3\}$	u	u	

Table 4.4 SEQ stability of the sustainable development game with simple preferences

State	$R_i^+(s)$		$R_i(s)$		SEQ stability		Equilibrium
	DM 1	DM 2	DM 1	DM 2	DM 1	DM 2	
s_1	∅	∅	$\{s_3\}$	$\{s_2\}$	s	s	Eq
s_2	∅	$\{s_1\}$	$\{s_4\}$	$\{s_1\}$	s	u	
s_3	$\{s_1\}$	∅	$\{s_1\}$	$\{s_4\}$	u	s	
s_4	$\{s_2\}$	$\{s_3\}$	$\{s_2\}$	$\{s_3\}$	u	u	

sustainable development model, s_3 is SEQ stable for DM 2. Let us analyze SEQ stability of s_3 for DM 1. DM 1 has a unilateral improvement from s_3 to s_1, but DM 2 has no any unilateral improvement from s_1. Hence, s_3 is GMR and SMR stable for DM 1 rather than SEQ stable. Similarly, one can assess whether other three states are SEQ stable for the two DMs by Definition 4.5. SEQ stability results are listed in Table 4.4 in which "s" indicates SEQ stable, "u" indicates SEQ unstable, and "Eq" indicates a SEQ equilibrium.

4.2.2 *Reachable Lists of a Coalition of Decision Makers*

Any nonempty subset H of DMs, $H \subseteq N$ and $H \neq \emptyset$, is called a coalition. If $|H| = 1$, then the coalition H is trivial; if $|H| > 1$, then the coalition H is nontrivial. (Here, $|H|$ denotes the cardinality of H.) Within an n-DM model ($n \geq 2$), an important coalition is the set of opponents of a fixed DM i, namely $N\backslash\{i\}$, where \ refers to "set subtraction". In order to analyze the stability of a state for DM $i \in N$, it is necessary to take into account possible responses by all other DMs $j \in N\backslash\{i\}$. The essential inputs of stability analysis are reachable lists of group $N\backslash\{i\}$ from state s, $R_{N\backslash\{i\}}(s)$ and $R^+_{N\backslash\{i\}}(s)$ for simple preference. For a two-DM model, DM i has only an opponent, DM j, so i's opponent's reachable lists from s are the states reachable by DM j's moves. In an n-DM model ($n > 2$), the opponents of a DM constitute a coalition of two or more DMs and the determination of their reachable lists is more subtle. The definition of a legal sequence of UMs for a nontrivial coalition is given first.

A legal sequence of UMs for a coalition of DMs is a sequence of states linked by unilateral moves by members of the coalition, in which a DM may move more than once, but not twice consecutively. In general, a DM's directed graph can be transitive or intransitive within the GMCR paradigm. When, for example, a DM can move from s_1 to s_2 and s_2 to s_3 in one step, moves are transitive if the DM can also move in one step from s_1 to s_3. If this is not possible, the move is intransitive. Hence, the restriction of non-successive-moves by the same DM means that GMCR can handle intransitive moves, in addition to transitive moves. Let the coalition $H \subseteq N$ satisfy $|H| \geq 2$ and let the status quo state be $s \in S$. Let $R_H(s) \subseteq S$ (defined formally below) denote the set of states that can be reached by any legal sequence of UMs, by some or all DMs in H, starting at state s. If $s_1 \in R_H(s)$, then $\Omega_H(s, s_1)$ (also defined formally below) denotes the set of all last DMs in legal sequences from s to s_1. The formal definition of $R_H(s) \subseteq S$ and $\Omega_H(s, s_1) \subseteq H$ for $s_1 \in R_H(s)$ is given as follows:

Definition 4.6 A unilateral move by H is a member of $R_H(s) \subseteq S$, defined inductively by

(1) assuming $\Omega_H(s, s_1) = \emptyset$ for all $s_1 \in S$;
(2) if $j \in H$ and $s_1 \in R_j(s)$, then $s_1 \in R_H(s)$ and $\Omega_H(s, s_1) = \Omega_H(s, s_1) \cup \{j\}$;
(3) if $s_1 \in R_H(s)$, $j \in H$, and $s_2 \in R_j(s_1)$, then, provided $\Omega_H(s, s_1) \neq \{j\}$, $s_2 \in R_H(s)$ and $\Omega_H(s, s_2) = \Omega_H(s, s_2) \cup \{j\}$.

Note that this definition is inductive: first, using (2), the states reachable from s are identified and added to $R_H(s)$; then, using (3), all states reachable from those states are identified and added to $R_H(s)$; then the process is repeated until no further states can be added to $R_H(s)$ and there is no change in $\Omega_H(s, s_2)$ for any $s_2 \in R_H(s)$. Because $R_H(s) \subseteq S$ and S is finite, this limit must be reached in finitely many steps.

To interpret Definition 4.6, note that if $s_1 \in R_H(s)$, then $\Omega_H(s, s_1) \subseteq H$ is the set of all last DMs in legal sequences from s to s_1. (If $s_1 \notin R_H(s)$, it can be assumed that $\Omega_H(s, s_1) = \emptyset$.) Suppose that $\Omega_H(s, s_1)$ contains only one DM, say $j \in N$. Then any move from s_1 to a subsequent state, say s_2, must be made by a member of H other than j; otherwise DM j would have to move twice in succession. On the other hand, if $|\Omega_H(s, s_1)| \geq 2$, any member of H who has a unilateral move from s_1 to s_2 may exercise it.

A legal sequence of UIs for a coalition can be defined similarly. Let $R_H^+(s) \subseteq S$ (defined formally below) denote the set of states that can be reached by any legal sequence of UIs, by some or all DMs in H, starting at state s. If $s_1 \in R_H^+(s)$, then $\Omega_H^+(s, s_1)$ (also defined formally below) denotes the set of all last DMs in legal sequences from s to s_1 by UIs. The formal definition of $R_H^+(s) \subseteq S$ and $\Omega_H^+(s, s_1) \subseteq H$ for $s_1 \in R_H^+(s)$ is given as follows:

Definition 4.7 Let $s \in S$, $H \subseteq N$, and $H \neq \emptyset$. A unilateral improvement by H is a member of $R_H^+(s) \subseteq S$, defined inductively by

(1) assuming $\Omega_H^+(s, s_1) = \emptyset$ for all $s_1 \in S$;
(2) if $j \in H$ and $s_1 \in R_j^+(s)$, then $s_1 \in R_H^+(s)$ and $\Omega_H^+(s, s_1) = \Omega_H^+(s, s_1) \cup \{j\}$;

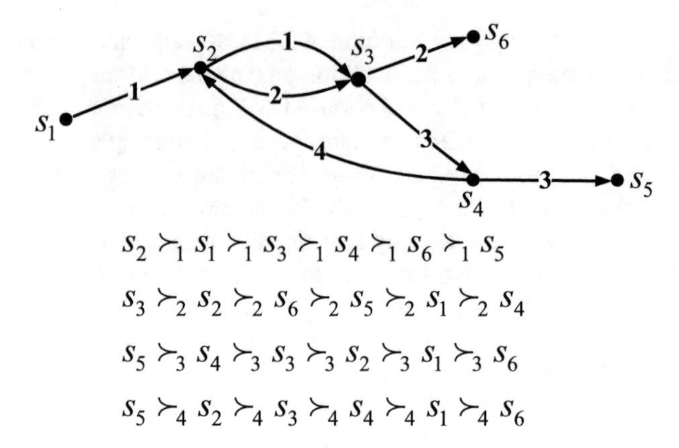

Fig. 4.3 Graph model with four DMs and six states

(3) if $s_1 \in R_H^+(s)$, $j \in H$, and $s_2 \in R_j^+(s_1)$, then, provided $\Omega_H^+(s, s_1) \neq \{j\}$, $s_2 \in R_H^+(s)$ and $\Omega_H^+(s, s_2) = \Omega_H^+(s, s_2) \cup \{j\}$.

Definition 4.7 is identical to Definition 4.6 except that all moves are required to be UIs, i.e. each move is to a state strictly preferred by the mover to the current state. Similarly, $\Omega_H^+(s, s_1)$ includes all last movers in UIs by H from state s to state s_1. An example that shows the procedures to construct the reachable lists of a group is presented as follows:

Example 4.6 (*Constructing Reachable Lists of a Coalition*) Figure 4.3 shows a graph model with DM set $N = \{1, 2, 3, 4\}$ and state set $S = \{s_1, s_2, s_3, s_4, s_5, s_6\}$. The labels on the arcs of the graph indicate the controlling DMs. Preference information is given below the directed graph. If $s = s_1$ is selected as the status quo state, then the reachable lists of $H = N$ from s, $R_N(s_1)$ and $R_N^+(s_1)$, can be constructed according to Definitions 4.6 and 4.7.

Constructing $R_N(s_1)$:

1. s_2 can be reached by DM 1 from s_1 by one step UM, so $s_2 \in R_N(s_1)$;
2. s_3 cannot be attained by DM 1 from s_2 since DM 1 cannot move twice consecutively;
3. s_3 can be reached by DM 2 from s_2 by one step UM, so $s_3 \in R_N(s_1)$;
4. s_6 cannot be attained by DM 2 from s_3 since DM 2 cannot move twice consecutively;
5. s_4 can be reached by DM 3 from s_3 by one step UM, so $s_4 \in R_N(s_1)$;
6. s_5 cannot be attained by DM 3 from s_4 since DM 3 cannot move twice consecutively;
7. s_2 is reachable again by DM 4 from s_4 by one step UM, then s_3 is reachable again by DM 1 from s_2, and s_6 is finally reachable by DM 2 from s_3, so $s_6 \in R_N(s_1)$.

Accordingly, $R_N(s_1) = \{s_2, s_3, s_4, s_6\}$.

Constructing $R_N^+(s_1)$:

From the preference information provided, $A_1^+ = \{(s_1, s_2)\}$, $A_2^+ = \{(s_2, s_3)\}$, $A_3^+ = \{(s_3, s_4), (s_4, s_5)\}$, and $A_4^+ = \{(s_4, s_2)\}$.

1. s_2 can be reached by DM 1 from s_1 by a UI, so $s_2 \in R_N^+(s_1)$;
2. s_3 can be reached by DM 2 from s_2 by a UI, so $s_3 \in R_N^+(s_1)$;
3. s_4 can be reached by DM 3 from s_3 by a UI, so $s_4 \in R_N^+(s_1)$;
4. s_5 cannot be attained by DM 3 from s_4 since DM 3 cannot move twice consecutively.

Therefore, $R_N^+(s_1) = \{s_2, s_3, s_4\}$.

The four basic stabilities of Nash, GMR, SMR, and SEQ with simple preference in two-DM models, described using logical representation in Sect. 4.2.1, can be extended to models including more than two DMs, which is the objective of the next subsection.

4.2.3 *n*-Decision Maker Case

In an n-DM model, where $n > 2$, the opponents of a DM can be thought of as the coalition of all other DMs. To calculate the stability of a state for DM $i \in N$, it is necessary to examine possible responses by this coalition, $N \setminus \{i\}$ from the states in $R_{N\setminus\{i\}}(s)$ or $R_{N\setminus\{i\}}^+(s)$. Let $i \in N$ and $s \in S$ in the following definitions.

Nash stability definition is identical for two-DM and n-DM models because this formal stability does not consider the opponents' responses.

Definition 4.8 State s is Nash stable for DM i, denoted by $s \in S_i^{Nash}$, iff $R_i^+(s) = \emptyset$.

For GMR stability, DM i expects that its opponents, $N \setminus \{i\}$, will respond to any unilateral improvement by i from s to s_1 with a sequence of legal unilateral moves to a state in $R_{N\setminus\{i\}}(s_1)$, so as to hurt i if possible. As before, i anticipates that the conflict will end after the opponents have responded.

Definition 4.9 State s is GMR for DM i, denoted by $s \in S_i^{GMR}$, iff for every $s_1 \in R_i^+(s)$ there exists at least one $s_2 \in R_{N\setminus\{i\}}(s_1)$ with $s \succeq_i s_2$.

As in the two-DM case, for SMR stability, DM i expects to have a chance to counterrespond (s_3) to its opponents' response (s_2) to i's original move.

Definition 4.10 State s is SMR for DM i, denoted by $s \in S_i^{SMR}$, iff for every $s_1 \in R_i^+(s)$ there exists at least one $s_2 \in R_{N\setminus\{i\}}(s_1)$ such that $s \succeq_i s_2$ and $s \succeq_i s_3$ for every $s_3 \in R_i(s_2)$.

A state is SEQ stable for a given DM iff the DM can be deterred by subsequent unilateral improvements by its opponents.

Definition 4.11 State s is SEQ for DM i, denoted by $s \in S_i^{SEQ}$, iff for every $s_1 \in R_i^+(s)$ there exists at least one $s_2 \in R_{N\setminus\{i\}}^+(s_1)$ with $s \succeq_i s_2$.

SEQ stability indicates that all UIs of the focal DM are sanctioned by a subsequent group unilateral improvement by the DM's opponents.

Definitions 4.8–4.11 cover Nash stability, GMR, SMR, and SEQ in the graph model for multiple-decision-maker conflict models (or, simply, n-DM models) with simple preference. These definitions retain the features of Definitions 4.2–4.5, in the two-DM case, except that DM i's opponents are a subset of N, instead of a single opponent. When $n = 2$, the DM set N is $\{i, j\}$, so that the reachable list of coalition $N \setminus \{i\}$ from s_1, $R_{N\setminus\{i\}}(s_1)$, reduces to DM j's reachable list from s_1, $R_j(s_1)$.

4.2.4 *Interrelationships Among Stability Definitions*

Fang et al. (1989, 1993) established general relationships among Nash, GMR, SMR, and SEQ solution concepts as shown in Fig. 4.4. The following theorem demonstrates that the same relationships hold for these solution concepts in both two-DM and n-DM models.

Theorem 4.1 *Let $i \in N$, $|N| = n$, and $n \geq 2$. Then the stable states under the four basic stability definitions satisfy*

$$S_i^{Nash} \subseteq S_i^{SMR} \subseteq S_i^{GMR} \tag{4.1}$$

and

$$S_i^{Nash} \subseteq S_i^{SEQ} \subseteq S_i^{GMR}. \tag{4.2}$$

Proof The inclusion relations presented in Eq. 4.1 will be proven. The proof for Eq. 4.2 is similar.

If $s \in S_i^{Nash}$, then $R_i^+(s) = \emptyset$, so using Definition 4.10, $s \in S_i^{SMR}$. Hence, $S_i^{Nash} \subseteq S_i^{SMR}$.

For any $s \in S_i^{SMR}$, if $R_i^+(s) = \emptyset$, then $s \in S_i^{GMR}$. Otherwise, for any $s_1 \in R_i^+(s)$ there exists at least one $s_2 \in R_{N\setminus\{i\}}(s_1)$ such that $s \succeq_i s_2$ and $s \succeq_i s_3$ for every

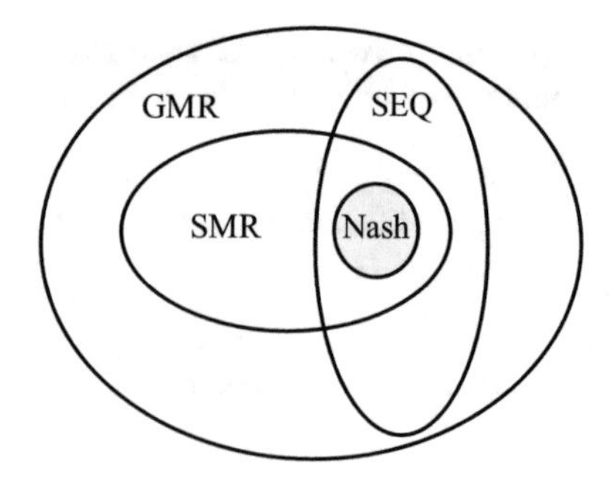

Fig. 4.4 Interrelationships among the solution concepts

Table 4.5 Summary of stability results for the sustainable development game with simple preferences

State number	Nash			GMR			SMR			SEQ		
	DM 1	DM 2	Eq	DM 1	DM 2	Eq	DM 1	DM 2	Eq	DM 1	DM 2	Eq
s_1	s	s	√	s	s	√	s	s	√	s	s	√
s_2	s	u		s	u		s	u		s	u	
s_3	u	s		s	s	√	s	s	√	u	s	
s_4	u	u		u	u		u	u		u	u	

$s_3 \in R_i(s_2)$, by Definition 4.10. Accordingly, for every $s_1 \in R_i^+(s)$ there exists $s_2 \in R_{N\setminus\{i\}}(s_1)$, such that $s \succeq_i s_2$. This implies that $S_i^{SMR} \subseteq S_i^{GMR}$. Therefore, the inclusion relations $S_i^{Nash} \subseteq S_i^{SMR} \subseteq S_i^{GMR}$ hold. □

There is no necessary inclusion relation between S_i^{SMR} and S_i^{SEQ}, i.e., it may be true that $S_i^{SMR} \supseteq S_i^{SEQ}$, or that $S_i^{SMR} \subseteq S_i^{SEQ}$, or neither.

The sustainable development model is utilized to illustrate the interrelationships among the four stabilities, Nash, GMR, SMR, and SEQ. Stability results for the four solution concepts are summarized in Table 4.5, where "Eq" means "equilibrium", "s" indicates stable, "u" indicates unstable, and "√" indicates an equilibrium for some solution concept. In fact, one can also utilize Theorem 4.1 to help determine some stabilities. For example, $R_i^+(s_1) = \emptyset$ for $i = 1, 2$, so s_1 is Nash stable for the two DMs. Therefore, s_1 is GMR, SMR, and SEQ for the two DMs by Eqs. 4.1 and 4.2 in Theorem 4.1. Because s_2 is GMR unstable for DM 2, then it is certain that s_2 is SMR unstable for DM 2 from the relation Eq. 4.1 in Theorem 4.1. In this case, s_3 is SMR but not SEQ for DM 2.

Theorem 4.1 examines the relationships of individual stability definitions from a single DM's viewpoint. Recall that a possible resolution or equilibrium of a graph model is a state that all DMs find stable under appropriate stability definitions. Hence, Theorem 4.1 implies that the same relationships hold for equilibria. Let $S^{Nash} = \bigcap_{i \in N} S_i^{Nash}$, $S^{GMR} = \bigcap_{i \in N} S_i^{GMR}$, $S^{SMR} = \bigcap_{i \in N} S_i^{SMR}$, and $S^{SEQ} = \bigcap_{i \in N} S_i^{SEQ}$ denote the equilibrium sets under the four stability definitions, respectively. The following theorem is immediate.

Theorem 4.2 *Let $i \in N$, $|N| = n$, and $n \geq 2$. Then the equilibria under the four basic stability definitions satisfy*

$$S^{Nash} \subseteq S^{SMR} \subseteq S^{GMR} \tag{4.3}$$

and

$$S^{Nash} \subseteq S^{SEQ} \subseteq S^{GMR}. \tag{4.4}$$

4.3 Matrix Representation of Stability Definitions

Stability definitions in the graph model are traditionally defined logically, in terms of the underlying graphs and preference relations, as in Sect. 4.2. However, as noted in the development of the DSS GMCR II (Fang et al. 2003a, b), the nature of logical representations makes coding difficult. The new preference structure proposed by Li et al. (2004) to represent uncertainty in DMs' preferences included some extensions of the four stability definitions, and algorithms were outlined but they have not been coded. The work of Hamouda et al. (2004, 2006) integrated strength of preference information into these four solution concepts but, again, it proved difficult to code and has not been integrated into GMCR II. Then, difficulties in coding, mainly because of the logical formulation, were the primary motivation for the development of explicit matrix representations of the graph model with simple preference. In the following subsection, matrix expressions are used to capture the relative preferences and reachable lists of a single DM, both by UMs and by UIs.

4.3.1 Preference Matrices and UM and UI Matrices

Let $m = |S|$ denote the number of states. For a graph model, several matrices can represent relative preference relations between two states.

Definition 4.12 For a graph model G, the preference matrix and the indifference matrix for DM i are $m \times m$ matrices, P_i^+ and $P_i^=$, with (s, q) entries

$$P_i^+(s,q) = \begin{cases} 1 & \text{if } q \succ_i s, \\ 0 & \text{otherwise}, \end{cases}$$

and

$$P_i^=(s,q) = \begin{cases} 1 & \text{if } q \sim_i s \text{ and } q \neq s, \\ 0 & \text{otherwise}. \end{cases}$$

A nonzero entry $P_i^+(s,q) = 1$ in the preference matrix indicates that DM i prefers state q to state s, while zero entry $P_i^+(s,q) = 0$ indicates that DM i either prefers s to q or is indifferent between s and q (by the property of preference completeness). Similarly, $P_i^=(s,q) = 1$ implies that DM i is indifferent between s and q while $P_i^=(s,q) = 0$ denotes that DM i prefers either s to q or q to s. It is convenient to define

$$P_i^{-,=} = E - I - P_i^+,$$

where E is the $m \times m$ unit matrix (all entries 1) and I is the $m \times m$ identity matrix. Note that $P_i^{-,=}(s,q) = 1$ means that DM i does not prefer state q to state s.

For $i \in N$ and $s \in S$, DM i's unilateral moves (UMs) and unilateral improvements (UIs) can be represented as follows:

Definition 4.13 For a graph model G, DM i's UM and UI matrices are the $m \times m$ matrices, J_i and J_i^+, with (s, q) entries

$$J_i(s,q) = \begin{cases} 1 & \text{if } (s,q) \in A_i, \\ 0 & \text{otherwise,} \end{cases}$$

and

$$J_i^+(s,q) = \begin{cases} 1 & \text{if } (s,q) \in A_i \text{ and } q \succ_i s, \\ 0 & \text{otherwise.} \end{cases}$$

Note that $J_i(s, q) = 1$ if and only if DM i can move from state s to state q (in one step). In other words, $(s, q) \in A_i$. Also, $J_i^+(s, q) = 1$ iff $J_i(s, q) = 1$ and DM i prefers q to s.

The set $R_i(s) = \{q \in S : J_i(s, q) = 1\}$ is DM i's *reachable list* from state s by UMs. It contains all states to which DM i can make unilateral moves from state s in one step. Similarly, $R_i^+(s) = \{q \in S : J_i^+(s, q) = 1\}$ is DM $i's$ reachable list from s by UIs. Clearly, $R_i^+(s)$ is identical to $R_i(s)$ except that all moves are required to be UIs. Note that, if $R_i(s)$ and $R_i^+(s)$ are written as 0–1 row vectors, then

$$R_i(s) = e_s^T \cdot J_i \text{ and } R_i^+(s) = e_s^T \cdot J_i^+,$$

where e_s^T denotes the transpose of the sth standard basis column vector of the m-dimensional Euclidean space, $\mathbb{R}^S$.

The definitions of DM i's UM matrix, J_i, UI matrix, J_i^+, and preference matrix, P_i^+, imply that

$$J_i^+ = J_i \circ P_i^+, \tag{4.5}$$

where "$\circ$" denotes the Hadamard product.

The objective of the next subsection is to develop an explicit algebraic version of a graph model, to facilitate stability calculations. The two-DM models are considered first. One will see that matrix representation of solution concepts (MRSC) is feasible for four graph model stability definitions in the two-DM graph model. Using explicit matrix formulations instead of graphical or logical representations makes MRSC more effective and convenient for calculating stabilities and identifying equilibria.

4.3.2 Two Decision Maker Case

The matrix representation of Nash, GMR, SMR, and SEQ stabilities in two-DM conflict models with simple preference is developed in this subsection. The system, called the MRSC method, incorporates a set of $m \times m$ matrices, M_i^{Nash}, M_i^{GMR},

M_i^{SMR}, and M_i^{SEQ}, to capture Nash, GMR, SMR, and SEQ for DM $i \in N$, where $m = |S|$. For now, $|N| = 2$.

From Definition 4.2, state s is Nash stable for DM i iff DM i cannot move from s to any state i prefers. Define DM i's Nash matrix as the $m \times m$ matrix $M_i^{Nash} = J_i^+ \cdot E$ ("E" denotes the $m \times m$ matrix with each entry being set to 1). The diagonal element of M_i^{Nash} matrix at (s, s) is

$$M_i^{Nash}(s, s) = e_s^T \cdot J_i^+ \cdot e, \tag{4.6}$$

for $s \in S$, and all off-diagonal entries zero. Here, e is the m-dimensional unit column vector (all elements 1). Then the following theorem shows how this matrix represents Nash stability.

Theorem 4.3 *State $s \in S$ is Nash stable for DM i iff $M_i^{Nash}(s, s) = 0$.*

Proof By Eq. 4.6, $M_i^{Nash}(s, s) = 0$ holds iff

$$e_s^T \cdot J_i^+ = 0^T.$$

According to the definition of DM i's UI matrix, $M_i^{Nash}(s, s) = 0$ iff $R_i^+(s) = \emptyset$, which is the definition of Nash stability for DM i given in Definition 4.2. □

Note that Theorem 4.3 provides a matrix method to assess whether state s is Nash stable for DM i by identifying the Nash matrix's diagonal entry $M_i^{Nash}(s, s)$. If the sth diagonal entry is zero, then s is Nash stable for DM i; otherwise s is Nash unstable for DM i. This matrix representation of Nash stability will be adapted to the other basic stability definitions.

Example 4.7 (*Matrix Representation of Nash Stability for the Sustainable Development Model*) The logical representation of Nash stability for the sustainable development game was presented in Example 4.2. For the graph model of the sustainable development conflict presented in Fig. 3.2, the UM matrices for DM 1 and DM 2 are

$$J_1 = \begin{pmatrix} 0 & 0 & 1 & 0 \\ 0 & 0 & 0 & 1 \\ 1 & 0 & 0 & 0 \\ 0 & 1 & 0 & 0 \end{pmatrix} \text{ and } J_2 = \begin{pmatrix} 0 & 1 & 0 & 0 \\ 1 & 0 & 0 & 0 \\ 0 & 0 & 0 & 1 \\ 0 & 0 & 1 & 0 \end{pmatrix}. \tag{4.7}$$

According to Definition 4.12, the preference matrices are

$$P_1^+ = \begin{pmatrix} 0 & 0 & 0 & 0 \\ 1 & 0 & 1 & 0 \\ 1 & 0 & 0 & 0 \\ 1 & 1 & 1 & 0 \end{pmatrix} \text{ and } P_2^+ = \begin{pmatrix} 0 & 0 & 1 & 0 \\ 1 & 0 & 1 & 1 \\ 0 & 0 & 0 & 0 \\ 1 & 0 & 1 & 0 \end{pmatrix}. \tag{4.8}$$

Accordingly, one uses $J_i^+ = J_i \circ P_i^+$, for $i = 1, 2$, to obtain

$$J_1^+ = \begin{pmatrix} 0 & 0 & 0 & 0 \\ 0 & 0 & 0 & 0 \\ 1 & 0 & 0 & 0 \\ 0 & 1 & 0 & 0 \end{pmatrix} \text{ and } J_2^+ = \begin{pmatrix} 0 & 0 & 0 & 0 \\ 1 & 0 & 0 & 0 \\ 0 & 0 & 0 & 0 \\ 0 & 0 & 1 & 0 \end{pmatrix}. \tag{4.9}$$

Then, from Eq. 4.6, the Nash matrices are

$$M_1^{Nash} = \begin{pmatrix} 0 & 0 & 0 & 0 \\ 0 & 0 & 0 & 0 \\ 0 & 0 & 1 & 0 \\ 0 & 0 & 0 & 1 \end{pmatrix} \text{ and } M_2^{Nash} = \begin{pmatrix} 0 & 0 & 0 & 0 \\ 0 & 1 & 0 & 0 \\ 0 & 0 & 0 & 0 \\ 0 & 0 & 0 & 1 \end{pmatrix}.$$

Since $M_1^{Nash}(1,1) = M_1^{Nash}(2,2) = 0$ and $M_1^{Nash}(3,3) = M_1^{Nash}(4,4) = 1$, then s_1 and s_2 are Nash stable, and s_3 and s_4 are Nash unstable for DM 1, according to Theorem 4.3. Similarly, because $M_2^{Nash}(1,1) = M_2^{Nash}(3,3) = 0$ and $M_2^{Nash}(2,2) = M_2^{Nash}(4,4) = 1$, s_1 and s_3 are Nash stable, and s_2 and s_4 are Nash unstable, for DM 2. The results are identical to those in Example 4.2 obtained by logical representation.

A state $s \in S$ is general metarational for DM i iff whenever DM i makes any UI from s, then its opponent can hurt i in response. Define DM i's $m \times m$ GMR matrix as

$$M_i^{GMR} = J_i^+ \cdot [E - sign\left(J_j \cdot (P_i^{-,=})^T\right)], \tag{4.10}$$

where $j \in N$, $j \neq i$. The following theorem establishes the matrix method to assess whether state s is GMR stable for a DM.

Theorem 4.4 *State $s \in S$ is GMR for DM i iff $M_i^{GMR}(s,s) = 0$.*

Proof Since

$$M_i^{GMR}(s,s) = (e_s^T \cdot J_i^+) \cdot [\left(E - sign\left(J_j \cdot (P_i^{-,=})^T\right)\right) \cdot e_s]$$

$$= \sum_{s_1=1}^{m} J_i^+(s,s_1) \cdot [1 - sign\left((e_{s_1}^T \cdot J_j) \cdot (e_s^T \cdot P_i^{-,=})^T\right)],$$

then $M_i^{GMR}(s,s) = 0$ holds iff

$$J_i^+(s,s_1) \cdot [1 - sign\left((e_{s_1}^T \cdot J_j) \cdot (e_s^T \cdot P_i^{-,=})^T\right)] = 0, \tag{4.11}$$

for every $s_1 \in S$. It is clear that Eq. 4.11 is equivalent to

$$(e_{s_1}^T \cdot J_j) \cdot (e_s^T \cdot P_i^{-,=})^T \neq 0,$$

for every $s_1 \in R_i^+(s)$. Therefore, for any $s_1 \in R_i^+(s)$, there exists at least one $s_2 \in R_j(s_1)$ with $s \succeq_i s_2$. According to Definition 4.3, $M_i^{GMR}(s, s) = 0$ implies that s is GMR stable for DM i. □

Theorem 4.4 proves that this matrix method, called matrix representation of GMR stability, is equivalent to the logical representation for two-DM GMR stability in Definition 4.3. To analyze GMR stability at s for DM i, one only needs to identify whether the diagonal entry $M_i^{GMR}(s, s)$ of i's GMR matrix is zero. If so, s is GMR stable for i; otherwise, s is GMR unstable for DM i. Note that all information about GMR stability is contained in the diagonal entries of the GMR matrix.

Example 4.8 (*Matrix Representation of GMR Stability for the Sustainable Development Model*) The logical representation of GMR stability for the sustainable development game was illustrated in Example 4.3. First, one uses $P_i^{-,=} = E - I - P_i^+$ for $i = 1, 2$, to obtain

$$P_1^{-,=} = \begin{pmatrix} 1 & 1 & 1 & 1 \\ 1 & 1 & 1 & 1 \\ 1 & 1 & 1 & 1 \\ 1 & 1 & 1 & 1 \end{pmatrix} - \begin{pmatrix} 1 & 0 & 0 & 0 \\ 0 & 1 & 0 & 0 \\ 0 & 0 & 1 & 0 \\ 0 & 0 & 0 & 1 \end{pmatrix} - \begin{pmatrix} 0 & 0 & 0 & 0 \\ 1 & 0 & 1 & 0 \\ 1 & 0 & 0 & 0 \\ 1 & 1 & 1 & 0 \end{pmatrix} = \begin{pmatrix} 0 & 1 & 1 & 1 \\ 0 & 0 & 0 & 1 \\ 0 & 1 & 0 & 1 \\ 0 & 0 & 0 & 0 \end{pmatrix}, \tag{4.12}$$

and

$$P_2^{-,=} = \begin{pmatrix} 1 & 1 & 1 & 1 \\ 1 & 1 & 1 & 1 \\ 1 & 1 & 1 & 1 \\ 1 & 1 & 1 & 1 \end{pmatrix} - \begin{pmatrix} 1 & 0 & 0 & 0 \\ 0 & 1 & 0 & 0 \\ 0 & 0 & 1 & 0 \\ 0 & 0 & 0 & 1 \end{pmatrix} - \begin{pmatrix} 0 & 0 & 1 & 0 \\ 1 & 0 & 1 & 1 \\ 0 & 0 & 0 & 0 \\ 1 & 0 & 1 & 0 \end{pmatrix} = \begin{pmatrix} 0 & 1 & 0 & 1 \\ 0 & 0 & 0 & 0 \\ 1 & 1 & 0 & 1 \\ 0 & 1 & 0 & 0 \end{pmatrix}. \tag{4.13}$$

From Eq. 4.10, DM i's GMR matrix is

$$M_i^{GMR} = J_i^+ \cdot \left(E - sign\left(J_j \cdot (P_i^{-,=})^T\right)\right),$$

where $i, j = 1, 2$. Therefore,

$$M_1^{GMR} = \begin{pmatrix} 0 & 0 & 0 & 0 \\ 0 & 0 & 0 & 0 \\ 1 & 0 & 0 & 0 \\ 0 & 1 & 0 & 0 \end{pmatrix} \cdot \left(\begin{pmatrix} 1 & 1 & 1 & 1 \\ 1 & 1 & 1 & 1 \\ 1 & 1 & 1 & 1 \\ 1 & 1 & 1 & 1 \end{pmatrix} - sign\left(\begin{pmatrix} 0 & 1 & 0 & 0 \\ 1 & 0 & 0 & 0 \\ 0 & 0 & 0 & 1 \\ 0 & 0 & 1 & 0 \end{pmatrix} \cdot \begin{pmatrix} 0 & 1 & 1 & 1 \\ 0 & 0 & 0 & 1 \\ 0 & 1 & 0 & 1 \\ 0 & 0 & 0 & 0 \end{pmatrix}^T \right) \right)$$

$$= \begin{pmatrix} 0 & 0 & 0 & 0 \\ 0 & 0 & 0 & 0 \\ 0 & 1 & 0 & 1 \\ 1 & 1 & 1 & 1 \end{pmatrix}.$$

Similarly, DM 2's GMR matrix is calculated by

$$M_2^{GMR} = \begin{pmatrix} 0 & 0 & 0 & 0 \\ 1 & 0 & 0 & 0 \\ 0 & 0 & 0 & 0 \\ 0 & 0 & 1 & 0 \end{pmatrix} \cdot \left(\begin{pmatrix} 1 & 1 & 1 & 1 \\ 1 & 1 & 1 & 1 \\ 1 & 1 & 1 & 1 \\ 1 & 1 & 1 & 1 \end{pmatrix} - sign \left(\begin{pmatrix} 0 & 0 & 1 & 0 \\ 0 & 0 & 0 & 1 \\ 1 & 0 & 0 & 0 \\ 0 & 1 & 0 & 0 \end{pmatrix} \cdot \begin{pmatrix} 0 & 1 & 0 & 1 \\ 0 & 0 & 0 & 0 \\ 1 & 1 & 0 & 1 \\ 0 & 1 & 0 & 0 \end{pmatrix}^T \right) \right)$$

$$= \begin{pmatrix} 0 & 0 & 0 & 0 \\ 1 & 1 & 1 & 1 \\ 0 & 0 & 0 & 0 \\ 1 & 1 & 0 & 1 \end{pmatrix}.$$

Since $M_1^{GMR}(1, 1) = M_1^{GMR}(2, 2) = M_1^{GMR}(3, 3) = 0$ and $M_1^{GMR}(4, 4) \neq 0$, then s_1, s_2, and s_3 are GMR stable, while s_4 is GMR unstable, for DM 1, according to Theorem 4.4. Similarly, because $M_2^{GMR}(1, 1) = M_2^{GMR}(3, 3) = 0$ and $M_2^{GMR}(2, 2) = M_2^{GMR}(4, 4) = 1$, s_1 and s_3 are GMR stable, and s_2 and s_4 are GMR unstable, for DM 2. These results are identical to those in Example 4.3 obtained by logical representation.

Symmetric metarationality is similar to general metarationality except that DM i expects to have a chance to counterrespond to its opponent j's response to i's original move. Define DM i's SMR $m \times m$ matrix as

$$M_i^{SMR} = J_i^+ \cdot [E - sign(Q)]$$

in which

$$Q = J_j \cdot [(P_i^{-,=})^T \circ \left(E - sign\left(J_i \cdot (P_i^+)^T\right)\right)],$$

for $j \in N$, $j \neq i$. The following theorem establishes the matrix method to determine whether state s is SMR stable for a DM.

Theorem 4.5 *State $s \in S$ is SMR for DM i iff $M_i^{SMR}(s, s) = 0$.*

Proof Since

$$M_i^{SMR}(s, s) = (e_s^T \cdot J_i^+) \cdot [(E - sign(Q)) \cdot e_s]$$

$$= \sum_{s_1=1}^{m} J_i^+(s, s_1)[1 - sign\,(Q(s_1, s))]$$

with

$$Q(s_1, s) = \sum_{s_2=1}^{m} J_j(s_1, s_2) \cdot W,$$

and

$$W = P_i^{-,=}(s, s_2) \cdot \left[1 - sign\left(\sum_{s_3=1}^{m}\left(J_i(s_2, s_3) \cdot P_i^{+}(s, s_3)\right)\right)\right],$$

then $M_i^{SMR}(s, s) = 0$ holds iff $Q(s_1, s) \neq 0$, for every $s_1 \in R_i^{+}(s)$, which is equivalent to the statement that, for every $s_1 \in R_i^{+}(s)$, there exists $s_2 \in R_j(s_1)$ such that

$$P_i^{-,=}(s, s_2) \neq 0, \tag{4.14}$$

and

$$\sum_{s_3=1}^{m} J_i(s_2, s_3) \cdot P_i^{+}(s, s_3) = 0. \tag{4.15}$$

Obviously, for every $s_1 \in R_i^{+}(s)$, there exists $s_2 \in R_j(s_1)$ such that Eqs. 4.14 and 4.15 hold iff for every $s_1 \in R_i^{+}(s)$ there exists $s_2 \in R_j(s_1)$ such that $s \succeq_i s_2$ and $s \succeq_i s_3$ for all $s_3 \in R_i(s_2)$. □

Theorem 4.5 proves that this matrix method, called matrix representation of SMR stability, is equivalent to the logical representation for two-DM SMR stability in Definition 4.4. To calculate SMR stability at s for DM i, one only needs to assess whether the diagonal entry $M_i^{SMR}(s, s)$ of i's SMR matrix is zero. If so, s is SMR stable for i; otherwise, s is SMR unstable for DM i.

Example 4.9 (*Matrix Representation of SMR Stability for the Sustainable Development Model*) The logical representation of SMR stability for the sustainable development game was described in Example 4.4. First, one uses Eq. 4.7 for J_i, Eq. 4.8 for P_i^{+}, Eq. 4.9 for J_i^{+}, and Eqs. 4.12 and 4.13 for $P_i^{-,=}$, for $i = 1, 2$. DM i's SMR matrix is

$$M_i^{SMR} = J_i^{+} \cdot [E - sign(Q)]$$

in which

$$Q = J_j \cdot [(P_i^{-,=})^T \circ \left(E - sign\left(J_i \cdot (P_i^{+})^T\right)\right)],$$

where $i, j = 1, 2$. Therefore,

$$M_1^{SMR} = \begin{pmatrix} 0 & 0 & 0 & 0 \\ 0 & 0 & 0 & 0 \\ 0 & 1 & 0 & 1 \\ 1 & 1 & 1 & 1 \end{pmatrix} \text{ and } M_2^{SMR} = \begin{pmatrix} 0 & 0 & 0 & 0 \\ 1 & 1 & 1 & 1 \\ 0 & 0 & 0 & 0 \\ 1 & 1 & 0 & 1 \end{pmatrix}.$$

Since $M_1^{SMR}(1, 1) = M_1^{SMR}(2, 2) = M_1^{SMR}(3, 3) = 0$ and $M_1^{SMR}(4, 4) \neq 0$, then s_1, s_2, and s_3 are SMR stable, while s_4 are SMR unstable, for DM 1, according to Theorem 4.5. Similarly, because $M_2^{SMR}(1, 1) = M_2^{SMR}(3, 3) = 0$ and $M_2^{SMR}(2, 2) = M_2^{SMR}(4, 4) = 1$, s_1 and s_3 are SMR stable, and s_2 and s_4 are SMR unstable, for DM 2. These results are identical to those in Example 4.4 obtained by logical representation.

Sequential stability is similar to general metarationality, but includes only those sanctions that are "credible". Define DM i's SEQ $m \times m$ matrix as

$$M_i^{SEQ} = J_i^+ \cdot [E - sign\left(J_j^+ \cdot (P_i^{-,=})^T\right)],$$

for $j \in N, j \neq i$. The following theorem provides the matrix method to analyze whether state s is SEQ stable for a DM.

Theorem 4.6 *State $s \in S$ is SEQ for DM i iff $M_i^{SEQ}(s, s) = 0$.*

Proof Since

$$M_i^{SEQ}(s, s) = (e_s^T J_i^+) \cdot [\left(E - sign(J_j^+ \cdot (P_i^{-,=})^T)\right) e_s]$$

$$= \sum_{s_1=1}^{|S|} J_i^+(s, s_1)[1 - sign((e_{s_1}^T J_j^+) \cdot (e_s^T P_i^{-,=})^T)],$$

then $M_i^{SEQ}(s, s) = 0$ holds iff

$$J_i^+(s, s_1)[1 - sign\left((e_{s_1}^T J_j^+) \cdot (e_s^T P_i^{-,=})^T\right)] = 0, \forall s_1 \in S. \qquad (4.16)$$

It is clear that Eq. 4.16 is equivalent to

$$(e_{s_1}^T J_j^+) \cdot (e_s^T P_i^{-,=})^T \neq 0, \forall s_1 \in R_i^+(s).$$

It implies that for any $s_1 \in R_i^+(s)$, there exists at least one $s_2 \in R_j^+(s_1)$ with $s \succeq_i s_2$. □

Note that the SEQ matrix is identical to the GMR matrix except that DM j's UM matrix J_j is replaced by the UI matrix J_j^+.

Theorem 4.6 proves that this matrix method, called matrix representation of SEQ stability, is equivalent to the logical representation for two-DM SEQ stability in Definition 4.5. To identify DM i's SEQ stability at s for DM i, one only needs to determine whether the diagonal entry $M_i^{SEQ}(s, s)$ of i's SEQ matrix is zero. If so, s is SEQ stable for i; otherwise, s is SEQ unstable for DM i.

Example 4.10 (Matrix Representation of SEQ Stability for the Sustainable Development Model) The logical representation of SEQ stability for the sustainable development game was presented in Example 4.5. First, one uses Eq. 4.9 to obtain J_i^+, and Eqs. 4.12 and 4.13 for $P_i^{-,=}$, for $i = 1, 2$. DM i's SEQ matrix is

$$J_i^+ \cdot [E - sign\left(J_j^+ \cdot (P_i^{-,=})^T\right)],$$

where $i, j = 1, 2$. Therefore,

$$M_1^{SEQ} = \begin{pmatrix} 0 & 0 & 0 & 0 \\ 0 & 0 & 0 & 0 \\ 1 & 1 & 1 & 1 \\ 1 & 1 & 1 & 1 \end{pmatrix} \text{ and } M_2^{SEQ} = \begin{pmatrix} 0 & 0 & 0 & 0 \\ 1 & 1 & 1 & 1 \\ 0 & 0 & 0 & 0 \\ 1 & 1 & 0 & 1 \end{pmatrix}.$$

Since $M_1^{SEQ}(1, 1) = M_1^{SEQ}(2, 2) = 0$ and $M_1^{SEQ}(3, 3) = M_1^{SEQ}(4, 4) = 1$, then s_1 and s_2 are SEQ stable, while s_3 and s_4 are SEQ unstable, for DM 1. Similarly, because $M_2^{SEQ}(1, 1) = M_2^{SEQ}(3, 3) = 0$ and $M_2^{SEQ}(2, 2) = M_2^{SEQ}(4, 4) = 1$, s_1 and s_3 are SEQ stable, and s_2 and s_4 are SEQ unstable, for DM 2. These results are identical to those in Example 4.5 obtained by logical representation.

4.3.3 *Matrices to Construct Reachable Lists of a Coalition*

The aim of a stability analysis is to find the states of a graph model that are stable for all DMs, under appropriate stability definitions, or equilibria. As discussed in Sect. 4.2.2, the reachable lists of coalition H by sequences of the legal UMs and the legal UIs, $R_H(s)$ and $R_H^+(s)$, are essential ingredients for stability analysis and the construction of these two sets is a complicated process. In this section, the reachability matrices M_H and M_H^+ are proposed to provide an algebraic method of constructing $R_H(s)$ and $R_H^+(s)$ (Xu et al. 2010b).

4.3.3.1 Several Extended Definitions in the Graph Model

First, the adjacency matrix and the incidence matrix of a graph (Godsil and Royle 2001) are extended to a graph model. Let $m = |S|$ and $l = |A|$.

Definition 4.14 For a graph model G, the **UM adjacency matrix** and the **UI adjacency matrix** for $H \subseteq N$ and $H \neq \emptyset$ are $m \times m$ matrices J_H and J_H^+ with (s, q) entries

$$J_H(s, q) = \begin{cases} 1 & \text{if } d_i(s, q) \in A \text{ for some } i \in H, \\ 0 & \text{otherwise,} \end{cases}$$

and

$$J_H^+(s,q) = \begin{cases} 1 & \text{if } d_i(s,q) \in A^+ \text{ for some } i \in H, \\ 0 & \text{otherwise,} \end{cases}$$

for $s, q \in S$ in which $d_i(s,q)$ denotes arc (s,q) controlled by some DM i.

The adjacency matrix for any coalition H has been defined. For example, if $H = i$, then $J_H(s,q)$ reduces to $J_i(s,q)$ that represents the adjacency relation between s and q in DM i's graph.

Definition 4.15 For the graph model, the **UM incidence matrix** and the **UI incidence matrix** are $m \times l$ matrices B and B^+, with (s,a) entries

$$B(s,a) = \begin{cases} -1 & \text{if } a = d_i(s,x) \in A \text{ for some } i \in N \text{ and some } x \in S, \\ 1 & \text{if } a = d_i(x,s) \in A \text{ for some } i \in N \text{ and some } x \in S, \\ 0 & \text{otherwise,} \end{cases}$$

and

$$B^+(s,a) = \begin{cases} -1 & \text{if } a = d_i(s,x) \in A^+ \text{ for some } i \in N \text{ and some } x \in S, \\ 1 & \text{if } a = d_i(x,s) \in A^+ \text{ for some } i \in N \text{ and some } x \in S, \\ 0 & \text{otherwise,} \end{cases}$$

where $s \in S$ and $a \in A$.

The extension of incidence matrix has two versions, both including and excluding preference information.

According to the signs of the entries, the UM incidence matrix can be separated into the UM in-incidence and out-incidence matrices.

Definition 4.16 For a graph model G, the **UM in-incidence matrix** and the **UM out-incidence matrix** are the $m \times l$ matrices B_{in} and B_{out} with (s,a) entries

$$B_{in}(s,a) = \begin{cases} 1 & \text{if } a = d_i(x,s) \in A \text{ for some } i \in N \text{ and some } x \in S, \\ 0 & \text{otherwise,} \end{cases}$$

and

$$B_{out}(s,a) = \begin{cases} 1 & \text{if } a = d_i(s,x) \in A \text{ for some } i \in N \text{ and some } x \in S, \\ 0 & \text{otherwise,} \end{cases}$$

where $s \in S$ and $a \in A$.

It is obvious that $B_{in} = (B + |B|)/2$ and $B_{out} = (|B| - B)/2$, where $|B|$ denotes the matrix in which each entry equals the absolute value of the corresponding entry of B. The **UI in-incidence matrix** B_{in}^+ and the **UI out-incidence matrix** B_{out}^+ can be defined similarly.

The relationships among the UM, UI adjacency matrices and the UM, UI in-incidence and out-incidence matrices are described as follows:

Theorem 4.7 *In a graph model G, J_H and J_H^+ denote the UM and the UI adjacency matrices for H, B_{in} and B_{out} denote the UM in-incidence and out-incidence matrices, and B_{in}^+ and B_{out}^+ indicate the UI in-incidence and out-incidence matrices. Then*

$$J_H = B_{out} \cdot I_H \cdot (B_{in})^T \text{ and } J_H^+ = B_{out}^+ \cdot I_H \cdot (B_{in}^+)^T,$$

where I_H is the $l \times l$ diagonal matrix in which $I_H(k,k) = 1$ if $a_k = d_i(s,q)$ and $i \in H$, otherwise $I_H(k,k) = 0$. Note that the diagonal matrix I_H has 1's as the (k,k) entry if and only if the arc a_k is controlled by DM i; otherwise all diagonal entries, and, of course, all non-diagonal entries are zeros.

From algebraic graph theory (Godsil and Royle 2001), Theorem 4.7 can follow easily. Two important matrices to link conflict evolution that will be introduced in Chap. 9 and conflict resolution in the graph model are proposed as follows:

Definition 4.17 For the graph model G, the **legal UM arc-incidence matrix** LJ_H and the **legal UI arc-incidence matrix** LJ_H^+ for coalition H are the $l \times l$ matrices with (a,b) entries

$$LJ_H(a,b) = \begin{cases} 1 & \text{if edge } a \text{ is incident on edge } b \text{ in } IG(G) \text{ for } a,b \in A, \\ & \text{and } a \text{ and } b \text{ controlled by different DMs in } H, \\ 0 & \text{otherwise,} \end{cases}$$

and

$$LJ_H^+(a,b) = \begin{cases} 1 & \text{if edge } a \text{ is incident on edge } b \text{ in } IG(G) \text{ for } a,b \in A^+, \\ & \text{and } a \text{ and } b \text{ controlled by different DMs in } H, \\ 0 & \text{otherwise.} \end{cases}$$

Note that if $H = N$, then LJ_N and LJ_N^+ are written as LJ and LJ^+, respectively.

Let D_i and D_i^+ denote the $l \times l$ diagonal matrices with (k,k) entries

$$D_i(k,k) = \begin{cases} 1 & \text{if } a_k = d_i(s,q) \text{ for } s,q \in S \text{ and } a_k \in A, \\ 0 & \text{otherwise,} \end{cases}$$

and

$$D_i^+(k,k) = \begin{cases} 1 & \text{if } a_k = d_i(s,q) \text{ for } s,q \in S \text{ and } a_k \in A^+, \\ 0 & \text{otherwise.} \end{cases}$$

Based on Definitions 4.16 and 4.17, the legal UM and the legal UI arc-incidence matrices can be obtained by the following theorem.

Theorem 4.8 *For the graph model G, let B_{in} and B_{out} be the UM in-incidence and the UM out-incidence matrices, and B^+_{in} and B^+_{out} denote the UI in-incidence and the UI out-incidence matrices. Then, the legal UM arc-incidence matrix LJ_H and the legal UI arc-incidence matrix LJ^+_H for coalition H satisfy that*

$$LJ_H = \bigvee_{i,j \in H, i \neq j} [(B_{in} \cdot D_i)^T \cdot (B_{out} \cdot D_j)],$$

and

$$LJ^+_H = \bigvee_{i,j \in H, i \neq j} [(B^+_{in} \cdot D^+_i)^T \cdot (B^+_{out} \cdot D^+_j)].$$

Proof Let $M = \bigvee_{i,j \in H, i \neq j} [(B_{in} \cdot D_i)^T \cdot \big(B_{out} \cdot D_j\big)]$. Thus, any entry (a_k, a_h) of matrix M can be expressed as

$$M(a_k, a_h) = sign[\sum_{i,j \in H, i \neq j} \sum_{q=1}^{m} (B_{in}(q, a_k) \cdot D_i(k, k) \cdot B_{out}(q, a_h) \cdot D_j(h, h))],$$

for $a_k, a_h \in A$ and $q \in S$.

Hence, $M(a_k, a_h) \neq 0$ iff $B_{in}(q, a_k) \cdot B_{out}(q, a_h) \neq 0$ for some $q \in S$ such that $a_k = d_i(s, q)$ and $a_h = d_j(q, u)$ for $s, u \in S$, and $i, j \in H$ and $i \neq j$. This implies that $M(a_k, a_h) \neq 0$ iff edge a_k is incident on edge a_h in $IG(G)$ and a_k and a_h are controlled by different DMs in H (see Fig. 4.5). Therefore, based on the definition of the matrix LJ, $M(a_k, a_h) \neq 0$ iff $LJ_H(a_k, a_h) \neq 0$. Since M and LJ_H are 0–1 matrices, then, $LJ_H = \bigvee_{i,j \in H, i \neq j} [(B_{in} \cdot D_i)^T \cdot (B_{out} \cdot D_j)]$ follows.

The proof of $LJ^+_H = \bigvee_{i,j \in H, i \neq j} [(B^+_{in} \cdot D^+_i)^T \cdot (B^+_{out} \cdot D^+_j)]$ is similar. □

It is obvious that unilateral moves on the branches of paths will end when the same arc appears twice. Generally, if there is no new appropriate arc produced, then the corresponding joint moves will stop. Therefore, the following Lemma 4.1 is obvious. Let $l = |A|$, $l^+ = |A^+|$ in the following lemma.

Lemma 4.1 *For the graph model G, let $H \subseteq N$. $R_H(s)$ and $R^+_H(s)$ are the reachable lists of H by the legal sequences of UMs and UIs from s. The δ_1 and δ_2 symbols are the numbers of iteration steps required to find $R_H(s)$ and $R^+_H(s)$, respectively. Then*

$$\delta_1 \leq l \quad and \quad \delta_2 \leq l^+.$$

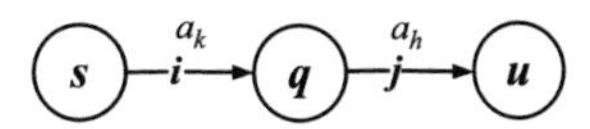

Fig. 4.5 a_k incident on a_h in $IG(G)$

Lemma 4.2 *For a graph model G, let t be a nonnegative integer, and fix $a, b \in A$. Then, $(LJ_H)^t(a, b)$, the (a, b) entry of matrix $(LJ_H)^t$ equals the number of legal UM arc-by-arc paths of length t in G for H, from edge a to edge b.*

Proof This Lemma is proved using induction on t.

When $t = 1$, the result is obvious. Note that $(LJ_H)^1(a, b)$ equals the number of arcs from a to b.

Assume that when $t = r$, the result holds. Then, when $t = r + 1$, $(LJ_H)^{r+1}(a_k, a_h) = \sum_{w=1}^{l}[(LJ_H)^r(a_k, a_w) \cdot LJ_H(a_w, a_h)]$. By the induction hypothesis, $(LJ_H)^r(a_k, a_w)$ denotes the number of legal paths by UMs from a_k to a_w with length r, and $LJ_H(a_w, a_h)$ equals the number of legal paths by UMs from a_w to a_h with length 1. Thus, $(LJ_H)^r(a_k, a_w) \cdot LJ_H(a_w, a_h)$ denotes the number of legal paths from a_k to a_h through a_w with length $r + 1$. Therefore, $\sum_{w=1}^{l}[(LJ_H)^r(a_k, a_w) \cdot LJ_H(a_w, a_h)]$ is the total number of legal paths from a_k to a_h by UMs with length $r + 1$.

Therefore, $(LJ_H)^t(a, b)$ equals the number of legal UM arc-by-arc paths for H from edge a to edge b with length t. □

Note that if $a = d_i(u, s)$ and $b = d_j(q, v)$ for $u, s, q, v \in S$ and $i, j \in H$, then the number of legal UM state-by-state paths for H from state u to state v of length $t + 1$ is at least $(LJ_H)^t(a, b)$. In fact, $(LJ_H)^t(a, b)$ is the number of legal paths of length t from u to v with initial edge a and terminal edge b. Similarly, $(LJ_H^+)^t(a, b)$ denotes the number of legal UI arc-by-arc paths for H in the G from edge a to edge b with length t. For example, Fig. 4.6a depicts an arc-by-arc path from arc a_1 to arc a_4 with length 5 in the graph model G presented in Fig. 4.3, where $a_1 = d_1(s_1, s_2)$ and $a_4 = d_2(s_3, s_6)$. Figure 4.6b presents the corresponding state-by-state path from s_1 to s_6 with initial edge a_1 and terminal edge a_4, which is of length 6.

The UM incidence matrix B and the UI incidence matrix B^+ depict unilateral move and unilateral improvement in one-step. The legal UM arc-incidence matrix LJ and the legal UI arc-incidence matrix LJ^+ can trace all evolutionary paths of length greater than 1 by UMs and UIs in a strategic conflict, respectively. The details of the evolution of a conflict will be discussed in Chap. 9.

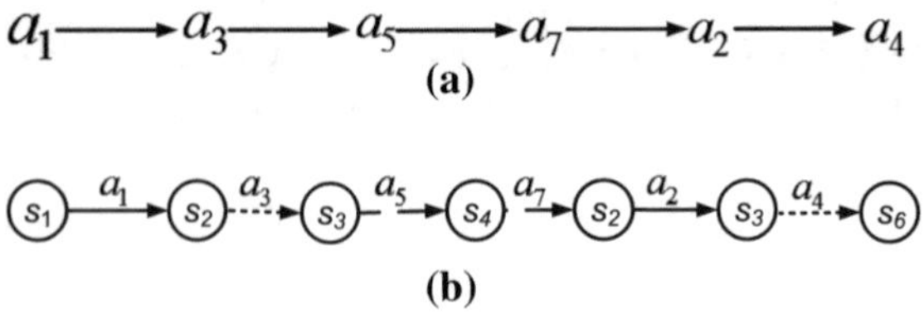

$a_1 \longrightarrow a_3 \longrightarrow a_5 \longrightarrow a_7 \longrightarrow a_2 \longrightarrow a_4$

(a)

$s_1 \xrightarrow{a_1} s_2 \xrightarrow{a_3} s_3 \xrightarrow{a_5} s_4 \xrightarrow{a_7} s_2 \xrightarrow{a_2} s_3 \xrightarrow{a_4} s_6$

(b)

Fig. 4.6 The arc-by-arc and the state-by-state UM paths

Example 4.11 Determine the legal UM and the legal UI arc-incidence matrices for the graph model G presented in Fig. 4.3 (Xu et al. 2010b).

Based on Fig. 4.3, the UM incidence matrix B is

$$B = \begin{pmatrix} -1 & 0 & 0 & 0 & 0 & 0 & 0 \\ 1 & -1 & -1 & 0 & 0 & 0 & 1 \\ 0 & 1 & 1 & -1 & -1 & 0 & 0 \\ 0 & 0 & 0 & 0 & 1 & -1 & -1 \\ 0 & 0 & 0 & 0 & 0 & 1 & 0 \\ 0 & 0 & 0 & 1 & 0 & 0 & 0 \end{pmatrix}.$$

Therefore,

$$B_{in} = \begin{pmatrix} 0 & 0 & 0 & 0 & 0 & 0 & 0 \\ 1 & 0 & 0 & 0 & 0 & 0 & 1 \\ 0 & 1 & 1 & 0 & 0 & 0 & 0 \\ 0 & 0 & 0 & 0 & 1 & 0 & 0 \\ 0 & 0 & 0 & 0 & 0 & 1 & 0 \\ 0 & 0 & 0 & 1 & 0 & 0 & 0 \end{pmatrix} \text{ and } \mathrm{B_{out}} = \begin{pmatrix} 1 & 0 & 0 & 0 & 0 & 0 & 0 \\ 0 & 1 & 1 & 0 & 0 & 0 & 0 \\ 0 & 0 & 0 & 1 & 1 & 0 & 0 \\ 0 & 0 & 0 & 0 & 0 & 1 & 1 \\ 0 & 0 & 0 & 0 & 0 & 0 & 0 \\ 0 & 0 & 0 & 0 & 0 & 0 & 0 \end{pmatrix}.$$

Then Theorem 4.8 implies that the legal UM arc-incidence and the legal UI arc-incidence matrices are

$$LJ = \begin{pmatrix} 0 & 0 & 1 & 0 & 0 & 0 & 0 \\ 0 & 0 & 0 & 1 & 1 & 0 & 0 \\ 0 & 0 & 0 & 0 & 1 & 0 & 0 \\ 0 & 0 & 0 & 0 & 0 & 0 & 0 \\ 0 & 0 & 0 & 0 & 0 & 0 & 1 \\ 0 & 0 & 0 & 0 & 0 & 0 & 0 \\ 0 & 1 & 1 & 0 & 0 & 0 & 0 \end{pmatrix} \text{ and } \mathrm{LJ}^{+} = \begin{pmatrix} 0 & 0 & 1 & 0 & 0 & 0 & 0 \\ 0 & 0 & 0 & 0 & 0 & 0 & 0 \\ 0 & 0 & 0 & 0 & 1 & 0 & 0 \\ 0 & 0 & 0 & 0 & 0 & 0 & 0 \\ 0 & 0 & 0 & 0 & 0 & 0 & 1 \\ 0 & 0 & 0 & 0 & 0 & 0 & 0 \\ 0 & 0 & 1 & 0 & 0 & 0 & 0 \end{pmatrix}.$$

Searching the nonzero entries of matrix LJ produces the UM arc-by-arc evolutionary path from a_1 to a_4 as presented in Fig. 4.7. Since there are two nonzero entries in the seventh row of matrix LJ, as seen in Fig. 4.7, branches a_2 and a_3 appear following a_7. However, arc a_3 has been passed in the path, so the branch with a_3 ends. Similarly, the branch following arc a_2 with arc a_5 stops. However, because a_4 is not a UI arc, a_4 cannot be reached by the legal UI paths so that s_6 is not reachable by the legal sequence of UIs from s_1.

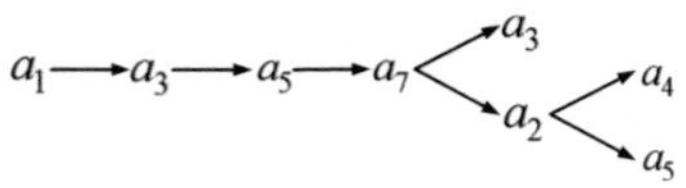

$$a_1 \rightarrow a_3 \rightarrow a_5 \rightarrow a_7 \begin{cases} \rightarrow a_3 \\ \rightarrow a_2 \begin{cases} \rightarrow a_4 \\ \rightarrow a_5 \end{cases} \end{cases}$$

Fig. 4.7 The arc-by-arc evolutionary paths from a_1 to a_4

4.3.3.2 Reachability Matrices to Construct Reachable Lists of a Coalition

Definition 4.18 For the graph model G, the t-**UM reachability matrix** and the t-**UI reachability matrix** for H, where $t = 1, 2, 3, \cdots$, are the $m \times m$ matrices with (s, q) entries

$$M_H^{(t)}(s, q) = \begin{cases} 1 & \text{if } q \in S \text{ is reachable by } H \text{ from } s \in S \text{ in exactly } t \text{ legal UMs,} \\ 0 & \text{otherwise,} \end{cases}$$

and

$$M_H^{(t,+)}(s, q) = \begin{cases} 1 & \text{if } q \in S \text{ is reachable by } H \text{ from } s \in S \text{ in exactly } t \text{ legal UIs,} \\ 0 & \text{otherwise.} \end{cases}$$

Obviously, $M_H^{(1)} = J_H$ and $M_H^{(1,+)} = J_H^+$. The t-UM and the t-UI reachability matrices for coalition H can be constructed by the following lemma.

Lemma 4.3 *For the graph model G, let B_{in} and B_{out} denote the in-incidence and out-incidence matrices, respectively. LJ_H and LJ_H^+ are the legal UM and the legal UI arc-incidence matrices for H. Then, for $t \geq 2$, the t-UM reachability and the t-UI reachability matrices for H can be expressed as*

$$M_H^{(t)} = sign[B_{out} \cdot (LJ_H)^{t-1} \cdot B_{in}^T] \text{ and } M_H^{(t,+)} = sign[B_{out}^+ \cdot (LJ_H^+)^{t-1} \cdot (B_{in}^+)^T].$$

Proof Based on Definition 4.18, $M_H^{(t)}(u, v) = 1$ iff state v is reachable by coalition H from state u in exactly t legal unilateral moves. Let $(LJ_H)^{t-1} = Q$ and $W = sign[B_{out} \cdot Q \cdot B_{in}^T]$. Then $W(u, v) \neq 0$ iff there exist $Q(a, b) \neq 0$ such that $a, b \in A$, $a = d_i(u, s)$, and $b = d_j(q, v)$ for $i, j \in H$, where $s, q, u, v \in S$. Using Lemma 4.2, $Q(a, b) \neq 0$ implies that state v can be attained by H from state u in exactly t legal UMs. Therefore, $M_H^{(t)}(u, v) = 1$ iff $W(u, v) \neq 0$. Since $M_H^{(t)}$ and W are 0–1 matrices, $M_H^{(t)} = sign[B_{out} \cdot (LJ)^{t-1} \cdot B_{in}^T]$.

The proof of $M_H^{(t,+)} = sign[B_{out}^+ \cdot (LJ_H^+)^{t-1} \cdot (B_{in}^+)^T]$ is similar. □

Definition 4.19 For the graph model G, the UM reachability matrix and the UI reachability matrix for H are the $m \times m$ matrices M_H and M_H^+ with (s, q) entries

$$M_H(s, q) = \begin{cases} 1 & \text{if } q \in R_H(s), \\ 0 & \text{otherwise,} \end{cases}$$

and

$$M_H^+(s, q) = \begin{cases} 1 & \text{if } q \in R_H^+(s), \\ 0 & \text{otherwise,} \end{cases}$$

respectively.

It is clear that $R_H(s) = \{q : M_H(s, q) = 1\}$ and $R_H^+(s) = \{q : M_H^+(s, q) = 1\}$. If $R_H(s)$ and $R_H^+(s)$ are written as 0–1 row vectors, then

$$R_H(s) = e_s^T \cdot M_H \text{ and } \mathrm{R_H^+(s) = e_s^T \cdot M_H^+},$$

where e_s^T denotes the transpose of the sth standard basis vector of the m-dimensional Euclidean space. Therefore, the reachability matrices for coalition H, M_H and M_H^+, can be used to construct the reachable lists of H from state s, $R_H(s)$ and $R_H^+(s)$.

The reachability matrices for coalition H can now be obtained by the following lemma.

Lemma 4.4 *For the graph model, let $M_H^{(t)}$ and $M_H^{(t,+)}$ be the t-UM and the t-UI reachability matrices. Then, the UM and the UI reachability matrices for H satisfy that*

$$M_H = \bigvee_{t=1}^{l} M_H^{(t)} \text{ and } M_H^+ = \bigvee_{t=1}^{l^+} M_H^{(t,+)}.$$

Proof Let $C = \bigvee_{t=1}^{l} M_H^{(t)}$. Based on the definition of M_H, $M_H(u, v) \neq 0$ iff v is reachable by H from u with a sequence of legal UMs. By Lemma 4.1, $l \geq \delta_1$. Hence, $M_H(u, v) \neq 0$ iff there exists $1 \leq t_0 \leq \delta_1 \leq l$ such that v is reachable by H from u with t_0 legal UMs. Based on Definition 4.18, this implies that $M_H^{(t_0)}(u, v) = 1$. Therefore, $M_H(u, v) \neq 0$ iff $C(u, v) \neq 0$. Since M_H and C are 0–1 matrices, $M_H = \bigvee_{t=1}^{l} M_H^{(t)}$ holds. The proof of $M_H^+ = \bigvee_{t=1}^{l^+} M_H^{(t,+)}$ can be carried out similarly. □

Lemma 4.5 *For the graph model G, LJ_H and LJ_H^+ denote the legal UM arc-incidence matrix and the legal UI arc-incidence matrix for H, respectively. Then*

(1) $(LJ_H + I)^n = \sum_{t=0}^{n} C_n^t \cdot (LJ_H)^t$,

(2) $(LJ_H^+ + I)^n = \sum_{t=0}^{n} C_n^t \cdot (LJ_H^+)^t$,

where the constant $C_n^t = \binom{n}{t} = \frac{n\cdot(n-1)\cdots(n-t+1)}{t!}$, $(LJ_H)^0 = (LJ_H^+)^0 = I_H$, *and I is the identity matrix.*

The above lemma is an obvious result of matrix theory. Based on the above discussions, the relations among the reachability matrices and the legal arc-incidence matrices for coalition H can now be established by the following theorem.

Theorem 4.9 *For the graph model G, LJ_H and LJ_H^+ are the legal UM and the legal UI arc-incidence matrices for H, respectively. The UM and UI reachability matrices for H, M_H and M_H^+, can be obtained by*

$M_H = sign[B_{out} \cdot (LJ_H + I)^{l-1} \cdot B_{in}^T]$ *and* $M_H^+ = sign[B_{out}^+ \cdot (LJ_H^+ + I)^{l^+-1} \cdot (B_{in}^+)^T]$,

where I *is the identity matrix.*

Proof Let $Q = sign[B_{out} \cdot (LJ_H + I)^{l-1} \cdot B_{in}^T]$. By Lemma 4.5 and $C_{l-1}^t > 0$, then

$$Q = sign[\sum_{t=0}^{l-1} C_{l-1}^t \cdot B_{out} \cdot (LJ_H)^t \cdot B_{in}^T] = (B_{out} \cdot I_H \cdot B_{in}^T) \bigvee sign[\sum_{t=1}^{l-1} B_{out} \cdot (LJ_H)^t \cdot B_{in}^T].$$

Based on Lemma 4.3 and Theorem 4.7,

$$Q = J_H \bigvee \left(\bigvee_{t=1}^{l-1} M_H^{(t+1)} \right) = \bigvee_{t=1}^{l} M_H^{(t)}.$$

Based on Lemma 4.4, $M_H = sign[B_{out} \cdot (LJ_H + I)^{l-1} \cdot B_{in}^T]$ follows.

The proof of $M_H^+ = sign[B_{out}^+ \cdot (LA^+ + I)^{l^+-1} \cdot (B_{in}^+)^T]$ is similar. □

The aim of a stability analysis is to find the equilibria of a graph model that are stable for all DMs under appropriate stability definitions. The reachable lists of coalition H by the sequences of the legal UMs and the legal UIs, $R_H(s)$ and $R_H^+(s)$, are essential components for stability analysis and the construction of the two state sets is a complicated process (Fang et al. 1993). An algebraic method for constructing $R_H(s)$ and $R_H^+(s)$ using the reachability matrices M_H and M_H^+ based on the incidence matrix B is developed here. In Chap. 5, another algebraic approach based on the adjacency matrix J is presented in Theorem 5.20.

Example 4.12 Fig. 4.3 shows a graph model with DM set $N = \{1, 2, 3, 4\}$ and state set $S = \{s_1, s_2, s_3, s_4, s_5, s_6\}$. The UM reachability matrix M_N is calculated according to Theorem 4.9 by

$$M_N = sign[B_{out} \cdot (LJ + I)^{l-1} \cdot B_{in}^T] = \begin{pmatrix} 0 & 1 & 1 & 1 & 0 & 1 \\ 0 & 0 & 1 & 1 & 0 & 1 \\ 0 & 1 & 0 & 1 & 0 & 1 \\ 0 & 1 & 1 & 0 & 1 & 1 \\ 0 & 0 & 0 & 0 & 0 & 0 \\ 0 & 0 & 0 & 0 & 0 & 0 \end{pmatrix},$$

where B_{out}, B_{in} and LJ are provided by Example 4.11. Similarly, the UI reachability matrix M_N^+ is obtained by

$$M_N^+ = \begin{pmatrix} 0 & 1 & 1 & 1 & 0 & 0 \\ 0 & 0 & 1 & 1 & 0 & 0 \\ 0 & 1 & 0 & 1 & 0 & 0 \\ 0 & 1 & 1 & 0 & 1 & 0 \\ 0 & 0 & 0 & 0 & 0 & 0 \\ 0 & 0 & 0 & 0 & 0 & 0 \end{pmatrix}.$$

If $s = s_1$ is selected as the status quo state, then the reachable lists of $H = N$ from s_1, $R_N(s_1)$ and $R_N^+(s_1)$, can be constructed by $R_H(s_1) = \{q : M_H(s_1, q) = 1\}$ and $R_H^+(s_1) = \{q : M_H^+(s_1, q) = 1\}$. Therefore, $R_N(s_1) = \{s_2, s_3, s_4, s_6\}$. $R_N^+(s_1) = \{s_2, s_3, s_4\}$.

Theorem 4.9 provides an algebraic method to construct the reachable lists of a coalition. The matrix representation of stability definitions can be extended to models including more than two DMs, which is the objective of the next subsection.

4.3.4 n-Decision Maker Case

Equivalent matrix representations of the logical definitions for Nash stability, GMR, SMR, and SEQ can be determined directly by using the relationship that has been established between matrix elements and the state set of a graph model, and by using preference relation matrices among the states.

Let $i \in N$, $|N| = n$, and $|S| = m$ in the following theorems. Nash stability in n-DM models is identical to two-DM cases because Nash stability does not consider opponents' responses.

It should be pointed out that the following stability matrices for n-DMs use the same notation as that presented in Sect. 4.3.2 for two-DMs. For general metarationality, DM i will take into account the opponents' possible responses, which are the legal sequence of UMs by members of $N\backslash\{i\}$. For $i \in N$, find DM i's UI adjacency matrix J_i^+ and the UM reachability matrix $M_{N\backslash\{i\}}$ using Theorem 4.9 for which $H = N\backslash\{i\}$. Define the $m \times m$ matrix M_i^{GMR} as

$$M_i^{GMR} = J_i^+ \cdot [E - sign(M_{N\backslash\{i\}} \cdot (P_i^{-,=})^T)].$$

Theorem 4.10 *State $s \in S$ is GMR for DM i, denoted by $s \in S_i^{GMR}$, iff $M_i^{GMR}(s, s) = 0$.*

Proof Since the diagonal element of matrix M_i^{GMR}

$$M_i^{GMR}(s, s) = \langle (J_i^+)^T e_s, \big(E - sign(M_{N\backslash\{i\}} \cdot (P_i^{-,=})^T)\big) e_s \rangle$$

$$= \sum_{s_1=1}^{m} J_i^+(s, s_1)[1 - sign(\langle (M_{N\backslash\{i\}})^T e_{s_1}, (P_i^{-,=})^T e_s \rangle)],$$

then $M_i^{GMR}(s, s) = 0$ iff

$$J_i^+(s, s_1)[1 - sign(\langle (M_{N\backslash\{i\}})^T e_{s_1}, (P_i^{-,=})^T e_s \rangle)] = 0, \forall s_1 \in S.$$

This implies that $M_i^{GMR}(s,s) = 0$ iff

$$(e_{s_1}^T M_{N\setminus\{i\}}) \cdot (e_s^T P_i^{-,=})^T \neq 0, \forall s_1 \in R_i^+(s). \quad (4.17)$$

Equation 4.17 means that, for any $s_1 \in R_i^+(s)$, there exists $s_2 \in S$ such that the m-dimensional row vector, $e_{s_1}^T \cdot M_{N\setminus\{i\}}$, with s_2th element 1 and the m-dimensional column vector, $(P_i^{-,=})^T \cdot e_s$, with s_2th element 1.

Therefore, $M_i^{GMR}(s,s) = 0$ iff for any $s_1 \in R_i^+(s)$, there exists at least one $s_2 \in R_{N\setminus\{i\}}(s_1)$ with $s \succeq_i s_2$. □

For symmetric metarationality, the n-DM model is similar to the two-DM model. The only modification is that responses come from DM i's opponents instead of from a single DM. Let

$$G = (P_i^{-,=})^T \circ [E - sign\big(J_i \cdot (P_i^+)^T\big)],$$

then define the $m \times m$ matrix M_i^{SMR} as

$$M_i^{SMR} = J_i^+ \cdot [E - sign(M_{N\setminus\{i\}} \cdot G)].$$

Theorem 4.11 *State $s \in S$ is SMR for DM i, denoted by $s \in S_i^{SMR}$, iff $M_i^{SMR}(s,s) = 0$.*

Proof Since the diagonal element of matrix M_i^{SMR}

$$M_i^{SMR}(s,s) = \langle (J_i^+)^T \cdot e_s, \big(E - sign(M_{N\setminus\{i\}} \cdot G)\big)e_s \rangle$$

$$= \sum_{s_1=1}^{m} J_i^+(s,s_1)[1 - sign\big(\langle (M_{N\setminus\{i\}})^T \cdot e_{s_1}, G \cdot e_s \rangle\big)],$$

then $M_i^{SMR}(s,s) = 0$ iff

$$J_i^+(s,s_1)[1 - sign\big(\langle (M_{N\setminus\{i\}})^T \cdot e_{s_1}, G \cdot e_s \rangle\big)] = 0, \forall s_1 \in S.$$

This means that $M_i^{SMR}(s,s) = 0$ iff

$$(e_{s_1}^T \cdot M_{N\setminus\{i\}}) \cdot (G \cdot e_s) \neq 0, \forall s_1 \in R_i^+(s). \quad (4.18)$$

Let $G(s_2,s)$ denote the (s_2,s) entry of matrix G. Since

$$(e_{s_1}^T M_{N\setminus\{i\}}) \cdot (G \cdot e_s) = \sum_{s_2=1}^{m} M_{N\setminus\{i\}}(s_1,s_2) \cdot G(s_2,s),$$

then Eq. 4.18 holds iff for any $s_1 \in R_i^+(s)$, there exists $s_2 \in R_{N\setminus\{i\}}(s_1)$ such that $G(s_2,s) \neq 0$.

Because $G(s_2, s) = P_i^{-,=}(s, s_2)[1 - sign(\sum_{s_3=1}^{m} J_i(s_2, s_3)P_i^+(s, s_3))]$, then $G(s_2, s) \neq 0$ implies that for $s_2 \in R_{N\backslash\{i\}}(s_1)$,

$$P_i^{-,=}(s, s_2) \neq 0 \tag{4.19}$$

and

$$\sum_{s_3=1}^{m} J_i(s_2, s_3)P_i^+(s, s_3) = 0. \tag{4.20}$$

Equation 4.19 is equivalent to the statement that, $\forall s_1 \in R_i^+(s)$, $\exists s_2 \in R_{N\backslash\{i\}}(s_1)$ such that $s \succeq_i s_2$. Equation 4.20 is the same as the statement that, $\forall s_1 \in R_i^+(s)$, $\exists s_2 \in R_{N\backslash\{i\}}(s_1)$ such that $P_i^+(s, s_3) = 0$ for $\forall s_3 \in R_i(s_2)$. Based on the definition of $m \times m$ matrix P_i^+, one knows that $P_i^+(s, s_3) = 0 \Longleftrightarrow s \succeq_i s_3$.

Therefore, the above discussion is concluded that $M_i^{SMR}(s, s) = 0$ iff for any $s_1 \in R_i^+(s)$, there exists at least one $s_2 \in R_{N\backslash\{i\}}(s_1)$ with $s \succeq_i s_2$ and $s \succeq_i s_3$ for all $s_3 \in R_i(s_2)$. □

Sequential stability examines the credibility of the sanctions by DM i's opponents. For $i \in N$, find the UI reachability matrix $M_{N\backslash\{i\}}^+$ using Theorem 4.9. Define the $m \times m$ matrix M_i^{SEQ} as

$$M_i^{SEQ} = J_i^+ \cdot [E - sign(M_{N\backslash\{i\}}^+ \cdot (P_i^{-,=})^T)].$$

Theorem 4.12 *State $s \in S$ is SEQ for DM i, denoted by $s \in S_i^{SEQ}$, iff M_i^{SEQ} $(s, s) = 0$.*

Proof Since the diagonal element of matrix M_i^{SEQ}

$$M_i^{SEQ}(s, s) = \langle (J_i^+)^T \cdot e_s, (E - sign(M_{N\backslash\{i\}}^+ \cdot (P_i^{-,=})^T))e_s \rangle$$

$$= \sum_{s_1=1}^{m} J_i^+(s, s_1)[1 - sign(\langle (M_{N\backslash\{i\}}^+)^T \cdot e_{s_1}, (P_i^{-,=})^T \cdot e_s \rangle)],$$

then $M_i^{SEQ}(s, s) = 0$ iff $J_i^+(s, s_1)[1 - sign(\langle (M_{N\backslash\{i\}}^+)^T \cdot e_{s_1}, (P_i^{-,=})^T \cdot e_s \rangle)] = 0, \forall s_1 \in S$. This implies that $M_i^{SEQ}(s, s) = 0$ iff

$$(e_{s_1}^T M_{N\backslash\{i\}}^+) \cdot (e_s^T \cdot P_i^{-,=})^T \neq 0, \forall s_1 \in R_i^+(s). \tag{4.21}$$

Equation 4.21 means that, for any $s_1 \in R_i^+(s)$, there exists $s_2 \in S$, such that the m-dimensional row vector, $e_{s_1}^T \cdot M_{N\backslash\{i\}}^+$, with s_2th element 1 and the m-dimensional column vector, $(P_i^{-,=})^T \cdot e_s$, with s_2th element 1.

Therefore, $M_i^{SEQ}(s, s) = 0$ iff for any $s_1 \in R_i^+(s)$, there exists at least one $s_2 \in R_{N\setminus\{i\}}^+(s_1)$ with $s \succeq_i s_2$. □

When $n = 2$, the DM set N becomes to $\{i, j\}$ in Theorems 4.10–4.12, and the reachable lists for $H = N \setminus \{i\}$ by legal sequences of UMs and UIs from s_1, $R_{N\setminus\{i\}}(s_1)$ and $R_{N\setminus\{i\}}^+(s_1)$, degenerate to $R_j(s_1)$ and $R_j^+(s_1)$, DM j's corresponding reachable lists from s_1. Thus, Theorems 4.10–4.12 are reduced to Theorems 4.4–4.6.

4.4 Computational Complexity

The proposed matrix method raises the question of computational complexity, which is the number of steps or arithmetic operations required to solve a computational problem. In this section, the computational complexities of MRSC and the graph model stability definitions are compared using general metarationality as an example.

4.4.1 Two Decision Maker Case

Recall the logical representation of GMR stability. State s is GMR for DM i iff for every $s_1 \in R_i^+(s)$ there exists at least one $s_2 \in R_j(s_1)$ with $s \succeq_i s_2$. Let $m = |S|$. The following procedures are utilized to calculate DM i's GMR stability.

- It takes at most m operations (or comparisons) to determine the state set $R_i^+(s)$;
- for every $s_1 \in R_i^+(s)$, it takes at most m operations (comparisons) to find $R_j(s_1)$;
- for $s_2 \in R_j(s_1)$, it makes at most $m - 1$ preference comparisons about states s and s_2.

Hence, using the logical definition to calculate DM i's GMR stability for state s will take at most $m + (m-1)(m+m-1) = 2m^2 - 2m + 1$ comparisons.

Recall DM i's matrix representation of GMR stability for state s. State s is GMR for DM i iff $M_i^{GMR}(s, s) = 0$, where $M_i^{GMR} = J_i^+ \left(E - sign\left(J_j \cdot (P_i^{-,=})^T\right)\right)$. By the proof of Theorem 4.4, one knows that $M_i^{GMR}(s, s) = 0$ iff

$$\sum_{s_1=1}^{m} J_i^+(s, s_1) \cdot \left(1 - sign((e_{s_1}^T \cdot J_j) \cdot (e_s^T \cdot P_i^{-,=})^T)\right) = 0. \tag{4.22}$$

It is easy to see that Eq. 4.22 takes $2m^2$ multiplication and addition operations. Comparing the computational complexities of these two methods to calculate GMR stability, one finds that their computational complexities are $O(m^2)$ at the same level. Note that the computational complexity of GMR stability is considered for the worst case. For logical representation of the GMR stability, the actual number of comparisons required is often smaller than $2m^2 - 2m + 1$. For matrix representation, the standard multiplication of two $m-$dimensional vectors is used, so it requires $O(m^2)$ arithmetic operations.

4.4.2 n-Decision Maker Case

In n-DM models, GMR stability is also selected as an example for analysis of the computational complexity of the proposed matrix method. According to Theorem 4.10, GMR stability definition is formulated using matrices as follows. State s is GMR for DM i iff $M_i^{GMR}(s,s)=0$, where $M_i^{GMR}=J_i^+\cdot[E-sign(M_{N\setminus\{i\}}\cdot(P_i^{-,=})^T)]$. By Theorem 4.9, one can estimate the computational complexity of the UM reachability matrix M_N. It is less than $\delta\cdot(n-1)\cdot O(m^3)$, where δ is the number of iterations, $n=|N|$ is the number of DMs, and m is the number of states. Let $l=|\bigcup_{i\in N} A_i|$. Then $\delta\leq l$ using Lemma 4.1. Therefore, the computational complexity to calculate DM i's GMR stability for state s in n-DM models is less than

$$l\cdot(n-1)\cdot O(m^3)+O(m^2)=l\cdot(n-1)\cdot O(m^3),$$

which presents a polynomial-time effective algorithm.

Many researchers are now attempting to develop faster algorithms for matrix operations. For example, for the multiplication of two $m\times m$ matrices, the standard method requires $O(m^3)$ arithmetic operations, but the Strassen algorithm (Strassen 1969) requires only $O(m^{2.807})$ operations. Coppersmith and Winograd's work (1990) shows that the computational complexity of matrix multiplication was decreased to $O(m^{2.376})$. In fact, some researchers believe that an optimal algorithm for multiplying $m\times m$ matrices will reduce the complexity to $O(m^2)$ (Cohn et al. 2005). Table 4.6 shows that the computational complexity of MRSC can be reduced using the Strassen or Coppersmith–Winograd algorithm. So far, the matrix representation of solution concepts has been established in multiple decision maker graph models for simple preference. As shown above, the matrix method for calculating the individual stability and equilibria is attractive from a computational point of view. Therefore, the proposed matrix method not only is propitious for theoretical analysis, but also has the potential to deal with large and complicated conflict problems.

In Sect. 4.3, matrix expressions are used to develop an explicit algebraic form conflict model that facilitates stability calculations. In following section, the efficiency of the matrix approach is illustrated using the Elmira conflict.

Table 4.6 The computational complexity of GMR stability using MRSC

Input	Output	Algorithm	Complexity
J_i, J_i^+, E, and $P_i^{-,=}$	$M_i^{GMR}(s,s)$	Standard matrix multiplication	$O(\delta\cdot n\cdot m^3)$
		Strassen algorithm	$O(\delta\cdot n\cdot m^{2.807})$
		Coppersmith–Winograd algorithm	$O(\delta\cdot n\cdot m^{2.376})$

4.5 Application: Elmira Conflict

As an introduction on how to formally investigate conflict taking place in the real-world, the Elmira groundwater contamination dispute was utilized in Sects. 1.2.2 (Modeling), 1.2.3 (Stability Analysis) and 1.2.4 (Follow-up Analysis). Here, as well as other sections in the book, this interesting dispute is utilized to explain and demonstrate technical definitions and concepts.

Briefly, Elmira, a small agricultural town renowned for its annual maple syrup festival, is located in southwestern Ontario, Canada. In 1989, the Ontario **Ministry of Environment (MoE)** tested the underground aquifer supplying water to Elmira, and determined that it was polluted by N-nitroso demethylamine (NDMA). A local pesticide and rubber manufacturer, **Uniroyal Chemical Ltd. (UR)**, was identified as the prime suspect, since NDMA was a by product of its production process. A Control Order was issued by MoE requiring UR to take expensive measures to remedy the contamination. UR immediately appealed the control order. The **Local Government (LG)**, consisting of the Regional Municipality of Waterloo and the Township of Woolwich, sided with MoE, but sought legal advice from independent consultants on its possible role in resolving this conflict (see Hipel et al. (1993) and Kilgour et al. (2001) for more details).

Hipel et al. (1993) established a graph model for this conflict, comprised of three DMs and five options, as follows:

- Ministry of Environment (MoE): its only option is to **modify** the Control Order to make it more acceptable to UR;
- Uniroyal Chemical Ltd. (UR): its options are to **delay** the appeal process, **accept** the Control Order in its current form, or **abandon** the Elmira operation; and
- Local Government (LG): its only option is to **insist** that the original Control Order be applied.

Given the five options in the model, there are 32 mathematically possible states. But many of them are infeasible for a variety of reasons; the nine feasible states are listed in Table 4.7 (where a "Y" indicates that an option is selected by the DM controlling it, an "N" means that the option is not chosen, and a dash "–" denotes that the entry

Table 4.7 Options and feasible states for the Elmira model

MoE									
1. Modify	N	Y	N	Y	N	Y	N	Y	–
UR									
2. Delay	Y	Y	N	N	Y	Y	N	N	–
3. Accept	N	N	Y	Y	N	N	Y	Y	–
4. Abandon	N	N	N	N	N	N	N	N	Y
LG									
5. Insist	N	N	N	N	Y	Y	Y	Y	–
State number	s_1	s_2	s_3	s_4	s_5	s_6	s_7	s_8	s_9

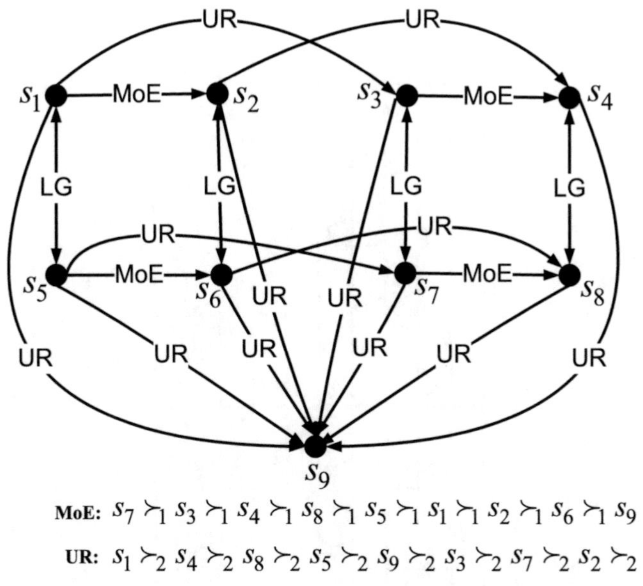

MoE: $s_7 \succ_1 s_3 \succ_1 s_4 \succ_1 s_8 \succ_1 s_5 \succ_1 s_1 \succ_1 s_2 \succ_1 s_6 \succ_1 s_9$

UR: $s_1 \succ_2 s_4 \succ_2 s_8 \succ_2 s_5 \succ_2 s_9 \succ_2 s_3 \succ_2 s_7 \succ_2 s_2 \succ_2 s_6$

LG: $s_7 \succ_3 s_3 \succ_3 s_5 \succ_3 s_1 \succ_3 s_8 \succ_3 s_6 \succ_3 s_4 \succ_3 s_2 \succ_3 s_9$

Fig. 4.8 Integrated graph model for the Elmira conflict

may be "Y" or "N"). The integrated graph model of the Elmira conflict is shown in Fig. 4.8, in which labels on the arcs indicate the DM controlling the move and preference information over the states is below the integrated graph.

4.5.1 *Procedures for Calculating Stability*

4.5.1.1 Finding Stable States from the Definitions

Let $N = \{1, 2, 3\}$ be the set of DMs (1 = MoE, 2 = UR, and 3 = LG). As an example, DM 3's SMR stability for state s_1 is analyzed using the logical representation presented in Definition 4.10. The procedures are as follows:

1. DM 3's reachable list from s_1 by UIs is $R_3^+(s_1) = \{s_5\}$;
2. The reachable list of coalition $H = N \setminus \{3\}$ from s_5 by UMs is $R_H(s_5) = \{s_6, s_7, s_8, s_9\}$;
3. $s_8 \in R_H(s_5)$ satisfies $s_1 \succ_3 s_8$; also $R_3(s_8) = \{s_4\}$ and $s_1 \succ_3 s_4$;
4. Therefore, s_1 is SMR stable for DM 3 by Definition 4.10.

Other cases can be analyzed similarly. Since the Elmira conflict is modeled as a standard graph model with simple preference, its stabilities can also be analyzed using DSS GMCR II (Fang et al. 2003a, b). The stability results of the Elmira conflict are presented in Table 4.8 in which "$\surd$" denotes that this state is stable for DM 1 (or

Table 4.8 Stability results of the Elmira conflict

State number	Nash				GMR				SMR				SEQ			
	MoE	UR	LG	Eq	MoE	UR	LG	Eq	MoE	UR	LG	Eq	MoE	UR	LG	Eq
s_1	√	√			√	√	√	√	√	√	√	√	√	√		
s_2	√				√		√		√		√		√		√	
s_3	√				√		√		√		√		√		√	
s_4	√	√			√	√	√	√	√	√	√	√	√	√		
s_5	√	√	√	√	√	√	√	√	√	√	√	√	√	√	√	√
s_6	√		√		√		√		√		√		√		√	
s_7	√		√		√		√		√		√		√		√	
s_8	√	√	√	√	√	√	√	√	√	√	√	√	√	√	√	√
s_9	√	√	√	√	√	√	√	√	√	√	√	√	√	√	√	√

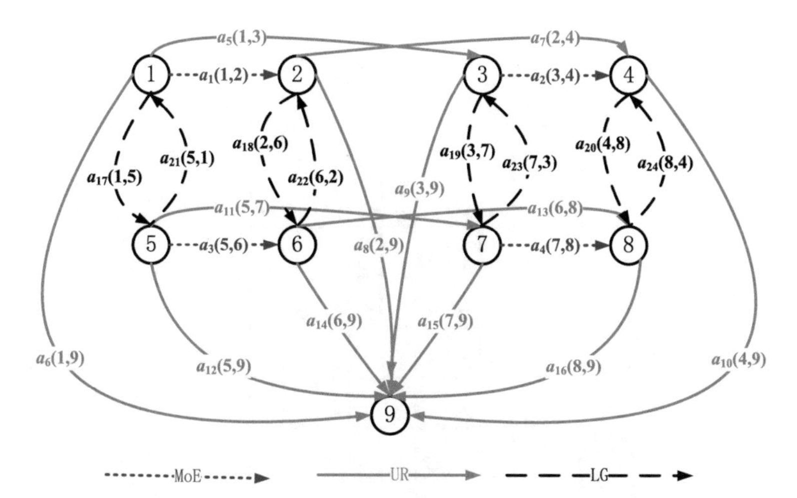

Fig. 4.9 The labeled graph for the Elmira conflict

MoE), DM 2 (or UR), or DM 3 (or LG) under the appropriate stability definitions, and "Eq" means an equilibrium that is stable for the three DMs.

4.5.1.2 Finding Stable States from Matrix Representation

The labeled graph of the Elmira conflict, determined according to the Rule of Priority presented in Sect. 3.3.2, is depicted in Fig. 4.9. The procedures to calculate the stabilities for the Elmira model using the matrix method are as follows:

1. For $i = 1, 2,$ and 3, using Fig. 4.8, determine DM i's adjacency matrix J_i and preference matrix P_i^+ as presented in Tables 4.9 and 4.10;
2. For $i = 1, 2,$ and 3, using $J_i^+ = J_i \circ P_i^+$, calculate the UI adjacency matrices;
3. For $i = 1, 2,$ and 3, using $P_i^{-,=} = E - I - P_i^+$, calculate the preference matrices $P_i^{-,=}$;
4. Construct the UM in-incidence and out-incidence matrices B_{in} and B_{out}, and the UI in-incidence and out-incidence matrices B_{in}^+ and B_{out}^+, based on the labeled graph in Fig. 4.9 and Definition 4.16;
5. Determine the UM arc-incidence and the UI arc-incidence matrices for H, LJ_H and LJ_H^+ using Theorem 4.8 by

$$LJ_H = \bigvee_{i,j \in H, i \neq j} [(B_{in} \cdot D_i)^T \cdot (B_{out} \cdot D_j)] \text{ and } LJ_H^+ = \bigvee_{i,j \in H, i \neq j} [(B_{in}^+ \cdot D_i^+)^T \cdot (B_{out}^+ \cdot D_j^+)];$$

6. Calculate the reachability matrices M_H and M_H^+ using Theorem 4.9 by

$$M_H = sign[B_{out} \cdot (LJ_H + I)^{l-1} \cdot B_{in}^T] \text{ and } M_H^+ = sign[B_{out}^+ \cdot (LJ_H^+ + I)^{l^+-1} \cdot (B_{in}^+)^T]$$

for $l = 24$ and $l^+ = 10$ as presented in Table 4.11;

Table 4.9 Adjacency matrices for the Elmira conflict

Matrix	J_1									J_2									J_3								
State	s_1	s_2	s_3	s_4	s_5	s_6	s_7	s_8	s_9	s_1	s_2	s_3	s_4	s_5	s_6	s_7	s_8	s_9	s_1	s_2	s_3	s_4	s_5	s_6	s_7	s_8	s_9
s_1	0	1	0	0	0	0	0	0	0	0	0	1	0	0	0	0	0	1	0	0	0	0	1	0	0	0	0
s_2	0	0	0	0	0	0	0	0	0	0	0	0	1	0	0	0	0	1	0	0	0	0	0	1	0	0	0
s_3	0	0	0	1	0	0	0	0	0	0	0	0	0	0	0	0	0	1	0	0	0	0	0	0	1	0	0
s_4	0	0	0	0	0	0	0	0	0	0	0	0	0	0	0	0	0	1	0	0	0	0	0	0	0	1	0
s_5	0	0	0	0	0	1	0	0	0	0	0	0	0	0	0	1	0	1	1	0	0	0	0	0	0	0	0
s_6	0	0	0	0	0	0	0	0	0	0	0	0	0	0	0	0	1	1	0	1	0	0	0	0	0	0	0
s_7	0	0	0	0	0	0	0	1	0	0	0	0	0	0	0	0	0	1	0	0	1	0	0	0	0	0	0
s_8	0	0	0	0	0	0	0	0	0	0	0	0	0	0	0	0	0	1	0	0	0	1	0	0	0	0	0
s_9	0	0	0	0	0	0	0	0	0	0	0	0	0	0	0	0	0	0	0	0	0	0	0	0	0	0	0

Table 4.10 Preference matrices for the Elmira conflict

Matrix	P_1^+									P_2^+									P_3^+								
State	s_1	s_2	s_3	s_4	s_5	s_6	s_7	s_8	s_9	s_1	s_2	s_3	s_4	s_5	s_6	s_7	s_8	s_9	s_1	s_2	s_3	s_4	s_5	s_6	s_7	s_8	s_9
s_1	0	0	1	1	1	0	1	1	0	0	0	0	0	0	0	0	0	0	0	0	1	0	1	0	1	0	0
s_2	1	0	1	1	1	0	1	1	0	1	0	1	1	1	0	1	1	1	1	0	1	1	1	1	1	1	0
s_3	0	0	0	0	0	0	1	0	0	1	0	0	1	1	0	0	1	1	0	0	0	0	0	0	1	0	0
s_4	0	0	1	0	0	0	1	0	0	1	0	0	0	0	0	0	0	0	1	0	1	0	1	1	1	1	0
s_5	0	0	1	1	0	0	1	1	0	1	0	0	1	0	0	0	1	0	0	0	1	0	0	0	1	0	0
s_6	1	1	1	1	1	0	1	1	0	1	1	1	1	1	0	1	1	1	1	0	1	0	1	0	1	1	0
s_7	0	0	0	0	0	0	0	0	0	1	0	1	1	1	0	0	1	1	0	0	0	0	0	0	0	0	0
s_8	0	0	1	1	0	0	1	0	0	1	0	0	1	0	0	0	0	0	1	0	1	0	1	0	1	0	0
s_9	1	1	1	1	1	1	1	1	0	1	0	0	1	1	0	0	1	0	1	1	1	1	1	1	1	1	0

Table 4.11 Reachability matrices for the Elmira conflict

Matrix	$M_{N\backslash\{1\}}$									$M_{N\backslash\{2\}}$									$M_{N\backslash\{3\}}$								
State	s_1	s_2	s_3	s_4	s_5	s_6	s_7	s_8	s_9	s_1	s_2	s_3	s_4	s_5	s_6	s_7	s_8	s_9	s_1	s_2	s_3	s_4	s_5	s_6	s_7	s_8	s_9
s_1	0	0	1	0	1	0	1	0	1	0	1	0	0	1	1	0	0	0	0	1	1	1	0	0	0	0	1
s_2	0	0	0	1	0	1	0	1	1	0	0	0	0	0	1	0	0	0	0	0	0	1	0	0	0	0	1
s_3	0	0	0	0	0	0	1	0	1	0	0	0	1	0	0	1	1	0	0	0	0	1	0	0	0	0	1
s_4	0	0	0	0	0	0	0	1	1	0	0	0	0	0	0	0	1	0	0	0	0	0	0	0	0	0	1
s_5	1	0	1	0	0	0	1	0	1	1	1	0	0	0	1	0	0	0	0	0	0	0	0	1	1	1	1
s_6	0	1	0	1	0	0	0	1	1	0	1	0	0	0	0	0	0	0	0	0	0	0	0	0	0	1	1
s_7	0	0	1	0	0	0	0	0	1	0	0	1	1	0	0	0	1	0	0	0	0	0	0	0	0	1	1
s_8	0	0	0	1	0	0	0	0	1	0	0	0	1	0	0	0	0	0	0	0	0	0	0	0	0	0	1
s_9	0	0	0	0	0	0	0	0	0	0	0	0	0	0	0	0	0	0	0	0	0	0	0	0	0	0	0
Matrix	$M^+_{N\backslash\{1\}}$									$M^+_{N\backslash\{2\}}$									$M^+_{N\backslash\{3\}}$								
State	s_1	s_2	s_3	s_4	s_5	s_6	s_7	s_8	s_9	s_1	s_2	s_3	s_4	s_5	s_6	s_7	s_8	s_9	s_1	s_2	s_3	s_4	s_5	s_6	s_7	s_8	s_9
s_1	0	0	0	0	1	0	0	0	0	0	0	0	0	1	0	0	0	0	0	0	0	0	0	0	0	0	0
s_2	0	0	0	1	0	1	0	1	1	0	0	0	0	0	1	0	0	0	0	0	0	1	0	0	0	0	1
s_3	0	0	0	0	0	0	1	0	1	0	0	0	0	0	0	1	0	0	0	0	0	0	0	0	0	0	1
s_4	0	0	0	0	0	0	0	1	0	0	0	0	0	0	0	0	1	0	0	0	0	0	0	0	0	0	0
s_5	0	0	0	0	0	0	0	0	0	0	0	0	0	0	0	0	0	0	0	0	0	0	0	0	0	0	0
s_6	0	0	0	0	0	0	0	1	1	0	0	0	0	0	0	0	0	0	0	0	0	0	0	0	0	1	1
s_7	0	0	0	0	0	0	0	0	1	0	0	0	0	0	0	0	0	0	0	0	0	0	0	0	0	0	1
s_8	0	0	0	0	0	0	0	0	0	0	0	0	0	0	0	0	0	0	0	0	0	0	0	0	0	0	0
s_9	0	0	0	0	0	0	0	0	0	0	0	0	0	0	0	0	0	0	0	0	0	0	0	0	0	0	0

Table 4.12 Stability matrices for the Elmira conflict

Stability matrices
$M_i^{Nash} = J_i^+ \cdot E$
$M_i^{GMR} = J_i^+ \cdot [E - sign\left(M_{N\setminus\{i\}} \cdot (P_i^{-,=})^T\right)]$
$M_i^{SMR} = J_i^+ \cdot [E - sign(M_{N\setminus\{i\}} \cdot Q)]$ with $Q = (P_i^{-,=})^T \circ [E - sign(J_i \cdot (P_i^+)^T)]$
$M_i^{SEQ} = J_i^+ \cdot [E - sign\left(M_{N\setminus\{i\}}^+ \cdot (P_i^{-,=})^T\right)]$

Table 4.13 Diagonal entries of stability matrices for the Elmira conflict

State number	Nash			GMR			SMR			SEQ		
	MoE	UR	LG	MoE	UR	LG	MoE	UR	LG	MoE	UR	LG
s_1	0	0	1	0	0	0	0	0	0	0	0	1
s_2	0	1	1	0	1	0	0	1	0	0	1	0
s_3	0	1	1	0	1	0	0	1	0	0	1	0
s_4	0	0	1	0	0	0	0	0	0	0	0	1
s_5	0	0	0	0	0	0	0	0	0	0	0	0
s_6	0	1	0	0	1	0	0	1	0	0	1	0
s_7	0	1	0	0	1	0	0	1	0	0	1	0
s_8	0	0	0	0	0	0	0	0	0	0	0	0
s_9	0	0	0	0	0	0	0	0	0	0	0	0

7. Calculate the stability matrices using the mathematical formulations in Table 4.12 and present their diagonal entries in Table 4.13; and
8. Analyze the stabilities of the conflict using Theorems 4.3 and 4.10–4.12 based on the information in Table 4.13.

The stability results using the matrix approach are identical to those obtained using logical definitions and presented in Table 4.8.

4.5.2 Analysis of Stability Results

The reachability matrices, M_H and M_H^+, are analyzed first. Using Table 4.11 with $H = N \setminus \{1\}$, one has:

$$e_4^T \cdot M_H = (0, 0, 0, 0, 0, 0, 0, 1, 1).$$

This means that $R_H(s_4) = \{s_8, s_9\}$, i.e. states s_8 and s_9 can be reached from the status quo $s = s_4$ by legal sequences of UMs by DMs in $H = \{2, 3\}$. Similarly,

$$e_4^T \cdot M_H^+ = (0, 0, 0, 0, 0, 0, 0, 1, 0),$$

which indicates that $R_H^+(s_4) = \{s_8\}$, i.e. s_8 can be reached from status quo $s = s_4$ by legal UI sequences for $H = \{2, 3\}$. It is obvious that if $R_H(s)$ and $R_H^+(s)$ are written as 0–1 row vectors, respectively, then

$$R_H(s) = e_s^T \cdot M_H \text{ and } R_H^+(s) = e_s^T \cdot M_H^+.$$

After the reachability matrices have been determined, stability analysis can be carried out using the stability matrices shown in Table 4.12. For example, the diagonal vector of DM 2's GMR stability matrix, $diag(M_2^{GMR}) = (0, 1, 1, 0, 0, 1, 1, 0, 0)^T$ indicates that states s_1, s_4, s_5, s_8, and s_9 are GMR stable for DM 2.

4.6 Important Ideas

The Graph Model for Conflict Resolution is a powerful tool to model, analyze, and understand strategic conflicts. In this chapter, logical and matrix representations of four basic stability definitions for simple preference are introduced for two-DM and multiple-DM conflicts. The graph model solution concepts discussed in Sect. 4.2 are expressed logically, making them difficult for computer implementation. But the matrix representation of solution concepts discussed in Sect. 4.3 handles this problem efficiently. In particular, the matrix method

- facilitates the development of improved algorithms to assess the stabilities of states,
- is ideally suited for the theoretical study of conflict problems,
- has the advantage of easy calculation and computer implementation, compared with the logical representation of solution concepts,
- provides explicit algebraic expressions that may be adapted for new solution concepts, and
- can be readily integrated into a decision support system as mentioned in Sect. 2.3.3 and explained in detail in Chap. 10.

Because of the nature of its explicit expressions, the matrix representation is easy to employ with different kinds of preference structures and associated modified solution concepts. For example, it could be extended to represent models with preference uncertainty or with multiple degrees of preference, and to determine stabilities in the graph model with these preference structures. The details are discussed in Chaps. 5 and 6, respectively.

4.7 Problems

4.7.1 In the tourism industry, two airlines are competing with each other by reducing the price to obtain more market share. Each airline company can either reduce the price (R) or do not reduce (D). The normal form of this conflict is as shown in Table 4.14.

Table 4.14 The airline conflict in normal form

		Air Cloud	
		Reduce Price (R)	**Do Not Reduce (D)**
Cloudways	**Reduce Price (R)**	s_1 **2, 2** **RR**	s_2 **4, 1** **RD**
	Do Not Reduce (D)	s_3 **1, 4** **DR**	s_4 **3, 3** **DD**

For this conflict, write the model in:

(a) Option form,
(b) Graph form, and
(c) Using logical form, calculate Nash, general metarational (GMR), symmetric metarational (SMR), and sequential (SEQ) stability for each state and DM. Indicate the equilibria and explain what they mean. Be sure to provide representative examples of stability calculations in normal, option and graph forms.

4.7.2 For the airline conflict provided in Problem 4.7.1, use the matrix formulation to carry out the stability calculations for each state and each DM for Nash, GMR, SMR, and SEQ stability. Determine the equilibria in this conflict and explain why they make sense.

4.7.3 The normal form of the game for Prisoner's Dilemma is given in Problem 3.5.1 in the previous chapter. Determine Nash stability for each of the four states and each of the two DMs. Does Nash stability predict a Nash equilibrium? What obvious equilibrium was missed? Howard (1971) as well as Fraser and Hipel (1979, 1984) refer to this as a breakdown of rationality. This breakdown provided the motivation for Howard to develop the solution concepts of GMR and SMR, and for Fraser and Hipel to propose the SEQ stability definition.

4.7.4 The normal form of the game of Chicken is presented in Problem 3.5.4. Determine which states are Nash stable for each of the two DMs. Are there any Nash equilibria? Which states do you think should be equilibria? The failure of not having a Nash equilibrium is referred to by Howard (1971) and also Fraser and Hipel (1979, 1984) as an example of the breakdown of rationality.

4.7.5 Using the logical form of the stability definitions, determine the stability of each of the four states and each of the two DMs in the game of Prisoner's Dilemma with respect to Nash, GMR, SMR, and SEQ stability. Which states are equilibria? Use the normal form of the game to explain your calculations and show the equilibria. How has the breakdown of rationality referred to in Problem 4.7.3 been resolved?

4.7.6 By employing the logical form of the four stability definitions given in this chapter, ascertain the stability of each of the four states for each of the two DMs in Prisoner's Dilemma. Show your calculations using the option form of the game. Point out which states are equilibria and explain why this is important.

4.7.7 Utilizing the logical form of the solution concepts consisting of Nash, GMR, SMR and SEQ stability, determine the stable states for each stability definition, DM and state for the game of Prisoner's Dilemma. Employ the graph form of the conflict to explain your calculations. Comment on the importance of the equilibria that you find.

4.7.8 In their 1984 book, Fraser and Hipel (1984) introduce Tableau Form to "graphically and intuitively" carry out stability calculations, especially for the case of Nash and SEQ stability. Recall that SEQ stability is especially well-designed because a DM will not harm himself or herself when levying a sanction against another DM's unilateral improvement (UI), since the move for the sanctioning DM must be a UI for him. Refer to Chaps. 2 and 3 in Fraser and Hipel's (1984) book to see how Tableau Form is written for the case of two and more than two DMs, respectively. Write Prisoner's Dilemma in Tableau Form and then carry out a stability analysis for Nash and SEQ stability. Notice the way the Tableau Form naturally portray how moves and countermoves work. How can Tableau Form be expanded to handle GMR and SMR stability?

4.7.9 Calculate Nash, GMR, SMR and SEQ stability using the matrix representation of GMCR for Prisoner's Dilemma for each DM and each of the four states.

4.7.10 For the game of Chicken shown in Problem 3.5.4, calculate Nash, GMR, SMR and SEQ stability for each state and DM using the matrix foumulation of GMCR. Which states are equilibria? Has the breakdown of rationality referred to in Problem 4.7.4 been overcome?

4.7.11 As is also described in Problem 3.5.10, a superpower nuclear confrontation (Fang et al. 1993) can be modeled using two DMs and six options shown in Table 4.15. These options determine the five feasible states as listed in Table 4.15. Note that state W represents a nuclear winter. The graph model for this dispute is displayed in Fig. 4.10.

(a) Analyze stabilities for this model using the logical representation of stability definitions;
(b) Analyze stabilities for this model employing the matrix representation of stability definitions.

4.7.12 The Rafferty-Alameda dams, in the Souris River basin in southern Saskatchewan, Canada, were planned for flood control, recreation and cooling the Shand generating plant (Roberts 1990). The **province of Saskatchewan** wanted to finish the project promptly, seeking a license from the Environment Minister of the

Table 4.15 Decision makers, options and feasible states for the superpower nuclear confrontation conflict

DM 1					
1. Peace (P)	Y	Y	N	N	N
2. Conventional attack (C)	N	N	Y	Y	N
3. Full nuclear attack (W)	N	N	N	N	Y
DM 2					
1. Peace (P)	Y	N	Y	N	N
2. Conventional attack (C)	N	Y	N	Y	N
3. Full nuclear attack (W)	N	N	N	N	Y
States	**PP**	**PC**	**CP**	**CC**	**W**

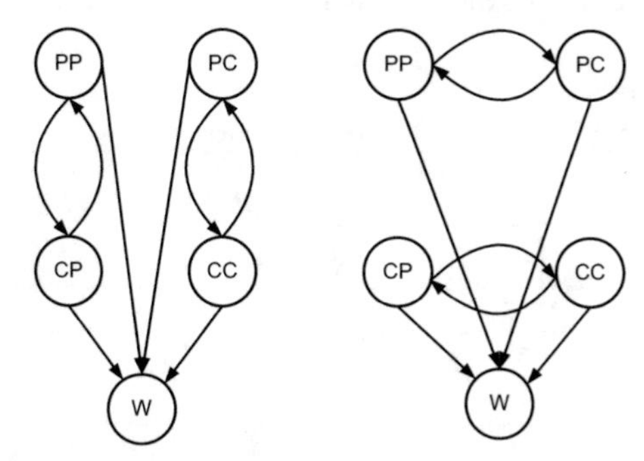

(a) Graph model for DM 1 (b) Graph model for DM 2

$DM1 : PP \succ_1 CP \succ_1 CC \succ_1 PC \succ_1 W$

$DM2 : PP \succ_2 PC \succ_2 CC \succ_2 CP \succ_2 W$

Fig. 4.10 The graph model of the superpower nuclear confrontation conflict

Federal Government. An **environmental group**, the Canadian Wildlife Federation, quickly petitioned against the license and argued that the provincial government had not respected regulations. The **federal court** sided with the environment group and ordered the suspension of the license, but later the license was reissued by a new federal environment minister. The environmental group petitioned again, and this time the federal court ordered the suspension of the license and the creation of a **review panel** to evaluate the project. However, construction of the dams continued during the review period, and the federal and provincial governments even reached an agreement that the project would continue while ten million dollars are set aside to alleviate any future environmental impacts. As the province had hoped, the project moved ahead at full speed, and the review panel resigned in protest. (See Hipel et al. (1991) for details.)

This conflict is modeled using four DMs: DM 1, **Federal (F)**; DM 2, **Saskatchewan (S)**; DM 3, **Groups (G)**; and DM 4, **Panel (P)**, each having some options. The following is a summary of the four DMs and their options:

- Federal Government (**Federal**): its options are to seek a court order to halt the project (Court Order) or to lift the license **(Lift)**,
- Province of Saskatchewan (**Saskatchewan**): its option is to go ahead at full speed **(Full speed)**,
- Environmental Groups (**Groups**): its option is to threaten court action to halt the project **(Court action)**, and
- Federal Environmental Review Panel (**Panel**): its option is to resign **(Resign)**.

Five options and ten feasible states of this model are presented in Table 4.16. The graph model of the Rafferty-Alameda dams conflict is shown in Fig. 4.11.

Table 4.16 Feasible states for the Rafferty-Alameda dams conflict

Federal										
1. Court order	–	N	Y	N	Y	N	Y	N	Y	N
2. Lift	–	N	N	N	N	N	N	N	N	Y
Saskatchewan										
3. Full speed	N	Y	Y	Y	Y	Y	Y	Y	Y	–
Groups										
4. Court action	–	N	N	Y	Y	N	N	Y	Y	–
Panel										
5. Resign	–	N	N	N	N	Y	Y	Y	Y	–
State number	s_1	s_2	s_3	s_4	s_5	s_6	s_7	s_8	s_9	s_{10}

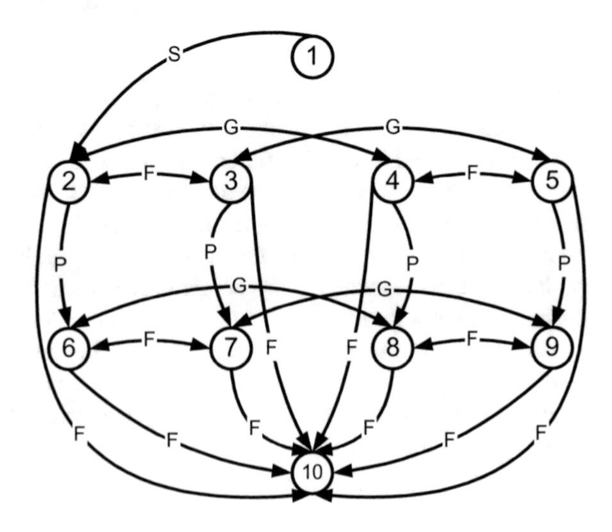

Fig. 4.11 The graph model of the Rafferty-Alameda dams conflict

The ordinal preferences for DMs 1, 2, 3, and 4 are

$$s_1 \succ_1 s_3 \succ_1 s_5 \succ_1 s_2 \succ_1 s_4 \succ_1 s_7 \succ_1 s_9 \succ_1 s_6 \succ_1 s_8 \succ_1 s_{10},$$

$$s_2 \succ_2 s_4 \succ_2 s_6 \succ_2 s_8 \succ_2 s_3 \succ_2 s_5 \succ_2 s_7 \succ_2 s_9 \succ_2 s_{10} \succ_2 s_1,$$

$$s_{10} \succ_3 s_1 \succ_3 s_7 \succ_3 s_3 \succ_3 s_6 \succ_3 s_2 \succ_3 s_9 \succ_3 s_5 \succ_3 s_8 \succ_3 s_4,$$

and

$$s_1 \succ_4 s_9 \succ_4 s_7 \succ_4 s_8 \succ_4 s_6 \succ_4 s_{10} \succ_4 s_5 \succ_4 s_3 \succ_4 s_4 \succ_4 s_2.$$

(a) Label the graph model in Fig. 4.11 according to the Rule of Priority and draw its labeled graph;
(b) Calculate the stabilities of Nash, GMR, SMR, and SEQ for the Rafferty-Alameda dams conflict using the matrix method.

References

Bennett, P. G. (1995). Modelling decision in international relations: Game theory and beyond. *Mershon International Studies Review*, *39*, 19–52. https://doi.org/10.2307/222691.

Cohn, H., Kleinberg, R., Szegedy, B., & Umans, C. (2005). Group-theoretic algorithms for matrix multiplication. *Proceedings of the 46th Annual Symposium on Foundations of Computer Science* (pp. 379–388). https://doi.org/10.1145/1145768.1145772.

Coppersmith, D., & Winograd, S. (1990). Matrix multiplication via arithmetic programming. *Journal of Symbolic Computation*, *9*(3), 251–280.

Fang, L., Hipel, K. W., & Kilgour, D. M. (1989). Conflict models in graph form: Solution concepts and their interrelationships. *European Journal of Operational Research*, *41*(1), 86–100. https://doi.org/10.1016/0377-2217(89)90041-6.

Fang, L., Hipel, K. W., & Kilgour, D. M. (1993). *Interactive decision making: The graph model for conflict resolution*. New York: Wiley. https://doi.org/10.2307/2583940.

Fang, L., Hipel, K. W., Kilgour, D. M., & Peng, X. (2003a). A decision support system for interactive decision making, part 1: Model formulation. *IEEE Transactions on Systems, Man and Cybernetics, Part C: Applications and Reviews*, *33*(1), 42–55. https://doi.org/10.1109/tsmcc.2003.809361.

Fang, L., Hipel, K. W., Kilgour, D. M., & Peng, X. (2003b). A decision support system for interactive decision making, part 2: Analysis and output interpretation. *IEEE Transactions on Systems, Man and Cybernetics, Part C: Applications and Reviews*, *33*(1), 56–66. https://doi.org/10.1109/tsmcc.2003.809360.

Fraser, N. M., & Hipel, K. W. (1979). Solving complex conflicts. *IEEE Transactions on Systems, Man, and Cybernetics*, *9*(12), 805–816. https://doi.org/10.1109/tsmc.1979.4310131.

Fraser, N. M., & Hipel, K. W. (1984). *Conflict analysis: Models and resolutions*. New York, North-Holland. https://doi.org/10.2307/2582031.

Godsil, C., & Royle, G. (2001). *Algebraic graph theory*. New York: Springer. https://doi.org/10.1007/978-1-4613-0163-9.

Hamouda, L., Kilgour, D. M., & Hipel, K. W. (2004). Strength of preference in the graph model for conflict resolution. *Group Decision and Negotiation*, *13*, 449–462. https://doi.org/10.1023/b:grup.0000045751.21207.35.

Hamouda, L., Kilgour, D. M., & Hipel, K. W. (2006). Strength of preference in graph models for multiple decision-maker conflicts. *Applied Mathematics and Computation*, *179*(1), 314–327. https://doi.org/10.1016/j.amc.2005.11.109.

Hipel, K. W. (2009). Conflict resolution: Theme overview paper in conflict resolution. *Encyclopedia of Life Support Systems (EOLSS)* (pp. 1–31). Oxford, U.K: EOLSS Publishers.

Hipel, K. W., Fang, L., & Kilgour, D. M. (1991). The graph model for conflicts in environmental planning. *Proceedings of the 1991 IEEE International Conference on Systems, Man, and Cybernetics* (Vol. 3, pp. 1997–2002). Charlottesville, Virginia, USA.

Hipel, K. W., Fang, L., Kilgour, D. M., & Haight, M. (1993). Environmental conflict resolution using the graph model. *Proceedings of the 1993 IEEE International Conference on Systems, Man, and Cybernetics* (Vol. 1, pp. 153–158). https://doi.org/10.1109/ICSMC.1993.384737.

Howard, N. (1971). *Paradoxes of rationality: Theory of metagames and political behavior*. Cambridge: MIT Press. https://doi.org/10.2307/1266876.

Howard, N., Bennett, P. G., Bryant, J. W., & Bradley, M. (1992). Manifesto for a theory of drama and irrational choice. *Journal of the Operational Research Society*, *44*, 99–103. https://doi.org/10.2307/2584447.

Kilgour, D. M., Hipel, K. W., & Fang, L. (1987). The graph model for conflicts. *Automatica*, *23*, 41–55. https://doi.org/10.1016/0005-1098(87)90117-8.

Kilgour, D. M., Hipel, K. W., Fang, L., & Peng, X. (2001). Coalition analysis in group decision support. *Group Decision and Negotiation*, *10*(2), 159–175.

Li, K. W., Hipel, K. W., Kilgour, D. M., & Fang, L. (2004). Preference uncertainty in the graph model for conflict resolution. *IEEE Transactions on Systems, Man, and Cybernetics, Part A, Systems and Humans*, *34*(4), 507–520. https://doi.org/10.1109/tsmca.2004.826282.

Nash, J. F. (1950). Equilibrium points in *n*-person games. *Proceedings of the National Academy of Sciences of the United States of America*, *36*(1), 48–49. https://doi.org/10.1073/pnas.36.1.48.

Nash, J. F. (1951). Noncooperative games. *Annals of Mathematics*, *54*(2):286–295. https://doi.org/10.1515/9781400884087-008.

Roberts, D. (1990). Rafferty-alameda: The tangled history of 2 dams projects. *The Globe and Mail, October 16* (pp. A4).

Saaty, T. L., & Alexander, J. M. (1989). *Conflict resolution: The analytic hierarchy approach*. New York: Praeger Publishers.

Strassen, V. (1969). Gaussian elimination is not optimal. *Numerische Mathematik*, *13*(4), 354–356. https://doi.org/10.1007/bf02165411.

Xu, H., Hipel, K. W., & Kilgour, D. M. (2007). Matrix representation of conflicts with two decision makers. *Proceedings of the 2007 IEEE International Conference on Systems, Man, and Cybernetics*, (pp. 1764–1769). Montreal, Canada. https://doi.org/10.1109/icsmc.2007.4413988.

Xu, H., Hipel, K. W., & Kilgour, D. M. (2009). Matrix representation of solution concepts in multiple decision maker graph models. *IEEE Transactions on Systems, Man, and Cybernetics, Part A: Systems and Humans*, *39*(1), 96–108. https://doi.org/10.1109/tsmca.2009.2007994.

Xu, H., Kilgour, D. M., & Hipel, K. W. (2010a). An integrated algebraic approach to conflict resolution with three-level preference. *Applied Mathematics and Computation*, *216*(3), 693–707. https://doi.org/10.1016/j.amc.2010.01.054.

Xu, H., Kilgour, D. M., Hipel, K. W., & Kemkes, G. (2010b). Using matrices to link conflict evolution and resolution in a graph model. *European Journal of Operational Research*, *207*, 318–329. https://doi.org/10.1016/j.ejor.2010.03.025.

Xu, H., Kilgour, D. M., & Hipel, K. W. (2011). Matrix representation of conflict resolution in multiple-decision-maker graph models with preference uncertainty. *Group Decision and Negotiation*, *20*(6), 755–779. https://doi.org/10.1007/s10726-010-9188-4.

Xu, H., Kilgour, D. M., Hipel, K. W., & McBean, E. A. (2014). Theory and implementation of coalitional analysis in cooperative decision making. *Theory and Decision*, *76*(2), 147–171. https://doi.org/10.1007/s11238-013-9363-6.

Chapter 5
Stability Definitions: Unknown Preference

Preferences play an important role in any type of decision model. Normally, for the graph model only a relative preference relationship is needed to represent a particular DM's preference for one state with respect to another, in terms of being more, equal or less preferred. This type of preference relation, called a simple preference structure, is discussed in Chap. 4, in which four stability definitions are provided for employment with simple preference. However, it is sometimes difficult to obtain accurate preference information. As pointed out by Fischer et al. (2000a, b), conflicts among the attributes or factors which may underlie the reasons for simple preference can create preference uncertainty. In other situations, there may be a scarcity of information or knowledge regarding preferences for one or more DMs involved in a dispute. Whatever the reason, uncertainty about preference may exist in a given conflict situation. To incorporate preference uncertainty into the graph model methodology, Li et al. (2004) proposed a new preference structure in which a DM's preferences include unknown preference. The main properties of this kind of preference structure are introduced in the next section. Additionally, in this chapter, the four basic stability definitions are extended to graph models with unknown preference including matrix representations of the four kinds of stabilities (Xu et al. 2007a, b, 2009a, 2011). Other approaches to taking into account preference uncertainty within the graph model structure are fuzzy (Hipel et al. 2011, Bashar et al. 2012, 2014, 2015, 2016, 2018), grey (Kuang et al. 2015, Zhao and Xu 2017), and probabilistic (Rego and dos Santos 2015) procedures.

5.1 Unknown Preference and Reachable Lists

To incorporate preference uncertainty into the graph model methodology, Li et al. (2004) proposed a new preference structure in which DM i's preferences are expressed by a triple of relations $\{\succ_i, \sim_i, U_i\}$ on S, where $s \succ_i q$ indicates strict

H. Xu et al., *Conflict Resolution Using the Graph Model: Strategic Interactions in Competition and Cooperation*, Studies in Systems, Decision and Control 153,
https://doi.org/10.1007/978-3-319-77670-5_5

preference, $s \sim_i q$ indicates indifference, and $s\ U_i\ q$ means DM i may prefer state s to state q, may prefer q to s, or may be indifferent between s and q.

In a graph model with uncertain preferences, the preferences of DM i over the set of states S can be expressed by a triple of relations $\{\succ_i, \sim_i, U_i\}$ on S. It is assumed that the preference relations of each DM $i \in N$ have the following properties:

(i) $\succ_i$ is asymmetric.
(ii) $\sim_i$ is reflexive and symmetric.
(iii) U_i is symmetric.
(iv) $\{\succ_i, \sim_i, U_i\}$ is strongly complete.

Property (iv) implies that, for any $s, t \in S$, exactly one of the following statements is true: $s \succ_i t$, $t \succ_i s$, $s \sim_i t$, or $s\ U_i\ t$.

If for any relation R and any states k, s, and q, $k\ R\ s$ and $s\ R\ q$ imply $k\ R\ q$, then R is transitive. For example, strict preference $\succ$ is transitive in many graph models. In the graph model, transitivity of preferences is not required, and all results hold whether preferences are transitive or intransitive. For example, the uncertain preference relation, U, is often intransitive. It means that although $s\ U_i\ q$ and $q\ U_i\ k$, DM i's preference between s and k may be certain.

5.1.1 *Reachable Lists of a Decision Maker*

Let S and N denote the state set and the DM set. The state set S can be partitioned into three subsets based on simple preference relative to a fixed state $s \in S$, which are $\Phi_i^+(s)$, $\Phi_i^=(s)$, and $\Phi_i^-(s)$ (see Chap. 4 for details). After uncertain preference is incorporated into the graph model, the state set S can be partitioned into four subsets, $\Phi_i^+(s)$, $\Phi_i^=(s)$, $\Phi_i^-(s)$, and $\Phi_i^U(s)$. The subset including preference uncertainty is described as follows:

Definition 5.1 For a graph model G, the subset of the state set S including uncertain preference relative to a fixed state $s \in S$ is

$$\Phi_i^U(s) = \{q : q\ U_i\ s\},$$

which contains the states uncertainly preferred to state s by DM i.

Therefore, $S = \Phi_i^+(s) \cup \Phi_i^=(s) \cup \Phi_i^-(s) \cup \Phi_i^U(s)$.

DM i's reachable lists from state $s \in S$ for simple preference including $R_i(s)$, $R_i^+(s)$, $R_i^=(s)$, and $R_i^-(s)$, are defined in Sect. 4.1.1. For a graph model with preference uncertainty, the set $R_i^U(s) = \{q : q \in R_i(s) \text{ and } q\ U_i\ s\}$ is the subset of DM i's unilateral moves from state s that result in a target state for which DM i's preference relative to s is uncertain. Similarly, $R_i^U(s)$ is called DM i's reachable list by uncertain unilateral moves (UUMs) from s. Therefore, DM i's unilateral moves from s in the graph model with uncertain preference is expressed as follows:

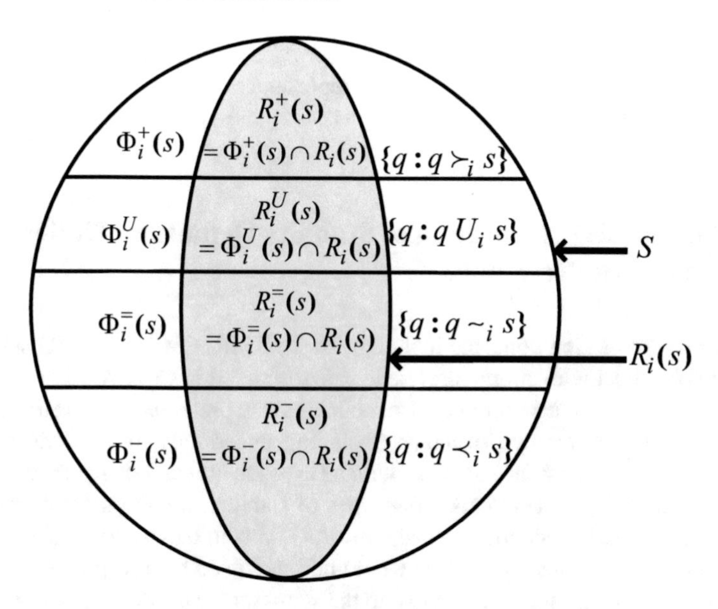

Fig. 5.1 Relations among subsets of S and reachable lists including preference uncertainty

$$R_i(s) = R_i^+(s) \cup R_i^=(s) \cup R_i^-(s) \cup R_i^U(s).$$

From the above definitions, the relations among the subsets of S and the corresponding reachable lists from state s for DM i in the graph model with preference uncertainty are depicted in Fig. 5.1, where the notation $\cap$ denotes the intersection operation.

DM i's oriented arcs A_i in the graph model with uncertain preference consist of UI arcs, A_i^+, equally preferred arcs, $A_i^=$, less preferred arcs, A_i^-, and unilateral uncertain move (UUM) arcs, A_i^U, where $A_i^U = \{(p, q) \in A_i : p\ U_i\ q\}$ is called the UUM arc set. DM i's UUM arcs from $s \in S$ are defined by $A_i^U(s) = \{(s, q) \in A_i : q \in R_i^U(s)\}$.

It is convenient to define

$$\Phi_i^{-,=,U}(s) = \Phi_i^-(s) \cup \Phi_i^=(s) \cup \Phi_i^U(s),$$

$$R_i^{+,U}(s) = R_i^+(s) \cup R_i^U(s), \text{ and}$$

$$A_i^{+,U}(s) = A_i^+(s) \cup A_i^U(s).$$

Note that $q \in \Phi_i^{-,=,U}(s)$ means that DM i does not prefer state q to state s or q is uncertainly preferred to s by DM i. DM i's reachable list $R_i^{+,U}(s)$ contains all states reachable from s by DM i by unilateral improvements in one step or uncertain unilateral moves in one step, so $R_i^{+,U}(s)$ is also called DM i's unilateral improvements or unilateral uncertain moves, denoted by unilateral improvements or uncertain moves (UIUMs), from s. The arc set $A_i^{+,U}(s)$ from s contains DM i's UI arcs and UUM arcs from s. Next, logical representation of stabilities is defined in the graph model with preference uncertainty.

5.2 Logical Representation of Stability Definitions Under Unknown Preference

The four basic solution concepts including Nash, GMR, SMR, and SEQ stabilities in the graph model with simple preference are discussed in Chap. 4. Li et al. (2004) extended these four solution concepts to models having preference uncertainty. Based on the extended preference structure (including uncertainty), they defined Nash, GMR, SMR, and SEQ stabilities to capture a DM's incentives to leave the status quo state and sensitivity to sanctions. Four types of stability definition were proposed, indexed a, b, c, and d, according to whether the DM would move to a state of uncertain preference and whether the DM would be sanctioned by a responding move to a state of uncertain preference, relative to the status quo. This range of extensions is needed, according to the work of Li et al. (2004), to address the diversity of possible risk profiles in face of uncertainty. A DM may be conservative or aggressive, avoiding or accepting states of uncertain preference, depending on the level of satisfaction with the current position.

Like all previous stability definitions in the graph model with simple preference, the four extensions were first defined logically, in terms of the underlying graphs. The four extended stability definitions are first introduced for two-DM models with preference uncertainty.

5.2.1 Two Decision Maker Case

5.2.1.1 Logical Representation of Stabilities Indexed *a*

In the definitions indexed a, DM i has an incentive to move to states with uncertain preferences relative to the status quo, but, when assessing possible sanctions, will not consider states with uncertain preferences (Li et al. 2004). Let $N = \{i, j\}, i \in N$, and $s \in S$ in the following definitions based on the research of Li et al. (2004).

State s is Nash indexed a stable (or, simply, $Nash_a$) for DM i if and only if the focal DM has no unilateral improvements or uncertain moves (UIUMs) from the status quo. Its formal definition is given as follows:

Definition 5.2 State s is $Nash_a$ stable for DM i, denoted by $s \in S_i^{Nash_a}$, iff $R_i^{+,U}(s) = \emptyset$.

Definition 5.2 implies that $Nash_a$ stability is identical for both two-DM and n-DM models because $Nash_a$ stability does not consider opponents' responses. Below, $Nash_a$ stability definition for n-DM models is not presented.

State $s \in S$ is GMR indexed a stable (or, simply, GMR_a) for DM i iff, whenever DM i makes any UIUM from s, then i's opponent can move to sanction i (that is, hurt i) in response.

Definition 5.3 State s is GMR_a stable for DM i, denoted by $s \in S_i^{GMR_a}$, iff for every $s_1 \in R_i^{+,U}(s)$ there exists at least one $s_2 \in R_j(s_1)$ with $s_2 \in \Phi_i^{-,=}(s)$ (or $s \succeq_i s_2$).

SMR indexed a stability (or, simply, SMR_a) is similar to GMR_a, but the focal DM considers not only the responses from opponents but also the DM's own counterresponses.

Definition 5.4 State s is SMR_a stable for DM i, denoted by $s \in S_i^{SMR_a}$, iff for every $s_1 \in R_i^{+,U}(s)$ there exists at least one $s_2 \in R_j(s_1)$ such that $s_2 \in \Phi_i^{-,=}(s)$ (or $s \succeq_i s_2$) and $s_3 \in \Phi_i^{-,=}(s)$ (or $s \succeq_i s_3$) for every $s_3 \in R_i(s_2)$.

SEQ indexed a stability (or, simply, SEQ_a) indicates that any UIUM of the focal DM is sanctioned by subsequent unilateral improvements or uncertain moves by opponents.

Definition 5.5 State s is SEQ_a stable for DM i, denoted by $s \in S_i^{SEQ_a}$, iff for every $s_1 \in R_i^{+,U}(s)$ there exists at least one $s_2 \in R_j^{+,U}(s_1)$ with $s_2 \in \Phi_i^{-,=}(s)$ (or $s \succeq_i s_2$).

Example 5.1 (*Stabilities indexed a for the Extended Sustainable Development Model*) Li et al. (2004) extended the sustainable development model discussed in Chap. 3 to include uncertain preference. Specifically, the conflict consists of two DMs: **an environmental agency** (DM 1: E) and **a developer** (DM 2: D); and a total of four options: DM 1 controls the two options of being proactive (labeled P) and being reactive (labeled R) in monitoring the developer's activities and their impacts on the environment, and DM 2 has the two options of practicing sustainable development (labeled S) and practicing unsustainable development (labeled U) for properly treating the environment. These options are combined to form four feasible states: s_1: PS, s_2: PU, s_3: RS, and s_4: RU. The four feasible states are listed in Table 5.1, where a "Y" indicates that an option is selected by the DM controlling it and an "N" means that the option is not chosen. The extended sustainable development model with preference uncertainty is used to illustrate how to determine the stabilities indexed a using Definitions 5.2–5.5.

The graph model of the conflict is shown in Fig. 5.2 in which DMs' preference information is below the two directed graphs G_1 and G_2. Consider analyzing $Nash_a$ stability of s_1 for DM 1. From Fig. 5.2, DM 1 has a unilateral move from s_1 to s_3. However, $s_1 \succ_1 s_3$ based on the preference information, so the move by DM 1 from s_1 to s_3 is not a unilateral improvement or a uncertain unilateral move. Therefore, state s_1 is $Nash_a$ stable for DM 1 according to Definition 5.2. Next, consider $Nash_a$ stability of s_1 for DM 2. Clearly, DM 2 has a unilateral move from s_1 to s_2. Because

Table 5.1 Options and feasible states for the extended sustainable development conflict

E: Environmental agency				
1. Proactive (labeled P)	Y	Y	N	N
2. Reactive (labeled R)	N	N	Y	Y
D: Developer				
3. Sustainable development (labeled S)	Y	N	Y	N
4. Unsustainable development (labeled U)	N	Y	N	Y
States	s_1	s_2	s_3	s_4

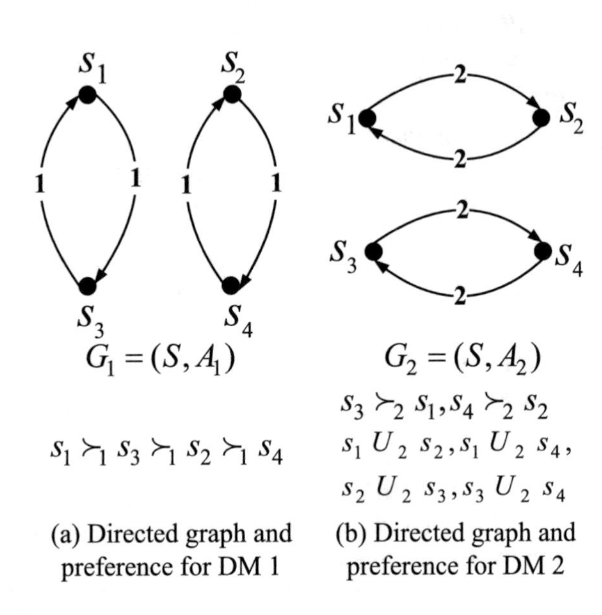

Fig. 5.2 Graph model for the extended sustainable development conflict

$s_1\ U_2\ s_2$, the move by DM 2 from s_1 to s_2 is not a UI but a UUM. Hence, state s_1 is $Nash_a$ unstable for DM 2. Similarly, the other three states can be assessed for $Nash_a$ stability.

By Definitions 5.2 and 5.3, one can see that if $R_i^{+,U}(s) = \emptyset$, then s is $Nash_a$ stable and GMR_a stable for DM i. Hence, for instance, s_1 is GMR_a stable for DM 1. One can assess whether s_3 is GMR_a stable for DM 1. DM 1 has a unilateral improvement from s_3 to s_1 and DM 2 has a unilateral move from s_1 to s_2. However, s_2 is less preferred than s_3 for DM 1, hence, s_3 is GMR_a stable for DM 1 according to Definition 5.3. The stabilities of other three states for the two DMs can be determined similarly.

By comparing Definitions 5.2–5.4, one can see that if $R_i^{+,U}(s) = \emptyset$, then s is $Nash_a$ stable, GMR_a stable, and SMR_a stable for DM i. Therefore, for instance, s_1 is SMR_a stable for DM 1. Next, considering analyzing SMR_a stability of s_3 for DM 1 using Definition 5.4. DM 1 has a UI from s_3 to s_1 and DM 2 has a UM from s_1 to s_2, then DM 1 has only a UM from s_2 to s_4. Because s_2 and s_4 are less preferred to

Table 5.2 Stabilities indexed a of the extended sustainable development game with uncertain preference

State	$Nash_a$			GMR_a			SMR_a			SEQ_a		
	1	2	Eq	1	2	Eq	1	2	Eq	1	2	Eq
s_1	√			√			√			√		
s_2	√			√			√			√		
s_3				√			√			√		
s_4												

s_3 by DM 1, so s_3 is SMR_a stable for DM 1. Using Definition 5.4, SMR_a stabilities for other states can be similarly determined.

Similar to GMR_a stability, if $R_i^{+,U}(s) = \emptyset$, then s is SEQ_a stable for DM i. Therefore, s_1 is SEQ_a stable for DM 1. Consider analyzing SEQ_a stability of s_3 for DM 1. DM 1 has a UI from s_3 to s_1. Although DM 2 has no UI from s_1, it can make a UUM from s_1 to s_2. However, s_2 is less preferred than s_3 for DM 1, hence, s_3 is SEQ_a stable for DM 1.

The results of the stabilities indexed a for the extended sustainable development model with uncertain preference are listed in Table 5.2, where "1" and "2" respectively denotes DM 1 and DM 2. "√" indicates a stable state for some solution concept and "Eq" means an equilibrium under some stability definition. From the above discussion, states s_1 and s_2 are stable for $Nash_a$, GMR_a, SMR_a, and SEQ_a for the extended sustainable development game with uncertain preference. Moreover, state s_3 is stable under GMR_a, SMR_a, and SEQ_a.

5.2.1.2 Logical Representation of Stabilities Indexed *b*

For the following definitions indexed b, DM i considers to leave a state or assesses sanctions, excluding uncertain preferences (Li et al. 2004). However, the definitions are different from those including simple preference only (presented in Chap. 4), since the current definitions are utilized to analyze conflict models including preference uncertainty.

Definition 5.6 State s is $Nash_b$ stable for DM i, denoted by $s \in S_i^{Nash_b}$, iff $R_i^+(s) = \emptyset$.

Definition 5.7 State s is GMR_b stable for DM i, denoted by $s \in S_i^{GMR_b}$, iff for every $s_1 \in R_i^+(s)$ there exists at least one $s_2 \in R_j(s_1)$ with $s_2 \in \Phi_i^{-,=}(s)$ (or $s \succeq_i s_2$).

Definition 5.8 State s is SMR_b stable for DM i, denoted by $s \in S_i^{SMR_b}$, iff for every $s_1 \in R_i^+(s)$ there exists at least one $s_2 \in R_j(s_1)$ such that $s_2 \in \Phi_i^{-,=}(s)$ (or $s \succeq_i s_2$) and $s_3 \in \Phi_i^{-,=}(s)$ (or $s \succeq_i s_3$) for every $s_3 \in R_i(s_2)$.

Definition 5.9 State s is SEQ_b stable for DM i, denoted by $s \in S_i^{SEQ_b}$, iff for every $s_1 \in R_i^+(s)$ there exists at least one $s_2 \in R_j^{+,U}(s_1)$ with $s_2 \in \Phi_i^{-,=}(s)$ (or $s \succeq_i s_2$).

Comparing stabilities indexed with a to those with b, the focal DM's attitudes to move from a starting state are different, because uncertain preference is allowed for a but is not permitted for b. In addition, their attitudes are identical for sanctions excluding uncertain preference.

5.2.1.3 Logical Representation of Stabilities Indexed *c*

For the extended definitions indexed c, DM i considers to leave a status quo state or evaluates sanctions including uncertain preference.

Definition 5.10 State s is $Nash_c$ stable for DM i, denoted by $s \in S_i^{Nash_c}$, iff $R_i^{+,U}(s) = \emptyset$.

Definition 5.11 State s is GMR_c stable for DM i, denoted by $s \in S_i^{GMR_c}$, iff for every $s_1 \in R_i^{+,U}(s)$ there exists at least one $s_2 \in R_j(s_1)$ with $s_2 \in \Phi_i^{-,=,U}(s)$ (or $s \succeq_i s_2$ or $s\ U_i\ s_2$).

Definition 5.12 State s is SMR_c stable for DM i, denoted by $s \in S_i^{SMR_c}$, iff for every $s_1 \in R_i^{+,U}(s)$ there exists at least one $s_2 \in R_j(s_1)$ such that $s_2 \in \Phi_i^{-,=,U}(s)$ (or $s \succeq_i s_2$ or $s\ U_i\ s_2$) and $s_3 \in \Phi_i^{-,=,U}(s)$ (or $s \succeq_i s_3$ or $s\ U_i\ s_3$) for every $s_3 \in R_i(s_2)$.

Definition 5.13 State s is SEQ_c stable for DM i, denoted by $s \in S_i^{SEQ_c}$, iff for every $s_1 \in R_i^{+,U}(s)$ there exists at least one $s_2 \in R_j^{+,U}(s_1)$ with $s_2 \in \Phi_i^{-,=,U}(s)$ (or $s \succeq_i s_2$ or $s\ U_i\ s_2$).

5.2.1.4 Logical Representation of Stabilities Indexed *d*

For the following definitions, indexed d, DM i would move only to preferred states from a status quo, but would be deterred by responses that result in states of uncertain preference.

Definition 5.14 State s is $Nash_d$ stable for DM i, denoted by $s \in S_i^{Nash_d}$, iff $R_i^+(s) = \emptyset$.

Definition 5.15 State s is GMR_d stable for DM i, denoted by $s \in S_i^{GMR_d}$, iff for every $s_1 \in R_i^+(s)$ there exists at least one $s_2 \in R_j(s_1)$ with $s_2 \in \Phi_i^{-,=,U}(s)$ (or $s \succeq_i s_2$ or $s\ U_i\ s_2$).

Definition 5.16 State s is SMR_d stable for DM i, denoted by $s \in S_i^{SMR_d}$, iff for every $s_1 \in R_i^+(s)$ there exists at least one $s_2 \in R_j(s_1)$ such that $s_2 \in \Phi_i^{-,=,U}(s)$ (or $s \succeq_i s_2$ or $s\ U_i\ s_2$) and $s_3 \in \Phi_i^{-,=,U}(s)$ (or $s \succeq_i s_3$ or $s\ U_i\ s_3$) for every $s_3 \in R_i(s_2)$.

Definition 5.17 State s is SEQ_d stable for DM i, denoted by $s \in S_i^{SEQ_d}$, iff for every $s_1 \in R_i^+(s)$ there exists at least one $s_2 \in R_j^{+,U}(s_1)$ with $s_2 \in \Phi_i^{-,=,U}(s)$ (or $s \succeq_i s_2$ or $s\ U_i\ s_2$).

Next the above definitions will be used to analyze stabilities for the extended sustainable development model with preference uncertainty.

Example 5.2 (*Stabilities for the Extended Sustainable Development Model with Preference Uncertainty*) The extended sustainable development model with preference uncertainty was described in Example 5.1. The graph model of this conflict is shown in Fig. 5.2 with the state set $S = \{s_1, s_2, s_3, s_4\}$ and the DM set $N = \{1, 2\}$. The stabilities indexed a were discussed in Example 5.1. In this example, it will be assessed whether the four states are stable for the stabilities indexed b, c, and d. Here, for instance, the following stabilities will be analyzed: s_1's $Nash_b$ stability and s_2's GMR_c and SEQ_c stabilities for DM 2; and s_3's SMR_d stability for DM 1. The analysis for the remaining cases is left as an exercise.

Since $R_2(s_1) = \{s_2\}$ and $s_1\ U_2\ s_2$, then $R_2^+(s_1) = \emptyset$. Accordingly, s_1 is $Nash_b$ stable, GMR_b stable, SMR_b stable, and SEQ_b stable for DM 2. One can assess whether s_2 is GMR_c or SEQ_c stable for DM 2. Due to $R_2(s_2) = \{s_1\}$ and $s_1\ U_2\ s_2$, then $R_2^{+,U}(s_2) = \{s_1\}$. Since $R_1(s_1) = \{s_3\}$ and $s_3\ U_2\ s_2$, then s_2 is GMR_c stable for DM 2 according to Definition 5.11. Similarly, for $s_1 \in R_2^{+,U}(s_2)$, $R_1^{+,U}(s_1) = \emptyset$, so s_2 is SEQ_c unstable for DM 2 according to Definition 5.13. Finally, it is determined whether s_3 is SMR_d stability for DM 1. Clearly, from Fig. 5.2, $R_1^+(s_3) = \{s_1\}$, $R_2(s_1) = \{s_2\}$, and $R_1(s_2) = \{s_4\}$. Since $s_3 \succ_1 s_2$ and $s_3 \succ_1 s_4$, then s_3 is SMR_d stable for DM 1 based on Definition 5.16.

Stability results of the extended sustainable development game with uncertain preference are listed in Table 5.3, where, as usual, "1" and "2" respectively denotes DM 1 and DM 2, "$\surd$" indicates a state stable for some stability definition, and "Eq" indicates an equilibrium under some stability definition.

Table 5.3 provides the stability results for the extended sustainable development game determined by logical representations for two-DM situations. Obviously, states s_1 and s_2 are equilibria for the four stabilities indexed b and indexed d in the extended sustainable development conflict, which means that they may be potential resolutions to solve this conflict.

5.2.2 *Reachable Lists of a Coalition*

To calculate the stability of a state for DM $i \in N$, it is necessary to examine possible responses by all other DMs $j \in N - \{i\}$, which may include sequential responses. To extend the graph model stability definitions to n-DM models with preference uncertainty, the definitions of a legal sequence of moves for simple preference must first be extended to take preference uncertainty into account. In order to achieve this, legal sequences of coalitional UIUMs must be defined first. Recall that a coalition is

Table 5.3 Stability results of the extended sustainable development game with uncertain preference

State		Nash			GMR			SMR			SEQ		
		1	2	Eq	1	2	Eq	1	2	Eq	1	2	Eq
a	s_1	√			√			√			√		
	s_2	√			√			√			√		
	s_3				√			√			√		
	s_4												
b	s_1	√	√	√	√	√	√	√	√	√	√	√	√
	s_2	√	√	√	√	√	√	√	√	√	√	√	√
	s_3		√		√	√	√	√	√	√	√	√	√
	s_4		√			√			√			√	
c	s_1	√			√	√	√	√			√		
	s_2	√			√	√	√	√			√		
	s_3				√	√	√	√	√	√	√	√	√
	s_4					√			√			√	
d	s_1	√	√	√	√	√	√	√	√	√	√	√	√
	s_2	√	√	√	√	√	√	√	√	√	√	√	√
	s_3		√		√	√	√	√	√	√	√	√	√
	s_4		√			√			√			√	

a nonempty subset of DMs, i.e., $H \subseteq N$ and $H \neq \emptyset$. A legal sequence of UIUMs is a sequence of allowable unilateral improvements or uncertain moves by a coalition, with the usual restriction that a member of the coalition may move more than once, but not twice consecutively. Let the coalition $H \subseteq N$ satisfy $|H| \geq 2$ and let the status quo state be $s \in S$. Let $R_H^{+,U}(s) \subseteq S$ (defined formally below) denote the set of states that can be reached by any legal sequence of UIUMs, by some or all DMs in H, starting at state s. If $s_1 \in R_H^{+,U}(s)$, then $\Omega_H^{+,U}(s, s_1)$ (also defined formally below) denotes the set of all last DMs in legal sequences from s to s_1 by UIUMs. The formal definitions of $R_H^{+,U}(s) \subseteq S$ and $\Omega_H^{+,U}(s, s_1) \subseteq H$ for $s_1 \in R_H^{+,U}(s)$ are given as follows:

Definition 5.18 A unilateral improvement or uncertain move (UIUM) by H is a member of $R_H^{+,U}(s) \subseteq S$, defined inductively by

(1) assuming $\Omega_H^{+,U}(s, s_1) = \emptyset$ for all $s_1 \in S$;
(2) if $j \in H$ and $s_1 \in R_j^{+,U}(s)$, then $s_1 \in R_H^{+,U}(s)$ and $\Omega_H^{+,U}(s, s_1) = \Omega_H^{+,U}(s, s_1) \cup \{j\}$;
(3) if $s_1 \in R_H^{+,U}(s)$, $j \in H$, and $s_2 \in R_j^{+,U}(s_1)$, then, provided $\Omega_H^{+,U}(s, s_1) \neq \{j\}$, $s_2 \in R_H^{+,U}(s)$ and $\Omega_H^{+,U}(s, s_2) = \Omega_H^{+,U}(s, s_2) \cup \{j\}$.

Note that if $s_1 \in R_H^{+,U}(s)$, then $\Omega_H^{+,U}(s, s_1) \subseteq H$ is the set of all last DMs in legal sequences of UIUMs from s to s_1. (If $s_1 \notin R_H^{+,U}(s)$, it is assumed that $\Omega_H^{+,U}(s, s_1) = \emptyset$.) Suppose that $\Omega_H^{+,U}(s, s_1)$ contains only one DM, say $j \in N$. Then any move

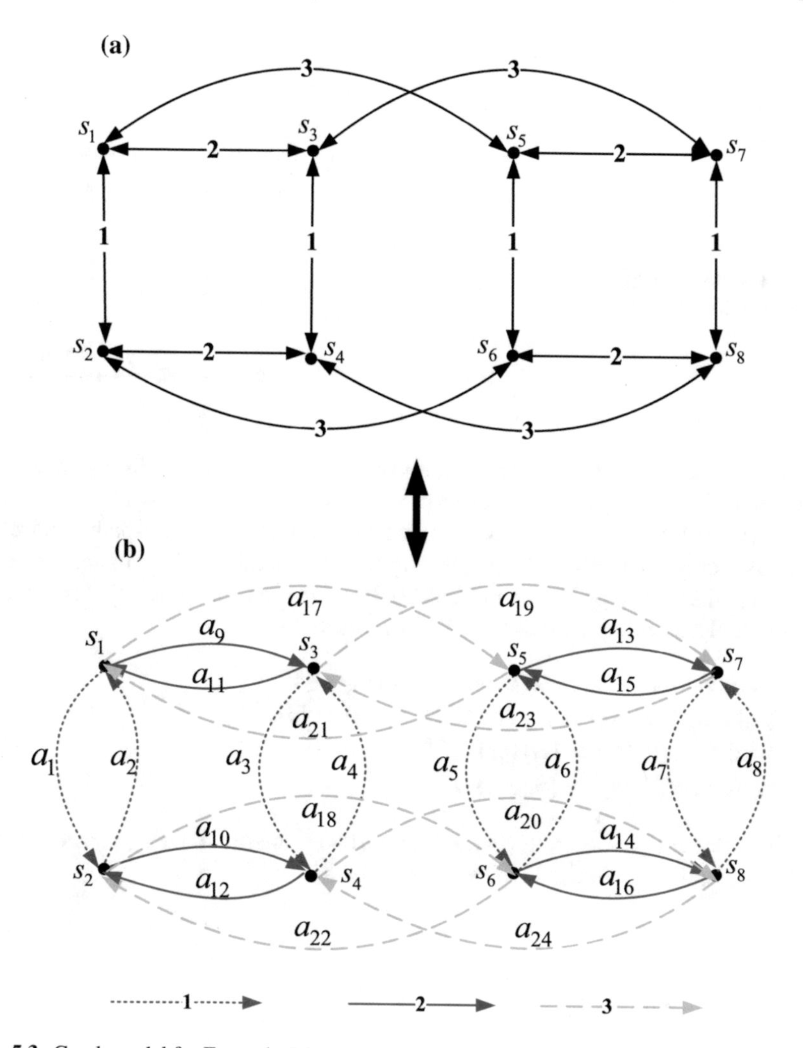

Fig. 5.3 Graph model for Example 5.3

from s_1 to a subsequent state, say s_2, must be made by a member of H other than j; otherwise DM j would have to move twice in succession. On the other hand, if $|\Omega_H^{+,U}(s, s_1)| \geq 2$, any member of H who has a unilateral improvement or a unilateral uncertain move from s_1 to s_2 may exercise it.

Example 5.3 (*Constructing Reachable Lists of a Coalition by UIUMs*) Figure 5.3a shows a graph model with DM set $N = \{1, 2, 3\}$ and state set $S = \{s_1, s_2, s_3, s_4, s_5, s_6, s_7, s_8\}$ (Xu et al. 2009b). The labels on the arcs of the graph indicate the controlling DMs. If DM i's oriented arcs are coded in color i, then, according to the *Rule of Priority* introduced in Sect. 3.3.2, Fig. 5.3a is converted to an edge labeled digraph as shown in Fig. 5.3b (The labeling process is left as an exercise). Preference

Table 5.4 Preference information for the graph model shown in Fig. 5.3

Colors	DMs	Certain preferences
Red	1	$s_2 \succ s_6 \succ s_4 \succ s_8 \succ s_1 \succ s_5 \succ s_3 \succ s_7$
Blue	2	$s_3 \succ s_7, s_4 \succ s_8, s_1 \succ s_5, s_2 \succ s_6$, only
Green	3	$s_3 \succ s_4 \succ s_7 \succ s_8 \succ s_5 \succ s_6 \succ s_1 \succ s_2$

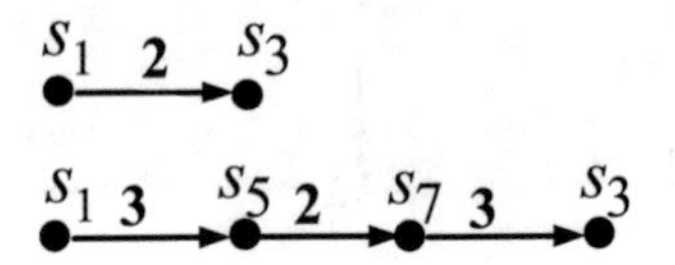

Fig. 5.4 Colored paths from s_1 by coalition $N \setminus \{1\}$

information is provided in Table 5.4. If $s = s_1$ is selected as the status quo state, then the reachable lists of $H = N \setminus \{1\}$ from s by unilateral improvements or uncertain moves (UIUMs), $R_H^{+,U}(s_1)$, can be obtained by searching all colored paths controlled by DMs 2 or 3 from s_1 by UIUMs (see Fig. 5.4. The details on how to determine all colored paths will be presented in Chap. 9). Here, the reachable list $R_H^{+,U}(s_1)$ can be constructed according to Definition 5.18 as follows:

1. Determine $R_2^{+,U}(s_1) = \{s_3\}$ and $R_3^{+,U}(s_1) = \{s_5\}$;
2. Obtain $\Omega_H^{+,U}(s_1, s_3) = \{2\}$ and $\Omega_H^{+,U}(s_1, s_5) = \{3\}$;
3. Calculate $R_2^{+,U}(s_5) = \{s_7\}$ and $R_3^{+,U}(s_7) = \{s_3\}$;
4. Construct $R_H^{+,U}(s_1) = \{s_3, s_5, s_7\}$.

The state set $R_H^{+,U}(s_1) = \{s_3, s_5, s_7\}$ means that states s_3, s_5, and s_7 can be reached by DMs 2 and 3 by the legal UIUMs from state s_1.

5.2.3 *Multiple Decision Maker Case*

In an n-DM model, where $n > 2$, the opponents of a DM can be thought of as a coalition of two or more DMs. The definitions of Nash, GMR, SMR, and SEQ stability in the graph model for multiple decision maker models with preference uncertainty were given by Li et al. (2004). They retain most features of the stability definitions in the 2-DM case, except that DM i's opponents are a subset of N, $N - \{i\}$, instead of a single opponent, j. Consequently, logical representation of solution concepts with preference uncertainty for 2-DM cases can be easily extended to that for n-DM situations.

5.2.3.1 Logical Representation of Stabilities Indexed *a*

First, extend the definitions indexed a for a two-DM case to an n-DM situation. Let $i \in N$ and $|N| = n$ in the following definitions.

Definition 5.19 State s is $Nash_a$ stable for DM i iff $R_i^{+,U}(s) = \emptyset$.

Definition 5.20 State s is GMR_a stable for DM i iff for every $s_1 \in R_i^{+,U}(s)$ there exists at least one $s_2 \in R_{N\backslash\{i\}}(s_1)$ with $s \succeq_i s_2$.

Definition 5.21 State s is SMR_a stable for DM i iff for every $s_1 \in R_i^{+,U}(s)$ there exists at least one $s_2 \in R_{N\backslash\{i\}}(s_1)$, such that $s \succeq_i s_2$ and $s \succeq_i s_3$ for any $s_3 \in R_i(s_2)$.

Definition 5.22 State s is SEQ_a stable for DM i iff for every $s_1 \in R_i^{+,U}(s)$ there exists at least one $s_2 \in R_{N\backslash\{i\}}^{+,U}(s_1)$ with $s \succeq_i s_2$.

5.2.3.2 Logical Representation of Stabilities Indexed *b*

For stabilities indexed b, DM i would move only to preferred states from a status quo and would be sanctioned only by less preferred or equally preferred states relative to the status quo.

Definition 5.23 State s is $Nash_b$ stable for DM i iff $R_i^{+}(s) = \emptyset$.

Definition 5.24 State s is GMR_b stable for DM i iff for every $s_1 \in R_i^{+}(s)$ there exists at least one $s_2 \in R_{N\backslash\{i\}}(s_1)$ with $s \succeq_i s_2$.

Definition 5.25 State s is SMR_b stable for DM i iff for every $s_1 \in R_i^{+}(s)$ there exists at least one $s_2 \in R_{N\backslash\{i\}}(s_1)$, such that $s \succeq_i s_2$ and $s \succeq_i s_3$ for any $s_3 \in R_i(s_2)$.

Definition 5.26 State s is SEQ_b stable for DM i iff for every $s_1 \in R_i^{+}(s)$ there exists at least one $s_2 \in R_{N\backslash\{i\}}^{+,U}(s_1)$ with $s \succeq_i s_2$.

5.2.3.3 Logical Representation of Stabilities Indexed *c*

For definitions indexed c, DM i would move to preferred states and states having uncertain preference relative to the starting state. With respect to sanctions, DM i does not want to end up at states that are less preferred or equally preferred relative to state s, and states having uncertain preference relative to state s.

Definition 5.27 State s is $Nash_c$ stable for DM i iff $R_i^{+,U}(s) = \emptyset$.

Definition 5.28 State s is GMR_c stable for DM i iff for every $s_1 \in R_i^{+,U}(s)$ there exists at least one $s_2 \in R_{N\backslash\{i\}}(s_1)$ with $s \succeq_i s_2$ or $s\ U_i\ s_2$.

Definition 5.29 State s is SMR_c stable for DM i iff for every $s_1 \in R_i^{+,U}(s)$ there exists at least one $s_2 \in R_{N\backslash\{i\}}(s_1)$, such that $s \succeq_i s_2$ or $s\ U_i\ s_2$ and $s \succeq_i s_3$ or $s\ U_i\ s_3$ for any $s_3 \in R_i(s_2)$.

Definition 5.30 State s is SEQ_c stable for DM i iff for every $s_1 \in R_i^{+,U}(s)$ there exists at least one $s_2 \in R_{N\backslash\{i\}}^{+,U}(s_1)$ with $s \succeq_i s_2$ or $s\ U_i\ s_2$.

5.2.3.4 Logical Representation of Stabilities Indexed *d*

For the last set of stabilities, indexed by d, a DM is not willing to move to a state with uncertain preference relative to the status quo, but is deterred by sanctions to states that have uncertain preference relative to the status quo.

Definition 5.31 State s is $Nash_d$ stable for DM i iff $R_i^+(s) = \emptyset$.

Definition 5.32 State s is GMR_d stable for DM i iff for every $s_1 \in R_i^+(s)$ there exists at least one $s_2 \in R_{N\backslash\{i\}}(s_1)$ with $s \succeq_i s_2$ or $s\ U_i\ s_2$.

Definition 5.33 State s is SMR_d stable for DM i iff for every $s_1 \in R_i^+(s)$ there exists at least one $s_2 \in R_{N\backslash\{i\}}(s_1)$, such that $s \succeq_i s_2$ or $s\ U_i\ s_2$ and $s \succeq_i s_3$ or $s\ U_i\ s_3$ for any $s_3 \in R_i(s_2)$.

Definition 5.34 State s is SEQ_d stable for DM i iff for every $s_1 \in R_i^+(s)$ there exists at least one $s_2 \in R_{N\backslash\{i\}}^{+,U}(s_1)$ with $s \succeq_i s_2$ or $s\ U_i\ s_2$.

When $n = 2$, the DM set N reduces to $\{i, j\}$ in Definitions 5.19–5.34. For example, the reachable list $R_{N\backslash\{i\}}^{+,U}(s_1)$ of $N\backslash\{i\}$ from s_1 by the legal sequences of UIUMs reduces to the reachable list $R_j^{+,U}(s_1)$ of j from s_1 by one step UIUMs.

From the solution concepts indexed a, b, c, and d presented above, it can be seen that a solution concept indexed a represents the stability for the most aggressive DMs. Firstly, the DM is aggressive in deciding whether to move from the status quo, thus being willing to accept the risk associated with moves to states of uncertain preference. In addition, when evaluating possible moves, the DM is deterred only by sanctions to states that are less preferred than the status quo and does not see states of uncertain preference (relative to the status quo) as sanctions. For the definitions indexed b, uncertainty in preferences is not considered by a DM. The definitions indexed c incorporate a mixed attitude toward the risk associated with states of uncertain preference. Specifically, the DM is aggressive in deciding whether to move from the status quo, but is conservative when evaluating possible moves, being deterred by sanctions to states that are less preferred or have uncertain preference relative to the status quo. Finally, the definition indexed d represents stability for the most conservative DMs, who would move only to preferred states from a status quo, but would be deterred by responses that result in states of uncertain preference (Li et al. 2004).

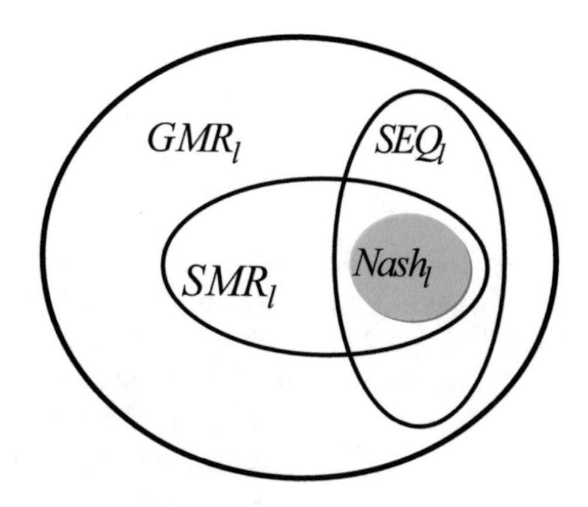

Fig. 5.5 Relationships among four stabilities indexed l

5.2.4 Relationships Among Stabilities in the Graph Model with Preference Uncertainty

Fang et al. (1993) determined relationships among Nash, GMR, SMR, and SEQ stabilities for the simple preference structure. Following this research direction, Li et al. (2004) established relationships among stability definitions with preference uncertainty.

Let l denote one of the four extensions indexed by a, b, c, and d, i.e., $l = a, b, c$, or d. In the following theorems, the symbol GS denotes a graph model stability, GMR, SMR, or SEQ. Then GS_l refers to the GS solution concept indexed l. The symbol $s \in S_i^{GS_l}$ denotes that $s \in S$ is stable for DM i according to stability GS indexed l. The relationships among the four stabilities of Nash, GMR, SMR, and SEQ, indexed l in the graph model with preference uncertainty are given next.

Theorem 5.1 *Let $l = a, b, c$, or d and $i \in N$. The relationships among the four stabilities indexed l are*

$$S_i^{Nash_l} \subseteq S_i^{SMR_l} \subseteq S_i^{GMR_l},$$

$$S_i^{Nash_l} \subseteq S_i^{SEQ_l} \subseteq S_i^{GMR_l}.$$

The proof of Theorem 5.1 easily follows from Definitions 5.19–5.34. Note that there is no necessary inclusion relationship between $S_i^{SMR_l}$ and $S_i^{SEQ_l}$, i.e., it may or may not be true that $S_i^{SMR_l} \supseteq S_i^{SEQ_l}$, or that $S_i^{SMR_l} \subseteq S_i^{SEQ_l}$. Then, the above inclusion relationships among the four stabilities indexed l are shown in Fig. 5.5.

Theorem 5.2 *The relationships among Nash stabilities indexed a, b, c, and d for DM i are*

$$S_i^{Nash_a} = S_i^{Nash_c},\ S_i^{Nash_b} = S_i^{Nash_d},$$

and

$$S_i^{Nash_a} \subseteq S_i^{Nash_b}.$$

This result is obvious from the above Nash stability definitions.

Theorem 5.3 *Let $i \in N$. The relationships among stabilities GS indexed a, b, c, and d are*

$$S_i^{GS_a} \subseteq S_i^{GS_b} \subseteq S_i^{GS_d},\ S_i^{GS_a} \subseteq S_i^{GS_c} \subseteq S_i^{GS_d}.$$

Proof The inclusion relations $S_i^{SMR_a} \subseteq S_i^{SMR_c} \subseteq S_i^{SMR_d}$ are proven first. If state $s \in S_i^{SMR_a}$, this implies that if $s_1 \in R_i^{+,U}(s)$, then there exists at least one $s_2 \in R_{N\setminus\{i\}}(s_1)$, such that $s_2 \in \Phi_i^{-,=}(s)$ and $s_3 \in \Phi_i^{-,=}(s)$ for every $s_3 \in R_i(s_2)$. Since $\Phi_i^{-,=}(s) \subseteq \Phi_i^{-,=,U}(s)$, then $s_2 \in \Phi_i^{-,=,U}(s)$ and $s_3 \in \Phi_i^{-,=,U}(s)$ for every $s_3 \in R_i(s_2)$. Therefore, if state $s \in S_i^{SMR_a}$, then state $s \in S_i^{SMR_c}$.

If state $s \in S_i^{SMR_c}$, this implies that if $s_1 \in R_i^{+,U}(s)$, then there exists at least one $s_2 \in R_{N\setminus\{i\}}(s_1)$, such that $s_2 \in \Phi_i^{-,=,U}(s)$ and $s_3 \in \Phi_i^{-,=,U}(s)$ for every $s_3 \in R_i(s_2)$. Since $R_i^{+}(s) \subseteq R_i^{+,U}(s)$, then $s \in S_i^{SMR_c}$ implies that if $s_1 \in R_i^{+}(s)$, then there exists at least one $s_2 \in R_{N\setminus\{i\}}(s_1)$, such that $s_2 \in \Phi_i^{-,=,U}(s)$ and $s_3 \in \Phi_i^{-,=,U}(s)$ for every $s_3 \in R_i(s_2)$. Therefore, $S_i^{SMR_c} \subseteq S_i^{SMR_d}$.

The inclusion relation $S_i^{SMR_a} \subseteq S_i^{SMR_c} \subseteq S_i^{SMR_d}$ is proved. Other inclusion relations about GMR and SEQ can be proved similarly. So

$$S_i^{GS_a} \subseteq S_i^{GS_c} \subseteq S_i^{GS_d}.$$

The proof of the inclusion relations

$$S_i^{GS_a} \subseteq S_i^{GS_b} \subseteq S_i^{GS_d}$$

can be similarly carried out. □

The above relationships among stabilities indexed a, b, c and d are shown in Fig. 5.6.

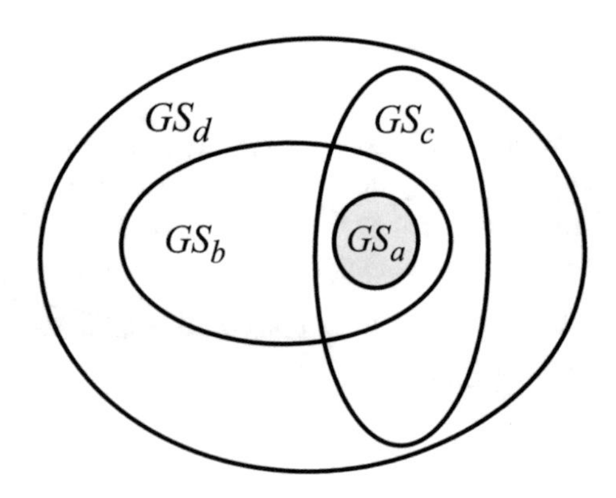

Fig. 5.6 Relationships among stabilities indexed a, b, c and d

5.3 Matrix Representation of Stability Definitions Under Unknown Preference

In this section, the matrix representation of solution concepts (MRSC) introduced in Sect. 4.3 is extended to conflicts with preference uncertainty. Below, procedures that define the four basic graph model stability definitions are applied to graph models with preference uncertainty using explicit matrix calculations. Matrix expressions are extended to define the reachable list of a DM, or a coalition, from a state by unilateral improvements or uncertain moves (UIUMs). These matrix expressions are used to develop an explicit algebraic form conflict model that facilitates stability calculations in both two-DM and n-DM ($n > 2$) models with preference uncertainty.

5.3.1 *Preference Matrices Including Uncertainty*

DM i's preference matrix P_i^+ and indifference matrix $P_i^=$ have been defined in Sect. 4.3.1.

Definition 5.35 For a graph model G, the uncertain preference matrix for DM i is the $m \times m$ matrix, P_i^U, with (s, q) entry $P_i^U(s,q) = \begin{cases} 1 \text{ if } s \; U_i \; q, \\ 0 \text{ otherwise.} \end{cases}$

Then define matrices $P_i^{-,=}$, $P_i^{+,U}$, and $P_i^{-,=,U}$ as

$$P_i^{-,=} = P_i^- \vee P_i^=,\; P_i^{+,U}(s,q) = E - I - P_i^{-,=}, \text{ and } P_i^{-,=,U}(s,q) = E - I - P_i^+.$$

The adjacency matrix is extended to a graph model with uncertain preference.

Definition 5.36 For a graph model G, DM i's UUM adjacency matrix is the $m \times m$ matrix, J_i^U, with (s, q) entries

$$J_i^U(s,q) = \begin{cases} 1 \text{ if } (s,q) \in A_i \text{ and } q \; U_i \; s, \\ 0 \text{ otherwise.} \end{cases}$$

Note that $J_i^U(s,q) = 1$ if and only if DM i can move from state s to state q by unilateral uncertain moves (in one step). It is convenient to define

$$J_i^{+,U}(s) = J_i^+ \bigvee J_i^U.$$

It follows that the UM adjacency matrix, UI adjacency matrix, UIUM adjacency matrix, and the preference matrices including uncertainty, satisfy the following relations:

$$J_i^+ = J_i \circ P_i^+, \quad \text{and} \quad J_i^{+,U} = J_i \circ P_i^{+,U}.$$

As has been observed previously, procedures to identify stable states based on logical representation definitions are difficult to code because of the nature of the logical representations. To overcome this limitation, the four stability definitions in two decision maker graph models with preference uncertainty are formulated explicitly in terms of matrices in the next subsection.

5.3.2 Two Decision Maker Case

Matrix representation of the four extensions of Nash, GMR, SMR, and SEQ stability definitions with preference uncertainty (MRSCU) in 2-DM conflict models is defined as follows. The system, called the MRSCU method, incorporates a set of $m \times m$ matrices, $M_i^{Nash_l}$, $M_i^{GMR_l}$, $M_i^{SMR_l}$, and $M_i^{SEQ_l}$, for $l \in \{a, b, c, d\}$, to capture $Nash_l$, GMR_l, SMR_l, and SEQ_l stabilities for DM $i \in N$, where $|N| = 2$, $m = |S|$, and DMs' preferences may include uncertainty.

5.3.2.1 Matrix Representation of Stabilities Indexed *a*

Define DM i's $m \times m$ $Nash_a$ stability matrix as

$$M_i^{Nash_a} = J_i^{+,U} \cdot E.$$

The following theorem establishes the matrix method to assess whether state s is $Nash_a$ stable for a DM.

Theorem 5.4 *State $s \in S$ is $Nash_a$ stable for DM i iff $M_i^{Nash_a}(s, s) = 0$.*

Proof It is obvious that $R_i^{+,U}(s) = \emptyset$ iff $e_s^T \cdot J_i^{+,U} = \overrightarrow{0}^T$. $e_s^T \cdot J_i^{+,U} = \overrightarrow{0}^T$ is equivalent to $M_i^{Nash_a}(s, s) = 0$. □

Define DM i's $m \times m$ GMR_a stability matrix as

$$M_i^{GMR_a} = J_i^{+,U} \cdot [E - sign\left(J_j \cdot (P_i^{-,=})^T\right)].$$

The following theorem provides a matrix method to calculate DM i's GMR_a stability.

Theorem 5.5 *State s is GMR_a stable for DM i iff*

$$M_i^{GMR_a}(s, s) = 0. \tag{5.1}$$

Proof Equation 5.1 is equivalent to

$$(e_s^T J_i^{+,U}) \cdot \left(\left(E - sign\left(J_j \cdot (P_i^{-,=})^T\right)\right) e_s\right) = 0.$$

Since

$$(e_s^T J_i^{+,U}) \cdot \left(\left(E - sign\left(J_j \cdot (P_i^{-,=})^T \right) \right) e_s \right)$$

$$= \sum_{s_1=1}^{m} J_i^{+,U}(s, s_1) \left(1 - sign\left((e_{s_1}^T J_j) \cdot (e_s^T P_i^{-,=})^T \right) \right),$$

then Eq. 5.1 holds iff

$$J_i^{+,U}(s, s_1)[1 - sign\left((e_{s_1}^T J_j) \cdot (e_s^T P_i^{-,=})^T \right)] = 0, \forall s_1 \in S. \tag{5.2}$$

It is clear that Eq. 5.2 is equivalent to

$$(e_{s_1}^T J_j) \cdot (e_s^T P_i^{-,=})^T \neq 0, \forall s_1 \in R_i^{+,U}(s),$$

which implies that, for any $s_1 \in R_i^{+,U}(s)$, there exists at least one $s_2 \in R_j(s_1)$ with $s \succeq_i s_2$. □

Define the $m \times m$ SMR_a stability matrix as

$$M_i^{SMR_a} = J_i^{+,U} \cdot [E - sign(Q)],$$

with

$$Q = J_j \cdot [(P_i^{-,=})^T \circ \left(E - sign\left(J_i \cdot (P_i^{+,U})^T \right) \right)], \text{ for } j \neq i.$$

Theorem 5.6 *State s is SMR_a stable for DM i iff*

$$M_i^{SMR_a}(s, s) = 0. \tag{5.3}$$

Proof Since

$$M_i^{SMR_a}(s, s) = (e_s^T J_i^{+,U}) \cdot [(E - sign(Q)) e_s]$$

$$= \sum_{s_1=1}^{m} J_i^{+,U}(s, s_1)[1 - sign\left(Q(s_1, s) \right)]$$

with

$$Q(s_1, s) = \sum_{s_2=1}^{m} J_j(s_1, s_2) \cdot W,$$

and

$$W = P_i^{-,=}(s, s_2)[1 - sign\left(\sum_{s_3=1}^{m} \left(J_i(s_2, s_3) P_i^{+,U}(s, s_3) \right) \right)],$$

then Eq. 5.3 holds iff $Q(s_1, s) \neq 0$, $\forall s_1 \in R_i^{+,U}(s)$, which is equivalent to the statement that, $\forall s_1 \in R_i^{+,U}(s)$, $\exists s_2 \in R_j(s_1)$ such that

$$P_i^{-,=}(s, s_2) \neq 0, \tag{5.4}$$

and

$$\sum_{s_3=1}^{m} \left(J_i(s_2, s_3) P_i^{+,U}(s, s_3) \right) = 0. \tag{5.5}$$

Obviously, $\forall s_1 \in R_i^{+,U}(s)$, $\exists s_2 \in R_j(s_1)$ such that Eqs. 5.4 and 5.5 hold iff for every $s_1 \in R_i^{+,U}(s)$ there exists $s_2 \in R_j(s_1)$ such that $s \succeq_i s_2$ and $s \succeq_i s_3$ for all $s_3 \in R_i(s_2)$. □

Define the $m \times m$ SEQ_a stability matrix as

$$M_i^{SEQ_a} = J_i^{+,U} \cdot [E - sign\left(J_j^{+,U} \cdot (P_i^{-,=})^T\right)].$$

Theorem 5.7 *State $s \in S$ is SEQ_a stable for DM i iff*

$$M_i^{SEQ_a}(s, s) = 0. \tag{5.6}$$

Proof Equation 5.6 is equivalent to

$$(e_s^T J_i^{+,U}) \cdot [\left(E - sign\left(J_j^{+,U} \cdot (P_i^{-,=})^T\right)\right) e_s] = 0.$$

Since

$$(e_s^T J_i^{+,U}) \cdot [\left(E - sign\left(J_j^{+,U} \cdot (P_i^{-,=})^T\right)\right) e_s]$$

$$= \sum_{s_1=1}^{m} J_i^{+,U}(s, s_1)[1 - sign\left((e_{s_1}^T J_j^{+,U}) \cdot (e_s^T P_i^{-,=})^T\right)],$$

then Eq. 5.6 holds iff

$$J_i^{+,U}(s, s_1)[1 - sign\left((e_{s_1}^T J_j^{+,U}) \cdot (e_s^T P_i^{-,=})^T\right)] = 0, \forall s_1 \in S. \tag{5.7}$$

It is clear that Eq. 5.7 is equivalent to

$$(e_{s_1}^T J_j^{+,U}) \cdot (e_s^T P_i^{-,=})^T \neq 0, \forall s_1 \in R_i^{+,U}(s).$$

It implies that for any $s_1 \in R_i^{+,U}(s)$, there exists at least one $s_2 \in R_j^{+,U}(s_1)$ with $s \succeq_i s_2$. □

5.3.2.2 Matrix Representation of Stabilities Indexed *b*

For the next definitions indexed using b for an uncertain situation, DM i considers to leave a state or assesses sanctions, excluding uncertain preferences.

Define the $m \times m$ $Nash_b$ stability matrix as

$$M_i^{Nash_b} = J_i^+ \cdot E.$$

Theorem 5.8 *State $s \in S$ is $Nash_b$ stable for DM i iff $M_i^{Nash_b}(s, s) = 0$.*

Define DM i's $m \times m$ GMR_b stability matrix as

$$M_i^{GMR_b} = J_i^+ \cdot [E - sign\left(J_j \cdot (P_i^{-,=})^T\right)].$$

Theorem 5.9 *State $s \in S$ is GMR_b stable* for *DM i iff $M_i^{GMR_b}(s, s) = 0$.*

Define DM i's $m \times m$ SMR_b stability matrix as

$$M_i^{SMR_b} = J_i^+ \cdot [E - sign(Q)],$$

with

$$Q = J_j \cdot [(P_i^{-,=})^T \circ \left(E - sign\left(J_i \cdot (P_i^{+,U})^T\right)\right)], \text{ for } j \neq i.$$

Theorem 5.10 *State s is SMR_b stable for DM i iff $M_i^{SMR_b}(s, s) = 0$.*

Define DM i's $m \times m$ SEQ_b matrix as

$$M_i^{SEQ_b} = J_i^+ \cdot [E - sign\left(J_j^{+,U} \cdot (P_i^{-,=})^T\right)].$$

Theorem 5.11 *State $s \in S$ is SEQ_b stable for DM i iff $M_i^{SEQ_b}(s, s) = 0$.*

The proofs of Theorems 5.8–5.11 are similar to those for theorems presented in Sect. 5.3.2.1, so they are left as an exercise. For the extended definitions indexed c, DM i has an incentive to move to states with uncertain preferences relative to the status quo, and will consider a move to a state with uncertain preference to be a sanction.

5.3.2.3 Matrix Representation of Stabilities Indexed *c*

Define DM i's $m \times m$ $Nash_c$ stability matrix as

$$M_i^{Nash_c} = J_i^{+,U} \cdot E.$$

Theorem 5.12 *State $s \in S$ is $Nash_c$ stable for DM i iff $M_i^{Nash_c}(s, s) = 0$.*

Define DM i's $m \times m$ GMR_c stability matrix as

$$M_i^{GMR_c} = J_i^{+,U} \cdot [E - sign\left(J_j \cdot (P_i^{-,=,U})^T\right)].$$

Theorem 5.13 *State s is GMR_c stable for DM i iff $M_i^{GMR_c}(s, s) = 0$.*

Define DM i's $m \times m$ SMR_c stability matrix as

$$M_i^{SMR_c} = J_i^{+,U} \cdot [E - sign(Q)],$$

in which

$$Q = J_j \cdot [(P_i^{-,=,U})^T \circ \left(E - sign\left(J_i \cdot (P_i^{+})^T\right)\right)], \text{ for } j \neq i.$$

Theorem 5.14 *State s is SMR_c stable for DM i iff $M_i^{SMR_c}(s, s) = 0$.*

Define DM i's $m \times m$ SEQ_c stability matrix as

$$M_i^{SEQ_c} = J_i^{+,U} \cdot [E - sign\left(J_j^{+,U} \cdot (P_i^{-,=,U})^T\right)].$$

Theorem 5.15 *State s is SEQ_c stable for DM i iff $M_i^{SEQ_c}(s, s) = 0$.*

5.3.2.4 Matrix Representation of Stabilities Indexed *d*

For the last definitions indexed d, DM i considers to leave a state, excluding preference uncertainty, but will consider a move to a state with uncertain preference to be a sanction.

Define DM i's $m \times m$ $Nash_d$ stability matrix as

$$M_i^{Nash_d} = J_i^{+} \cdot E.$$

Theorem 5.16 *State s is $Nash_d$ stable for DM i iff $M_i^{Nash_d}(s, s) = 0$.*

Define DM i's $m \times m$ GMR_d stability matrix as

$$M_i^{GMR_d} = J_i^+ \cdot [E - sign\left(J_j \cdot (P_i^{-,=,U})^T\right)].$$

Theorem 5.17 *State s is GMR_d stable for DM i iff* $M_i^{GMR_d}(s, s) = 0$.

Define DM i's $m \times m$ SMR_d stability matrix as

$$M_i^{SMR_d} = J_i^+ \cdot [E - sign(Q)],$$

in which

$$Q = J_j \cdot [(P_i^{-,=,U})^T \circ \left(E - sign\left(J_i \cdot (P_i^+)^T\right)\right)], \text{ for } j \neq i.$$

Theorem 5.18 *State s is SMR_d stable for DM i iff* $M_i^{SMR_d}(s, s) = 0$.

Define DM i's $m \times m$ SEQ_d stability matrix as

$$M_i^{SEQ_d} = J_i^+ \cdot [E - sign\left(J_j^{+,U} \cdot (P_i^{-,=,U})^T\right)].$$

Theorem 5.19 *State s is SEQ_d stable for DM i iff* $M_i^{SEQ_d}(s, s) = 0$.

Example 5.4 (Matrix Representation for the Extended Sustainable Development Model with Preference Uncertainty) The extended sustainable development model with preference uncertainty was described in Examples 5.1 and 5.2. The graph model of this conflict is shown in Fig. 5.2 with the state set $S = \{s_1, s_2, s_3, s_4\}$ and the DM set $N = \{1, 2\}$. In this example, it will be assessed whether the four states are stable for the stabilities indexed a, b, c, and d using the proposed matrix method. The stability matrices are summarized in Table 5.5. The procedures to calculate stabilities for the extended sustainable development model with preference uncertainty using the proposed matrix method are as follows:

1. Using Fig. 5.2, the UM adjacency matrices for DM 1 and DM 2 are obtained by
$$J_1 = \begin{pmatrix} 0\,0\,1\,0 \\ 0\,0\,0\,1 \\ 1\,0\,0\,0 \\ 0\,1\,0\,0 \end{pmatrix} \text{ and } J_2 = \begin{pmatrix} 0\,1\,0\,0 \\ 1\,0\,0\,0 \\ 0\,0\,0\,1 \\ 0\,0\,1\,0 \end{pmatrix}.$$
2. Using preference information provided by the graph model in Fig. 5.2, preference matrices for DM 1 and DM 2 respectively are
$$P_1^+ = \begin{pmatrix} 0\,0\,0\,0 \\ 1\,0\,1\,0 \\ 1\,0\,0\,0 \\ 1\,1\,1\,0 \end{pmatrix}, P_1^- = \begin{pmatrix} 0\,1\,1\,1 \\ 0\,0\,0\,1 \\ 0\,1\,0\,1 \\ 0\,0\,0\,0 \end{pmatrix}, P_1^U = \begin{pmatrix} 0\,0\,0\,0 \\ 0\,0\,0\,0 \\ 0\,0\,0\,0 \\ 0\,0\,0\,0 \end{pmatrix}, \text{ and}$$

Table 5.5 Stability matrices for two-DM conflicts with preference uncertainty

Preference	Sets of definitions	Stability matrices
Including uncertainty	a	$M_i^{Nash_a} = J_i^{+,U} \cdot E$
		$M_i^{GMR_a} = J_i^{+,U} \cdot [E - sign\left(J_j \cdot (P_i^{-,=})^T\right)]$
		$M_i^{SMR_a} = J_i^{+,U} \cdot [E - sign(J_j \cdot Q)]$, with $Q = (P_i^{-,=})^T \circ [E - sign(J_i \cdot (P_i^{+,U})^T)]$
		$M_i^{SEQ_a} = J_i^{+,U} \cdot [E - sign\left(J_j^{+,U} \cdot (P_i^{-,=})^T\right)]$
	b	$M_i^{Nash_b} = J_i^{+} \cdot E$
		$M_i^{GMR_b} = J_i^{+} \cdot [E - sign\left(J_j \cdot (P_i^{-,=})^T\right)]$
		$M_i^{SMR_b} = J_i^{+} \cdot [E - sign(J_j \cdot Q)]$, with $Q = (P_i^{-,=})^T \circ [E - sign(J_i \cdot (P_i^{+,U})^T)]$
		$M_i^{SEQ_b} = J_i^{+} \cdot [E - sign\left(J_j^{+,U} \cdot (P_i^{-,=})^T\right)]$
	c	$M_i^{Nash_c} = J_i^{+,U} \cdot E$
		$M_i^{GMR_c} = J_i^{+,U} \cdot [E - sign\left(J_j \cdot (P_i^{-,=,U})^T\right)]$
		$M_i^{SMR_c} = J_i^{+,U} \cdot [E - sign(J_j \cdot Q)]$, with $Q = (P_i^{-,=,U})^T \circ [E - sign\left(J_i \cdot (P_i^{+})^T\right)]$
		$M_i^{SEQ_c} = J_i^{+,U} \cdot [E - sign\left(J_j^{+,U} \cdot (P_i^{-,=,U})^T\right)]$
	d	$M_i^{Nash_d} = J_i^{+} \cdot E$
		$M_i^{GMR_d} = J_i^{+} \cdot [E - sign\left(J_j \cdot (P_i^{-,=,U})^T\right)]$
		$M_i^{SMR_d} = J_i^{+} \cdot [E - sign(J_j \cdot Q)]$, with $Q = (P_i^{-,=,U})^T \circ [E - sign\left(J_i \cdot (P_i^{+})^T\right)]$
		$M_i^{SEQ_d} = J_i^{+} \cdot [E - sign\left(J_j^{+,U} \cdot (P_i^{-,=,U})^T\right)]$

$$P_2^{+} = \begin{pmatrix} 0\,0\,1\,0 \\ 0\,0\,0\,1 \\ 0\,0\,0\,0 \\ 0\,0\,0\,0 \end{pmatrix}, \; P_2^{-} = \begin{pmatrix} 0\,0\,1\,0 \\ 0\,0\,0\,1 \\ 1\,0\,0\,0 \\ 0\,1\,0\,0 \end{pmatrix}, \; P_2^{U} = \begin{pmatrix} 0\,1\,0\,1 \\ 1\,0\,1\,0 \\ 0\,1\,0\,1 \\ 1\,0\,1\,0 \end{pmatrix}.$$

3. Other preference matrices are calculated using $P_i^{+,U} = P_i^{+} \bigvee P_i^{U}$, $P_i^{-,=} = E - I - P_i^{+,U}$, and $P_i^{-,=,U} = E - I - P_i^{+}$ for $i = 1, 2$.
4. Using the mathematical formulation of stability matrices presented in Table 5.5, all diagonal entries of these stability matrices are calculated and given in Table 5.6.

Note that in the second column of Table 5.6, 1, 2, 3, and 4 denote state numbers and in the second row 1 and 2 indicate DM 1 and DM 2. The column vector $(0, 0, 1, 1)^T$ corresponding to the extension a and Nash stability for DM 1 is the diagonal vector of DM 1's $Nash_a$ stability matrix $M_1^{Nash_a}$, i.e., $diag(M_1^{Nash_a}) = (0, 0, 1, 1)^T$.

Table 5.6 Diagonal entries of stability matrices for the extended sustainable development game with uncertain preference

State		Nash		GMR		SMR		SEQ	
		1	2	1	2	1	2	1	2
a	1	0	1	0	1	0	1	0	1
	2	0	1	0	1	0	1	0	1
	3	1	1	0	1	0	1	0	1
	4	1	1	1	1	1	1	1	1
b	1	0	0	0	0	0	0	0	0
	2	0	0	0	0	0	0	0	0
	3	1	0	0	0	0	0	0	0
	4	1	0	1	0	1	0	1	0
c	1	0	1	0	0	0	1	0	1
	2	0	1	0	0	0	1	0	1
	3	1	1	0	0	0	0	0	0
	4	1	1	1	0	1	0	1	0
d	1	0	0	0	0	0	0	0	0
	2	0	0	0	0	0	0	0	0
	3	1	0	0	0	0	0	0	0
	4	1	0	1	0	1	0	1	0

Based on Theorem 5.4, since $M_1^{Nash_a}(s_1, s_1) = M_1^{Nash_a}(s_2, s_2) = 0$, then s_1 and s_2 are $Nash_a$ stable for DM 1. The column vector $(1, 1, 0, 0)^T$ corresponding to the extension c and SMR stability for DM 2 is the diagonal vector of DM 2's SMR_c stability matrix $M_2^{SMR_c}$, i.e., $diag(M_2^{SMR_c}) = (1, 1, 0, 0)^T$. Therefore, $M_1^{SMR_c}(s_3, s_3) = M_2^{SMR_c}(s_4, s_4) = 0$, which imply that s_3 and s_4 are SMR_c stable for DM 2 by Theorem 5.14. Other stabilities can be similarly analyzed. It is easy to see that the results of the stabilities are precisely the same as those presented in Table 5.3 which is obtained using the logical method.

Theorems 5.4–5.19 prove that the proposed matrix representation of solution concepts are equivalent to the solution concepts for two-DM conflicts defined in Sect. 5.2.1. The matrix representation can be extended to models including more than two DMs, which is the objective of the next subsection.

5.3.3 Reachability Matrices for a Coalition

Fix $H \subseteq N$ such that $|H| \geq 2$, and let $s \in S$. Now it is demonstrated how to find matrices corresponding to $R_H(s)$, the reachable list of H from s by legal sequences of UMs, $R_H^+(s)$, the reachable list of H from s by legal sequences of UIs, and $R_H^{+,U}$, the reachable list of H from s by legal sequences of UIUMs.

Definition 5.37 For $i \in H$, $H \subseteq N$, and $t = 1, 2, 3, \ldots$, define the $m \times m$ matrices $M_i^{(t)}$, $M_i^{(t,+)}$, and $M_i^{(t,+,U)}$ with (s, q) entries as follows:

$$M_i^{(t)}(s,q) = \begin{cases} 1 \text{ if } q \in S \text{ is reachable from } s \in S \text{ in exactly} \\ \quad t \text{ legal UMs by } H \text{ with last mover DM } i, \\ 0 \text{ otherwise,} \end{cases}$$

$$M_i^{(t,+)}(s,q) = \begin{cases} 1 \text{ if } q \in S \text{ is reachable from } s \in S \text{ in exactly} \\ \quad t \text{ legal UIs by } H \text{ with last mover DM } i, \\ 0 \text{ otherwise.} \end{cases}$$

and

$$M_i^{(t,+,U)}(s,q) = \begin{cases} 1 \text{ if } q \in S \text{ is reachable from } s \in S \text{ in exactly} \\ \quad t \text{ legal UIUMs by } H \text{ with last mover DM } i, \\ 0 \text{ otherwise.} \end{cases}$$

Based on Definition 5.37, one has:

Lemma 5.1 *For $i \in N$ and $H \subseteq N$, the three matrices $M_i^{(t)}$, $M_i^{(t,+)}$ and $M_i^{(t,+,U)}$ satisfy*

$$M_i^{(1)}(s,q) = J_i(s,q) \text{ and, for } t = 2, 3, \ldots, M_i^{(t)} = sign[(\bigvee_{j \in H-\{i\}} M_j^{(t-1)}) \cdot J_i], \tag{5.8}$$

$$M_i^{(1,+)}(s,q) = J_i^+(s,q) \text{ and, for } t = 2, 3, \ldots, M_i^{(t,+)} = sign[(\bigvee_{j \in H-\{i\}} M_j^{(t-1,+)}) \cdot J_i^+], \tag{5.9}$$

$$M_i^{(1,+,U)}(s,q) = J_i^{+,U}(s,q) \text{ and, for } t = 2, 3, \ldots, M_i^{(t,+,U)} = sign[(\bigvee_{j \in H-\{i\}} M_j^{(t-1,+,U)}) \cdot J_i^{+,U}]. \tag{5.10}$$

Proof The verification of Eqs. 5.8 and 5.9 is similar to Eq. 5.10. Here, Eq. 5.10 is verified. For $t = 2$, the definition of matrix multiplication shows that $G(s, q)$, the (s, q) entry of the matrix $G = (\bigvee_{j \in H-\{i\}} J_j^{+,U}) \cdot J_i^{+,U}$, is nonzero iff state q is reachable from state s in exactly two UIUMs, with last mover DM i. The condition $j \in H - \{i\}$ implies that DM i does not make two moves consecutively. Hence, $G(s, q) \neq 0$ iff state q is reachable from state s in exactly two legal UIUMs. Then

$$sign[(\bigvee_{j \in H-\{i\}} J_j^{+,U}) \cdot J_i^{+,U}] = sign[(\bigvee_{j \in H-\{i\}} M_j^{(1,+,U)}) \cdot J_i^{+,U}] = M_i^{(2,+,U)}.$$

Now suppose that $t > 2$. Since

$$M_j^{(t-1,+,U)}(s,q) = \begin{cases} 1 \text{ if } q \in S \text{ is reachable from } s \in S \text{ in exactly } t-1 \\ \quad \text{legal UIUMs by } H \text{ with last mover DM } j, \\ 0 \text{ otherwise,} \end{cases}$$

using matrix multiplication, matrix $B = sign[(\bigvee_{j\in H-\{i\}} M_j^{(t-1,+,U)}) \cdot J_i^{+,U}]$ has (s,q) entry

$$B(s,q) = \begin{cases} 1 \text{ if } q \in S \text{ is reachable from } s \in S \text{ in exactly } t \\ \quad \text{legal UIUMs by } H \text{ with last mover DM } i, \\ 0 \text{ otherwise,} \end{cases}$$

which confirms Eq. 5.10. □

The UM and UI reachability matrices have been defined by Definition 4.19 in Chap. 4. The two matrices are extended to the graph model including uncertain preference.

Definition 5.38 For the graph model G, the UIUM reachability matrix for H is the $m \times m$ matrix $M_H^{+,U}$ with (s,q) entry

$$M_H^{+,U}(s,q) = \begin{cases} 1 & \text{if } q \in R_H^{+,U}(s), \\ 0 & \text{otherwise.} \end{cases}$$

Obviously, $R_H^{+,U}(s) = \{q : M_H^{+,U}(s,q) = 1\}$. If $R_H^{+,U}(s)$ is written as a 0–1 row vector, then

$$R_H^{+,U}(s) = e_s^T \cdot M_H^{+,U},$$

where e_s^T denotes the transpose of the sth standard basis vector of the m-dimensional Euclidean space. Therefore, the UIUM reachability matrix for coalition H, M_H, can be used to construct the reachable lists of H from state s by the legal sequence of UIUMs, $R_H^{+,U}(s)$.

Recall that A_i is DM i's arc set in a graph model. Let A_i^+ and $A_i^{+,U}$ denote i's UI arc set and UIUM arc set, respectively. For $s \in S$, let $A_i(s)$, $A_i^+(s)$, and $A_i^{+,U}(s)$ denote the respective subsets of these three sets with initial state s. Therefore, these arc sets are related by $A_i = \bigcup_{s\in S} A_i(s)$, $A_i^+ = \bigcup_{s\in S} A_i^+(s)$, and $A_i^{+,U} = \bigcup_{s\in S} A_i^{+,U}(s)$.

It is obvious that unilateral moves on the branches of paths will end when the same arc appears twice. Generally, if there is no new appropriate arc produced, then the corresponding joint moves will stop. Therefore, Lemma 5.1 can be extended to the following cases. Let $l_3 = |A_H^{+,U}|$ in the following lemma.

Lemma 5.2 *For a graph model G, let $H \subseteq N$. $R_H^{+,U}(s)$ is the reachable list of H by the legal sequences of UIUMs from s. δ_3 is the number of iteration steps required to find $R_H^{+,U}(s)$. Then $\delta_3 \le l_3$.*

Then the following theorem can be derived using Lemmas 5.1 and 5.2.

Theorem 5.20 *Let $s \in S$, $H \subseteq N$, and $H \neq \emptyset$. The reachability matrices M_H, M_H^+ and $M_H^{+,U}$ by H can be respectively expressed by*

$$M_H = \bigvee_{t=1}^{l_1} \bigvee_{i\in H} M_i^{(t)}, \tag{5.11}$$

$$M_H^+ = \bigvee_{t=1}^{l_2} \bigvee_{i \in H} M_i^{(t,+)}, \tag{5.12}$$

and

$$M_H^{+,U} = \bigvee_{t=1}^{l_3} \bigvee_{i \in H} M_i^{(t,+,U)}. \tag{5.13}$$

Proof The proofs of Eqs. 5.11 and 5.12 are left as exercises. Here, only Eq. 5.13 is proven. Assume that $C = \bigvee_{t=1}^{l_3} \bigvee_{i \in H} M_i^{(t,+,U)}$. Using the definition for matrix $M_H^{+,U}$, $M_H^{+,U}(s,q) = 1$ iff $q \in R_H^{+,U}(s)$. Since $l_3 = |\bigcup_{i \in N} A_i^{+,U}|$, then, by Lemma 5.2, $l_3 \geq \delta_3$. Therefore, by Definition 5.37, $q \in R_H^{+,U}(s)$ implies that there exists $1 \leq t_0 \leq \delta_3$ and $i_0 \in H$ such that $M_{i_0}^{(t_0,+,U)}(s,q) = 1$. This implies that matrix C has (s,q) entry 1. Therefore, $M_H^{+,U}(s,q) = 1$ iff $C(s,q) = 1$. Since $M_H^{+,U}$ and C are 0–1 matrices, it follows that $M_H^{+,U} = C$. □

Compared to the approach presented in Theorem 4.9 based on the incidence matrix, the algebraic method founded on the adjacency matrix developed here is easier for calculation purposes. However, the incidence matrix-based method can trace multiple edges.

5.3.4 *Multiple Decision Maker Case*

The logical definitions of Nash, GMR, SMR, and SEQ stabilities in the graph model for multiple decision maker conflict models with preference uncertainty are described in Sect. 5.2.3. They retain most features of the stability definitions in the 2-DM case, except that DM i's opponents are a subset of N, $N - \{i\}$, instead of a single opponent, j. It is obvious that in the n-DM case, the algebraic characterizations of stabilities are similar to those presented in Sect. 5.3.2. Consequently, matrix representation of solution concepts with preference uncertainty for 2-DM cases can be easily extended to that for n-DM situations. Let $i \in N$ and $|N| = n$ in the following theorems.

5.3.4.1 Matrix Representation of Stabilities Indexed *a* for Preference with Uncertainty

In the definitions indexed a, DM i has an incentive to move to states with uncertain preferences relative to the status quo, but, when assessing possible sanctions, will not consider states with uncertain preferences (Li et al. 2004).

Theorem 5.21 *State $s \in S$ is $Nash_a$ stable for DM i iff* $(e_s)^T \cdot J_i^{+,U} \cdot e = 0$.

Theorem 5.21 implies that Nash stability definitions are identical for both 2-DM and n-DM models with preference uncertainty because Nash stability does not consider opponents' responses.

For GMR, DM i considers the opponents' responses, which are reachable states $R_{N\backslash\{i\}}$ of coalition $H = N\backslash\{i\}$ by the legal UM sequences. First, the matrix $M_{N\backslash\{i\}}$ using Theorem 5.20 can be constructed for $H = N \setminus \{i\}$. Define the $m \times m$ GMR_a stability matrix as

$$M_i^{GMR_a} = J_i^{+,U} \cdot [E - sign\left(M_{N\backslash\{i\}} \cdot (P_i^{-,=})^T\right)].$$

Then the following theorem provides a matrix method to calculate GMR_a stability.

Theorem 5.22 *State $s \in S$ is GMR_a stable for DM i, denoted by $s \in S_i^{GMR_a}$, iff* $M_i^{GMR_a}(s, s) = 0$.

Proof Since the diagonal entry of matrix $M_i^{GMR_a}$

$$M_i^{GMR_a}(s, s) = (e_s^T \cdot J_i^{+,U}) \cdot [\left(E - sign\left(M_{N\backslash\{i\}} \cdot (P_i^{-,=})^T\right)\right) \cdot e_s]$$

$$= \sum_{s_1=1}^{m} J_i^{+,U}(s, s_1)[1 - sign\left((e_{s_1}^T \cdot M_{N\backslash\{i\}}) \cdot (e_s^T \cdot P_i^{-,=})^T\right)],$$

then $M_i^{GMR_a}(s, s) = 0$ iff $J_i^{+,U}(s, s_1)[1 - sign\left((e_{s_1}^T \cdot M_{N\backslash\{i\}}) \cdot (e_s^T \cdot P_i^{-,=})^T\right)] = 0$ for any $s_1 \in S$. This implies that $M_i^{GMR_a}(s, s) = 0$ iff

$$(e_{s_1}^T \cdot M_{N\backslash\{i\}}) \cdot (e_s^T \cdot P_i^{-,=})^T \neq 0 \quad \text{for any } s_1 \in R_i^{+,U}(s). \tag{5.14}$$

By Eq. 5.14, for any $s_1 \in R_i^{+,U}(s)$, there exists $s_2 \in S$, such that the m-dimensional row vector, $e_{s_1}^T \cdot M_{N\backslash\{i\}}$, has the s_2th element 1 and the m-dimensional column vector, $(P_i^{-,=})^T \cdot e_s$, has the s_2th element 1.

Therefore, $M_i^{GMR_a}(s, s) = 0$ iff for any the $s_1 \in R_i^{+,U}(s)$, there exists at least one the $s_2 \in R_{N\backslash\{i\}}(s_1)$ with $s \succeq_i s_2$. □

Symmetric metarationality in the n-DM model is similar to in the 2-DM model. The only modification is that responses are from DM i's opponents instead of a single DM. Let $D = (P_i^{-,=})^T \circ [E - sign\left(J_i \cdot (P_i^{+,U})^T\right)]$, then define the $m \times m$ SMR_a stability matrix as

$$M_i^{SMR_a} = J_i^{+,U} \cdot [E - sign(M_{N\backslash\{i\}} \cdot D)].$$

Thus, the following theorem provides a matrix method to calculate SMR_a stability.

Theorem 5.23 *State $s \in S$ is SMR_a stable for DM i, denoted by $s \in S_i^{SMR_a}$, iff $M_i^{SMR_a}(s, s) = 0$.*

Proof Let $G = M_{N\backslash\{i\}} \cdot D$. Since the diagonal element of matrix $M_i^{SMR_a}$

$$M_i^{SMR_a}(s, s) = (e_s^T \cdot J_i^{+,U}) \cdot [(E - sign(G)) \cdot e_s]$$

$$= \sum_{s_1=1}^{m} J_i^{+,U}(s, s_1)[1 - sign\,(G(s_1, s))]$$

with

$$G(s_1, s) = \sum_{s_2=1}^{m} M_{N\backslash\{i\}}(s_1, s_2) \cdot P_i^{-,=}(s, s_2)[1 - sign\left(\sum_{s_3=1}^{m}\left(J_i(s_2, s_3)P_i^{+,U}(s, s_3)\right)\right)],$$

thus, $M_i^{SMR_a}(s, s) = 0$ holds iff $G(s_1, s) \neq 0$ for any $s_1 \in R_i^{+,U}(s)$, which is equivalent to the statement that, for any $s_1 \in R_i^{+,U}(s)$ there exists $s_2 \in R_{N\backslash\{i\}}(s_1)$ such that

$$P_i^{-,=}(s, s_2) \neq 0, \quad \text{and} \quad \sum_{s_3=1}^{m}\left(J_i(s_2, s_3)P_i^{+,U}(s, s_3)\right) = 0. \tag{5.15}$$

Obviously, for any $s_1 \in R_i^{+,U}(s)$ there exists $s_2 \in R_{N\backslash\{i\}}(s_1)$ such that Eq. 5.15 holds iff for every $s_1 \in R_i^{+,U}(s)$ there exists $s_2 \in R_{N\backslash\{i\}}(s_1)$ such that $s \succeq_i s_2$ and $s \succeq_i s_3$ for all $s_3 \in R_i(s_2)$. □

SEQ is similar to GMR, but includes only those sanctions that are "credible" (unilaterally improved) or uncertain moves, i.e., SEQ examines the credibility and uncertainty of the sanctions by DM i's opponents. First, the matrix $M_{N\backslash\{i\}}^{+,U}$ using Theorem 5.20 can be built. Define the $m \times m$ SEQ_a stability matrix $M_i^{SEQ_a}$ as

$$M_i^{SEQ_a} = J_i^{+,U} \cdot [E - sign\left(M_{N\backslash\{i\}}^{+,U} \cdot (P_i^{-,=})^T\right)].$$

Thus the following theorem provides a matrix method to calculate SEQ_a stability.

Theorem 5.24 *State $s \in S$ is SEQ_a stable for DM i, denoted by $s \in S_i^{SEQ_a}$, iff $M_i^{SEQ_a}(s, s) = 0$.*

Proof Since the diagonal element of matrix $M_i^{SEQ_a}$

$$M_i^{SEQ_a}(s, s) = (e_s^T \cdot J_i^{+,U}) \cdot [\left(E - sign\left(M_{N\backslash\{i\}}^{+,U} \cdot (P_i^{-,=})^T\right)\right) \cdot e_s]$$

$$= \sum_{s_1=1}^{m} J_i^{+,U}(s, s_1)[1 - sign\left((e_{s_1}^T \cdot M_{N\setminus\{i\}}^{+,U}) \cdot (e_s^T \cdot P_i^{-,=})^T\right)],$$

then $M_i^{SEQ_a}(s, s) = 0$ iff $J_i^{+,U}(s, s_1)[1 - sign\left((e_{s_1}^T \cdot M_{N\setminus\{i\}}^{+,U}) \cdot (e_s^T \cdot P_i^{-,=})^T\right)] = 0$ for any $s_1 \in S$. This implies that $M_i^{SEQ_a}(s, s) = 0$ iff

$$(e_{s_1}^T \cdot M_{N\setminus\{i\}}^{+,U}) \cdot (e_s^T \cdot P_i^{-,=})^T \neq 0 \text{ for any } s_1 \in R_i^{+,U}(s). \tag{5.16}$$

By Eq. 5.16, for any $s_1 \in R_i^{+,U}(s)$, there exists $s_2 \in S$ such that the m-dimensional row vector, $e_{s_1}^T \cdot M_{N\setminus\{i\}}^{+,U}$, has the s_2th element 1 and the m-dimensional column vector, $(P_i^{-,=})^T \cdot e_s$, has the s_2th element 1.

Therefore, $M_i^{SEQ_a}(s, s) = 0$ iff for any $s_1 \in R_i^{+,U}(s)$, there exists at least one $s_2 \in R_{N\setminus\{i\}}^{+,U}(s_1)$ with $s \succeq_i s_2$. □

$Nash_a$ stability means that the focal DM has no unilateral improvements or uncertain moves (UIUMs). GMR_a denotes that the UIUMs of the focal DM are sanctioned by subsequent unilateral moves by the opponents of the focal DM. SMR_a is similar to GMR_a, but the focal DM considers not only the responses from his opponents but also his own counterresponses. SEQ_a indicates that UIUMs of the focal DM are sanctioned by subsequent unilateral improvements or uncertain moves by the opponents of the focal DM.

For the next set of definitions indexed b, DM i considers to leave a state or assesses sanctions, excluding uncertain preferences (Li et al. 2004). However, the definitions are different from those including simple preference only (Fang et al. 1993), since the current definitions are utilized to analyze conflict models including preference uncertainty. Following theorems can be similarly verified as the above theorems.

5.3.4.2 Matrix Representation of Stabilities Indexed b for Preference with Uncertainty

Theorem 5.25 *State $s \in S$ is $Nash_b$ stable for DM i iff $(e_s)^T \cdot J_i^+ \cdot e = 0$.*

Define the $m \times m$ stability matrix $M_i^{GMR_b}$ as

$$M_i^{GMR_b} = J_i^+ \cdot [E - sign\left(M_{N\setminus\{i\}} \cdot (P_i^{-,=})^T\right)].$$

Theorem 5.26 *State $s \in S$ is GMR_b stable for DM i iff $M_i^{GMR_b}(s, s) = 0$.*

Define the $m \times m$ stability matrix $M_i^{SMR_b} = J_i^+ \cdot [E - sign(G)]$, with

$$G = M_{N\setminus\{i\}} \cdot [(P_i^{-,=})^T \circ \left(E - sign\left(J_i \cdot (P_i^{+,U})^T\right)\right)].$$

Theorem 5.27 *State $s \in S$ is SMR_b stable for DM i iff $M_i^{SMR_b}(s,s) = 0$.*

Define the $m \times m$ stability matrix $M_i^{SEQ_b}$ as

$$M_i^{SEQ_b} = J_i^{+} \cdot [E - sign\left(M_{N\setminus\{i\}}^{+,U} \cdot (P_i^{-,=})^T\right)].$$

Theorem 5.28 *State $s \in S$ is SEQ_b stable for DM i iff $M_i^{SEQ_b}(s,s) = 0$.*

5.3.4.3 Matrix Representation of Stabilities Indexed c for Preference with Uncertainty

For the extended definitions indexed c, DM i considers moving from a status quo state or evaluating sanctions including uncertain preferences.

Theorem 5.29 *State $s \in S$ is $Nash_c$ stable for DM i iff $(e_s)^T \cdot J_i^{+,U} \cdot e = 0$.*

Define the $m \times m$ stability matrix $M_i^{GMR_c}$ as

$$M_i^{GMR_c} = J_i^{+,U} \cdot [E - sign\left(M_{N\setminus\{i\}} \cdot (P_i^{-,=,U})^T\right)].$$

Theorem 5.30 *State $s \in S$ is GMR_c stable for DM i iff $M_i^{GMR_c}(s,s) = 0$.*

Define the $m \times m$ stability matrix $M_i^{SMR_c}$ as

$$M_i^{SMR_c} = J_i^{+,U} \cdot [E - sign(M_{N\setminus\{i\}} \cdot D)],$$

in which

$$D = (P_i^{-,=,U})^T \circ [E - sign\left(J_i \cdot (P_i^{+})^T\right)].$$

Theorem 5.31 *State $s \in S$ is SMR_c stable for DM i iff $M_i^{SMR_c}(s,s) = 0$.*

Define the $m \times m$ stability matrix $M_i^{SEQ_c}$ as

$$M_i^{SEQ_c} = J_i^{+,U} \cdot [E - sign\left(M_{N\setminus\{i\}}^{+,U} \cdot (P_i^{-,=,U})^T\right)].$$

Theorem 5.32 *State $s \in S$ is SEQ_c stable for DM i iff $M_i^{SEQ_c}(s,s) = 0$.*

5.3.4.4 Matrix Representation of Stabilities Indexed d for Preference with Uncertainty

For the last definitions, indexed d, a DM is not motivated to leave the status quo to move to states with uncertain preference, but will consider moving to states with uncertain preference to be a sanction.

Theorem 5.33 *State $s \in S$ is $Nash_d$ stable for DM i iff* $(e_s)^T \cdot J_i^+ \cdot e = 0$.

Define the $m \times m$ stability matrix as

$$M_i^{GMR_d} = J_i^+ \cdot [E - sign\left(M_{N\setminus\{i\}} \cdot (P_i^{-,=,U})^T\right)].$$

Theorem 5.34 *State $s \in S$ is GMR_d stable for DM i iff* $M_i^{GMR_d}(s,s) = 0$.

Define the $m \times m$ stability matrix $M_i^{SMR_d} = J_i^+ \cdot [E - sign(M_{N\setminus\{i\}} \cdot D)]$, in which $D = (P_i^{-,=,U})^T \circ [E - sign\left(J_i \cdot (P_i^+)^T\right)]$.

Theorem 5.35 *State $s \in S$ is SMR_d stable for DM i iff* $M_i^{SMR_d}(s,s) = 0$.

Define the $m \times m$ stability matrix $M_i^{SEQ_d} = J_i^+ \cdot [E - sign\left(M_{N\setminus\{i\}}^{+,U} \cdot (P_i^{-,=,U})^T\right)]$.

Theorem 5.36 *State $s \in S$ is SEQ_d stable for DM i iff* $M_i^{SEQ_d}(s,s) = 0$.

When $n = 2$, the DM set N becomes $\{i, j\}$ and Theorems 5.21–5.36 are reduced to Theorems 5.4–5.19.

5.3.5 *Computational Complexity*

A graph model structure can handle any finite number of states and DMs, each of whom controls any finite number of options. As pointed out by Fang et al. (2003a, b), an available decision support system (DSS) for stability analysis of a graph model for simple preference works well. In Sect. 4.4, the computational complexities of the matrix method and the graph model stability definitions for 2-DM models are compared, using GMR stability as an example. Both methods have complexity of $O(m^2)$, where m is the number of states. Now compare the GMR stability matrix $J_i^+ \cdot [E - sign\left(J_j \cdot (P_i^{-,=})^T\right)]$ and the GMR_a stability matrix $J_i^{+,U} \cdot [E - sign\left(J_j \cdot (P_i^{-,=})^T\right)]$ in the two-DM graph model for simple preference and for preference with uncertainty. Since $J_i^{+,U} = J_i^+ \vee J_i^U$, then in two-DM conflicts, MRSC and MRSCU have the same level of computational complexity.

In n-DM models, GMR stability was also selected as an example for analysis of the computational complexity of MRSC (see Sect. 4.4). The computational complexity of GMR stability for state s in n-DM models is less than $l \cdot (n-1) \cdot O(m^3) + O(m^2) = l \cdot (n-1) \cdot O(m^3)$, where $l = |\bigcup_{i \in N} A_i|$. This observation implies that the MRSC method is a polynomial-time effective algorithm. By comparing the GMR stability matrix $J_i^+ \cdot [E - sign\left(M_{N\setminus\{i\}} \cdot (P_i^{-,=})^T\right)]$ for simple preference with the GMR_a stability matrix $J_i^{+,U} \cdot [E - sign\left(M_{N\setminus\{i\}} \cdot (P_i^{-,=})^T\right)]$ for preference with uncertainty, it is not difficult to determine that the MRSCU method possesses a similar property.

5.4 Application: Lake Gisborne Conflict

In this section, the matrix method is demonstrated by using a practical problem. Lake Gisborne is located near the south coast of the Canadian Atlantic province of Newfoundland and Labrador. In June 1995, a project to export bulk water from Lake Gisborne to foreign markets was proposed by a division of the McCurdy Group of Companies, Canada Wet Incorporated. On December 5, 1996, the government of Newfoundland and Labrador approved this project because of its potential economic benefits. However, due to the risk of harmful impacts on local environment, a wide variety of lobby groups opposed the proposal. The Federal Government of Canada supported the opponents and introduced a policy to forbid bulk water export from major drainage basins in Canada. In response to this pressure, the government of Newfoundland and Labrador introduced a new bill to ban bulk water export from the province, forcing Canada Wet to abandon the Gisborne Water Export project (see details in Fang et al. (2002) and Li et al. (2004)).

Since several groups supported the project, an economical-oriented provincial government might have considered supporting it because of the urgent need for cash. However, an environmental-oriented provincial government might have opposed it because of the possibility of devastating environmental consequences. In 1999, it was unclear which of these two different attitudes described the provincial government's thinking, resulting in uncertainty in preferences in the Gisborne conflict. The details can be found in Li et al. (2004). A model of this conflict featuring uncertain preference uses three DMs: DM 1, **Federal (Fe)**, DM 2, **Provincial (Pr)**, and DM 3, **Support (Su)**, and a total of three options, which are presented in Table 5.7. The following is a summary of the three DMs and their options (Li et al. 2004):

- Federal government of Canada (**Federal**): its only option is to continue a Canada wide accord on the prohibition of bulk water exports (**Continue**), or not,
- Provincial government of Newfoundland and Labrador (**Provincial**): its only option is to lift the ban on bulk water exports (**Lift**), or not, and
- Support groups (**Support**): its only option is to appeal for continuation of the Gisborne project (**Appeal**), or not.

In the Lake Gisborne model, the three options are combined to form eight feasible states listed in Table 5.7, where a "Y" indicates that an option is selected by the DM

Table 5.7 Options and feasible states for the Gisborne conflict

Federal								
1. Continue	N	Y	N	Y	N	Y	N	Y
Provincial								
2. Lift	N	N	Y	Y	N	N	Y	Y
Support								
3. Appeal	N	N	N	N	Y	Y	Y	Y
State number	s_1	s_2	s_3	s_4	s_5	s_6	s_7	s_8

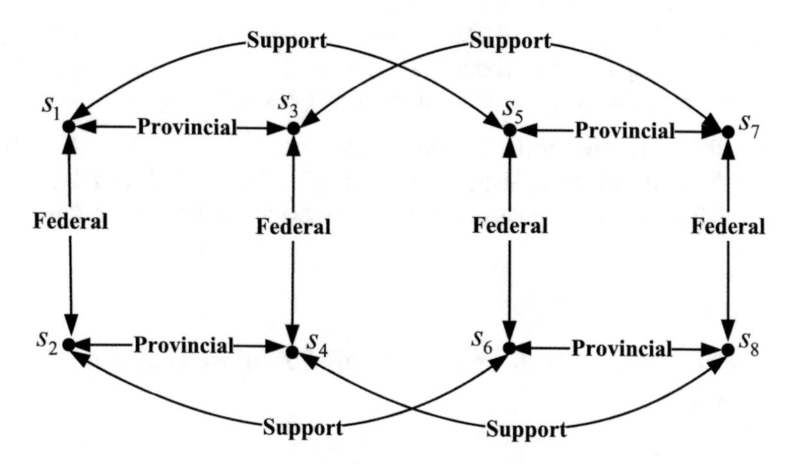

Fig. 5.7 Graph model for the Gisborne conflict

Table 5.8 Certain preference information for the Gisborne model

DMs	Certain preferences
Federal	$s_2 \succ s_6 \succ s_4 \succ s_8 \succ s_1 \succ s_5 \succ s_3 \succ s_7$
Provincial	$s_3 \succ s_7$, $s_4 \succ s_8$, $s_1 \succ s_5$, $s_2 \succ s_6$, only
Support	$s_3 \succ s_4 \succ s_7 \succ s_8 \succ s_5 \succ s_6 \succ s_1 \succ s_2$

controlling it and an "N" means that the option is not chosen. The graph model of the Lake Gisborne conflict is shown in Fig. 5.7. The labels on the arcs of the graph indicate the DM who can make the move. The certain preferences over states of DMs are given in Table 5.8, where $\succ$ denotes the strict preference and is transitive. Consequently, the preference information of DM **Federal** and DM **Support** is certain, but DM **Provincial** only knows that it prefers state s_3 to s_7, state s_4 to s_8, state s_1 to s_5, and state s_2 to s_6. The preference of DM **Provincial** is uncertain between other pair of states such as states s_3 and s_6. It is obvious that DM **Provincial**'s preference information includes uncertainty.

5.4.1 Procedures for Calculating Stability with Unknown Preference

The labeled graph of the Gisborne model, determined according to the Rule of Priority presented in Sect. 3.3.2, is depicted in Fig. 5.3b. To use the Gisborne model as an example to demonstrate how the matrix representation of the four basic solution concepts with preference uncertainty works, one can adhere to the following steps:

- Construct matrices, J_i, J_i^+, $J_i^{+,U}$, P_i^+, and $P_i^{-,=}$, for $i = 1, 2,$ and 3, using information provided by Fig. 5.7 and Table 5.8;
- Calculate the UM, UI, and UIUM reachability matrices by $H = N - \{i\}$ for $i = 1, 2,$ and 3, using inductive formulations provided by Theorem 5.20; and
- Analyze the stabilities of the model using Theorems 5.21–5.36.

At the first step, one constructs DM i's UM matrix, J_i, UI matrix, J_i^+, UIUM matrix, $J_i^{+,U}$, and preference matrices, P_i^+ and $P_i^{-,=}$, for $i = 1, 2,$ and 3 (Xu et al. 2007a, b). This and the previous chapters provide details on how to construct these matrices.

5.4.2 Reachability Matrices of a Coalition in the Gisborne Model

In multiple decision maker graph models, the UM reachability matrix, the UI reachability matrix, and the UIUM reachability matrix by a coalition are essential components for the MRSCU method. Using the Gisborne model as an example, the construction of the reachability matrices is shown and their results are analyzed next.

Let $N = \{1, 2, 3\}$ and $H = N - \{i\}, i = 1, 2,$ and 3. Let $L_1 = |\bigcup_{i \in N} A_i| = 24$, $L_2 = |\bigcup_{i \in N} A_i^+| = 8$, and $L_3 = |\bigcup_{i \in N} A_i^{+,U}| = 16$. Thus, the reachability matrices by $N - \{i\}$, $M_{N-\{i\}}$, $M_{N-\{i\}}^+$, and $M_{N-\{i\}}^{+,U}$, are calculated using the following inductive formulations:

$$M_i^{(1)} = J_i,\ M_i^{(1,+)} = J_i^+, \text{ and } M_i^{(1,+,U)} = J_i^{+,U} \text{for } i \in N;$$

$$M_j^{(t)} = sign[(\bigvee_{p \in N-\{i,j\}} M_p^{(t-1)}) \cdot J_j],$$

$$M_j^{(t,+)} = sign[(\bigvee_{p \in N-\{i,j\}} M_p^{(t-1,+)}) \cdot J_j^+], \text{ and}$$

$$M_j^{(t,+,U)} = sign[(\bigvee_{p \in N-\{i,j\}} M_p^{(t-1,+,U)}) \cdot J_j^{+,U}] \text{ for } j \in N - \{i\}.$$

Finally,

$$M_{N-\{i\}} = \bigvee_{t=1}^{L_1} \bigvee_{j \in N-\{i\}} M_j^{(t)},$$

$$M_{N-\{i\}}^{+} = \bigvee_{t=1}^{L_2} \bigvee_{j \in N-\{i\}} M_j^{(t,+)}, \text{ and}$$

$$M_{N-\{i\}}^{+,U} = \bigvee_{t=1}^{L_3} \bigvee_{j \in N-\{i\}} M_j^{(t,+,U)} \text{ for } i = 1, 2, 3.$$

Let the state set $S = \{s_1, s_2, s_3, s_4, s_5, s_6, s_7, s_8\}$. Tables 5.9, 5.10 and 5.11 show the results for the construction of the reachability matrices. It is clear that if $R_H(s)$, $R_H^+(s)$, and $R_H^{+,U}(s)$ are written as 0-1 row vectors, then

$$R_H(s) = e_s^T \cdot M_H, \; R_H^+(s) = e_s^T \cdot M_H^+, \text{ and } R_H^{+,U}(s) = e_s^T \cdot M_H^{+,U} \text{for any } s \in S.$$

For example, using Table 5.11, one has:

$$e_2^T \cdot M_{N-\{1\}}^{+,U} = (0, 0, 0, 1, 0, 1, 0, 1),$$

which indicates that the reachable list of $N - \{1\}$ by the legal UIUMs from state s_2, $R_{N-\{1\}}^{+,U}(s_2) = \{s_4, s_6, s_8\}$, i.e., states s_4, s_6, and s_8 can be reached by any legal UIUM sequence, by DM **Provincial** and DM **Support**, from the status quo $s = s_2$. Consequently, the construction of the reachability matrices provides an algebraic method for constructing $R_H(s)$, $R_H^+(s)$, and $R_H^{+,U}(s)$, the reachable lists of H by the legal sequences of UMs, UIs, and UIUMs, for any $s \in S$.

5.4.3 Analysis of Stability Results for the Gisborne Model

For analyzing the stabilities of the Gisborne conflict, one can utilize either the logical representation or the matrix approach. Let $N = \{1, 2, 3\}$ be the set of DMs (1 = Federal (or Fe), 2 = Provincial (or Pr), and 3 = Support (or Su)).

As an example, DM 1's SEQ_c stability for state s_1 can be analyzed using the logical representation presented in Definition 5.30. The procedures are as follows:

1. DM 1's reachable list from s_1 by UIs is $R_1^+(s_1) = \{s_2\}$;
2. the reachable list of coalition $H = N - \{1\}$ from s_2 by UIUMs is $R_H^{+,U}(s_2) = \{s_4, s_6, s_8\}$;

Table 5.9 UM reachability matrices for the Gisborne model

Matrix	$M_{N-\{1\}}$								$M_{N-\{2\}}$								$M_{N-\{3\}}$							
State	1	2	3	4	5	6	7	8	1	2	3	4	5	6	7	8	1	2	3	4	5	6	7	8
1	0	0	1	0	1	0	1	0	0	1	0	0	1	1	0	0	0	1	1	1	0	0	0	0
2	0	0	0	1	0	1	0	1	1	0	0	0	1	1	0	0	1	0	1	1	0	0	0	0
3	1	0	0	0	1	0	1	0	0	0	0	1	0	0	1	1	1	1	0	1	0	0	0	0
4	0	1	0	0	0	1	0	1	0	0	1	0	0	0	1	1	1	1	1	0	0	0	0	0
5	1	0	1	0	0	0	1	0	1	1	0	0	0	1	0	0	0	0	0	0	0	1	1	1
6	0	1	0	1	0	0	0	1	1	1	0	0	1	0	0	0	0	0	0	0	1	0	1	1
7	1	0	1	0	1	0	0	0	0	0	1	1	0	0	0	1	0	0	0	0	1	1	0	1
8	0	1	0	1	0	1	0	0	0	0	1	1	0	0	1	0	0	0	0	0	1	1	1	0

Table 5.10 UI reachability matrices for the Gisborne model

Matrix	$M^+_{N-\{1\}}$								$M^+_{N-\{2\}}$								$M^+_{N-\{3\}}$							
State	1	2	3	4	5	6	7	8	1	2	3	4	5	6	7	8	1	2	3	4	5	6	7	8
1	0	0	0	0	1	0	0	0	0	1	0	0	1	1	0	0	0	1	0	0	0	0	0	0
2	0	0	0	0	0	1	0	0	0	0	0	0	0	1	0	0	0	0	0	0	0	0	0	0
3	0	0	0	0	0	0	0	0	0	0	0	1	0	0	0	0	0	0	0	1	0	0	0	0
4	0	0	0	0	0	0	0	0	0	0	0	0	0	0	0	0	0	0	0	0	0	0	0	0
5	0	0	0	0	0	0	0	0	0	0	0	0	0	1	0	0	0	0	0	0	0	1	0	0
6	0	0	0	0	0	0	0	0	0	0	0	0	0	0	0	0	0	0	0	0	0	0	0	0
7	0	0	1	0	0	0	0	0	0	0	1	1	0	0	0	1	0	0	0	0	0	0	0	1
8	0	0	0	1	0	0	0	0	0	0	0	1	0	0	0	0	0	0	0	0	0	0	0	0

3. $s_4 \in R_H^{+,U}(s_2)$ satisfies $s_1 \succ_1 s_4$;
4. therefore, s_1 is SEQ_c stable for DM 1 by Definition 5.30.

Other cases can be analyzed similarly (see details in Li et al. (2004)). Furthermore, the matrix representations in Theorems 5.21–5.36 can also be employed to calculate the stabilities of the Gisborne conflict.

The stable states and equilibria under the four sets of definitions (indexed a, b, c, and d) for the four basic solution concepts, Nash, GMR, SMR and SEQ, are summarized in Table 5.12, in which "√" indicates stable for a DM—Fe, Pr, or Su—and "Eq" indicates an equilibrium. For instance, states s_4 and s_6 are equilibria for all b and d solution concepts. If the provincial government is economical-oriented and has complete preference information ($s_3 \succ s_7 \succ s_4 \succ s_8 \succ s_1 \succ s_5 \succ s_2 \succ s_6$), then this is a standard graph model, and the likely resolution is state s_4, as can be demonstrated using DSS GMCR II (Fang et al. 2003a, b). Similarly, an environmental-oriented provincial government, with preferences $s_2 \succ s_6 \succ s_1 \succ s_5 \succ s_4 \succ s_8 \succ s_3 \succ s_7$, state s_6 is the likely resolution. From the above discussions and Table 5.7, it is known that the outcome of the conflict depends on the provincial government's

Table 5.11 UIUM reachability matrices for the Gisborne model

Matrix	$M_{N-\{1\}}^{+,U}$								$M_{N-\{2\}}^{+,U}$								$M_{N-\{3\}}^{+,U}$							
State	1	2	3	4	5	6	7	8	1	2	3	4	5	6	7	8	1	2	3	4	5	6	7	8
1	0	0	1	0	1	0	1	0	0	1	0	0	1	1	0	0	0	1	1	1	0	0	0	0
2	0	0	0	1	0	1	0	1	0	0	0	0	0	1	0	0	0	0	0	1	0	0	0	0
3	1	0	0	0	1	0	1	0	0	0	0	1	0	0	0	0	1	1	0	1	0	0	0	0
4	0	1	0	0	0	1	0	1	0	0	0	0	0	0	0	0	0	1	0	0	0	0	0	0
5	1	0	1	0	0	0	1	0	0	0	0	0	0	1	0	0	0	0	0	0	0	1	1	1
6	0	1	0	1	0	0	0	1	0	0	0	0	0	0	0	0	0	0	0	0	0	0	0	1
7	1	0	1	0	1	0	0	0	0	0	1	1	0	0	0	1	0	0	0	0	1	1	0	1
8	0	1	0	1	0	1	0	0	0	0	0	1	0	0	0	0	0	0	0	0	0	1	0	0

attitude. If the support group convinces the provincial government of the urgent need for cash, state s_4 is selected as a resolution for the Gisborne conflict, which means that the economical-oriented provincial government will lift the ban on bulk water export. On the other hand, for the environmental-oriented provincial government, the resolution for the Gisborne conflict is likely to be state s_6, which means that the provincial government will not lift the ban.

5.5 Important Ideas

In this chapter, the logical and matrix representations of four basic solution concepts for simple preference for two-DM and multiple-DM conflicts from Chap. 4 are expanded to handle preference uncertainty. Although the graph model solution concepts are extended to models with preference uncertainty in Sect. 5.2, procedures to identify stable states based on these solution concepts are not easy to code because of the nature of their logical representations, which may explain why implementation algorithms for these solution concepts have not been developed. The matrix representation of solution concepts in the graph model with preference uncertainty in Sect. 5.3 handles this problem efficiently, and, therefore, facilitates the development of improved algorithms to assess the stabilities of states within a well-designed decision support system as explained in Sect. 10.2. More specifically, using GMR stability as an example, Table 5.13 shows how the procedures for matrix representation is extended from 2-DM models to n-DM models, as well as how matrix formulation is expanded for addressing uncertainty. Consequently, a key advantage of this matrix method is that it can be easily adapted to new solution concepts because of the nature of explicit matrix expressions. Hence, unknown preference can be readily operationalized by embedding it within a well-designed decision support system for GMCR.

Table 5.12 Stability analysis of the Gisborne model

State		Nash				GMR				SMR				SEQ			
		Fe	Pr	Su	Eq	Fe	Pr	Su	Eq	Fe	Pr	Su	Eq	Fe	Pr	Su	Eq
s_1	a																
	b		√				√				√				√		
	c						√				√				√		
	d		√				√				√				√		
s_2	a	√				√				√				√			
	b	√	√			√	√			√	√			√	√		
	c	√				√	√			√	√			√			
	d	√	√			√	√			√	√			√	√		
s_3	a			√				√				√				√	
	b		√	√			√	√			√	√			√	√	
	c			√			√	√			√	√			√	√	
	d		√	√			√	√			√	√			√	√	
s_4	a	√		√		√		√		√		√		√		√	
	b	√	√	√	√	√	√	√	√	√	√	√	√	√	√	√	√
	c	√		√		√	√	√	√	√	√	√	√	√	√	√	√
	d	√	√	√	√	√	√	√	√	√	√	√	√	√	√	√	√
s_5	a			√				√				√				√	
	b		√	√			√	√			√	√			√	√	
	c			√			√	√			√	√			√	√	
	d		√	√			√	√			√	√			√	√	
s_6	a	√		√		√		√		√		√		√		√	
	b	√	√	√	√	√	√	√	√	√	√	√	√	√	√	√	√
	c	√		√		√	√	√	√	√	√	√	√	√	√	√	√
	d	√	√	√	√	√	√	√	√	√	√	√	√	√	√	√	√
s_7	a							√				√				√	
	b		√				√	√			√	√			√	√	
	c						√	√			√	√			√	√	
	d		√				√	√			√	√			√	√	
s_8	a	√				√		√		√		√		√		√	
	b	√	√			√	√	√	√	√	√	√	√	√	√	√	√
	c	√				√	√	√	√	√	√	√	√	√		√	
	d	√	√			√	√	√	√	√	√	√	√	√	√	√	√

5.6 Problems

5.6.1 In your own words, explain three reasons why you think uncertainty over preferences may occur in a conflict situation.

Table 5.13 The features of the explicit matrix method

Preference	The number of DMs	GMR matrix or GMR_a matrix	Extend 2-DM models to n-DM models	Extend simple preference to preference with uncertainty
Simple preference	Two	$J_i^+ \cdot [E - sign\left(J_j \cdot (P_i^{-,=})^T\right)]$	J_j is replaced with $M_{N-\{i\}}$	J_i^+ is replaced with $J_i^{+,U}$
	n	$J_i^+ \cdot [E - sign\left(M_{N-\{i\}} \cdot (P_i^{-,=})^T\right)]$		
Uncertain preference	Two	$J_i^{+,U} \cdot [E - sign\left(J_j \cdot (P_i^{-,=})^T\right)]$		
	n	$J_i^{+,U} \cdot [E - sign\left(M_{N-\{i\}} \cdot (P_i^{-,=})^T\right)]$		

5.6.2 Based on a real-world conflict which is of direct interest to you, explain why uncertainty of preference for one or more DMs may arise. For background information to your conflict, you may wish to refer to a newspaper, magazine or journal article.

5.6.3 Outline how the fuzzy approach to capturing preference uncertainty works within GMCR by referring to the articles by Hipel et al. (2011) and Bashar et al. (2012). Qualitatively, compare the kind of preference uncertainties reflected in the unknown preference approach of this chapter to that of the fuzzy preference procedure.

5.6.4 Briefly explain how the grey preference method proposed by Kuang et al. (2015) works. In a qualitative sense, compare the type of preference uncertainty modeled by grey preference to that of unknown preference.

5.6.5 Succinctly describe how the probabilistic preference method of Rego and dos Santos (2015) works. Explain how their probabilistic technique mimics the key steps that Bashar et al. (2012) employ in their fuzzy preference technique.

5.6.6 For the Prisoner's Dilemma game put forward in Problem 3.5.1, introduce unknown preference into this conflict in a reasonable way. Using the logical stability definitions given in Sect. 5.2, carry out a stability analysis for both Nash and sequential stability. Comment upon any insights that you gained from your investigation.

5.6.7 Using the Prisoner's Dilemma dispute explained in Problem 3.5.1, describe how unknown preference could arise within this dispute. Employing the matrix representation of unknown preference, execute a stability analysis to determine individual stability for each state with respect to Nash and sequential stability for each of the DMs, as well as the equilibrium results. If you already completed Problem 5.6.6, you may wish to assume the same unknown preference that you did in Problem 5.6.6.

5.6.8 With respect to the Game of Chicken explained in Problem 3.5.4, make a reasonable assumption as to how unknown preference may arise. Using either the logical approach of Sect. 5.2 or the matrix procedure of Sect. 5.3, carry out a stability analysis of the Chicken game for Nash, GMR, SMR, and SEQ stability. Comment upon any interesting strategic insights that you find.

5.6.9 The Sustainable Development Conflict is displayed in normal form in Table 3.2. Explain why and how unknown preference could occur in this conflict. Carry out a stability analysis of the Sustainable Development game for this situation for Nash, GMR, SMR, and SEQ stability. Comment upon any strategic insights you may find.

5.6.10 The Elmira groundwater contamination dispute is modeled in option form in Sect. 1.2.2 and Tables 1.2 and 3.5. A stability analysis under simple preference is presented in Sect. 4.5. Explain why and how unknown preference could be present in this dispute. Show detailed calculations for carrying out a stability analysis for Nash and SEQ stability for the Elmira conflict. Point out any interesting strategic insights you find.

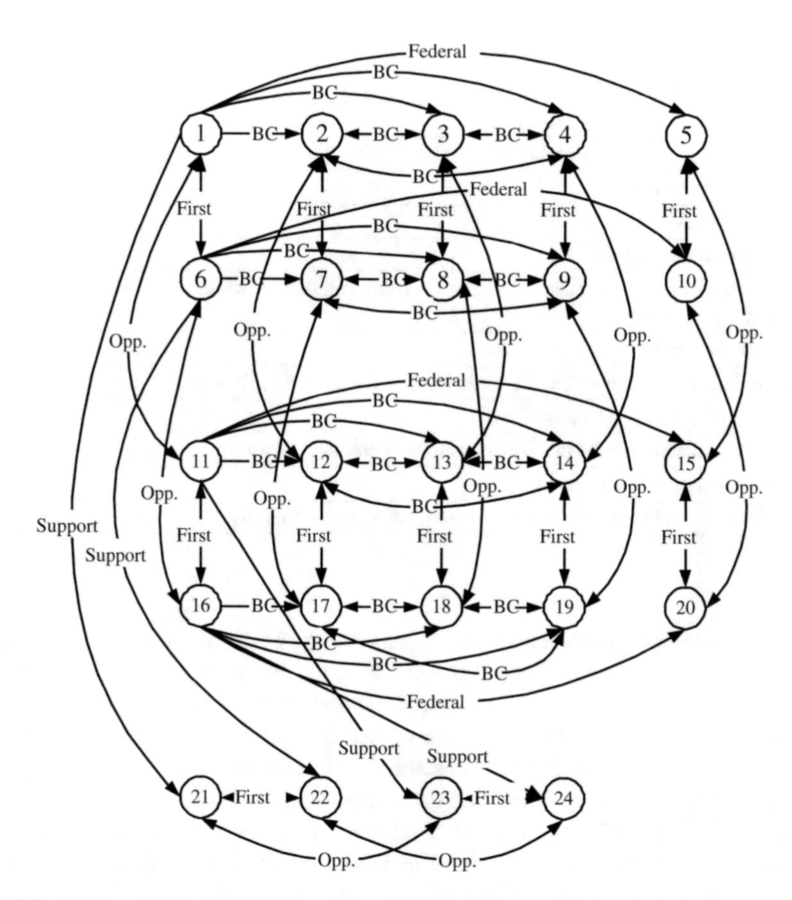

Fig. 5.8 Graph model for the BC salmon aquaculture conflict

5.6.11 In Sect. 5.4, the Lake Gisborne conflict over proposed bulk water exports is used as an example for carrying out stability calculations for unknown preference using a matrix formulation. Based on the logical interpretation of unknown preference, execute a stability analysis for Nash and sequential stability.

5.6.12 In the past decades, the salmon farming industry expanded rapidly in the province of British Columbia (BC) in Canada to satisfy the demand for high quality fish products. However, several groups, such as the fishing industry, First Nations, and others, strongly opposed the increased farming of salmon in order to protect wild salmon. In 1995, the BC government imposed a temporary moratorium on the expansion of salmon farming and required the BC Environmental Assessment Office to evaluate the potential negative impacts of salmon aquaculture on the environment. In August 1997, after discussions with a number of interest groups and evaluation of the environmental and socioeconomic aspects of the industry, the BC Environmental Assessment Office released its Salmon Aquaculture Review (SAR). This

Table 5.14 DMs and options of the BC salmon aquaculture conflict

DMs and options	Status quo
Provincial government of BC (**BC**)	
1. Adopt: Adopt the recommendations	N
2. Lift: Lift the moratorium	N
3. Partial: Partially lift the moratorium	N
Federal government (**Federal**)	
4. Unilateral: Unilateral implementation of the recommendations within Federal jurisdiction	N
First Nations (**First**)	
5. Legal: Legal action under the Fisheries Act and Canadian Constitution	N
Aquaculture opposition (**Opposition**)	
6. Legal: Legal action under the Fisheries Act and political pressure	N
Aquaculture support (**Support**)	
7. Implement: Implement some of the recommendations voluntarily	N

report provided 49 recommendations to improve the environmental performance of salmon farming. However, until October 1999, the BC government took no steps to implement the recommendations, and the moratorium on salmon farming expansion continued in force. This situation caused a conflict (see Li et al. (2005) for more details). Noakes et al. (2003) established the graph model shown in Fig. 5.8 for this conflict, which contains five decision makers and seven options presented in Table 5.14. The seven options are combined to form 24 feasible states listed in Table 5.15. The preference relations of the DMs over the feasible states are listed in Table 5.16, where $\succ$ denotes strict preference and is transitive. Preference information for DM **Federal**, DM **First**, DM **Opposition**, and DM **Support** is certain, but some of DM **BC**'s preferences are certain, and others are uncertain.

1. Label the graph model according to the Rule of Priority;
2. Calculate the stabilities of Nash, GMR, SMR, and SEQ with $a, b, c, andd$ for the BC salmon aquaculture conflict using the MRSCU method.

5.6.13 Theorem 4.9 in Chap. 4 provides a method to construct the UM and the UI reachability matrices by

$$M_H = sign[B_{out} \cdot (LJ_H + I)^{l-1} \cdot B_{in}^T] \text{ and } M_H^+ = sign[B_{out}^+ \cdot (LJ_H^+ + I)^{l^+-1} \cdot (B_{in}^+)^T].$$

Theorem 5.20 provides another method to calculate the two matrices by

$$M_H = \bigvee_{t=1}^{l_1} \bigvee_{i \in H} M_i^{(t)} \text{ and } M_H^+ = \bigvee_{t=1}^{l_2} \bigvee_{i \in H} M_i^{(t,+)}.$$

Table 5.15 Feasible state for the BC salmon aquaculture conflict

DM/Option	1	2	3	4	5	6	7	8	9	10	11	12	13	14	15	16	17	18	19	20	21	22	23	24
BC																								
Adopt	N	Y	Y	Y	N	N	Y	Y	Y	N	N	Y	Y	Y	N	N	Y	Y	Y	N	N	N	N	N
Lift	N	N	Y	N	N	N	N	Y	N	N	N	N	Y	N	N	N	N	Y	N	N	N	N	N	N
Partial	N	N	N	Y	N	N	N	N	Y	N	N	N	N	Y	N	N	N	N	Y	N	N	N	N	N
Federal																								
Unilateral	N	N	N	N	Y	N	N	N	N	Y	N	N	N	N	Y	N	N	N	N	Y	N	N	N	N
First																								
Legal	N	N	N	N	N	Y	Y	Y	Y	Y	N	N	N	N	N	Y	Y	Y	Y	Y	N	Y	N	Y
Opposition																								
Legal	N	N	N	N	N	N	N	N	N	N	Y	Y	Y	Y	Y	Y	Y	Y	Y	Y	N	N	Y	Y
Support																								
Implement	N	N	N	N	N	N	N	N	N	N	N	N	N	N	N	N	N	N	N	N	Y	Y	Y	Y

Table 5.16 Relative preference for DMs in the BC salmon aquaculture conflict

DM	Preference
BC	[3 ≻ 13 ≻ 8 ≻ 18], [4 ≻ 14 ≻ 9 ≻ 19], [21 ≻ 23 ≻ 22 ≻ 24], [5 ≻ 15 ≻ 10 ≻ 20], [2 ≻ 12 ≻ 7 ≻ 17], [1 ≻ 11 ≻ 6 ≻ 16]
Federal	21 ≻ 23 ≻ 22 ≻ 24 ≻ 20 ≻ 10 ≻ 15 ≻ 5 ≻ 16 ≻ 6 ≻ 11 ≻ 1 ≻ 18 ≻ 8 ≻ 13 ≻ 3 ≻ 19 ≻ 9 ≻ 14 ≻ 4 ≻ 17 ≻ 7 ≻ 12 ≻ 2
First	17 ≻ 7 ≻ 2 ≻ 12 ≻ 21 ≻ 23 ≻ 24 ≻ 22 ≻ 20 ≻ 10 ≻ 5 ≻ 15 ≻ 16 ≻ 6 ≻ 1 ≻ 11 ≻ 19 ≻ 9 ≻ 4 ≻ 14 ≻ 18 ≻ 8 ≻ 3 ≻ 13
Opposition	17 ≻ 12 ≻ 24 ≻ 20 ≻ 16 ≻ 23 ≻ 15 ≻ 11 ≻ 7 ≻ 2 ≻ 22 ≻ 10 ≻ 6 ≻ 21 ≻ 5 ≻ 1 ≻ 19 ≻ 14 ≻ 9 ≻ 4 ≻ 18 ≻ 13 ≻ 8 ≻ 3
Support	3 ≻ 13 ≻ 8 ≻ 18 ≻ 4 ≻ 14 ≻ 9 ≻ 19 ≻ 21 ≻ 23 ≻ 22 ≻ 24 ≻ 5 ≻ 15 ≻ 10 ≻ 20 ≻ 2 ≻ 12 ≻ 7 ≻ 17 ≻ 1 ≻ 11 ≻ 6 ≻ 16

1. Prove the two methods are equivalent;
2. Compare the advantages and disadvantages of the two methods.

References

Bashar, M. A., Kilgour, D. M., & Hipel, K. W. (2012). Fuzzy preferences in the graph model for conflict resolution. *IEEE Transactions on Fuzzy Systems*, *20*(4), 760–770. https://doi.org/10.1109/tfuzz.2012.2183603.

Bashar, M. A., Kilgour, D. M., & Hipel, K. W. (2014). Fuzzy option prioritization for the graph model for conflict resolution. *Fuzzy Sets and Systems*, *26*(2014), 34–48. https://doi.org/10.1016/j.fss.2014.02.011.

Bashar, M. A., Hipel, K. W., Kilgour, D. M., & Obeidi, A. (2015). Coalition fuzzy stability analysis in the graph model for conflict resolution. *Journal of Intelligent and Fuzzy Systems*, *29*(2), 593–607. https://doi.org/10.3233/ifs-141336.

Bashar, M. A., Obeidi, A., Kilgour, D. M., & Hipel, K. W. (2016). Modeling fuzzy and interval fuzzy preferences within a graph model framework. *IEEE Transactions on Fuzzy Systems*, *24*(4), 765–778. https://doi.org/10.1109/tfuzz.2015.2446536.

Bashar, M. A., Hipel, K. W., Kilgour, D. M., & Obeidi, A. (2018). Interval fuzzy preferences in the graph model for conflict resolution. *Fuzzy Optimization and Decision Making*. https://doi.org/10.1007/s10700-017-9279-7.

Fang, L., Hipel, K. W., & Kilgour, D. M. (1993). *Interactive decision making: The graph model for conflict resolution*. New York: Wiley. https://doi.org/10.2307/2583940.

Fischer, G. W., Jia, J., & Luce, M. F. (2000a). Attribute conflict and preference uncertainty: The RandMAU model. *Management Science*, *46*(5), 669–684. https://doi.org/10.1287/mnsc.46.5.669.12051.

Fischer, G. W., Luce, M. F., & Jia, J. (2000b). Attribute conflict and preference uncertainty: Effects on judgment time and error. *Management Science*, *46*(1), 88–103. https://doi.org/10.1287/mnsc.46.1.88.15131.

Fang, L., Hipel, K. W., & Wang, L. (2002). Gisborne water export conflict study. In Schmiz, G. H. (Ed.), *Proceedings of the Third International Conference on Water Resources and Environment Research* (Vol. 1, pp. 432–436). Dresden, Germany.

Fang, L., Hipel, K. W., Kilgour, D. M., & Peng, X. (2003a). A decision support system for interactive decision making, part 1: Model formulation. *IEEE Transactions on Systems, Man and*

Cybernetics, Part C: Applications and Reviews, *33*(1), 42–55. https://doi.org/10.1109/tsmcc.2003.809361.

Fang, L., Hipel, K. W., Kilgour, D. M., & Peng, X. (2003b). A decision support system for interactive decision making, part 2: Analysis and output interpretation. *IEEE Transactions on Systems, Man and Cybernetics, Part C: Applications and Reviews*, *33*(1), 56–66. https://doi.org/10.1109/tsmcc.2003.809360.

Hipel, K. W., Kilgour, D. M., & Bashar, M. A. (2011). Fuzzy preferences in multiple participant decision making. *Scientia Iranica, Transactions D: Computer Science and Engineering and Electrical Engineering*, *18*(3(D1)), 627–638. https://doi.org/10.1016/j.scient.2011.04.016.

Kuang, H., Bashar, M. A., Hipel, K. W., & Kilgour, D. M. (2015). Grey-based preference in a graph model for conflict resolution with multiple decision makers. *IEEE Transactions on Systems, Man and Cybernetics: Systems*, *45*(9), 1254–1267. https://doi.org/10.1109/tsmc.2014.2387096.

Li, K. W., Hipel, K. W., Kilgour, D. M., & Fang, L. (2004). Preference uncertainty in the graph model for conflict resolution. *IEEE Transactions on Systems, Man, and Cybernetics, Part A, Systems and Humans*, *34*(4), 507–520. https://doi.org/10.1109/tsmca.2004.826282.

Li, K. W., Hipel, K. W., Kilgour, D. M., & Noakes, D. J. (2005). Integrating uncertain preferences into status quo analysis with application to an environmental conflict. *Group Decision and Negotiation*, *14*(6), 461–479. https://doi.org/10.1007/s10726-005-9003-9.

Noakes, D. J., Fang, L., Hipel, K. W., & Kilgour, D. M. (2003). An examination of the salmon aquaculture conflict in British Columbia using the graph model for conflict resolution. *Fisheries Management and Ecology*, *10*, 123–137. https://doi.org/10.1046/j.1365-2400.2003.00336.x.

Rego, L. C., & dos Santos, A. M. (2015). Probabilistic preferences in the graph model for conflict resolution. *IEEE Transactions on Systems, Man, and Cybernetics: Systems*, *45*(4), 595–608. https://doi.org/10.1109/tsmc.2014.2379626.

Xu, H., Hipel, K. W., & Kilgour, D. M. (2007a). Matrix representation of conflicts with two decision makers. In *Proceedings of the 2007 IEEE International Conference on Systems, Man, and Cybernetics* (pp. 1764–1769). Montreal, Canada. https://doi.org/10.1109/icsmc.2007.4413988.

Xu, H., Kilgour, D. M., & Hipel, K. W. (2007b). Matrix representation of solution concepts in graph models for two decision-makers with preference uncertainty. *Dynamics of Continuous, Discrete and Impulsive Systems*, *14*(S1), 703–707.

Xu, H., Hipel, K. W., & Kilgour, D. M. (2009a). Matrix representation of solution concepts in multiple decision maker graph models. *IEEE Transactions on Systems, Man, and Cybernetics, Part A: Systems and Humans*, *39*(1), 96–108. https://doi.org/10.1109/tsmca.2009.2007994.

Xu, H., Li, K. W., Kilgour, D. M., & Hipel, K. W. (2009b). A matrix-based approach to searching colored paths in a weighted colored multidigraph. *Applied Mathematics and Computation*, *215*, 353–366. https://doi.org/10.1016/j.amc.2009.04.086.

Xu, H., Kilgour, D. M., & Hipel, K. W. (2011). Matrix representation of conflict resolution in multiple-decision-maker graph models with preference uncertainty. *Group Decision and Negotiation*, *20*(6), 755–779. https://doi.org/10.1007/s10726-010-9188-4.

Zhao, S., & Xu, H. (2017). Grey option prioritization for the graph model for conflict resolution. *Journal of Grey System*, *29*(3), 14–25.

Chapter 6
Stability Definitions: Degrees of Preference

In a water quality dispute, an environmental agency may greatly prefer that an industrial enterprise does not seriously pollute a nearby river into which it discharges wastes. The purpose of this chapter is to present a formal methodology that can handle this type of "degree", "strength", or "level" of preference, which often arises in practice, in order to determine its strategic consequences. More specifically, a multiple-degree preference structure is developed within the paradigm of the Graph Model for Conflict Resolution (GMCR) in conjunction with associated stability definitions for determining individual stability of each state from a given decision maker's (DM's) viewpoint as well as the overall equilibria (Hamouda et al. 2004, 2006, Xu et al. 2009, 2010, 2011). Within this structure, a DM may have multiple degrees of preference when comparing pairs of states. For example, if state a is preferred to state b, it may be mildly preferred at degree 1 ($d = 1$), more strongly preferred at degree 2 ($d = 2$), ..., or maximally preferred at degree r ($d = r$), where $r > 0$ is a fixed parameter. The number of degrees, r, is unrestricted in this system, thereby extending the earlier simple preference structure having two types of preferences consisting of equally preferred (degree zero) and more preferred (degree one) in Chap. 4 and the special case of three kinds of preferences (equally preferred, mildly preferred, and greatly more preferred) discussed in detail in this chapter.

The main properties of the preference structure according to degree are introduced in Sect. 6.1 in this chapter. Because DMs make moves and countermoves when interacting with one another under conflict, reachable lists are defined in Sect. 6.2 to keep track of the possible unilateral movements in one step from a given state for a particular DM with respect to multiple types of preference. When considering stability definitions for more than two DMs, coalition moves are defined since two or more DMs can participate in blocking a unilateral improvement by another DM. Subsequently, multiple-degree versions of four stability definitions consisting of Nash

H. Xu et al., *Conflict Resolution Using the Graph Model: Strategic Interactions in Competition and Cooperation*, Studies in Systems, Decision and Control 153,
https://doi.org/10.1007/978-3-319-77670-5_6

stability, general metarationality, symmetric metarationality, and sequential stability, are defined for the graph model with this extended preference structure and the relationships among them are investigated. Additionally, in this chapter, matrix representations of the four stabilities are presented for graph models having a preference structure of up to degree 3.

6.1 Multiple Degrees of Preference

The simple preference structure discussed in Chaps. 3 and 4 contains two types of preferences: indifference, in which a DM is indifferent between, or equally prefers, two states, and strict preference, in which a DM prefers one state more than another. The third kind of preference can be added by allowing a DM to greatly prefer one state over another. Hence, an expanded preference structure for a given DM can have two states being equally preferred (called preference of degree zero, or simply $d = 0$), one state being more or mildly preferred over another (degree $d = 1$), or one state being greatly more preferred than another ($d = 2$). In fact, one can extend two degrees of preference to an unlimited number. Below, preference structures having preferences of up to two degrees and the general case of having any number of degrees are discussed in Sects. 6.1.1 and 6.1.2, respectively.

6.1.1 Three Types of Preference

A triplet relation on S that expresses strength of preference according to indifferent, mild, or strong preference, was developed by Hamouda et al. (2004, 2006). For states $s, q \in S$, the preference relation $s \sim_i q$ indicates that DM i is indifferent between states s and q, the relation $s >_i q$ means that DM i mildly prefers s to q, and $s \gg_i q$ denotes that DM i strongly prefers s to q. Similar to the properties for simple preference given in Sect. 3.2.4, the characteristics of the preference structure, $\{\sim_i, >_i, \gg_i\}$, containing three kinds of preference for each DM $i \in N$, are as follows:

(i) $\sim_i$ is reflexive and symmetric;
(ii) $>_i$ and $\gg_i$ are asymmetric; and
(iii) $\{\sim_i, >_i, \gg_i\}$ is strongly complete.

Note that $\{\sim_i, >_i, \gg_i\}$ is strongly complete. Hence, if $s, q \in S$, then exactly one of the following relations holds: $s \sim_i q$, $s >_i q$, $s \gg_i q$, $q >_i s$, or $q \gg_i s$. Also, it is assumed that, for any $s, q \in S$, $s >_i q$ is equivalent to $q <_i s$. The preference type "$\gg_i$" has similar properties to "$>_i$".

Table 6.1 Subsets of S with respect to three degrees of preference for DM i

Subsets of S	Descriptions
$\Phi_i^{++}(s) = \{q : q \gg_i s\}$	States strongly preferred to state s by DM i
$\Phi_i^{+m}(s) = \{q : q >_i s\}$	States mildly preferred to state s by DM i
$\Phi_i^{=}(s) = \{q : q \sim_i s\}$	States equally preferred to state s by DM i
$\Phi_i^{-m}(s) = \{q : s >_i q\}$	States mildly less preferred than state s for DM i
$\Phi_i^{--}(s) = \{q : s \gg_i q\}$	States strongly less preferred to state s by DM i

The set of feasible states, S, can be partitioned or divided into a set of non-overlapping or disjoint subsets based on the types of preference relative to a specific state $s \in S$. These categorizations of preferences are needed for carrying out stability analyses according to different kinds of human behavior under conflict as explicitly defined in Sect. 6.3. For example, a DM may be tempted to unilaterally move to a mildly preferred state which can be blocked by another DM moving to a state which is greatly less preferred by the original DM. The descriptions of these different classifications of preferences are presented in Table 6.1.

Let $s \in S$ and $i \in N$. Based on different structures of preferences, DM i can identify different subsets of S. For simple preference, DM i can identify three subsets of S with respect to a state s: the set of states more preferred by DM i than state s (denoted by $\Phi_i^{+}(s)$); the set of states equally preferred to state s by DM i ($\Phi_i^{=}(s)$); and the set of states less preferred by DM i to state s ($\Phi_i^{-}(s)$) (see Sect. 4.1 for details). For the three types of preference, DM i can identify five subsets of S: $\Phi_i^{++}(s)$, $\Phi_i^{+m}(s)$, $\Phi_i^{=}(s)$, $\Phi_i^{-m}(s)$, and $\Phi_i^{--}(s)$, which are explained in Table 6.1. Notice that in this table that the set of states mildly preferred to state s by DM i, given by $\Phi_i^{+m}(s)$, have an "m" in the superscript in order to distinguish this set from $\Phi_i^{+}(s)$ for the case of a simple preference structure in which $\Phi_i^{++}(s)$ does not exist. Therefore, all states that are more preferred to state s by DM i would be included in $\Phi_i^{+}(s)$ for a simple preference structure. Similar comments hold for the set $\Phi_i^{-m}(s)$ in Table 6.1.

In Sect. 6.3, a given DM can levy a sanction against a unilateral improvement by DM i from state s if the sanctioning DM can put DM i in either a less preferred or equally preferred state relative to state s. Therefore, the set of states given by $\Phi_i^{--,-,=}(s) = \Phi_i^{--}(s) \cup \Phi_i^{-m}(s) \cup \Phi_i^{=}(s)$, where $\cup$ denotes the union operation, is important in various stability definitions. Note that in the graph model with strength of preference, $s \succ_i q$ iff either $s >_i q$ or $s \gg_i q$. Hence, the three types of preference structure expand simple preference.

The simple preference structure having the set of binary relations given as $\{\sim, \succ\}$, and the expanded preference structure with strength of preference, which has the set of binary relations $\{\sim, >, \gg\}$, are referred to as having two types of preferences

Table 6.2 Degree of relative preference

Degree of strength	Description	Notation
$d = r$	Preferred at degree r	$\overbrace{> \cdots >}^{r}$
......	......	......
$d = 3$	Very strongly preferred	$\ggg$
$d = 2$	Strongly preferred	$\gg$
$d = 1$	Moderately preferred	$>$
$d = 0$	Equally preferred	$\sim$

and three kinds of preferences, respectively. The existing two preference structures in the graph model are extended to the general case of multiple types of preference structures with any specified degree in the next section (Xu et al. 2009).

6.1.2 Multiple Degrees of Preference

A set of new and more general binary relations $\overbrace{> \cdots >}^{d}$ for $d = 1, 2, \ldots, r$, as listed in Table 6.2, are introduced in this section to represent DM i's preference at each degree d. With the introduction of these new binary relations, the three types of preference structures in the graph model are extended from a triplet of relations, to an $r + 1$ types of preference relations for DM i over the set of states, which is expressed as $\{\sim_i, >_i, \gg_i, \ldots, \overset{r}{\succ}_i\}$ on S, where $\overset{r}{\succ}_i$ denotes $\overbrace{> \cdots >}^{r}_i$, i.e., DM i has preference at degree r for comparing states with respect to preference. For instance, $s \ggg_i q$ means that DM i very strongly prefers state s to state q. Similar to the case for simple preference as described in Sect. 3.2.4, it is assumed that the preference relations of each DM $i \in N$ have the following properties:

(i) $\sim_i$ is reflexive and symmetric (i.e., $\forall s, q \in S$, $s \sim_i s$, and if $s \sim_i q$, then $q \sim_i s$);

(ii) $\overset{d}{\succ}_i$ for $d = 1, 2, \ldots, r$, is asymmetric (i.e., $s \overset{r}{\succ}_i q$ and $q \overset{r}{\succ}_i s$ cannot occur simultaneously); and

(iii) $\{\sim_i, >_i, \gg_i, \ldots, \overset{r}{\succ}_i\}$ is strongly complete (i.e. if $s, q \in S$, then exactly one of the following relations holds: $s \sim_i q$, $s \overset{d}{\succ}_i q$, or $q \overset{d}{\succ}_i s$ for $d = 1, 2, \ldots, r$).

Preference information can be either transitive or intransitive. For states $k, s, q \in S$, if $k \overset{d}{\succ}_i s$ and $s \overset{d}{\succ}_i q$ imply $k \overset{d}{\succ}_i q$, then the preference $\overset{d}{\succ}_i$ is transitive. Otherwise, the preferences are called intransitive. Note that the assumption of transitivity of preferences is not required in the following definitions so that the results in this chapter hold for both transitive and intransitive preferences. When all of the preferences for

Table 6.3 Subsets of S for DM i with respect to multiple degrees of preference

Degree of strength	Subsets of S	Description
$d = r$	$\Phi_i^{+(r)}(s) = \{q : q \overbrace{> \cdots >_i}^{r} s\}$	States preferred to state s at degree r by DM i
	$\Phi_i^{-(r)}(s) = \{q : s \overbrace{> \cdots >_i}^{r} q\}$	States less preferred to state s at degree r by DM i
$\vdots$		
$\vdots$		
$d = 3$	$\Phi_i^{+(3)}(s) = \{q : q \ggg_i s\}$	States very strongly preferred to state s by DM i
	$\Phi_i^{-(3)}(s) = \{q : s \ggg_i q\}$	States very strongly less preferred to state s by DM i
$d = 2$	$\Phi_i^{+(2)}(s) = \{q : q \gg_i s\}$	States strongly preferred to state s by DM i
	$\Phi_i^{-(2)}(s) = \{q : s \gg_i q\}$	States strongly less preferred to state s by DM i
$d = 1$	$\Phi_i^{+(1)}(s) = \{q : q >_i s\}$	States moderately preferred to state s by DM i
	$\Phi_i^{-(1)}(s) = \{q : s >_i q\}$	States moderately less preferred to state s by DM i
$d = 0$	$\Phi_i^{(0)}(s) = \Phi_i^{=}(s) = \{q : q \sim_i s\}$	States equally preferred to state s by DM i

a given DM i are transitive, the preferences are said to be ordinal and, hence, the states in a conflict can be ordered or ranked from most to least preferred, where ties are allowed. Sometimes this ranking of states according to preference is referred to as a "preference ranking".

A list and associated descriptions for the range of subsets of S with respect to multiple types of preference are presented in Table 6.3. Starting at the bottom of the table at degree 0, the notation for the states equally preferred to state s by DM i is given as $\Phi_i^{(0)}(s)$ or $\Phi_i^{=}(s)$. Notice that for degree of strength $d = 1, \ldots, r$, two subsets of states are given for each degree as $\Phi_i^{+(d)}(s)$ and $\Phi_i^{-(d)}(s)$, to indicate subsets of states preferred to state s at degree d by DM i, and states less preferred to state s at degree d by DM i, respectively. Hence, overall there is a total of $2r + 1$ subsets of S when considering multiple degrees of preference. A diagram displaying these degrees of preference for DM i is furnished later in Sect. 6.2.2 as the left side of Fig. 6.1.

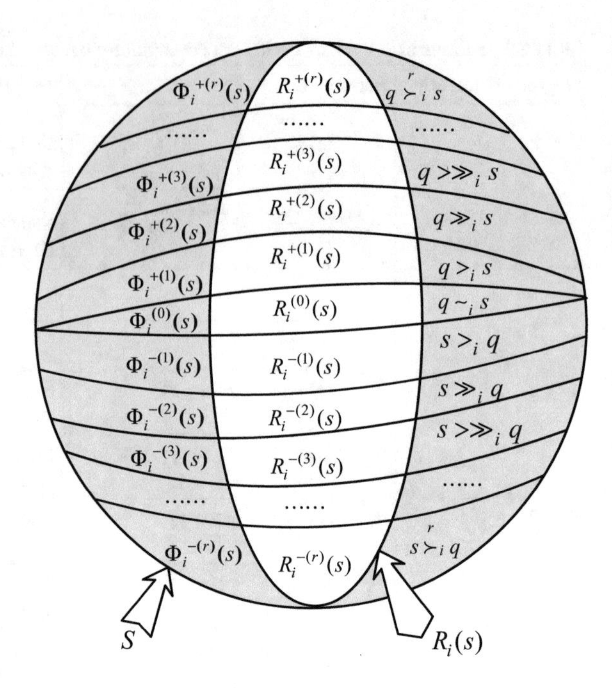

Fig. 6.1 Relationships among subsets of S and reachable lists from s

6.2 Reachable Lists of a Decision Maker

In addition to preference, one must be aware of the moves DMs control when ascertaining stability. Accordingly, in this section, moves unilaterally controlled by a DM in one step are defined as reachable lists for the cases of three types and multiple kinds of preferences in Sects. 6.2.1 and 6.2.2, respectively. Potential moves by a DM in the face of simple preference are defined using reachable lists in Sect. 4.1.1. In the upcoming two subsections, let $i \in N$, $s \in S$, and $m = |S|$ be the number of the states in S. The notation given by $\cap$ denotes the intersection operation while $\cup$ is the union operation. Recall that each arc of $A_i \subseteq S \times S$ indicates that DM i can make a unilateral move (in one step) from the initial state to the terminal state of the arc.

6.2.1 Reachable Lists for Three Degrees of Preference

The reachable lists of a DM for three types of preference are defined as follows:

(i) $R_i^{++}(s) = \{q \in S : (s, q) \in A_i \text{ and } q \gg_i s\}$ stands for DM i's reachable list from state s by a strong unilateral improvement. This set contains all states q which are strongly preferred by DM i to state s and can be reached in one step from s;

Table 6.4 Unilateral movements for DM i in the three types of preference structure

Type of movements	Description
$R_i^{++}(s) = R_i(s) \cap \Phi_i^{++}(s)$	All strong unilateral improvements from state s for DM i
$R_i^{+m}(s) = R_i(s) \cap \Phi_i^{+m}(s)$	All mild unilateral improvements from state s for DM i
$R_i^{=}(s) = R_i(s) \cap \Phi_i^{=}(s)$	All equally preferred states reachable from state s by DM i
$R_i^{-m}(s) = R_i(s) \cap \Phi_i^{-m}(s)$	All mild unilateral disimprovements from state s for DM i
$R_i^{--}(s) = R_i(s) \cap \Phi_i^{--}(s)$	All strong unilateral disimprovements from state s for DM i

(ii) $R_i^{+m}(s) = \{q \in S : (s, q) \in A_i \text{ and } q >_i s\}$ denotes DM i's reachable list from state s by a mild unilateral improvement;
(iii) $R_i^{-m}(s) = \{q \in S : (s, q) \in A_i \text{ and } s >_i q\}$ denotes DM i's reachable list from state s by a mild unilateral disimprovement;
(iv) $R_i^{--}(s) = \{q \in S : (s, q) \in A_i \text{ and } s \gg_i q\}$ is DM i's reachable list from state s by a strong unilateral disimprovement;
(v) $R_i^{+,++}(s) = R_i^{+m}(s) \cup R_i^{++}(s) = \{q \in S : (s, q) \in A_i \text{ and } q >_i s \text{ or } q \gg_i s\}$ denotes DM i's reachable list from state s by a mild unilateral move or strong unilateral move.

From the above definitions, these reachable lists from state s by DM i can be summarized as presented in Table 6.4. As discussed in Sect. 4.1.1, DM i's reachable list from state s, $R_i(s)$, represents DM $i's$ unilateral moves (UMs). $R_i(s)$ is partitioned according to the three kinds of preference structure as $R_i(s) = R_i^{++}(s) \cup R_i^{+m}(s) \cup R_i^{=}(s) \cup R_i^{-m}(s) \cup R_i^{--}(s)$.

6.2.2 *Reachable Lists for Multiple Degrees of Preference*

The set $R_i(s)$ denotes the unilateral moves (UMs) of DM i from $s \in S$, and is also called i's reachable list from s. It contains all states to which DM i can move, unilaterally and in one step, from state s. Similarly, the set $R_i^+(s) = \{q \in S : q \in R_i(s) \text{ and } q \overset{d}{\succ_i} s \text{ for } d = 1, 2, \ldots, r\}$ contains DM i's unilateral improvements (UIs) from state s for all degrees of preference. Note that although the same notation "$R_i^+(s)$" is used in Sect. 4.1.1 to represent DM i's unilateral improvements from state s at degree 1, the meaning of $R_i^+(s)$ here differs from that: there, it denotes all unilateral improvements, which can only be of degree 1 from s by DM i, whereas here, it includes all unilateral improvements, no matter what degree. All reachable lists from state s at each degree of preference for DM i are expressed by $R_i^{+(r)}(s), \ldots,$

Table 6.5 Reachable lists of DM i at some degree of preference

Type of movement	Description
$R_i^{+(d)}(s) = R_i(s) \cap \Phi_i^{+(d)}(s)$ $(d = 1, 2, \ldots, r)$	All unilateral improvements of degree d from state s for DM i
$R_i^{(0)}(s) = R_i^{=}(s) = R_i(s) \cap \Phi_i^{=}(s)$ $(d = 0)$	All equally preferred states reachable from state s by DM i
$R_i(s)^{-(d)}(s) = R_i(s) \cap \Phi_i^{-(d)}(s)$ $(d = 1, 2, \ldots, r)$	All unilateral disimprovements of degree d from state s for DM i

$R_i^{+(1)}(s)$, $R_i^{(0)}(s)$, $R_i^{-(1)}(s), \ldots,$ and $R_i^{-(r)}(s)$. Let $R_i(s) = \bigcup_{d=0}^{r}(R_i^{-(d)}(s) \cup R_i^{+(d)}(s))$ and $R_i^{+}(s) = \bigcup_{d=1}^{r} R_i^{+(d)}(s)$, where $R_i^{+(d)}(s)$ and $R_i^{-(d)}(s)$ for $d = 0, 1, \ldots, r$, are described in Table 6.5. Additionally, the relations among the subsets of S, $\Phi_i^{+(d)}(s)$ and $\Phi_i^{-(d)}(s)$ for $d = 0, 1, \ldots, r$, and the corresponding reachable lists from state s for DM i, $R_i^{+(d)}(s)$ and $R_i^{-(d)}(s)$ for $d = 0, 1, \ldots, r$, are depicted in Fig. 6.1.

Incorporating these extended multiple kinds of preference into the Graph Model for Conflict Resolution results in multi-degree versions of the four basic solution concepts presented in Sect. 6.4. The stability definitions for three types of preference are presented in next section.

6.3 Logical Representation of Stabilities for Three Types of Preference

Three types of preference including strength of preference are integrated into the Graph Model for Conflict Resolution to extend the four basic solution concepts in order to ascertain their strategic impacts. Recall that the three types of preference are equally preferred ($\sim$), mildly preferred ($>$), and strongly preferred ($\gg$) which together form the preference structure denoted as $\{\sim, >, \gg\}$. The four stability definitions given in the next subsection recognize two cases in which the degree of strength in the three kinds of preference are distinguished. Firstly, general stabilities are defined, and then the two subclasses, strong and weak, are determined. Stabilities of the first kind are referred to as general because they are in essence the same as the stability definitions using simple preference, as defined in Sect. 4.2. Stability definitions are called strong or weak stabilities in order to reflect the additional preference information contained in the strength of the preference relation. These more sophisticated definitions furnish expanded strategic insights into a conflict model that handles strength of preference. Sections 6.3.1 and 6.3.3 furnish the above stability definitions for 2-DM and n-DM ($n \geq 2$), respectively, while Sect. 6.3.2 presents definitions for reachable lists of a coalition of DMs required in the n-DM stability definitions.

6.3.1 Two Decision Maker Case

In order to calculate the stability of a state for a given DM $i \in N$, it is necessary to examine possible responses by all other DMs $j \in N \setminus \{i\}$. In a two-DM model, the only opponent of DM i is the remaining DM j. For all of the definitions given in next section, assume that $N = \{i, j\}$ and $s \in S$.

6.3.1.1 Logical Representation of General Stabilities

Four general solution concepts are given below in which strength of preference is not considered in sanctioning. However, the general stabilities are different from those defined in Sect. 4.2 for simple preference, because the stability definitions for simple preference do not directly take into account degree or strength of preference.

Definition 6.1 State s is Nash stable for DM i, denoted by $s \in S_i^{Nash}$, iff $R_i^{+,++}(s) = \emptyset$.

Definition 6.2 State s is general GMR (GGMR) for DM i, denoted by $s \in S_i^{GGMR}$, iff for every $s_1 \in R_i^{+,++}(s)$ there exists at least one $s_2 \in R_j(s_1)$ such that $s_2 \in \Phi_i^{--,-,=}(s)$.

Definition 6.3 State s is general SMR (GSMR) for DM i, denoted by $s \in S_i^{GSMR}$, iff for every $s_1 \in R_i^{+,++}(s)$ there exists at least one $s_2 \in R_j(s_1)$, such that $s_2 \in \Phi_i^{--,-,=}(s)$ and $s_3 \in \Phi_i^{--,-,=}(s)$ for any $s_3 \in R_i(s_2)$.

Definition 6.4 State s is general SEQ (GSEQ) for DM i, denoted by $s \in S_i^{GSEQ}$, iff for every $s_1 \in R_i^{+,++}(s)$ there exists at least one $s_2 \in R_j^{+,++}(s_1)$ such that $s_2 \in \Phi_i^{--,-,=}(s)$.

6.3.1.2 Logical Representation of Strong and Weak Stabilities

When strength of preference is introduced into the graph model, stability definitions can be strong or weak, according to the degree of sanctioning. For three kinds of preference, stabilities are divided into strongly and weakly stable with respect to the strength of possible sanctions. Hence, if a particular state s is general stable, then s is either strongly stable or weakly stable. Strong and weak stabilities only include GMR, SMR, and SEQ because Nash stability does not involve sanctions.

Definition 6.5 State s is strongly GMR (SGMR) for DM i, denoted by $s \in S_i^{SGMR}$, iff for every $s_1 \in R_i^{+,++}(s)$ there exists at least one $s_2 \in R_j(s_1)$ such that $s_2 \in \Phi_i^{--}(s)$.

Definition 6.6 State s is strongly SMR (SSMR) for DM i, denoted by $s \in S_i^{SSMR}$, iff for every $s_1 \in R_i^{+,++}(s)$ there exists at least one $s_2 \in R_j(s_1)$, such that $s_2 \in \Phi_i^{--}(s)$ and $s_3 \in \Phi_i^{--}(s)$ for all $s_3 \in R_i(s_2)$.

Definition 6.7 State s is strongly SEQ (SSEQ) for DM i, denoted by $s \in S_i^{SSEQ}$, iff for every $s_1 \in R_i^{+,++}(s)$ there exists at least one $s_2 \in R_j^{+,++}(s_1)$ such that $s_2 \in \Phi_i^{--}(s)$.

Definition 6.8 Let $s \in S$ and $i \in N$. State s is weakly stable for DM i iff s is general stable, but not strongly stable for some stability definition.

Example 6.1 (Stabilities for the Extended Sustainable Development Model under Three-degree Preference) The sustainable development conflict is introduced in Example 3.1. Here, this conflict is expanded to include three degrees of preference for the two-DM case. Specifically, the conflict consists of two DMs: an environmental agency (DM 1: E) and a developer (DM 2: D); and two options: DM 1 controls the option of being proactive (labeled P) and DM 2 has the option of practicing sustainable development (labeled SD) for properly treating the environment. The two options are combined to form four feasible states: s_1, s_2, s_3, and s_4. These results are listed in Table 6.6, where a "Y" indicates that an option is selected by the DM controlling it and an "N" means that the option is not chosen.

The preference information for each DM among the four states is provided at the bottom of Table 6.6. As can be seen for the case of DM 1, this DM prefers s_1 over s_3, greatly prefers s_3 to s_2, which is equally preferred to s_4. Notice that DM 2 greatly prefers s_1 to s_4. The graph model for the extended sustainable development conflict is presented in Fig. 6.2. One can see, for instance from DM 1's directed graph on the left side of Fig. 6.2, that this DM controls the movement between states s_1 and s_3 as well as s_2 and s_4.

The extended sustainable development model with three degrees of preference is used to illustrate how to determine general and strong stabilities under the three-degree version using Definitions 6.1–6.8. In particular, first consider analyzing

Table 6.6 Extended sustainable development game in option form under three-degree preference

DM 1: Environmental agency				
1. Proactive (P)	Y	Y	N	N
DM 2: Developer				
2. Sustainable development (SD)	Y	N	Y	N
States	s_1	s_2	s_3	s_4

Preferences $s_1 >_1 s_3 \gg_1 s_2 \sim_1 s_4$ for DM 1 and
$s_3 >_2 s_1 \gg_2 s_4 \sim_2 s_2$ for DM 2

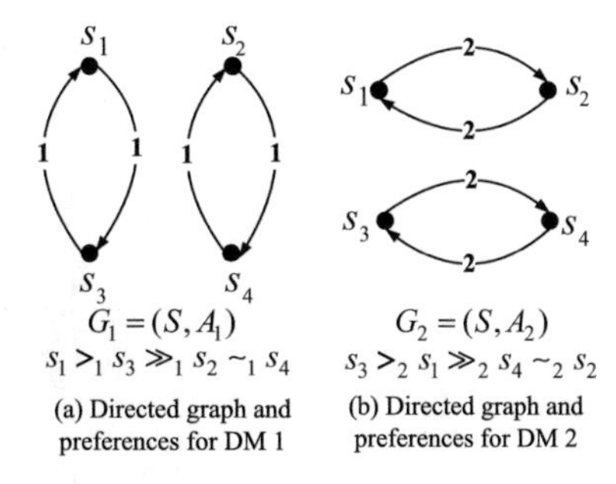

Fig. 6.2 Graph model for the extended sustainable development conflict under three-degree preference

state s_1 with respect to general Nash stability for DM 1. From Fig. 6.2, DM 1 has a unilateral move (UM) from s_1 to s_3. However, s_1 is mildly more preferred to s_3 by DM 1 because the move from s_1 to s_3 does not fall into the category of a mild or strong unilateral improvement. Therefore, state s_1 is general Nash stable for DM 1 according to Definition 6.1. Moreover, s_1 is also general GMR, SMR, and SEQ stable for DM 1.

Next, one can assess whether s_3 is general GMR stable for DM 1. From Fig. 6.2, DM 1 has a mild unilateral improvement from s_3 to s_1 and DM 2 has a unilateral move from s_1 to s_2. However, since s_2 is strongly less preferred than s_3 for DM 1, state s_3 is general GMR stable for DM 1 according to Definition 6.2. The stabilities of the other three states for the two DMs can be determined in a similar fashion.

Now, consider analyzing state s_3 from DM 1's viewpoint for GSMR stability using Definition 6.3. As can be seen from DM 1's directed graph in Fig. 6.2a, DM 1 has a mild UI from s_3 to s_1 and DM 2 has a UM from s_1 to s_2, from which DM 1 has only a UM from s_2 to s_4. Because DM 1 is indifferent between s_2 and s_4, which are greatly less preferred to state s_3, s_3 is GSMR stable for DM 1 using Definition 6.3. General SMR stability for other states can be calculated in a similar way.

To explain how general SEQ stability is calculated, consider state s_3 from DM 1's perspective. Because DM 2's possible countermove from s_1 to s_2 is, in fact, a move to a greatly less preferred state, this DM has no credible sanction to stop DM 1 from taking advantage of its UI from s_3 to s_1. Accordingly, state s_3 is not general SEQ stable for DM 1.

One could also provide an explanation for determining strong or weak stabilities using Definitions 6.5–6.8 for the extended sustainable development conflict. The discussion would be quite similar to the general stabilities.

The stable states and equilibria for the extended sustainable development conflict under three-degree preference are summarized in Table 6.7, in which "$\surd$" for a given state means that this state is general, strong, or weak stable for DM 1 or DM 2 and "Eq" is an equilibrium for an appropriate solution concept. The results provided by Table 6.7 show that state s_1 is a strong equilibrium for the four basic stabilities. State s_3 is strongly stable for GMR and SMR for all DMs. Hence, s_1 and s_3 are better choices for decision makers.

Table 6.7 Stability results of the extended sustainable development conflict under three-degree preference

State	Nash			GGMR			GSMR			GSEQ			SGMR			SSMR			SSEQ			WGMR			WSMR			WSEQ		
	1	2	Eq	1	2	Eq	1	2	Eq	1	2	Eq	1	2	Eq	1	2	Eq	1	2	Eq	1	2	Eq	1	2	Eq	1	2	Eq
s_1	√	√	√	√	√	√	√	√	√	√	√	√	√	√	√	√	√	√	√	√	√									
s_2	√			√			√			√			√			√			√											
s_3		√		√	√	√	√	√	√		√		√	√	√	√	√	√		√										
s_4	√			√			√			√			√			√			√											

6.3.2 Reachable Lists of a Coalition of Decision Makers

To extend the definitions of the reachable lists for a coalition to take three kinds of preference ($\sim, >, \gg$) into account, a legal sequence of coalitional mild or strong unilateral improvements (MSUIs) must be defined first. The reachable lists of coalition H from state s by the legal sequences of UMs and UIs are defined in Sect. 4.2.2 for simple preference. The reachable lists of coalition H are expanded to three kinds of preference in this section. A legal sequence of MSUIs is a sequence of allowable mild unilateral improvements or strong unilateral improvements by a coalition, with the same restriction that any member in the coalition may move more than once, but not twice consecutively. The formal definition for reachable lists of coalition H by the legal sequence of MSUIs is presented as follows.

Definition 6.9 Let $s \in S$, $H \subseteq N$, and $H \neq \emptyset$. A mild or strong unilateral improvement (MSUI) by H is a member of $R_H^{+,++}(s) \subseteq S$, defined inductively by

(1) assuming $\Omega_H^{+,++}(s, s_1) = \emptyset$ for all $s_1 \in S$;
(2) if $j \in H$ and $s_1 \in R_j^{+,++}(s)$, then $s_1 \in R_H^{+,++}(s)$ and $\Omega_H^{+,++}(s, s_1) = \Omega_H^{+,++}(s, s_1) \cup \{j\}$;
(3) if $s_1 \in R_H^{+,++}(s)$, $j \in H$, and $s_2 \in R_j^{+,++}(s_1)$, then, provided $\Omega_H^{+,++}(s, s_1) \neq \{j\}$, $s_2 \in R_H^{+,++}(s)$ and $\Omega_H^{+,++}(s, s_2) = \Omega_H^{+,++}(s, s_2) \cup \{j\}$.

Definition 6.9 is similar to Definition 4.7 for simple preference in Sect. 4.2.2 and Definition 5.18 for unknown preference in Sect. 5.2.2. It is also an inductive definition. By (2) in Definition 6.9, the states reachable from s are identified and added to the set $R_H^{+,++}(s)$; then, using (3), all states reachable from those states are identified and added to $R_H^{+,++}(s)$; afterwards the process is repeated in finitely many steps until no further states are added to the coalitional reachable list by legal sequences of mild or strong unilateral improvements, $R_H^{+,++}(s)$. For $\Omega_H^{+,++}(s, s_1)$, if $s_1 \in R_H^{+,++}(s)$, then $\Omega_H^{+,++}(s, s_1) \subseteq H$ is the set of all last DMs in legal MSUI sequences from s to s_1. Suppose that $\Omega_H^{+,++}(s, s_1)$ contains only one DM $j \in N$. Then any move from s_1 to a subsequent state s_2 must be made by a member of H other than j; otherwise DM j would have to move twice in succession.

6.3.3 n-Decision Maker Case

Within an n-DM model ($n \geq 2$) for three degrees of preference structure, DM i's opponents, $N \setminus \{i\}$, consist of a group of one or more DMs. In order to analyze the stability of a state for DM $i \in N$, it is necessary to take into account possible responses by all other DMs $j \in N \setminus \{i\}$. The key components in stability definitions for three degrees of preference are reachable lists of coalition $N \setminus \{i\}$ from state s, $R_{N\setminus\{i\}}(s)$ and $R_{N\setminus\{i\}}^{+,++}(s)$, discussed above. The stability definitions for two DM cases presented in Sect. 6.3.1 are extended to general n-DM models next.

6.3.3.1 Logical Representation of General Stabilities

Four standard solution concepts are given below in which strength of preference is not considered in sanctioning. However, the general stabilities are different from those defined in Sect. 4.2.3 for simple preference, because stability definitions for simple preference cannot analyze conflict models having strength of preference. Let $i \in N$ and $s \in S$ for the following definitions.

Definition 6.10 State s is Nash stable for DM i, denoted by $s \in S_i^{Nash}$, iff $R_i^{+,++}(s) = \emptyset$.

Nash stability definitions are identical for both the 2-DM and the n-DM models because Nash stability does not consider opponents' responses.

Definition 6.11 State s is general GMR (GGMR) for DM i, denoted by $s \in S_i^{GGMR}$, iff for every $s_1 \in R_i^{+,++}(s)$ there exists at least one $s_2 \in R_{N\backslash\{i\}}(s_1)$ such that $s_2 \in \Phi_i^{--,-,=}(s)$.

Definition 6.12 State s is general SMR (GSMR) for DM i, denoted by $s \in S_i^{GSMR}$, iff for every $s_1 \in R_i^{+,++}(s)$ there exists at least one $s_2 \in R_{N\backslash\{i\}}(s_1)$, such that $s_2 \in \Phi_i^{--,-,=}(s)$ and $s_3 \in \Phi_i^{--,-,=}(s)$ for any $s_3 \in R_i(s_2)$.

Definition 6.13 State s is general SEQ (GSEQ) for DM i, denoted by $s \in S_i^{GSEQ}$, iff for every $s_1 \in R_i^{+,++}(s)$ there exists at least one $s_2 \in R_{N\backslash\{i\}}^{+,++}(s_1)$ such that $s_2 \in \Phi_i^{--,-,=}(s)$.

Similar to 2-DM case, general stabilities for n-DM models are partitioned into strong or weak stabilities according to the level of sanctioning. Strong and weak stabilities only include GMR, SMR, and SEQ because Nash stability does not involve sanctions.

6.3.3.2 Logical Representation of Strong and Weak Stabilities

Definition 6.14 State s is strongly GMR (SGMR) for DM i, denoted by $s \in S_i^{SGMR}$, iff for every $s_1 \in R_i^{+,++}(s)$ there exists at least one $s_2 \in R_{N\backslash\{i\}}(s_1)$ such that $s_2 \in \Phi_i^{--}(s)$.

Definition 6.15 State s is strongly SMR (SSMR) for DM i, denoted by $s \in S_i^{SSMR}$, iff for every $s_1 \in R_i^{+,++}(s)$ there exists at least one $s_2 \in R_{N\backslash\{i\}}(s_1)$, such that $s_2 \in \Phi_i^{--}(s)$ and $s_3 \in \Phi_i^{--}(s)$ for all $s_3 \in R_i(s_2)$.

Definition 6.16 State s is strongly SEQ (SSEQ) for DM i, denoted by $s \in S_i^{SSEQ}$, iff for every $s_1 \in R_i^{+,++}(s)$ there exists at least one $s_2 \in R_{N\backslash\{i\}}^{+,++}(s_1)$ such that $s_2 \in \Phi_i^{--}(s)$.

The important components, $R_{N\backslash\{i\}}(s_1)$ and $R_{N\backslash\{i\}}^{+,++}(s_1)$, in Definitions 6.14–6.16 are defined in Sects. 4.2.2 and 6.3.2, respectively, for $H = N\backslash\{i\}$. The definition of weak stability is presented next.

Definition 6.17 Let $s \in S$ and $i \in N$. State s is weakly stable for DM i iff s is general stable, but not strongly stable for some stability definition.

6.4 Logical Representation of Stabilities for Multiple Degrees of Preferences

The following stability definitions for multiple kinds of preference are analogous to the concepts for three types of preference presented in Sect. 6.3. The multiple-degree preference is included into the Graph Model for Conflict Resolution resulting in multilevel versions of the four basic solution concepts, $Nash_k$, GMR_k, SMR_k, and SEQ_k for $k = 0, 1, \ldots, r$. The stability definitions in a 2-DM conflict model are presented next.

6.4.1 *Two Decision Maker Case*

6.4.1.1 Logical Representation of General Stabilities

Definition 6.18 State s is **general Nash stable** (GNash) for DM i, denoted by $s \in S_i^{GNash}$, iff $R_i^+(s) = \emptyset$.

Definition 6.19 State s is **general GMR** (GGMR) for DM i, denoted by $s \in S_i^{GGMR}$, iff for every $s_1 \in R_i^+(s)$ there exists at least one $s_2 \in R_j(s_1)$ with $s_2 \in \bigcup_{d=0}^{r} \Phi_i^{-(d)}(s)$.

Definition 6.20 State s is **general SMR** (GSMR) for DM i, denoted by $s \in S_i^{GSMR}$, iff for every $s_1 \in R_i^+(s)$ there exists at least one $s_2 \in R_j(s_1)$ with $s_2 \in \bigcup_{d=0}^{r} \Phi_i^{-(d)}(s)$ and $s_3 \in \bigcup_{d=0}^{r} \Phi_i^{-(d)}(s)$ for all $s_3 \in R_i(s_2)$.

Definition 6.21 State s is **general SEQ** (GSEQ) for DM i, denoted by $s \in S_i^{GSEQ}$, iff for every $s_1 \in R_i^+(s)$ there exists at least one $s_2 \in R_j^+(s_1)$ with $s_2 \in \bigcup_{d=0}^{r} \Phi_i^{-(d)}(s)$.

Note that in this section the meaning of $R_i^+(s)$ differs from that in Sect. 4.1.1 to represent DM i's UI from state s for simple preference; there, it denotes all one-degree unilateral improvements from s by DM i, whereas here, it includes all unilateral improvements, no matter how many degrees of preference. For three degrees of preference discussed above, general stabilities are divided into strongly and weakly stable according to the strength of the possible sanction, i.e., if a particular state s is general stable, then s is either strongly stable or weakly stable. Within multiple degrees of preference, the general stabilities are constituted by stabilities at each level of preference.

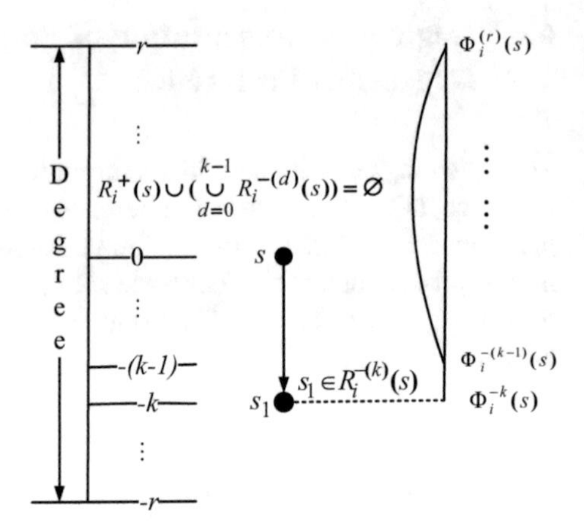

Fig. 6.3 Nash stability at degree k for DM i

6.4.1.2 Logical Representation of Stabilities at Degree k

Firstly, definitions are now given for different strengths of Nash stability. Even though unilateral improvements do not exist under Nash stability, the idea of strength of stability can still be captured using the degree of preference for the most preferred states to which the DM could unilaterally move. All these states must be less preferred than the initial state. A special definition is required for the case in which no movements of any type exist for the DM. In particular, if DM i has no unilateral move at all degrees of preference from state s, state s is extremely stable. The stability is proposed next.

Definition 6.22 If $R_i(s) = \emptyset$, then state s is **super stable** for DM i at any degree of preference, denoted by $s \in S_i^{Super}$.

Definition 6.23 State s is **Nash stable** ($Nash_0$) at degree 0 for DM i, denoted by $s \in S_i^{Nash_0}$, iff $R_i^+(s) = \emptyset$ and $R_i^{(0)}(s) \neq \emptyset$.

Notice in the definition of $Nash_0$ that no unilateral improvements by DM i from state s exist but an equally preferred state must be present.

Definition 6.24 For $1 \leq k \leq r$, state s is **Nash stable** ($Nash_k$) at degree k for DM i, denoted by $s \in S_i^{Nash_k}$, iff $R_i^+(s) \cup (\bigcup_{d=0}^{k-1} R_i^{-(d)}(s)) = \emptyset$ and $R_i^{-(k)}(s) \neq \emptyset$.

For $Nash_k$ stability, the most preferred state to which DM i can unilaterally move from s is located at degree $-k$ (below degree 0). The kth degree Nash stability is depicted in Fig. 6.3. The super stability is referred to as Nash stability at the highest degree, because no unilateral moves exist for DM i from s.

When multiple-degree preference is incorporated into the graph model, GMR, SMR, and SEQ stabilities at different degrees can be distinguished according to the strength of the sanctions.

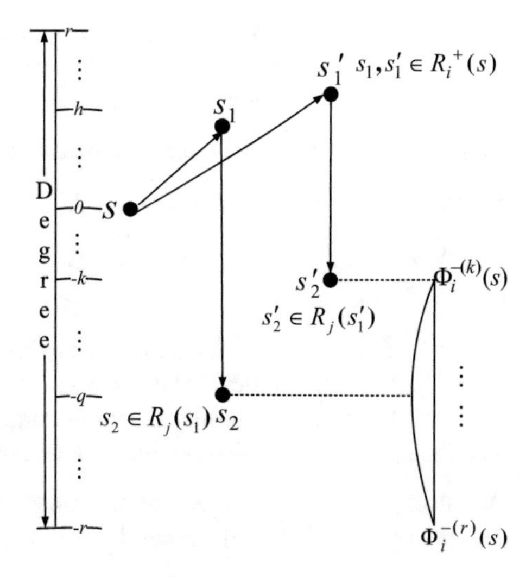

Fig. 6.4 GMR stability at degree k for DM i

Definition 6.25 State s is **general metarational** (GMR_0) at degree 0 for DM i, denoted by $s \in S_i^{GMR_0}$, iff either $R_i^+(s) = \emptyset$ and $R_i^{(0)}(s) \neq \emptyset$, or $R_i^+(s) \neq \emptyset$ and for every $s_1 \in R_i^+(s)$ there exists at least one $s_2 \in R_j(s_1)$ with $s_2 \in \bigcup_{d=0}^{r} \Phi_i^{-(d)}(s)$ and there exists at least one $s'_1 \in R_i^+(s)$ and $s'_2 \in R_j(s'_1)$ such that $s'_2 \in \Phi_i^{(0)}(s)$ and $R_j(s'_1) \bigcap (\bigcup_{d=1}^{r} \Phi_i^{-(d)}(s)) = \emptyset$.

Based on Definition 6.25, when DM i has no UIs from state s and it is $Nash_0$ stable, as in Definition 6.24, then state s is also GMR stable at degree 0.

Definition 6.26 For $1 \leq k \leq r-1$, state s is **general metarational** (GMR_k) at degree k for DM i, denoted by $s \in S_i^{GMR_k}$, iff either $\bigcup_{d=0}^{k-1} R_i^{-(d)}(s) \cup R_i^+(s) = \emptyset$ and $R_i^{-(k)}(s) \neq \emptyset$, or $R_i^+(s) \neq \emptyset$ and for every $s_1 \in R_i^+(s)$ there exists at least one $s_2 \in R_j(s_1)$ with $s_2 \in \bigcup_{d=k}^{r} \Phi_i^{-(d)}(s)$ and there exists at least one $s'_1 \in R_i^+(s)$ and $s'_2 \in R_j(s'_1)$ such that $s'_2 \in \Phi_i^{-(k)}(s)$ and $R_j(s'_1) \bigcap (\bigcup_{d=k+1}^{r} \Phi_i^{-(d)}(s)) = \emptyset$.

Figure 6.4 contains a specific example to explain the meaning of Definition 6.26. Notice that DM i has UIs from state s to states s_1 and s'_1, each of which can be at any degree from 1 to r. From state s_1, DM j, who is DM i's opponent, has one unilateral move to state s_2 (labeled $R_j(s_1)$), which is as shown on the degree axis to be of degree $-q$, where q can range from k to r relative to s. With respect to state s'_1, DM j can move to state s'_2, which is only located at degree $-k$ relative to state s. Therefore, state s for DM i possesses general meterational stability at degree k for which $0 < k < r$ according to Definition 6.26.

If all of DM i's UIs from a state are sanctioned at the highest degree r (exactly r levels below the state), then the state is called general metarational at degree r. Its formal definition is given below.

Definition 6.27 State s is **general metarational** (GMR_r) at degree r for DM i, denoted by $s \in S_i^{GMR_r}$, iff either $\bigcup_{d=0}^{r-1} R_i^{-(d)}(s) \cup R_i^+(s) = \emptyset$ and $R_i^{-(r)}(s) \neq \emptyset$, or $R_i^+(s) \neq \emptyset$ and for every $s_1 \in R_i^+(s)$ there exists at least one $s_2 \in R_j(s_1)$ with $s_2 \in \Phi_i^{-(r)}(s)$.

For DM i, if a UI from a state is sanctioned at degree k below the state and all other UIs from the particular state are sanctioned at a degree of at least k below the state, and these corresponding sanctions cannot be avoided by any counterresponse, then the state is called SMR stable at degree k. Its formal definition is given below.

Definition 6.28 State s is **symmetric metarational** (SMR_0) at degree 0 for DM i, denoted by $s \in S_i^{SMR_0}$, iff either $R_i^+(s) = \emptyset$ and $R_i^{(0)}(s) \neq \emptyset$, or $R_i^+(s) \neq \emptyset$ and for every $s_1 \in R_i^+(s)$ there exists at least one $s_2 \in R_j(s_1)$ with $s_2 \in \bigcup_{d=0}^{r} \Phi_i^{-(d)}(s)$ and there exists at least one $s_1' \in R_i^+(s)$ and $s_2' \in R_j(s_1')$ such that $s_2' \in \Phi_i^{(0)}(s)$ and $R_j(s_1') \bigcap (\bigcup_{d=1}^{r} \Phi_i^{-(d)}(s)) = \emptyset$, as well as $s_3 \in \bigcup_{d=0}^{r} \Phi_i^{-(d)}(s)$ for any $s_3 \in R_i(s_2) \cup R_i(s_2')$.

Symmetric metarationality at degree k ($0 < k \leq r$) for DM i consists of SMR_{k^+} and SMR_{k^-} that are defined next.

Definition 6.29 For $1 \leq k \leq r - 1$, state s is **symmetric metarational** (SMR_{k^+}) at degree k for DM i, denoted by $s \in S_i^{SMR_{k^+}}$, iff either $\bigcup_{d=0}^{k-1} R_i^{-(d)}(s) \cup R_i^+(s) = \emptyset$ and $R_i^{-(k)}(s) \neq \emptyset$, or $R_i^+(s) \neq \emptyset$ and for every $s_1 \in R_i^+(s)$ there exists at least one $s_2 \in R_j(s_1)$ with $s_2 \in \bigcup_{d=k}^{r} \Phi_i^{-(d)}(s)$ and there exists at least one $s_1' \in R_i^+(s)$ and $s_2' \in R_j(s_1')$ such that $s_2' \in \Phi_i^{-(k)}(s)$ and $R_j(s_1') \bigcap (\bigcup_{d=k+1}^{r} \Phi_i^{-(d)}(s)) = \emptyset$, as well as $s_3 \in \bigcup_{d=k}^{r} \Phi_i^{-(d)}(s)$ for any $s_3 \in R_i(s_2) \cup R_i(s_2')$.

Figure 6.5 vividly illustrates the SMR stability at k^+ for DM i. Stability SMR_{k^-} is defined by $S_i^{SMR_{k^-}} = S_i^{GSMR} \cap S_i^{GMR_k} - S_i^{SMR_{k^+}}$. Equivalently,

Definition 6.30 For $1 \leq k \leq r - 1$, state s is **symmetric metarational** (SMR_{k^-}) at degree k for DM i, denoted by $s \in S_i^{SMR_{k^-}}$, iff $s \in S_i^{GMR_k}$ and $R_i^+(s) \neq \emptyset$ and for every $s_1 \in R_i^+(s)$ there exists at least one $s_2 \in R_j(s_1)$ with $s_2 \in \bigcup_{d=k}^{r} \Phi_i^{-(d)}(s)$ and $s_3 \in \bigcup_{d=0}^{r} \Phi_i^{-(d)}(s)$ for all $s_3 \in R_i(s_2)$, as well as there exists $s_1' \in R_i^+(s)$ and

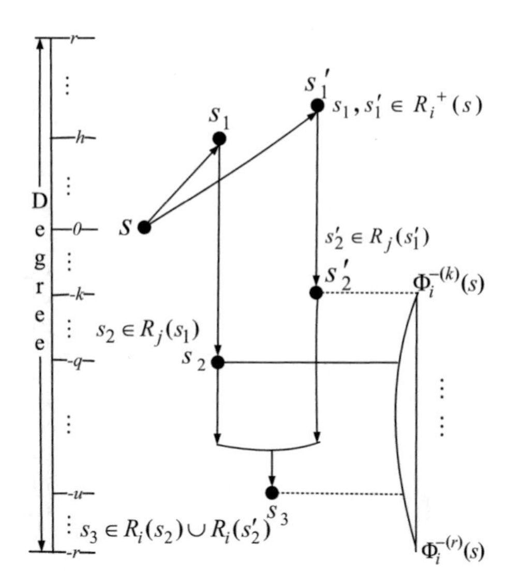

Fig. 6.5 SMR stability at degree k^+ for DM i

for every $s_2' \in R_j(s_1') \cap (\bigcup_{d=k}^{r} \Phi_i^{-(d)}(s))$, $R_i(s_2') \cap \Phi_i^{(-d)}(s) \neq \emptyset$ for at least one $d \in \{0, \ldots, (k-1)\}$.

Definition 6.31 State s is **symmetric metarational** (SMR_{r^+}) at degree r for DM i, denoted by $s \in S_i^{SMR_{r^+}}$, iff either $\bigcup_{d=0}^{r-1} R_i^{-(d)}(s) \cup R_i^{+}(s) = \emptyset$ and $R_i^{-(r)}(s) \neq \emptyset$, or $R_i^{+}(s) \neq \emptyset$ and for every $s_1 \in R_i^{+}(s)$ there exists at least one $s_2 \in R_j(s_1)$ with $s_2 \in \Phi_i^{-(r)}(s)$ and $s_3 \in \Phi_i^{-(r)}(s)$ for any $s_3 \in R_i(s_2)$.

Definition 6.32 State s is **symmetric metarational** (SMR_{r^-}) at degree r for DM i, denoted by $s \in S_i^{SMR_{r^-}}$, iff $R_i^{+}(s) \neq \emptyset$ and for every $s_1 \in R_i^{+}(s)$ there exists at least one $s_2 \in R_j(s_1)$ with $s_2 \in \Phi_i^{-(r)}(s)$ and $s_3 \in \bigcup_{d=0}^{r} \Phi_i^{-(d)}(s)$ for all $s_3 \in R_i(s_2)$, as well as there exists $s_1' \in R_i^{+}(s)$ and for every $s_2' \in R_j(s_1) \cap \Phi_i^{-(r)}(s)$, $R_i(s_2') \cap \Phi_i^{(-d)}(s) \neq \emptyset$ for at least one $d \in \{0, \ldots, (r-1)\}$.

Sequential stability at degree k is similar to the stability of GMR at the same degree. The only modification is that all DM i's UIs are subject to credible sanctions by DM i's opponent. Its formal definition is given below.

Definition 6.33 State s is **sequentially stable** (SEQ_0) at degree 0 for DM i, denoted by $s \in S_i^{SEQ_0}$, iff either $R_i^{+}(s) = \emptyset$ and $R_i^{(0)}(s) \neq \emptyset$, or $R_i^{+}(s) \neq \emptyset$ and for every $s_1 \in R_i^{+}(s)$ there exists at least one $s_2 \in R_j^{+}(s_1)$ with $s_2 \in \bigcup_{d=0}^{r} \Phi_i^{-(d)}(s)$ and there exists at least one $s_1' \in R_i^{+}(s)$ and $s_2' \in R_j^{+}(s_1')$ such that $s_2' \in \Phi_i^{(0)}(s)$ and $R_j^{+}(s_1') \bigcap (\bigcup_{d=1}^{r} \Phi_i^{-(d)}(s)) = \emptyset$.

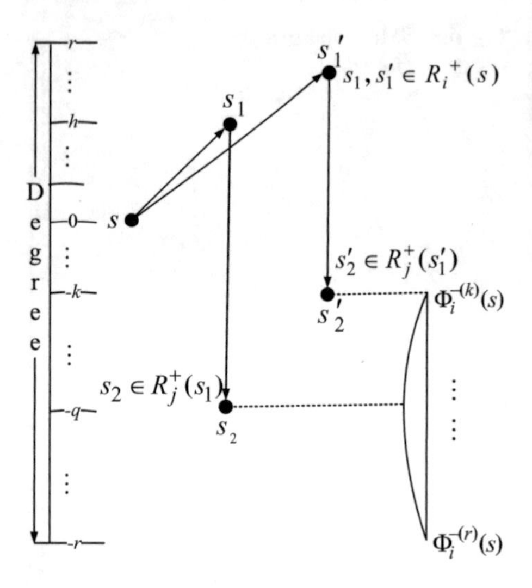

Fig. 6.6 SEQ stability at degree k for DM i

Definition 6.34 For $1 \le k \le r-1$, state s is **sequentially stable** (SEQ_k) at degree k for DM i, denoted by $s \in S_i^{SEQ_k}$, iff either $\bigcup_{d=0}^{k-1} R_i^{-(d)}(s) \cup R_i^+(s) = \emptyset$ and $R_i^{-(k)}(s) \neq \emptyset$, or $R_i^+(s) \neq \emptyset$ and for every $s_1 \in R_i^+(s)$ there exists at least one $s_2 \in R_j^+(s_1)$ with $s_2 \in \bigcup_{d=k}^{r} \Phi_i^{-(d)}(s)$ and there exists at least one $s_1' \in R_i^+(s)$ and $s_2' \in R_j^+(s_1')$ such that $s_2' \in \Phi_i^{-(k)}(s)$ and $R_j^+(s_1') \bigcap (\bigcup_{d=k+1}^{r} \Phi_i^{-(d)}(s)) = \emptyset$.

Figure 6.6 can be used to explain the meaning of Definition 6.34. In fact, Definition 6.34 is similar to Definition 6.29. The only difference is that DM j, who is DM i's opponent, has one unilateral improvement to state s_2 at a degree ranged from k to r relative to s. With respect to state s_1', DM j has a unilateral improvement s_2', which is only located at degree $-k$ relative to state s. Therefore, state s for DM i is sequentially stable at degree k for which $0 < k < r$ according to Definition 6.34.

Definition 6.35 State s is **sequentially stable** (SEQ_r) at degree r for DM i, denoted by $s \in S_i^{SEQ_r}$, iff either $\bigcup_{d=0}^{r-1} R_i^{-(d)}(s) \cup R_i^+(s) = \emptyset$ and $R_i^{-(r)}(s) \neq \emptyset$, or $R_i^+(s) \neq \emptyset$ and for every $s_1 \in R_i^+(s)$ there exists at least one $s_2 \in R_j^+(s_1)$ with $s_2 \in \Phi_i^{-(r)}(s)$.

In an n-DM model, where $n \geq 2$, the opponents of a DM can be thought of as a coalition of one or more DMs. To extend the graph model stability definitions to stability definitions in n-DM models with multiple degrees of preference, the definitions of a legal sequence of moves for three degrees of preference presented in Sect. 6.3.2 must first be extended to take multiple degrees of preference into account.

6.4.2 Reachable Lists of a Coalition of Decision Makers

A legal sequence of UMs in a graph model with multiple degrees of preference for a coalition of DMs is a sequence of states linked by unilateral moves controlled by members of the coalition, in which a DM may move more than once, but not twice in succession. As explained in Sect. 4.2.2 before Definition 4.6, this rule allows the GMCR methodology to handle intransitive moves, in addition to transitive moves. When $H = \{i\}$, a legal sequence of UMs for the coalition H reduces to a unilateral move of DM i.

Let the coalition $H \subseteq N$ satisfy $|H| \geq 2$ and let the status quo state be $s \in S$. Define $R_H(s) \subseteq S$, the reachable list of coalition H from state s by a legal sequence of UMs in a graph model with multiple degrees of preference. The following definitions are adapted from Fang et al. (1993) and Hamouda et al. (2006):

Definition 6.36 Let $s \in S$, $H \subseteq N$, and $H \neq \emptyset$. Here, $R_j(s) = \bigcup_{d=0}^{r}(R_j^{-(d)}(s) \cup R_j^{+(d)}(s))$ for any $j \in H$. A unilateral move by H is a member of $R_H(s) \subseteq S$, defined inductively by:

(1) if $j \in H$ and $s_1 \in R_j(s)$, then $s_1 \in R_H(s)$ and $\Omega_H(s, s_1) = \Omega_H(s, s_1) \cup \{j\}$;
(2) if $s_1 \in R_H(s)$, $j \in H$ and $s_2 \in R_j(s_1)$, then, provided $\Omega_H(s, s_1) \neq \{j\}$, $s_2 \in R_H(s)$ and $\Omega_H(s, s_2) = \Omega_H(s, s_2) \cup \{j\}$.

Note that Definition 6.36 is analogous to Definition 4.6, but, here, unilateral moves include the states that are reachable from state s by multiple degrees of preference (may have more than three degrees) listed in Table 6.5.

In a graph model with multiple degrees of preference, a legal sequence of UIs for coalition H is a sequence of states linked by unilateral improvements including each-degree UIs controlled by members of the coalition H with the usual restriction that a member of the coalition may move more than once, but not twice consecutively. The formal definition is given below.

Definition 6.37 Let $R_j^+(s) = \bigcup_{d=1}^{r} R_j^{+(d)}(s)$ for any $j \in H$. A unilateral improvement by H is a member of $R_H^+(s) \subseteq S$, defined inductively by:

(1) if $j \in H$ and $s_1 \in \bigcup_{d=1}^{r} R_j^{+(d)}(s)$, then $s_1 \in R_H^+(s)$ and $\Omega_H^+(s)(s, s_1) = \Omega_H^+(s)(s, s_1) \cup \{j\}$;
(2) if $s_1 \in R_H^+(s)$, $j \in H$ and $s_2 \in \bigcup_{d=1}^{r} R_j^{+(d)}(s_1)$, then, provided $\Omega_H^+(s)(s, s_1) \neq \{j\}$, $s_2 \in R_H^+(s)$ and $\Omega_H^+(s, s_2) = \Omega_H^+(s, s_2) \cup \{j\}$.

Definition 6.37 is identical to Definition 6.36 except that each move is to a state strictly preferred with some degree of preference by the mover to the current state. Similarly, $\Omega_H^+(s, s_1)$ includes all last movers in a legal sequence of UIs by coalition

H from state s to state s_1. Specifically, this definition is inductive: first, using (1), the states reachable by a single DM in H from s by one step UIs in multiple levels of preference are identified and added to $R_H^+(s)$; then, using (2), all states reachable from those states are identified and added to $R_H^+(s)$; afterwards the process is repeated until no further states are added to $R_H^+(s)$ by repeating (2). Because $R_H^+(s) \subseteq S$, and S is finite, this limit must be reached in finitely many steps.

6.4.3 n-Decision Maker Case

6.4.3.1 Logical Representation of General Stabilities

Super stability and Nash stability definitions are identical for both the 2-DM and the n-DM models because these stabilities do not consider the opponents' responses. Let $i \in N$ and $s \in S$ for the following definitions.

Definition 6.38 State $s \in S$ is GGMR for DM i, denoted by $s \in S_i^{GGMR}$, iff for every $s_1 \in R_i^+(s)$ there exists at least one $s_2 \in R_{N\setminus\{i\}}(s_1)$ with $s_2 \in \bigcup_{d=0}^{r} \Phi_i^{-(d)}(s)$.

Definition 6.39 State $s \in S$ is GSMR for DM i, denoted by $s \in S_i^{GSMR}$, iff for every $s_1 \in R_i^+(s)$ there exists at least one $s_2 \in R_{N\setminus\{i\}}(s_1)$ with $s_2 \in \bigcup_{d=0}^{r} \Phi_i^{-(d)}(s)$ and $s_3 \in \bigcup_{d=0}^{r} \Phi_i^{-(d)}(s)$ for all $s_3 \in R_i(s_2)$.

Definition 6.40 State $s \in S$ is GSEQ for DM i, denoted by $s \in S_i^{GSEQ}$, iff for every $s_1 \in R_i^+(s)$ there exists at least one $s_2 \in R_{N\setminus\{i\}}^+(s_1)$ with $s_2 \in \bigcup_{d=0}^{r} \Phi_i^{-(d)}(s)$.

6.4.3.2 Logical Representation of Stabilities at k Degree

Similar to 2-DM conflicts, solution concepts for n-DM conflicts can be defined as different-degree stabilities, according to degrees of preference. Nash stability definitions in multiple DM conflicts are the same as those in 2-DM cases. Therefore, only the extended GMR, SMR, and SEQ are defined here. For DM i, if a UI from state s is sanctioned by the legal sequence of UMs of i's opponents in exactly k degrees below s and all other UIs from state s are sanctioned in at least k degrees below s, then the status quo s is called general metarational at degree k. The process is portrayed in Fig. 6.7 and the formal definition is given below.

Definition 6.41 State s is GMR_0 for DM i, denoted by $s \in S_i^{GMR_0}$, iff either $R_i^+(s) = \emptyset$ and $R_i^{(0)}(s) \neq \emptyset$, or $R_i^+(s) \neq \emptyset$ and for every $s_1 \in R_i^+(s)$ there exists

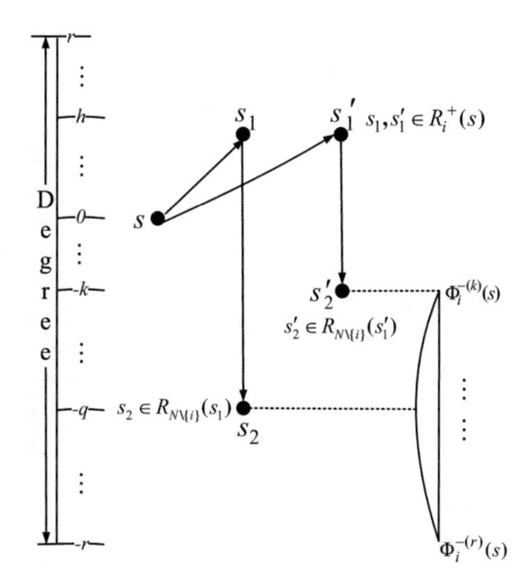

Fig. 6.7 General metarationality at degree k for DM i

at least one $s_2 \in R_{N\backslash\{i\}}(s_1)$ with $s_2 \in \bigcup_{d=0}^{r} \Phi_i^{-(d)}(s)$ and there exists at least one $s_1' \in R_i^+(s)$ and $s_2' \in R_{N\backslash\{i\}}(s_1')$ such that $s_2' \in \Phi_i^{(0)}(s)$ and $R_{N\backslash\{i\}}(s_1') \bigcap (\bigcup_{d=1}^{r} \Phi_i^{-(d)}(s)) = \emptyset$.

Definition 6.42 For $1 \leq k \leq r-1$, state s is GMR_k for DM i, denoted by $s \in S_i^{GMR_k}$, iff either $\bigcup_{d=0}^{k-1} R_i^{-(d)}(s) \cup R_i^+(s) = \emptyset$ and $R_i^{-(k)}(s) \neq \emptyset$, or $R_i^+(s) \neq \emptyset$ and for every $s_1 \in R_i^+(s)$ there exists at least one $s_2 \in R_{N\backslash\{i\}}(s_1)$ with $s_2 \in \bigcup_{d=k}^{r} \Phi_i^{-(d)}(s)$ and there exists at least one $s_1' \in R_i^+(s)$ and $s_2' \in R_{N\backslash\{i\}}(s_1')$ such that $s_2' \in \Phi_i^{-(k)}(s)$ and $R_{N\backslash\{i\}}(s_1') \bigcap (\bigcup_{d=k+1}^{r} \Phi_i^{-(d)}(s)) = \emptyset$.

If all of DM i's UIs from a state are sanctioned at exactly r degrees below the state, then the state is called general metarational at degree r. Its formal definition is given below.

Definition 6.43 State s is GMR_r for DM i, denoted by $s \in S_i^{GMR_r}$, iff either $\bigcup_{d=0}^{r-1} R_i^{-(d)}(s) \cup R_i^+(s) = \emptyset$ and $R_i^{-(r)}(s) \neq \emptyset$, or $R_i^+(s) \neq \emptyset$ and for every $s_1 \in R_i^+(s)$ there exists at least one $s_2 \in R_{N\backslash\{i\}}(s_1)$ with $s_2 \in \Phi_i^{-(r)}(s)$.

For DM i, if a UI from a state is sanctioned by the legal sequence of UMs of DM i's opponents at degree k and all other UIs from the particular state are sanctioned at degree at least k, and these corresponding sanctions cannot be avoided by any counterresponse, then the state is called symmetric metarational at degree k. The

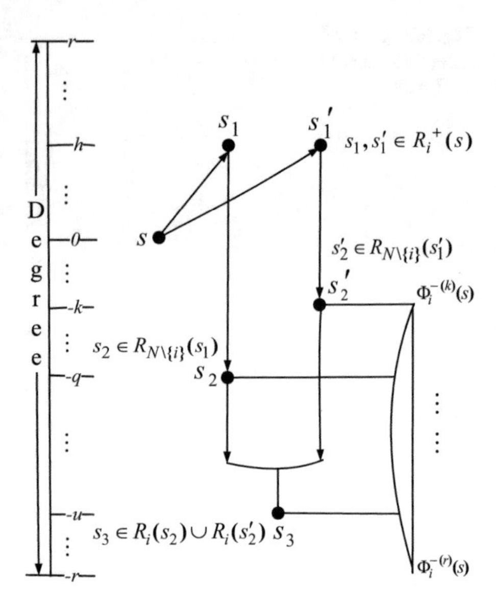

Fig. 6.8 Symmetric metarationality at degree k^+ for DM i

stability of SMR at degree k is portrayed in Fig. 6.8 and the formal definition is given below.

Definition 6.44 State s is SMR_0 for DM i, denoted by $s \in S_i^{SMR_0}$, iff either $R_i^+(s) = \emptyset$ and $R_i^{(0)}(s) \neq \emptyset$, or $R_i^+(s) \neq \emptyset$ and for every $s_1 \in R_i^+(s)$ there exists at least one $s_2 \in R_{N\setminus\{i\}}(s_1)$ with $s_2 \in \bigcup_{d=0}^{r} \Phi_i^{-(d)}(s)$ and there exists at least one $s_1' \in R_i^+(s)$ and $s_2' \in R_{N\setminus\{i\}}(s_1')$ such that $s_2' \in \Phi_i^{(0)}(s)$ and $R_{N\setminus\{i\}}(s_1') \bigcap (\bigcup_{d=1}^{r} \Phi_i^{-(d)}(s)) = \emptyset$, as well as $s_3 \in \bigcup_{d=0}^{r} \Phi_i^{-(d)}(s)$ for any $s_3 \in R_i(s_2) \cup R_i(s_2')$.

Symmetric metarationality at degree k $(0 < k \leq r)$ for DM i consists of SMR_{k^+} and SMR_{k^-} that are defined next.

Definition 6.45 For $1 \leq k \leq r-1$, state s is SMR_{k^+} for DM i, denoted by $s \in S_i^{SMR_{k^+}}$, iff either $\bigcup_{d=0}^{k-1} R_i^{-(d)}(s) \cup R_i^+(s) = \emptyset$ and $R_i^{-(k)}(s) \neq \emptyset$, or $R_i^+(s) \neq \emptyset$ and for every $s_1 \in R_i^+(s)$ there exists at least one $s_2 \in R_{N\setminus\{i\}}(s_1)$ with $s_2 \in \bigcup_{d=k}^{r} \Phi_i^{-(d)}(s)$ and there exists at least one $s_1' \in R_i^+(s)$ and $s_2' \in R_{N\setminus\{i\}}(s_1')$ such that $s_2' \in \Phi_i^{-(k)}(s)$ and $R_{N\setminus\{i\}}(s_1') \bigcap (\bigcup_{d=k+1}^{r} \Phi_i^{-(d)}(s)) = \emptyset$, as well as $s_3 \in \bigcup_{d=k}^{r} \Phi_i^{-(d)}(s)$ for any $s_3 \in R_i(s_2) \cup R_i(s_2')$.

Stability SMR_k- is defined by $S_i^{SMR_{k-}} = S_i^{GSMR} \cap S_i^{GMR_k} - S_i^{SMR_k}$. Equivalently,

Definition 6.46 For $1 \leq k \leq r-1$, state s is SMR_k- for DM i, denoted by $s \in S_i^{SMR_{k-}}$, iff $s \in S_i^{GMR_k}$ and $R_i^+(s) \neq \emptyset$, and for every $s_1 \in R_i^+(s)$ there exists at least one $s_2 \in R_{N\setminus\{i\}}(s_1)$ with $s_2 \in \bigcup_{d=k}^{r} \Phi_i^{-(d)}(s)$ and $s_3 \in \bigcup_{d=0}^{r} \Phi_i^{-(d)}(s)$ for all $s_3 \in R_i(s_2)$, as well as there exists $s_1' \in R_i^+(s)$ and for every $s_2' \in R_{N\setminus\{i\}}(s_1') \cap (\bigcup_{d=k}^{r} \Phi_i^{-(d)}(s))$, $R_i(s_2') \cap \Phi_i^{(-d)}(s) \neq \emptyset$ for at least one $d \in \{0, \ldots, (k-1)\}$.

Definition 6.47 State s is SMR_{r^+} for DM i, denoted by $s \in S_i^{SMR_{r+}}$, iff either $\bigcup_{d=0}^{r-1} R_i^{-(d)}(s) \cup R_i^+(s) = \emptyset$ and $R_i^{-(r)}(s) \neq \emptyset$, or $R_i^+(s) \neq \emptyset$ and for every $s_1 \in R_i^+(s)$ there exists at least one $s_2 \in R_{N\setminus\{i\}}(s_1)$ with $s_2 \in \Phi_i^{-(r)}(s)$ and $s_3 \in \Phi_i^{-(r)}(s)$ for any $s_3 \in R_i(s_2)$.

Definition 6.48 State s is SMR_{r^-} for DM i, denoted by $s \in S_i^{SMR_{r-}}$, iff $R_i^+(s) \neq \emptyset$ and for every $s_1 \in R_i^+(s)$ there exists at least one $s_2 \in R_{N\setminus\{i\}}(s_1)$ with $s_2 \in \Phi_i^{-(r)}(s)$ and $s_3 \in \bigcup_{d=0}^{r} \Phi_i^{-(d)}(s)$ for all $s_3 \in R_i(s_2)$, as well as there exists $s_1' \in R_i^+(s)$ and for every $s_2' \in R_{N\setminus\{i\}}(s_1) \cap \Phi_i^{-(r)}(s)$, $R_i(s_2') \cap \Phi_i^{(-d)}(s) \neq \emptyset$ for at least one $d \in \{0, \ldots, (r-1)\}$.

The only modification between GMR_k and SEQ_k is that all DM i's UIs are subject to credible sanctions by the legal sequence of UIs of DM i's opponents. Figure 6.9 depicts sequential stability at degree k. Its formal definition is given below.

Definition 6.49 State s is **sequentially stable** (SEQ_0) at level 0 for DM i, denoted by $s \in S_i^{SEQ_0}$, iff either $R_i^+(s) = \emptyset$ and $R_i^{(0)}(s) \neq \emptyset$, or $R_i^+(s) \neq \emptyset$ and for every $s_1 \in R_i^+(s)$ there exists at least one $s_2 \in R_{N\setminus\{i\}}^+(s_1)$ with $s_2 \in \bigcup_{d=0}^{r} \Phi_i^{-(d)}(s)$ and there exists at least one $s_1' \in R_i^+(s)$ and $s_2' \in R_{N\setminus\{i\}}^+(s_1')$ such that $s_2' \in \Phi_i^{(0)}(s)$ and $R_{N\setminus\{i\}}^+(s_1') \bigcap (\bigcup_{d=1}^{r} \Phi_i^{-(d)}(s)) = \emptyset$.

Definition 6.50 For $1 \leq k \leq r-1$, state s is **sequentially stable** (SEQ_k) at level k for DM i, denoted by $s \in S_i^{SEQ_k}$, iff either $\bigcup_{d=0}^{k-1} R_i^{-(d)}(s) \cup R_i^+(s) = \emptyset$ and $R_i^{-(k)}(s) \neq \emptyset$, or $R_i^+(s) \neq \emptyset$ and for every $s_1 \in R_i^+(s)$ there exists at least one $s_2 \in R_{N\setminus\{i\}}^+(s_1)$ with $s_2 \in \bigcup_{d=k}^{r} \Phi_i^{-(d)}(s)$ and there exists at least one $s_1' \in R_i^+(s)$ and $s_2' \in R_{N\setminus\{i\}}^+(s_1')$ such that $s_2' \in \Phi_i^{-(k)}(s)$ and $R_{N\setminus\{i\}}^+(s_1') \bigcap (\bigcup_{d=k+1}^{r} \Phi_i^{-(d)}(s)) = \emptyset$.

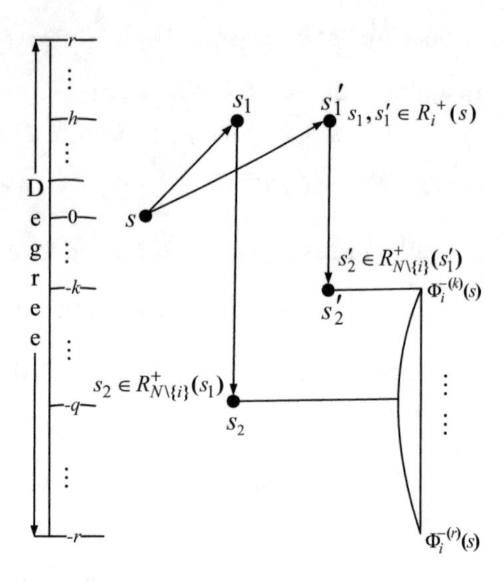

Fig. 6.9 Sequential stability at degree k for DM i

Definition 6.51 State s is **sequentially stable** (SEQ_r) at level r for DM i, denoted by $s \in S_i^{SEQ_r}$, iff either $\bigcup_{d=0}^{r-1} R_i^{-(d)}(s) \cup R_i^+(s) = \emptyset$ and $R_i^{-(r)}(s) \neq \emptyset$, or $R_i^+(s) \neq \emptyset$ and for every $s_1 \in R_i^+(s)$ there exists at least one $s_2 \in R^+_{N\backslash\{i\}}(s_1)$ with $s_2 \in \Phi_i^{-(r)}(s)$.

When $n = 2$, the DM set N becomes $\{i, j\}$ in Definitions 6.41–6.51, and the reachable lists for $H = N \setminus \{i\}$ by legal sequences of UMs and UIs from s_1, $R_{N\backslash\{i\}}(s_1)$ and $R^+_{N\backslash\{i\}}(s_1)$, degenerate to $R_j(s_1)$ and $R_j^+(s_1)$, DM j's corresponding reachable lists from s_1. Obviously, Definitions 6.25–6.35 are special cases of Definitions 6.41–6.51, so the same notation is used for two DM cases and n-DM situations.

6.4.4 Interrelationship Among Stability Definitions for Multiple Degrees of Preference

In Sect. 4.2.4, relationships among the four basic stabilities consisting of Nash, GMR, SMR, and SEQ are presented for two types of preference (or simple preference). Within Sect. 6.3.3, stabilities under three kinds of preference are defined. In Sect. 6.4.1.2, the four stability definitions at degree k are formally defined. In the following five theorems, a range of theoretical relationships among and within stability definitions for different degrees of preference are proven.

Theorem 6.1 *The interrelationships among the four basic stabilities at degree k are*

$$S_i^{Nash_k} \subseteq S_i^{SMR_{k+}} \subseteq S_i^{GMR_k},\ S_i^{SMR_{k-}} \subseteq S_i^{GMR_k},\ \textit{and}\ S_i^{Nash_k} \subseteq S_i^{SEQ_k} \subseteq S_i^{GMR_k},$$

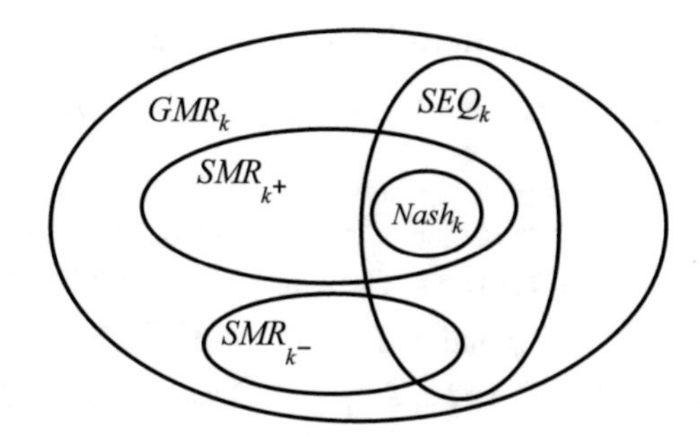

Fig. 6.10 Interrelationships among four stabilities at level k

for $0 \le k \le r$.

Proof When $k = 0$, the results are obvious, since there are no unilateral improvements by DM i relative to state s, but there exist equally preferred states. Assume that $0 < k \le r$. If $s \in S_i^{Nash_k}$, then $\bigcup_{d=0}^{k-1} R_i^{-(d)}(s) \cup R_i^{+}(s) = \emptyset$ and $R_i^{-(k)}(s) \neq \emptyset$. This implies that state $s \in S_i^{SMR_{k^+}}$ using Definitions 6.45 and 6.47. Hence, if $s \in S_i^{Nash_k}$ for $0 \le k \le r$, then $s \in S_i^{SMR_{k^+}}$, which implies $S_i^{Nash_k} \subseteq S_i^{SMR_{k^+}}$.

Using Definitions 6.41–6.47, if $s \in S_i^{SMR_{k^+}}$, it is obvious that $s \in S_i^{GMR_k}$ for $0 \le k \le r$. Therefore, the inclusion relations $S_i^{Nash_k} \subseteq S_i^{SMR_{k^+}} \subseteq S_i^{GMR_k}$ now follow.

Based on Definitions 6.46 and 6.48, the relation $S_i^{SMR_{k^-}} \subseteq S_i^{GMR_k}$ is obvious. Relations $S_i^{Nash_k} \subseteq S_i^{SEQ_k} \subseteq S_i^{GMR_k}$ can be similarly verified. □

Let $0 \le k \le r$. The inclusion relationships presented by Theorem 6.1 are depicted in Fig. 6.10. One should keep in mind that these relationships among stabilities are valid for the situations in which all stabilities being compared have the same degree. As can be clearly seen in this diagram, for example, if a state is $Nash_k$, it is also GMR_k, SMR_{k^+}, and SEQ_k, which is similar to the finding in Fig. 4.4 for the case of simple preference.

The next theorem confirms the relationship that exists for a specific stability definition at two different degrees. In particular, for each of the stability definitions, there are no common stable states when the preferences are different degrees.

Theorem 6.2 *Let* $0 \le h, q \le r$. *When* $h \neq q$, *the relationships between stabilities at h degree and at q degree are*

$$S_i^{Nash_h} \cap S_i^{Nash_q} = \emptyset, \tag{6.1}$$

$$S_i^{GMR_h} \cap S_i^{GMR_q} = \emptyset, \tag{6.2}$$

$$S_i^{SMR_{h^+}} \cap S_i^{SMR_{q^+}} = \emptyset,\ S_i^{SMR_{h^-}} \cap S_i^{SMR_{q^-}} = \emptyset,\ S_i^{SMR_{h^+}} \cap S_i^{SMR_{h^-}} = \emptyset, \text{ and} \tag{6.3}$$

$$S_i^{SEQ_h} \cap S_i^{SEQ_q} = \emptyset. \tag{6.4}$$

Proof First, Eq. 6.1 is proven. Assume that $h > q$. If there exists $s \in S_i^{Nash_h} \cap S_i^{Nash_q}$, then $s \in S_i^{Nash_h}$ and $s \in S_i^{Nash_q}$. Therefore, $R_i^+(s) \cup (\bigcup_{d=0}^{h-1} R_i^{-(d)}(s)) = \emptyset$ and $R_i^{-(h)}(s) \neq \emptyset$ as s is $Nash_h$ stable. Since $h-1 \geq q$, $R_i^{-(q)}(s) = \emptyset$. This contradicts the hypothesis that s is $Nash_q$ stable. Therefore, Eq. 6.1 holds.

Now, Eq. 6.2 is verified. If $s \in (S_i^{Nash_h} \cup S_i^{Nash_q})$, Eq. 6.2 is obvious. Assume that $h > q$ and $s \notin (S_i^{Nash_h} \cup S_i^{Nash_q})$. If there exists $s \in S_i^{GMR_h} \cap S_i^{GMR_q}$, then $s \in S_i^{GMR_h}$ and $s \in S_i^{GMR_q}$. Because s is GMR_q stable, $R_i^+(s) \neq \emptyset$ and for every $s_1 \in R_i^+(s)$ there exists at least one $s_2 \in R_{N\backslash\{i\}}(s_1)$ with $s_2 \in \bigcup_{d=q}^{r} \Phi_i^{-(d)}(s)$ and there exists at least one $s_1' \in R_i^+(s)$ and $s_2' \in R_{N\backslash\{i\}}(s_1')$ such that $s_2' \in \Phi_i^{-(q)}(s)$ and $R_{N\backslash\{i\}}(s_1') \bigcap (\bigcup_{d=q+1}^{r} \Phi_i^{-(d)}(s)) = \emptyset$. This implies that for all $s_2' \in R_{N\backslash\{i\}}(s_1')$, $s_2' \in \bigcup_{d=0}^{q} \Phi_i^{-(d)}(s)$ which means $s_2' \notin \bigcup_{d=h}^{r} \Phi_i^{-(d)}(s)$ as $h > q$. This contradicts with the hypothesis that s is GMR_h stable. Therefore, Eq. 6.2 follows.

Finally, the verification of Eqs. 6.3 and 6.4 can be similarly carried out using contradiction. □

The interrelationships among general stabilities, super stability, and stabilities at each degree are presented in the following theorem. Specifically, for each of the stability definitions, the set of stable states over the general stabilities is the same as the union of all of the stable states over all of the degrees of preference plus the super stable states.

Theorem 6.3 *The interrelationships among general stabilities, super stability, and stabilities at each level are*

$$S_i^{GNash} = (S_i^{Super}) \cup \left(\bigcup_{d=0}^{r} S_i^{Nash_d} \right), \tag{6.5}$$

$$S_i^{GGMR} = (S_i^{Super}) \cup \left(\bigcup_{d=0}^{r} S_i^{GMR_d} \right), \tag{6.6}$$

$$S_i^{GSMR} = (S_i^{Super}) \cup \left(\bigcup_{d=0}^{r} (S_i^{SMR_{d+}} \cup S_i^{SMR_{d-}}) \right), \textit{ and} \tag{6.7}$$

$$S_i^{GSEQ} = (S_i^{Super}) \cup \left(\bigcup_{d=0}^{r} S_i^{SEQ_d} \right). \tag{6.8}$$

Proof Equation 6.5 is derived directly from Definitions 6.22–6.24. Now consider the proof for Eq. 6.6. The inclusion relation $S_i^{GGMR} \supseteq (S_i^{Super}) \cup (\bigcup_{d=0}^{r} S_i^{GMR_d})$ is

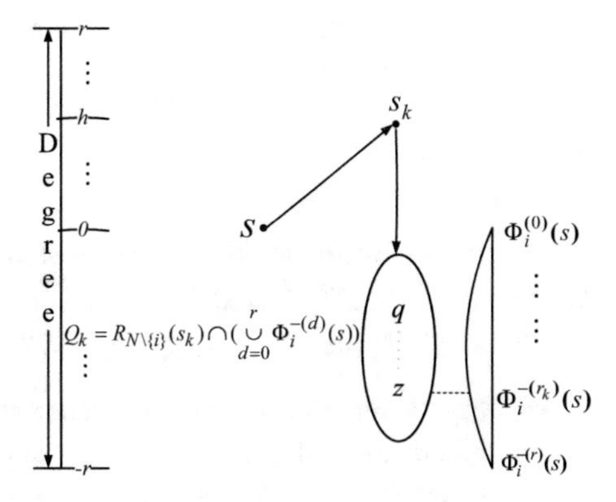

Fig. 6.11 The legal sequence of UM from state s_k

obvious from Definitions 6.41–6.43. It will be proved that the inclusion relation $S_i^{GGMR} \subseteq (S_i^{Super}) \cup (\bigcup_{d=0}^{r} S_i^{GMR_d})$ holds. The two cases will be respectively proved when $s \in (S_i^{Super} \cup S_i^{GNash})$ and $s \notin (S_i^{Super} \cup S_i^{GNash})$. For any $s \in S_i^{GGMR}$, based on Definition 6.38, if $s \in (S_i^{Super} \cup S_i^{GNash})$, then the above inclusion relation must be true.

Next, assume that $s \notin (S_i^{Super} \cup S_i^{GNash})$. Let $|R_i^+(s)| = l$ denote the cardinality of $R_i^+(s)$. Then, for any $s \in S_i^{GGMR}$, $R_i^+(s) \neq \emptyset$ and for every $s_k \in R_i^+(s)$ $(k = 1, \ldots, l)$, there exists at least one $s_k' \in R_{N\setminus\{i\}}(s_k)$ with $s_k' \in \bigcup_{d=0}^{r} \Phi_i^{-(d)}(s)$. Let $Q_k = \{q : q \in R_{N\setminus\{i\}}(s_k) \cap \bigcup_{d=0}^{r} \Phi_i^{-(d)}(s)\}$. It is obvious that $s_k' \in Q_k$. Hence, $Q_k \neq \emptyset$. Let $z \in Q_k$ and be DM i's least preferred in the state set Q_k. Since $z \in R_{N\setminus\{i\}}(s_k) \cap (\bigcup_{d=0}^{r} \Phi_i^{-(d)}(s))$, there exists $0 \leq r_k \leq r$ such that $z \in \Phi_i^{-(r_k)}(s)$ for $k = 1, \ldots, l$. Therefore, either $r_k = r$ or $R_{N\setminus\{i\}}(s_k) \cap (\bigcup_{d=r_k+1}^{r} \Phi_i^{-(d)}(s)) = \emptyset$. This process is portrayed in Fig. 6.11.

Let $r_m = \min\{r_k : k = 1, \ldots, l\}$. Then, $0 \leq r_m \leq r$. It is easy to follow that if $s \in S_i^{GGMR}$ and $R_i^+(s) \neq \emptyset$, then $s \in S_i^{GMR_{r_m}}$. In fact, for every $s_k \in R_i^+(s)$, there exists at least one $s_k' \in R_{N\setminus\{i\}}(s_k)$ with $s_k' \in \Phi_i^{-(r_k)}(s)$. Since $0 \leq r_m \leq r_k$, then $s_k' \in \bigcup_{d=r_m}^{r} \Phi_i^{-(d)}(s)$, and $s_m' \in R_{N\setminus\{i\}}(s_m)$ with $s_m' \in \Phi_i^{-(r_m)}(s)$. Based on the rule of selecting r_m, either $r_m = r$ so that $s \in S_i^{GMR_r}$, or $R_{N\setminus\{i\}}(s_m) \cap (\bigcup_{d=r_m+1}^{r} \Phi_i^{-(d)}(s)) = \emptyset$ so that $s \in S_i^{GMR_{r_m}}$. From the above discussion, Eq. 6.6 is proven.

Equations 6.7 and 6.8 can be proven in a fashion similar to that just presented for Eq. 6.6. □

Let S_i^{Nash}, S_i^{GMR}, S_i^{SMR}, and S_i^{SEQ} denote the set of stable states for DM i for Nash, GMR, SMR, and SEQ stability, respectively, in the graph model for simple preference presented in Sect. 4.2.3. When $r = 1$, stabilities having multiple-degree preference degenerate to the stabilities presented in Sect. 4.2.3, which includes two types of preference. Specifically,

Theorem 6.4 *For the multiple levels of preference, when* $r = 1$, $S_i^{Super} \cup S_i^{Nash_0} \cup S_i^{Nash_1} = S_i^{Nash}$, $S_i^{Super} \cup S_i^{GMR_0} \cup S_i^{GMR_1} = S_i^{GMR}$, $S_i^{Super} \cup S_i^{SMR_0} \cup S_i^{SMR_{1+}} \cup S_i^{SMR_{1-}} = S_i^{SMR}$, *and* $S_i^{Super} \cup S_i^{SEQ_0} \cup S_i^{SEQ_1} = S_i^{SEQ}$.

Stability calculations for the preference structure for $r = 2$ in the graph model for multiple degrees of preference produces the same stability findings as found for the three types of preference or strength of preference framework. More specifically, let S_i^{SGMR}, S_i^{SSMR}, and S_i^{SSEQ} denote the set of strong stable states for strongly GMR, SMR, and SEQ stability, respectively, presented in Sect. 6.3.3.2. The stabilities at degree 2 in the graph model with three kinds of preference degenerate to the corresponding strong stabilities presented in Sect. 6.3.3.2, except for the states that are Nash stable, because Nash stable states are not considered in strong GMR, SMR, and SEQ stability in Sect. 6.3.3.2. Formally, this is expressed in the next theorem.

Theorem 6.5 *For the multiple degrees of preference, when* $r = 2$, $S_i^{GMR_2} \backslash S_i^{Nash_2} = S_i^{SGMR}$, $S_i^{SMR_{2+}} \backslash S_i^{Nash_2} = S_i^{SSMR}$, *and* $S_i^{SEQ_2} \backslash S_i^{Nash_2} = S_i^{SSEQ}$.

The previous two theorems can be easily proven using the appropriate stability definitions.

6.5 Matrix Representation of Stability Definitions for Three Degrees of Preference

The matrix representations for conflict resolution for simple preference and unknown preference are presented in Sects. 4.3 and 5.3, respectively. It is natural to extend the logical form for conflict resolution under the three types of preference presented in Sect. 6.3 to the matrix representation. Following definitions for preference matrices and a reachability matrix in Sect. 6.5.1, matrix representation for various stability definitions for the two DM and n-DM cases for both general and strong stabilities are presented in Sects. 6.5.2 and 6.5.4, respectively, for three degrees of preference.

6.5.1 *Preference Matrices Including Strength of Preference*

Preference information is an important component in the Graph Model for Conflict Resolution under the three types of preference. Preference matrices corresponding to the preference information are constructed now.

Let $m = |S|$ denote the number of states. For DM i, a mild or strong unilateral improvement matrix (MSUI matrix) $J_i^{+,++}$ is an $m \times m$ matrix defined by

$$J_i^{+,++}(s,q) = \begin{cases} 1 & \text{if } q \in R_i^{+,++}(s), \\ 0 & \text{otherwise,} \end{cases} \tag{6.9}$$

where $R_i^{+,++}(s)$ stands for DM i's reachable list from states s by a MSUI as defined at the start of Sect. 6.2.1. Because the elements of the vector e_s are assigned a value of zero except for the element connected to state s where a value of 1 is given, $R_i^{+,++}(s) = e_s^T \cdot J_i^{+,++}$, if $R_i^{+,++}(s)$ is written as a 0-1 row vector, where a "1" at the jth element indicates DM i has a MSUI from s to s_j. The MSUI matrix $J_i^{+,++}$ depicts DM i's mild or strong unilateral improvements in one step. To carry out a stability analysis, a set of matrices corresponding to strength of preference is constructed next. Specifically,

$$P_i^{++}(s,q) = \begin{cases} 1 & \text{if } q \gg_i s, \\ 0 & \text{otherwise,} \end{cases}$$

and

$$P_i^{--}(s,q) = \begin{cases} 1 & \text{if } s \gg_i q, \\ 0 & \text{otherwise.} \end{cases}$$

Therefore, $(P_i^{++})^T = P_i^{--}$, where T denotes the transpose of a matrix.

$$P_i^{--,-,=}(s,q) = \begin{cases} 1 & \text{if } q \ll_i s,\ q <_i s, \text{ or } (q \sim_i s \text{ and } q \neq s), \\ 0 & \text{otherwise.} \end{cases}$$

and

$$P_i^{+,++}(s,q) = \begin{cases} 1 & \text{if } q >_i s \text{ or } q \gg_i s, \\ 0 & \text{otherwise.} \end{cases}$$

For three-degree preference, $P_i^{--,-,=}(s,q) = 1 - P_i^{+,++}(s,q)$ for $s, q \in S$ and $s \neq q$.

Based on the aforementioned definitions, for DM i, a set of adjacency matrices, J_i and $J_i^{+,++}$, and preference matrix $P_i^{+,++}$ have the following relationship between them:

$$J_i^{+,++} = J_i \circ P_i^{+,++},$$

where "$\circ$" denotes the Hadamard product given in Definition 3.15.

6.5.2 Two Decision Maker Case

6.5.2.1 Matrix Representation of General Stabilities

Equivalent matrix representations of the aforementioned logical definitions for Nash stability, GGMR, GSMR, and GSEQ in a two-DM graph model can be determined directly by using the matrices containing information regarding possible moves such as $J_i^{+,++}$ and those keeping track of preferences. Let $i \in N$, $|N| = 2$, and $m = |S|$.

Let E denote an $m \times m$ matrix with each entry equal to 1. Define the $m \times m$ Nash matrix M_i^{Nash} as

$$M_i^{Nash} = J_i^{+,++} \cdot E.$$

Theorem 6.6 *State $s \in S$ is Nash stable for DM i iff $M_i^{Nash}(s, s) = 0$.*

Note that Theorem 6.6 provides a matrix method to assess whether state s is Nash stable for DM i by identifying the Nash matrix's diagonal entry $M_i^{Nash}(s, s)$.

For the case of general GMR stability, define the $m \times m$ matrix M_i^{GGMR} as

$$M_i^{GGMR} = J_i^{+,++} \cdot [E - sign\left(J_j \cdot (P_i^{--,-,=})^T\right)],$$

where E is an $m \times m$ matrix with each entry having a value of 1.

Theorem 6.7 *State s is general GMR (GGMR) for DM i iff $M_i^{GGMR}(s, s) = 0$.*

Proof Since $M_i^{GGMR}(s, s) = (e_s^T \cdot J_i^{+,++}) \cdot [\left(E - sign\left(J_j \cdot (P_i^{--,-,=})^T\right)\right) \cdot e_s]$

$$= \sum_{s_1=1}^{m} J_i^{+,++}(s, s_1)[1 - sign\left((e_{s_1}^T \cdot J_j) \cdot (e_s^T \cdot P_i^{--,-,=})^T\right)],$$

then $M_i^{GGMR}(s, s) = 0$ iff $J_i^{+,++}(s, s_1)[1 - sign\left((e_{s_1}^T \cdot J_j) \cdot (e_s^T \cdot P_i^{--,-,=})^T\right)] = 0$, for $\forall s_1 \in S$. This implies that $M_i^{GGMR}(s, s) = 0$ iff

$$(e_{s_1}^T \cdot J_j) \cdot (e_s^T \cdot P_i^{--,-,=})^T \neq 0, \forall s_1 \in R_i^{+,++}(s). \tag{6.10}$$

From Eq. 6.10, for any $s_1 \in R_i^{+,++}(s)$, there exists $s_2 \in S$ such that the m-dimensional row vector $e_{s_1}^T \cdot J_j$ has a value of 1 for the s_2th element and the m-dimensional column vector $(P_i^{--,-,=})^T \cdot e_s$ has an entry of 1 for the s_2th element.

Therefore, $M_i^{GGMR}(s, s) = 0$ iff for any $s_1 \in R_i^{+,++}(s)$, there exists at least one $s_2 \in R_j(s_1)$ with $s_2 \in \Phi_i^{--,-,=}(s)$. □

In order to consider the general SMR stability, define the $m \times m$ matrix M_i^{GSMR} as $M_i^{GSMR} = J_i^{+,++} \cdot [E - sign(F)]$, with

$$F = J_j \cdot [(P_i^{--,-,=})^T \circ \left(E - sign\left(J_i \cdot (P_i^{+,++})^T\right)\right)].$$

Theorem 6.8 *State s is general SMR (GSMR) for DM i iff* $M_i^{GSMR}(s,s) = 0$.

Proof Let $G = (P_i^{--,-,=})^T \circ \left(E - sign\left(J_i \cdot (P_i^{+,++})^T\right)\right)$.

Since $M_i^{GSMR}(s,s) = (e_s^T \cdot J_i^{+,++}) \cdot [(E - sign(F)) \cdot e_s]$

$$= \sum_{s_1=1}^{m} J_i^{+,++}(s, s_1)[1 - sign\,(F(s_1, s))]$$

with

$$F(s_1, s) = \sum_{s_2=1}^{m} J_j(s_1, s_2) \cdot G(s_2, s),$$

and $G(s_2, s) = P_i^{--,-,=}(s, s_2)[1 - sign\left(\sum_{s_3=1}^{m}\left(J_i(s_2, s_3) P_i^{+,+}(s, s_3)\right)\right)]$, thus, $M_i^{GSMR}(s,s) = 0$ holds iff $F(s_1, s) \neq 0$, $\forall s_1 \in R_i^{+,++}(s)$, which is equivalent to the statement that, $\forall s_1 \in R_i^{+,++}(s)$, $\exists s_2 \in R_j(s_1)$ such that

$$P_i^{--,-,=}(s, s_2) \neq 0, \tag{6.11}$$

and

$$\sum_{s_3=1}^{m}\left(J_i(s_2, s_3) P_i^{+,++}(s, s_3)\right) = 0. \tag{6.12}$$

Obviously, for $\forall s_1 \in R_i^{+,++}(s)$, $\exists s_2 \in R_j(s_1)$, Eq. 6.11 holds iff $s_2 \in \Phi_i^{--,-,=}(s)$. For $\forall s_1 \in R_i^{+,++}(s)$, $\exists s_2 \in R_j(s_1)$, Eq. 6.12 holds iff for all $s_3 \in R_i(s_2)$, $P_i^{+,++}(s, s_3) = 0$ which implies $s_3 \in \Phi_i^{--,-,=}(s)$.

Therefore, $M_i^{GSMR} = 0$ iff for every $s_1 \in R_i^{+,++}(s)$ there exists $s_2 \in R_j(s_1)$ such that $s_2 \in \Phi_i^{--,-,=}(s)$ and $s_3 \in \Phi_i^{--,-,=}(s)$ for all $s_3 \in R_i(s_2)$. □

In order to analyze general SEQ stability using matrix approach, define the $m \times m$ matrix M_i^{GSEQ} as

$$M_i^{GSEQ} = J_i^{+,++} \cdot [E - sign\left(J_j^{+,++} \cdot (P_i^{--,-,=})^T\right)].$$

Theorem 6.9 *State s is general SEQ (GSEQ) for DM i iff* $M_i^{GSEQ}(s,s) = 0$.

Proof Since $M_i^{GSEQ}(s,s) = (e_s^T \cdot J_i^{+,++}) \cdot [\left(E - sign\left(J_j^{+,++} \cdot (P_i^{--,-,=})^T\right)\right) \cdot e_s]$

$$= \sum_{s_1=1}^{m} J_i^{+,++}(s, s_1)[1 - sign\left((e_{s_1}^T \cdot J_j^{+,++}) \cdot (e_s^T \cdot P_i^{--,-,=})^T\right)],$$

then $M_i^{GSEQ}(s, s) = 0$ iff for any $s_1 \in S$,

$$J_i^{+,++}(s, s_1)[1 - sign\left((e_{s_1}^T \cdot J_j^{+,++}) \cdot (e_s^T \cdot P_i^{--,-,=})^T\right)] = 0.$$

This implies that $M_i^{GSEQ}(s, s) = 0$ iff

$$(e_{s_1}^T \cdot J_j^{+,++}) \cdot (e_s^T \cdot P_i^{--,-,=})^T \neq 0, \forall s_1 \in R_i^{+,++}(s). \quad (6.13)$$

By Eq. 6.13, for any $s_1 \in R_i^{+,++}(s)$, there exists $s_2 \in S$ such that the m-dimensional row vector $e_{s_1}^T \cdot J_j^{+,++}$ has the s_2th element 1 and the m-dimensional column vector $(P_i^{--,-,=})^T \cdot e_s$ has the s_2th element 1.

Therefore, $M_i^{GSEQ}(s, s) = 0$ iff for any $s_1 \in R_i^{+,++}(s)$, there exists at least one $s_2 \in R_j^{+,++}(s_1)$ with $s_2 \in \Phi_i^{--,-,=}(s)$. □

6.5.2.2 Matrix Representation of Strong Stabilities

Corresponding to the logical representation of strong stabilities for three degrees of preference, matrix representation of strong GMR, SMR, and SEQ stabilities are presented below according to the degree of sanctioning. For three kinds of preference, these stabilities are divided into strongly and weakly stable with respect to the strength of possible sanctions. Hence, if a particular state s is general stable, then s is either strongly stable or weakly stable. Strong and weak stabilities only include GMR, SMR, and SEQ because Nash stability does not involve sanctions.

In the upcoming theorems, let $i \in N$, $|N| = 2$, and $m = |S|$. To consider strong GMR stability, define the $m \times m$ matrix M_i^{SGMR} as

$$M_i^{SGMR} = J_i^{+,++} \cdot [E - sign\left(J_j \cdot (P_i^{--})^T\right)].$$

Theorem 6.10 *State $s \in S$ is strong general metarational ($SGMR$) for DM i iff $M_i^{SGMR}(s, s) = 0$.*

In order to analyze strong SMR stability, define the $m \times m$ matrix M_i^{SSMR} as

$$M_i^{SSMR} = J_i^{+,++} \cdot [E - sign(J_j \cdot F)],$$

with

$$F = (P_i^{++}) \circ [E - sign\left(J_i \cdot (E - P_i^{++})\right)].$$

Theorem 6.11 *State $s \in S$ is strong symmetric metarational ($SSMR$) for DM i iff $M_i^{SSMR}(s, s) = 0$.*

In order to calculate strong SEQ, define the $m \times m$ matrix M_i^{SSEQ} as

$$M_i^{SSEQ} = J_i^{+,++} \cdot [E - sign\left(J_j^{+,++} \cdot (P_i^{--})^T\right)].$$

Theorem 6.12 *State $s \in S$ is strong sequentially stable (SSEQ) for DM i iff* $M_i^{SSEQ}(s, s) = 0$.

The proofs of Theorems 6.10–6.12 are similar to those of the three general stabilities presented in Theorems 6.7–6.9, respectively, in Sect. 6.5.2.1.

Let GS denote a graph model stability, GMR, SMR, or SEQ. The symbols GGS, SGS, and WGS respectively represent a general stability, GGMR, GSMR, or GSEQ, the strong stability, SGMR, SSMR, or SSEQ, and the weak stability, WGMR, WSMR, or WSEQ, under three degrees of preference. M_i^{GGS} and M_i^{SGS} denote DM i's general stability matrix, M_i^{GGMR}, M_i^{GSMR}, or M_i^{GSEQ}, and DM i's strong stability matrix, M_i^{SGMR}, M_i^{SSMR}, or M_i^{SSEQ}, respectively. Based on the notation, one has the following theorem.

Theorem 6.13 *State $s \in S$ is weak stable (WGS) for DM i iff $M_i^{GGS}(s, s) = 0$, but* $M_i^{SGS}(s, s) \neq 0$.

Theorem 6.13 means that if s is general stable, but not strong stable for a GS stability, then s is weak stable for the GS stability.

Example 6.2 (Stabilities for the Extended Sustainable Development Model under Three-degree Preference by using Matrix Representation) The sustainable development conflict is explained in Example 3.1. In this illustration, this conflict is expended to include three degrees of preference for the two-DM case. Specifically, the conflict consists of two DMs: an environmental agency (DM 1: E) and a developer (DM 2: D); and two options: DM 1 controls the option of being proactive (labeled P) and DM 2 has the option of practicing sustainable development (labeled SD) for properly treating the environment. The two options are combined to form four feasible states: s_1, s_2, s_3, and s_4. These results are listed in Table 6.8, where a "Y" indicates that an option is selected by the DM controlling it and an "N" means that the option is not chosen.

From Table 6.8, DM 1 and DM 2's preference information includes strength. The graph model for the extended sustainable development conflict is presented in Fig. 6.12.

From the graph model, the UM adjacency matrices for each DM are constructed by

$$J_1 = \begin{pmatrix} 0\,0\,1\,0 \\ 0\,0\,0\,1 \\ 1\,0\,0\,0 \\ 0\,1\,0\,0 \end{pmatrix} \text{ and } J_2 = \begin{pmatrix} 0\,1\,0\,0 \\ 1\,0\,0\,0 \\ 0\,0\,0\,1 \\ 0\,0\,1\,0 \end{pmatrix}.$$

Table 6.8 Extended sustainable development game in option form

DM 1: Environmental agency				
1. Proactive (P)	Y	Y	N	N
DM 2: Developer				
2. Sustainable development (SD)	Y	N	Y	N
States	s_1	s_2	s_3	s_4

Preferences $s_1 >_1 s_3 \gg_1 s_2 \sim_1 s_4$ for DM 1 and
$s_3 >_2 s_1 \gg_2 s_4 \sim_2 s_2$ for DM 2.

The preference matrices for the DMs 1 and 2 are given by

$$P_1^{++} = \begin{pmatrix} 0\,0\,0\,0 \\ 1\,0\,1\,0 \\ 0\,0\,0\,0 \\ 1\,0\,1\,0 \end{pmatrix}, \; P_1^{+,++} = \begin{pmatrix} 0\,0\,0\,0 \\ 1\,0\,1\,0 \\ 1\,0\,0\,0 \\ 1\,0\,1\,0 \end{pmatrix},$$

$$P_2^{++} = \begin{pmatrix} 0\,0\,1\,0 \\ 0\,0\,1\,0 \\ 0\,0\,0\,0 \\ 0\,0\,1\,0 \end{pmatrix}, \text{ and } P_2^{+,++} = \begin{pmatrix} 0\,0\,0\,0 \\ 1\,0\,1\,1 \\ 0\,0\,0\,0 \\ 1\,0\,1\,0 \end{pmatrix}.$$

Therefore, $J_i^{+,++} = J_i \circ P_i^{+,++}$, $P_i^{--,-,=} = E - I - P_i^{+,++}$, and $P_i^{--} = (P_i^{++})^T$ for $i = 1, 2$.

The stability matrices used by Theorems 6.6–6.13 are included in Table 6.9, which are employed to calculate the general stabilities of Nash, GMR, SMR, and SEQ, as well as the strong stabilities of SGMR, SSMR, and SSEQ for two-DM conflicts, respectively.

The stable states and equilibria for the sustainable development conflict are summarized in Table 6.10, in which "$\surd$" for a given state means that this state is stable for DM 1 or DM 2 and "Eq" is an equilibrium for an appropriate solution concept.

The results provided by Table 6.10 show that state s_1 is strong equilibrium for the four basic stabilities. State s_3 is strongly stable for GMR and SMR. Hence, s_1 and s_3 are better choices for decision makers.

6.5.3 Reachability Matrix Under Strength of Preference

An important matrix corresponding to the reachable list under three degrees of preference is now defined. Fix coalition $H \subseteq N$ such that $|H| \geq 2$, and let $s \in S$. In order to construct the reachability matrix corresponding to $R_H^{+,++}(s)$ presented in Definition 6.9, the reachable list of H from s by legal sequences of MSUIs, the t-step reachability matrix is defined as follows.

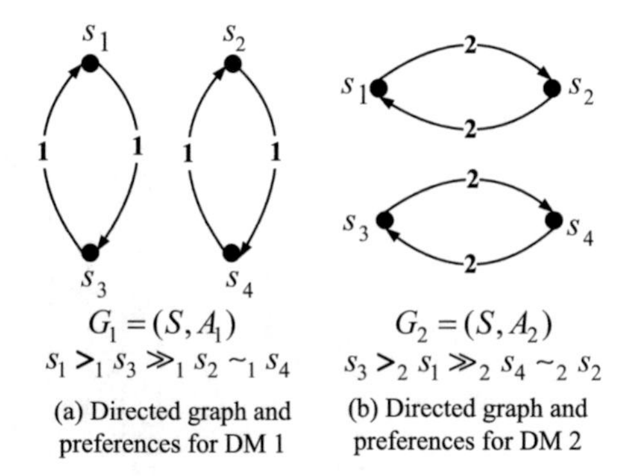

Fig. 6.12 Graph model for the extended sustainable development conflict under three-degree preference

Table 6.9 Stability matrices under three degrees of preference in two decision maker case

Category	Stability matrices
General stabilities	$M_i^{Nash} = J_i^{+,++} \cdot E$
	$M_i^{GGMR} = J_i^{+,++} \cdot [E - sign\left(J_j \cdot (P_i^{--,-,=})^T\right)]$
	$M_i^{GSMR} = J_i^{+,++} \cdot [E - sign(F)]$ with $F = J_j \cdot [(P_i^{--,-,=})^T \circ \left(E - sign\left(J_i \cdot (P_i^{+,++})^T\right)\right)]$
	$M_i^{GSEQ} = J_i^{+,++} \cdot [E - sign\left(J_j^{+,++} \cdot (P_i^{--,-,=})^T\right)]$
Strong stabilities	$M_i^{SGMR} = J_i^{+,++} \cdot [E - sign\left(J_j \cdot (P_i^{--})^T\right)]$
	$M_i^{SSMR} = J_i^{+,++} \cdot [E - sign(J_j \cdot F)]$ with $F = (P_i^{++}) \circ [E - sign\left(J_i \cdot (E - P_i^{++})\right)]$
	$M_i^{SSEQ} = J_i^{+,++} \cdot [E - sign\left(J_j^{+,++} \cdot (P_i^{--})^T\right)]$
Weak stabilities	$S_i^{WGS} = S_i^{GGS} - S_i^{SGS}$

Definition 6.52 For $i \in H$, $H \subseteq N$, and $t = 1, 2, 3, \ldots$, define the $m \times m$ matrix $M_i^{(H,t,+,++)}$ with (s, q) entries as follows:

$$M_i^{(H,t,+,++)}(s,q) = \begin{cases} 1 & \text{if } q \in S \text{ is reachable from } s \in S \text{ in exactly} \\ & t \text{ legal MSUIs by } H \text{ with last mover DM } i, \\ 0 & \text{otherwise.} \end{cases}$$

Similar to Lemma 5.1, one has

Lemma 6.1 *For $i \in H$ and $H \subseteq N$, the matrix $M_i^{(H,t,+,++)}$ satisfies that*

$$for\ t = 2, 3, \ldots,\ M_i^{(H,t,+,++)} = sign[(\bigvee_{j \in H-\{i\}} M_j^{(H,t-1,+,++)}) \cdot J_i^{+,++}].$$

Table 6.10 Stability results of the sustainable development conflict with strength of preference

State	Nash			GGMR			GSMR			GSEQ			SGMR			SSMR			SSEQ			WGMR			WSMR			WSEQ		
	1	2	Eq	1	2	Eq	1	2	Eq	1	2	Eq	1	2	Eq	1	2	Eq	1	2	Eq	1	2	Eq	1	2	Eq	1	2	Eq
s_1	√	√	√	√	√	√	√	√	√	√	√	√	√	√	√	√	√	√	√	√	√									
s_2	√			√			√			√			√			√			√											
s_3		√		√	√	√	√	√	√		√		√	√	√	√	√	√		√										
s_4	√			√			√			√			√			√			√											

with $M_i^{(H,1,+,++)}(s,q) = J_i^{+,++}(s,q)$.

The proof of this lemma is similar to that of Lemma 5.1.

The UM and UI reachability matrices are given in Definition 4.19 in Chap. 4. The MSUI reachability matrix is now similarly defined for a graph model having three degrees of preference.

Definition 6.53 For the graph model G, the MSUI reachability matrix for H is the $m \times m$ matrix $M_H^{+,++}$ with (s,q) entry

$$M_H^{+,++}(s,q) = \begin{cases} 1 & \text{if } q \in R_H^{+,++}(s), \\ 0 & \text{otherwise.} \end{cases}$$

Obviously, $R_H^{+,++}(s) = \{q : M_H^{+,++}(s,q) = 1\}$. If $R_H^{+,++}(s)$ is written as a 0-1 row vector, then

$$R_H^{+,++}(s) = e_s^T \cdot M_H^{+,++},$$

where e_s^T denotes the transpose of the sth standard basis vector of the m-dimensional Euclidean space. Therefore, the MSUI reachability matrix for coalition H, $M_H^{+,++}$, can be used to calculate the reachable lists of H from state s by the legal sequence of MSUIs, $R_H^{+,++}(s)$.

Let $L_4 = |\bigcup_{i \in H} A_i^{+,++}|$, where $A_i^{+,++}$ is DM i's MSUI oriented arcs, representing mild or strong unilateral improvements by DM i in coalition H. Then the following theorem can be derived using Lemma 6.1.

Theorem 6.14 *Let* $L_4 = |\bigcup_{i \in N} A_i^{+,++}|$, $s \in S$, $H \subseteq N$, *and* $H \neq \emptyset$. *The MSUI reachability matrix* $M_H^{+,++}$ *by* H *can be expressed as*

$$M_H^{+,++} = \bigvee_{t=1}^{L_4} \bigvee_{i \in H} M_i^{(H,t,+,++)}. \tag{6.14}$$

Proof To prove Eq. 6.14, assume that $C = \bigvee_{t=1}^{L_4} \bigvee_{i \in H} M_i^{(H,t,+,++)}$. Using the definition for matrix $M_H^{+,++}$, $M_H^{+,++}(s,q) = 1$ iff $q \in R_H^{+,++}(s)$. Using Definition 6.9, $q \in R_H^{+,++}(s)$ implies that there exists $1 \leq t_0 \leq L_4$ and $i_0 \in H$ such that $M_{i_0}^{(H,t_0,+,++)}(s,q) = 1$. This implies that matrix C has (s,q) entry 1. Therefore, $M_H^{+,++}(s,q) = 1$ iff $C(s,q) = 1$. Since $M_H^{+,++}$ and C are 0-1 matrices, it follows that $M_H^{+,++} = C$. □

6.5.4 n-Decision Maker Case

6.5.4.1 Matrix Representation of General Stabilities

Matrix representations of solution concepts with three degrees of preference for 2-DM cases presented in Sect. 6.5.2 are now extended to n-DM situations. Nash stability definitions are identical for both the 2-DM and the n-DM models, because Nash stability does not consider opponents' responses. In this subsection, let $i \in N$ and $|N| = n$.

In order to consider general GMR stability for n-DMs, define the $m \times m$ general GMR matrix M_i^{GGMR} as

$$M_i^{GGMR} = J_i^{+,++} \cdot [E - sign\left(M_{N\setminus\{i\}} \cdot (P_i^{--,-,=})^T\right)].$$

In order to avoid using a complex notation, the symbol used for the general GMR matrix representation for the n-DM situation is the same as that employed for the 2-DM case in Sect. 6.5.2.1. The context in which the definition is being utilized will clearly indicate whether it is for the 2-DM or n-DM situation. The same comments hold for the other definitions given in this section as well as Sect. 6.5.4.2. Because the proofs of the next three theorems are similar to the GMR, SMR, and SEQ stabilities presented in Sect. 6.5.2.2 for 2-DM models, the proofs are not given for the n-DM case.

Theorem 6.15 *State s is general GMR for DM i iff $M_i^{GGMR}(s, s) = 0$.*

The above matrix method, called matrix representation of general GMR stability, is equivalent to the logical representation for general GMR stability given in Definition 6.11. To analyze general GMR stability of state s for DM i, one only needs to check if the entry, $M_i^{GMR}(s, s)$, in the GMR matrix is zero. If so, state s is general GMR stable for i; otherwise, s is general GMR unstable for DM i. Note that all information about general GMR stability is contained in the diagonal entries of the general GMR matrix.

To analyze general SMR stability, define the $m \times m$ general SMR matrix M_i^{GSMR} as $M_i^{GSMR} = J_i^{+,++} \cdot [E - sign(Q)]$, with

$$Q = M_{N\setminus\{i\}} \cdot [(P_i^{--,-,=})^T \circ \left(E - sign\left(J_i \cdot (P_i^{+,++})^T\right)\right)].$$

Theorem 6.16 *State s is general SMR for DM i iff $M_i^{GSMR}(s, s) = 0$.*

Theorem 6.16 indicates that the matrix representation of general SMR stability is equivalent to the logical representation for general SMR stability presented in Definition 6.12. To calculate general SMR stability of state s for DM i, one only has to assess whether the diagonal entry, $M_i^{GSMR}(s, s)$, of DM i's general SMR matrix is zero. If so, state s is general SMR stable for i; otherwise, s is general SMR unstable for DM i.

General sequential stability is similar to general GMR stability, but includes only those sanctions that are "credible". Define the $m \times m$ general SEQ matrix M_i^{GSEQ} as

$$M_i^{GSEQ} = J_i^{+,++} \cdot [E - sign\left(M_{N\backslash\{i\}}^{+,++} \cdot (P_i^{--,-,=})^T\right)].$$

Theorem 6.17 *State s is general SEQ for DM i iff* $M_i^{GSEQ}(s, s) = 0$.

Similar to the previous two theorems, the matrix representation of SEQ stability is equivalent to the logical version given in Definition 6.13. When the diagonal entry at (s, s) is zero, the state s under consideration is SEQ stable for DM i.

6.5.4.2 Matrix Representation of Strong Stabilities

Similar to the two-DM case, matrix representations of general stabilities under the three degrees of preference for n-DMs include matrix versions of strong or weak stability. First, construct matrices J_i and $J_i^{+,++}$ using Definition 4.13 and Eq. 6.9. The matrices M_H and $M_H^{+,++}$ are calculated utilizing Theorems 4.9 and 7.5, for which $H = N\backslash\{i\}$. For convenience, the same notation employed for the two-DM situation in Sect. 6.5.2.2 is used for the n-DM case.

Define the $m \times m$ strong GMR matrix M_i^{SGMR} for DM i as

$$M_i^{SGMR} = J_i^{+,++} \cdot [E - sign\big(M_{N\backslash\{i\}} \cdot (P_i^{--})^T\big)].$$

Theorem 6.18 *State* $s \in S$ *is strong GMR (SGMR) for DM i, denoted by* $s \in S_i^{SGMR}$, *iff* $M_i^{SGMR}(s, s) = 0$.

Proof Since $M_i^{SGMR}(s, s) = (e_s^T \cdot J_i^{+,++}) \cdot [(E - sign\left(M_{N\backslash\{i\}} \cdot (P_i^{--})^T\right)) \cdot e_s]$

$$= \sum_{s_1=1}^{m} J_i^{+,++}(s, s_1)[1 - sign\left((e_{s_1}^T \cdot M_{N\backslash\{i\}}) \cdot (e_s^T \cdot P_i^{--})^T\right)],$$

then

$$M_i^{SGMR}(s, s) = 0 \Leftrightarrow J_i^{+,++}(s, s_1)[1 - sign\left((e_{s_1}^T \cdot M_{N\backslash\{i\}}) \cdot (e_s^T \cdot P_i^{--})^T\right)] = 0, \forall s_1 \in S.$$

This implies that $M_i^{SSGM}(s, s) = 0$ iff

$$(e_{s_1}^T \cdot M_{N\backslash\{i\}}) \cdot (e_s^T \cdot P_i^{--})^T \neq 0, \forall s_1 \in R_i^{+,++}(s). \tag{6.15}$$

By Eq. 6.15, for any $s_1 \in R_i^{+,++}(s)$, there exists $s_2 \in S$ such that the m-dimensional row vector, $e_{s_1}^T \cdot M_{N\backslash\{i\}}$, has the s_2th element 1 and the m-dimensional column vector, $(P_i^{--})^T \cdot e_s$, has the s_2th element 1.

Therefore, $M_i^{SGMR}(s, s) = 0$ iff for any $s_1 \in R_i^{+,++}(s)$, there exists at least one $s_2 \in R_{N\backslash\{i\}}(s_1)$ with $s \gg_i s_2$. □

For strong SMR, the n-DM model is similar to the two-DM case. The only modification is that responses to block improvements by DM i can come from more than one of DM i's opponents instead of from a single DM.

If $F = (P_i^{++}) \circ [E - sign(J_i \cdot (E - P_i^{++}))]$, then one can define the $m \times m$ strong SMR matrix M_i^{SSMR} for DM i as

$$M_i^{SSMR} = J_i^{+,++} \cdot [E - sign(M_{N\backslash\{i\}} \cdot F)].$$

Theorem 6.19 *State $s \in S$ is strong SMR (SSMR) for DM i, denoted by $s \in S_i^{SSMR}$, iff $M_i^{SSMR}(s, s) = 0$.*

Proof Let $Q = M_{N\backslash\{i\}} \cdot F$. Since $M_i^{SSMR}(s, s) = (e_s^T \cdot J_i^{+,++}) \cdot [(E - sign(Q)) \cdot e_s]$

$$= \sum_{s_1=1}^{m} J_i^{+,++}(s, s_1)[1 - sign(Q(s_1, s))]$$

then $M_i^{SSMR}(s, s) = 0$ iff $J_i^{+,++}(s, s_1)[1 - sign(Q(s_1, s))] = 0$, for any $s_1 \in S$. This means that $M_i^{SSMR}(s, s) = 0$ iff

$$(e_{s_1}^T \cdot M_{N\backslash\{i\}}) \cdot (F \cdot e_s) \neq 0, \forall s_1 \in R_i^{+,++}(s). \tag{6.16}$$

Since $(e_{s_1}^T \cdot M_{N\backslash\{i\}}) \cdot (F \cdot e_s) = \sum_{s_2=1}^{m} M_{N\backslash\{i\}}(s_1, s_2) \cdot F(s_2, s)$, then Eq. 6.16 holds iff for any $s_1 \in R_i^{+,++}(s)$, there exists $s_2 \in R_{N\backslash\{i\}}(s_1)$ such that $F(s_2, s) \neq 0$.

Because $F(s_2, s) = P_i^{++}(s_2, s) \cdot [1 - sign(\sum_{s_3=1}^{m} J_i(s_2, s_3)(1 - P_i^{++}(s_3, s)))]$, $F(s_2, s) \neq 0$ implies that for $s_2 \in R_{N\backslash\{i\}}(s_1)$,

$$P_i^{++}(s_2, s) \neq 0 \tag{6.17}$$

and

$$\sum_{s_3=1}^{m} J_i(s_2, s_3)(1 - P_i^{++}(s_3, s)) = 0. \tag{6.18}$$

Equation 6.17 is equivalent to the statement that, $\forall s_1 \in R_i^{+,++}(s)$, $\exists s_2 \in R_{N\backslash\{i\}}(s_1)$ such that $s \gg_i s_2$. Equation 6.18 is the same as the statement that, $\forall s_1 \in R_i^{+,++}(s)$, $\exists s_2 \in R_{N\backslash\{i\}}(s_1)$ such that $P_i^{++}(s_3, s) \neq 0$, for $\forall s_3 \in R_i(s_2)$. Based on the definition of the $m \times m$ preference matrix P_i^{++}, one knows that $P_i^{++}(s_3, s) \neq 0 \Leftrightarrow s \gg_i s_3$.

Therefore, one can conclude from the above discussion that $M_i^{SMR}(s, s) = 0$ iff for any $s_1 \in R_i^{+,++}(s)$, there exists at least one $s_2 \in R_{N\backslash\{i\}}(s_1)$ with $s \gg_i s_2$ and $s \gg_i s_3$ for all $s_3 \in R_i(s_2)$. □

Strong sequential stability examines the credibility of the sanctions by DM i's opponents, in the sense that opponents will not move to less preferred situations to block improvements by DM i.

First, find matrix $M^{+,++}_{N\backslash\{i\}}$ using Theorem 7.5 for $H = N\backslash\{i\}$. Define the $m \times m$ strong SEQ matrix M_i^{SSEQ} for DM i as

$$M_i^{SSEQ} = J_i^{+,++} \cdot [E - sign(M^{+,++}_{N\backslash\{i\}} \cdot (P_i^{--})^T)].$$

Theorem 6.20 *State $s \in S$ is strong SEQ (SSEQ) for DM i, denoted by $s \in S_i^{SSEQ}$, iff $M_i^{SSEQ}(s, s) = 0$.*

Proof Since $M_i^{SSEQ}(s, s) = (e_s^T \cdot J_i^{+,++}) \cdot [\left(E - sign\left(M^{+,++}_{N\backslash\{i\}} \cdot (P_i^{--})^T\right)\right) \cdot e_s]$

$$= \sum_{s_1=1}^{m} J_i^{+,++}(s, s_1)[1 - sign\left(e_{s_1}^T \cdot M^{+,++}_{N\backslash\{i\}} \cdot (e_s^T \cdot P_i^{--})^T\right)],$$

then

$$M_i^{SSEQ}(s, s) = 0 \Leftrightarrow J_i^{+,++}(s, s_1)[1 - sign\left((e_{s_1}^T \cdot M^{+,++}_{N\backslash\{i\}}) \cdot (e_s^T \cdot P_i^{--})^T\right)] = 0, \forall s_1 \in S.$$

This implies that $M_i^{SSEQ}(s, s) = 0$ iff

$$(e_{s_1}^T \cdot M^{+,++}_{N\backslash\{i\}}) \cdot (e_s^T \cdot P_i^{--})^T \neq 0, \forall s_1 \in R_i^{+,++}(s). \tag{6.19}$$

By Eq. 6.19, for any $s_1 \in R_i^{+,++}(s)$, there exists $s_2 \in S$ such that the m-dimensional row vector, $e_{s_1}^T \cdot M^{+,++}_{N\backslash\{i\}}$, has the s_2th element 1 and the m-dimensional column vector, $(P_i^{--})^T \cdot e_s$, has the s_2th element 1.

Therefore, $M_i^{SSEQ}(s, s) = 0$ iff for any $s_1 \in R_i^{+,++}(s)$, there exists at least one $s_2 \in R^{+,++}_{N\backslash\{i\}}(s_1)$ with $s \gg_i s_2$. □

Theorems 6.18–6.20 indicate that the matrix representation of strong solution concepts are equivalent to the strong stability definitions in the logical forms presented in Sect. 6.3.3.2. When $n = 2$, Theorems 6.18–6.20 are reduced to those theorems presented in Sect. 6.5.2.2.

6.6 Application: The Garrison Diversion Unit (GDU) Conflict

In this section, the four-degree version of stability definitions presented in Sect. 6.4 is applied to the Garrison Diversion Unit (GDU) conflict to illustrate how the procedure works. For combination with a brief overview of this international environmental dispute between Canada and United States, a conflict model in terms of DMs, options,

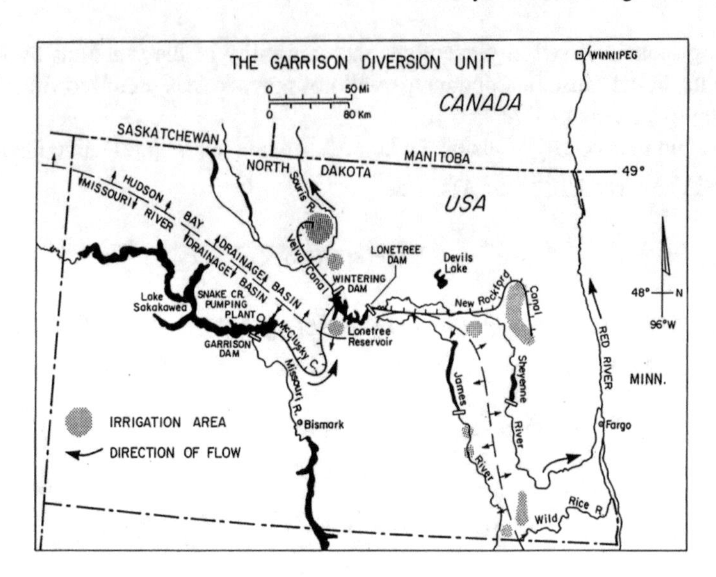

Fig. 6.13 Garrison Diversion Unit (GDU)

and preferences is constructed in the next subsection. Subsequently, a stability analysis is executed for four degrees of preference utilizing the calibrated model and insights regarding the stability results are discussed.

6.6.1 Model of the GDU Conflict

The history of the GDU conflict dates back to the nineteenth century. In order to irrigate land in the northeastern region of the American State of North Dakota, an irrigation project was proposed by the **United States Support (USS)** regarding construction of the McClusky Canal to transfer an immense amount of water from the Missouri River Basin to the Hudson Bay Basin as depicted in Fig. 6.13, which originally appeared in Fraser and Hipel (1984). From the Lonetree Reservoir, water can be conveyed to the planned irrigation areas marked on the map. Eventually, the irrigation runoff would flow into the Canadian province of Manitoba via the Red and Souris rivers. This irrigation initiative is called the Garrison Diversion Unit project. Among other problems, biologists were concerned that foreign biota from the Missouri River Basin could adversely affect biological species in the Hudson Bay Drainage Basin and could, for example, decimate fish species in Lake Winnipeg and thereby destroy the fishing industry. The GDU conflict was strategically analyzed using metagame analysis, conflict analysis and the graph model by Hipel and Fraser (1980), Fraser and Hipel (1984), and Fang et al. (1993) using two-degree preference. Later, Hamouda et al. (2006) examined a simplified version of the GDU dispute for three degrees of preference. In this simpler conflict, the **Canadian Opposition**

Table 6.11 Feasible states for the GDU model

USS									
1. Proceed	Y	Y	N	Y	N	Y	N	Y	N
2. Modify	N	N	Y	N	Y	N	Y	N	Y
CDO									
3. Legal	N	N	N	Y	Y	N	N	Y	Y
IJC									
4. Completion	N	Y	Y	Y	Y	N	N	N	N
5. Modification	N	N	N	N	N	Y	Y	Y	Y
State number	s_1	s_2	s_3	s_4	s_5	s_6	s_7	s_8	s_9

(CDO) was considering whether or not to oppose the project because of the potential negative environmental impacts that Canada would suffer. Based on the Boundary Water Treaty of 1909 between Canada and the United States, the **International Joint Commission (IJC)** consisting of representatives from the governments of the USA and Canada was called upon by both nations to carry out unbiased studies and make recommendations regarding the proposed GDU project.

The graph model for the simplified GDU conflict is comprised of three DMs: 1. **USS**, 2. **CDO**, and 3. **IJC**; and five options: 1. **Proceed**–Proceed with the project regardless of Canada's concerns; 2. **Modify**–Modify the project to reduce impacts on Canada; 3. **Legal**–Legal action by CDO based on the Boundary Waters Treaty; 4. **Completion**–IJC recommends completion of the project as originally planned; and 5. **Modification**–IJC stipulates modification of the project to reduce environmental impacts on Canada. Each of these three DMs followed by the option or options under its control are listed as the left column in Table 6.11. As explained in Sect. 3.1.2, when using the option form, a state is defined as a selection of options for each DM. Since there are five options in the GDU dispute, a total of $2^5 = 32$ states is mathematically possible. However, some states can be removed because they cannot possibly occur in reality. For instance, because options 4 and 5 are mutually exclusive, these two options cannot be selected together. Likewise, options 1 and 2 are mutually exclusive for USS. Moreover, it is assumed that the USS will do something and thereby choose one of its options. After all of the infeasible states are eliminated, only nine states are identified as being feasible. The feasible states are designated as columns of Ys and Ns on the right side of Table 6.11 in which a "Y" indicates that an option is selected by the DM controlling it and an "N" means that the option is not chosen.

The integrated graph model of the GDU conflict is shown in Fig. 6.14, in which a label on an arc indicates the DM who controls the move. Notice, for instance, that USS controls movement between states s_2 and s_3. From Table 6.11, one can see that for states s_2 and s_3 the option selections for USS change while the option choices by the other DMs, consisting of CDO and IJC, remain the same. All that is still required for a graph model is knowledge of each DM's preferences over the feasible states for the situation of four-degree preference in the GDU conflict. The preference

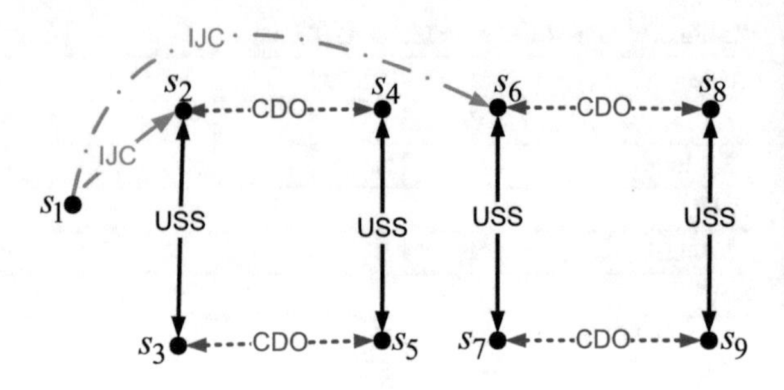

Fig. 6.14 The integrated graph model for movement in the GDU conflict

Table 6.12 Four-degree preferences for DMs in the GDU conflict

DM	Preference
USS	$s_2 > s_4 > s_3 > s_5 > s_1 > s_6 > s_9 > s_7 \ggg s_8$
CDO	$\{s_3 \sim s_7\} > \{s_5 \sim s_9\} > \{s_4 \sim s_8\} \ggg \{s_1 \sim s_2 \sim s_6\}$
IJC	$\{s_2 \sim s_3 \sim s_4 \sim s_5 \sim s_6 \sim s_7 \sim s_8 \sim s_9\} \gg s_1$

information for this conflict over the feasible states is given in Table 6.12, where $>$, $\gg$, and $\ggg$ mean more preferred, strongly preferred, and very strongly preferred, respectively, and equally preferred states are given in brackets and connected using the symbol $\sim$. The fact that states are ranked from most preferred on the left to least preferred on the right, where ties are allowed, indicates that the preferences are transitive for this application. One can see that state s_8 is very strongly less preferred to all other states for USS, because at state s_8 the USS is proceeding to construct the full project while IJC recommends a modified version and CDO is taking legal action based on the Boundary Waters Treaty. The DM CDO considers states s_1, s_2, and s_6 to be equally preferred and very strongly less preferred relative to all other states. Note that this representation of preference information presented in Table 6.12 implies that the preferred relations, $>$, $\gg$, and $\ggg$ are transitive. For instance, since $s_9 > s_7$ and $s_7 \ggg s_8$, then $s_9 \ggg s_8$ for USS. However, in general, the preference structure presented in this book does not require the transitivity of preference relations, and hence can handle intransitive preferences.

6.6.2 Stability Analysis Under Four-Degree Preference

Formally, in a stability analysis, one determines the stability of each state for each DM for various solution concepts. Here, four-degree versions of five stability definitions consisting of super stability; Nash stability, $Nash_k$; general metarationality, GMR_k; symmetric metarationality, SMR_k; and sequential stability, SEQ_k, for $k = 0, 1, 2,$ and 3, are employed to obtain stability results for the GDU conflict. An equilibrium for degree k for a specific solution concept represents a likely resolution to the conflict, since it is stable for every DM according to the stability definition under consideration. Note that the super stable states are treated as Nash stable at the highest level when determining an equilibrium in the graph model with multiple degrees of preference.

To explain how a stability calculation is carried out, consider SMR_k stability for state s_5 from DM 2's viewpoint for $k = 0, 1, 2,$ and 3. Using the definition of a reachable list presented in Sect. 6.3.2 and Table 6.12, $R_2^+(s_5) = \{s_3\}$ and $R_{N\setminus 2}(s_3) = \{s_2\}$ with $s_5 \ggg_2 s_2$ and $s_5 >_2 s_4$ for $R_2(s_2) = \{s_4\}$. Therefore, according to Definition 6.48 state s_5 is stable for SMR_{3^-}. Other cases can be analyzed similarly. The stability results for the GDU conflict are summarized in Table 6.13, in which "$\surd$" for a given state under a DM means that this state is stable at a given degree for the particular DM; "$\surd^{k^+}$" and "$\surd^{k^-}$" for a given state under a DM means that this state is SMR_{k^+} or SMR_{k^-} stable for the specified DM; and "$\surd^{k}$" for a state under "Eq" signifies that this state is an equilibrium for a corresponding solution concept at degree k. Note that U, C, and I displayed in Table 6.13 denote the three DMs, USS, CDO, and IJC, respectively.

A state that is not an equilibrium has no long-term stability because there is at least one individual DM who has an incentive to move to a more preferred state and thereby not permit an equilibrium to form. Table 6.14 provides stability results for different versions of preference. In particular, when stabilities are analyzed using two degrees of preference introduced in Sect. 4.2, states s_4, s_7, and s_9 are equilibria; if preference information is provided using three degrees of preference, then states s_7 and s_9 are equilibria using stability definitions presented in Sect. 6.3; there is only one equilibrium state s_9 for four degrees of preference. If state s_4 were the resolution for the GDU conflict, this would mean that IJC recommends to complete the GDU project regardless of Canada's concerns, so USS proceeds with this project. It is obvious that this resolution cannot settle this conflict in the long term. State s_7 means that the USS follows the IJC recommendation to modify this project, but Canada does not take legal action based on the Boundary Waters Treaty. State s_9 is the same as state s_7 except that Canada chooses legal procedures. When comparing states s_7 and s_9, equilibrium s_9 is a more reasonable resolution for solving this conflict. Therefore, a multiple-degree version of a stability analysis provides new insights and valuable guidance for decision analysts.

Although the example of the GDU conflict shown in Table 6.11 and Fig. 6.14 is a relatively small model having three DMs, five options, and nine feasible states, a graph model structure can handle any finite number of states and DMs, each of whom can control any finite number of options. As pointed out by Fang et al. (2003a, b), an available decision support system (DSS) for stability analysis of a graph model with two degrees of preference can work well. Theorem 6.4 reveals the relationship of stabilities between two degrees of preference presented in Chap. 4 and multiple degrees of preference. This theorem indicates the possibility of developing an effective algorithm to implement the multilevel versions of the four stabilities within a DSS, which would be essential if the proposed stability analysis were applied to larger practical problems. In fact, a DSS based on the matrix version will be designed in Chap. 10.

Table 6.13 Stability results of the GDU conflict for the graph model with four levels of preference

State	Super				Level(k)	Nash				GMR				SMR				SEQ			
	U	C	I	Eq		U	C	I	Eq	U	C	I	Eq	U	C	I	Eq	U	C	I	Eq
s_1	√	√			0																
					1																
					2																
					3	√	√			√	√			$\surd^{3+}$	$\surd^{3+}$			√	√		
s_2			√		0																
					1	√				√				$\surd^{1+}$				√			
					2																
					3			√				√				$\surd^{3+}$				√	
s_3			√		0																
					1		√				√				$\surd^{1+}$				√		
					2																
					3			√				√				$\surd^{3+}$				√	
s_4			√		0																
					1	√				√				$\surd^{1+}$				√			
					2																
					3		√	√			√	√			$\surd^{3+}$	$\surd^{3+}$			√	√	
s_5			√		0																
					1																
					2																
					3			√			√	√			$\surd^{3-}$	$\surd^{3+}$			√	√	
s_6			√		0																
					1	√				√				$\surd^{1+}$				√			
					2																
					3			√				√				$\surd^{3+}$				√	
s_7			√		0																
					1		√				√				$\surd^{1+}$				√		
					2																
					3			√		√		√				$\surd^{3+}$		√		√	
s_8			√		0																
					1																
					2																
					3		√	√			√	√			$\surd^{3+}$	$\surd^{3+}$			√	√	
s_9			√		0																
					1																
					2																
					3	√		√		√	√	√	$\surd^{3}$	$\surd^{3+}$	$\surd^{3-}$	$\surd^{3+}$		√	√	√	$\surd^{3}$

Table 6.14 The comparison of stability results for three versions of preference

Version of preference	Equilibria	Analysis method
Two degrees of preference	s_4, s_7, s_9	See Sect. 4.2
Three degrees of preference	s_7, s_9	See Sect. 6.3
Four degrees of preference	s_9	See Sect. 6.4.3.2

6.7 Important Ideas

In this chapter, a multiple-degree preference framework is developed for the graph model methodology to handle multiple degrees of preference, which lie between relative and cardinal preferences in terms of information content. Multilevel versions of the four solution concepts consisting of Nash, GMR, SMR, and SEQ are defined in the graph model for multiple degrees of preference. Specifically, solution concepts at degree k are defined for $Nash_k$, GMR_k, SMR_k, and SEQ_k for $k = 1, 2, \ldots, r$, where r is the maximum number of degrees of preference between two states. The proposed stability definitions extend existing definitions based on two degrees and three degrees of preference, so that more practical and complicated problems can be analyzed at greater depth.

The algebraic system to ease the coding of logically-defined stability definitions proposed in Chaps. 4 and 5 for simple preference and unknown preference, respectively, is extended in this chapter in a similar way to handle three degrees of preference. The algebraic method is developed to represent general, strong, and weak graph model stability definitions based on strength of preference using explicit matrix formulations instead of graphical or logical representations. These explicit algebraic formulations allow algorithms to assess rapidly the stabilities of states, and to be applied to large and complicated conflict models, using an advanced decision support system (DSS) like the one designed in Chap. 10. Because of the flexible nature of these explicit expressions, the matrix representations introduced here can be used as a solid framework for incorporating new solution concepts reflecting human behavior and novel theoretical constructs for handling different kinds of conflict situations, into the basic GMCR paradigm.

6.8 Problems

6.8.1 The concept of degree of preference constitutes a procedure for internalizing the psychological phenomenon of emotion. For instance, an environmentalist greatly prefers that a company not pollute the surrounding environment via discharges of gas, liquid and solid wastes. Describe two types of real-world disputes in which emotions are present and hence must be taken into account.

6.8.2 Attitudes can play a role in how people may behave in a conflict situation. Based on the research of Inohara et al. (2007) and Bernath Walker et al. (2009, 2012a), outline how attitude is operationalized within GMCR. Qualitatively, what connections do you see between degrees of preference and attitudes? Do you think they could be combined?

6.8.3 The concept of dominating attitudes within GMCR is put forward by Bernath Walker et al. (2012b). Briefly explain how this approach works and discuss its links to degrees of preference.

6.8.4 In the Prisoner's Dilemma conflict described in Problem 3.5.1, suppose that both DMs greatly prefer state s_1, in which they both cooperate, over state s_4, in which they do not cooperate with one another. Carry out a complete stability analysis following a logical interpretation using the four solution concepts utilized in this chapter as well as other chapters in the text. Did you gain additional strategic insights using this approach over the situation in which degree of preference is not present? Justify your response.

6.8.5 For the question involving Prisoner's Dilemma in Problem 6.8.4, execute the stability analysis using the matrix or algebraic formulation rather than the logical form.

6.8.6 For the game of Chicken in Problem 3.5.4, in which both drivers who are driving at high speed towards one another in their cars, carry out a stability analysis in which both drivers greatly prefer not to have a head-on crash over all of the other scenarios. Use the logical form of the four stability definitions for calculating individual stability and the associated equilibria. Explain why your findings make sense.

6.8.7 For the game of Chicken in Problem 3.5.4, in which both drivers are racing towards each other at high speed, execute a stability analysis for which one of the two drivers greatly prefers the situation in which they do not crash over all of the others states. Comment about the strategic meaning of your stability results.

6.8.8 For the Elmira conflict described and modeled in option form in Sect. 1.2.2 and analyzed for the case of simple preference in Sect. 4.5, suggest a reasonable model containing strength of preference such as when the Ministry of the Environment greatly prefers that Uniroyal not close down its chemical plant over situations in which it does terminates its operations. Carry out a complete stability analysis in which the DMs behave according to Nash or SEQ stability. Explain whether or not your findings make strategic sense.

6.8.9 Three degrees of preference often occur in practice. However, four degrees of preference may not take place as often. Explain a conflict situation in which it makes sense to entertain four degrees of preference in a conflict investigation. Provide references to support your claim.

6.8.10 The Gisborne conflict arising over the export of fresh water in bulk quantities is studied in Sect. 5.4. Provide a version of this conflict model in which it is reasonable to consider three degrees of preference for at least one of the DMs. Carry out a stability analysis of this conflict model for the case of Nash and SEQ stability. Discuss interesting stability results that you found.

References

Bernath Walker, S., Hipel, K. W., & Inohara, T. (2009). Strategic decision making for improved environmental security: Coalitions and attitudes in the graph model for conflict resolution. *Journal of Systems Science and Systems Engineering*, *18*(4), 461–476. https://doi.org/10.1007/s11518-009-5119-9.

Bernath Walker, S., Hipel, K. W., & Inohara, T. (2012a). Attitudes and preferences: Approaches to representing decision maker desires. *Applied Mathematics and Computation*, *218*(12), 6637–6647. https://doi.org/10.1016/j.amc.2011.11.102.

Bernath Walker, S., Hipel, K. W., & Inohara, T. (2012b). Dominating attitudes in the graph model for conflict resolution. *Journal of Systems Science and Systems Engineering*, *21*(3), 316–336. https://doi.org/10.1007/s11518-012-5198-x.

Fang, L., Hipel, K. W., & Kilgour, D. M. (1993). *Interactive decision making: The graph model for conflict resolution*. New York: Wiley. https://doi.org/10.2307/2583940.

Fang, L., Hipel, K. W., Kilgour, D. M., & Peng, X. (2003a). A decision support system for interactive decision making, part 1: Model formulation. *IEEE Transactions on Systems, Man and Cybernetics, Part C: Applications and Reviews*, *33*(1), 42–55. https://doi.org/10.1109/tsmcc.2003.809361.

Fang, L., Hipel, K. W., Kilgour, D. M., & Peng, X. (2003b). A decision support system for interactive decision making, part 2: Analysis and output interpretation. *IEEE Transactions on Systems, Man and Cybernetics, Part C: Applications and Reviews*, *33*(1), 56–66. https://doi.org/10.1109/tsmcc.2003.809360.

Fraser, N. M., & Hipel, K. W. (1984). *Conflict analysis: Models and resolutions*. New York: North-Holland Publishing Company. https://doi.org/10.2307/2582031.

Hamouda, L., Kilgour, D. M., & Hipel, K. W. (2004). Strength of preference in the graph model for conflict resolution. *Group Decision and Negotiation*, *13*, 449–462. https://doi.org/10.1023/b:grup.0000045751.21207.35.

Hamouda, L., Kilgour, D. M., & Hipel, K. W. (2006). Strength of preference in graph models for multiple decision-maker conflicts. *Applied Mathematics and Computation*, *179*(1), 314–327. https://doi.org/10.1016/j.amc.2005.11.109.

Hipel, K. W., & Fraser, N. M. (1980). Metagame alaysis of the Garrison conflict. *Water Resources Research*, *16*(4), 629–637. https://doi.org/10.1029/wr016i004p00629.

Inohara, T., Hipel, K. W., & Bernath Walker, S. (2007). Conflict analysis approaches for investigating attitudes and misperceptions in the War of 1812. *Journal of Systems Science and Systems Engineering*, *16*(2), 181–201. https://doi.org/10.1007/s11518-007-5042-x.

Xu, H., Hipel, K. W., & Kilgour, D. M. (2009). Multiple levels of preference in interactive strategic decisions. *Discrete Applied Mathematics*, *157*(15), 3300–3313. https://doi.org/10.1016/j.dam.2009.06.032.

Xu, H., Kilgour, D. M., & Hipel, K. W. (2010). An integrated algebraic approach to conflict resolution with three-level preference. *Applied Mathematics and Computation*, *216*(3), 693–707. https://doi.org/10.1016/j.amc.2010.01.054.

Xu, H., Kilgour, D. M., & Hipel, K. W. (2011). Matrix representation of conflict resolution in multiple-decision-maker graph models with preference uncertainty. *Group Decision and Negotiation*, *20*(6), 755–779. https://doi.org/10.1007/s10726-010-9188-4.

Chapter 7
Stability Definitions: Hybrid Preference

In the previous three chapters, different types of preference structures are presented and integrated into the Graph Model for Conflict Resolution (GMCR): simple preference in Chap. 4, unknown preference in Chap. 5, and degrees of preference in Chap. 6. A hybrid preference framework is proposed in this chapter for strategic conflict analysis to incorporate both unknown preference and three degrees of preference into the paradigm of GMCR under multiple decision makers (DMs). This structure offers DMs and analysts a more flexible mechanism for preference expression, which can include a strong or mild relative preference of one state over another, equally preferred states, and unknown preference between two states (Li et al. 2004, Hamouda et al. 2006, Xu et al. 2008, 2010a, b, c, 2011, 2013).

The main properties of hybrid preference structure, as well as the reachable lists of a DM to keep track of the DM's moves and countermoves, are introduced in Sect. 7.1 in this chapter. When considering stability definitions for more than two DMs, coalition moves are defined since two or more DMs can participate in blocking a unilateral improvement by another DM. Subsequently, the logical representations of four stability definitions consisting of Nash stability, general metarationality (GMR), symmetric metarationality (SMR), and sequential stability (SEQ), are defined for the graph model with hybrid preference in Sect. 7.2 while the relationships among them are investigated in Sect. 7.2.4. Additionally, in this chapter, matrix representations of hybrid preference and the four stabilities with hybrid preference are presented in Sects. 7.3 and 7.4, respectively. A case study is described in Sect. 7.5 to demonstrate how the stability definitions work under hybrid preference. Specifically, the water export conflict discussed in Sect. 5.4 is utilized to show how this methodology works for the situations in which two or more than two DMs are present.

H. Xu et al., *Conflict Resolution Using the Graph Model: Strategic Interactions in Competition and Cooperation*, Studies in Systems, Decision and Control 153,
https://doi.org/10.1007/978-3-319-77670-5_7

7.1 Hybrid Preference and Reachable Lists

The hybrid preference is defined using a quadruple relation $\{\sim_i, \gg_i, >_i, U_i\}$ in a graph model for DM i. The hybrid structure is complete, i.e. if $s, q \in S$, then exactly one of the following relations holds: $s \sim_i q, s \gg_i q, q \gg_i s, s >_i q, q >_i s$, and $s\, U_i\, q$. For hybrid preference, DM i can control six corresponding reachable lists from state s which are $R_i^{++}(s)$, $R_i^{+}(s)$, $R_i^{U}(s)$, $R_i^{=}(s)$, $R_i^{-}(s)$, and $R_i^{--}(s)$. A_i indicates that DM i can make a unilateral move (in one step) from the initial state to the terminal state of the arc. The reachable lists of DM i under hybrid preference are defined as follows:

(i) $R_i^{++}(s) = \{q \in S : (s, q) \in A_i$ and $q \gg_i s\}$ stands for DM i's reachable list from state s by a strong unilateral improvement. This set contains all states q which are strongly preferred by DM i to state s and can be reached in one step from s;
(ii) $R_i^{+}(s) = \{q \in S : (s, q) \in A_i$ and $q >_i s\}$ denotes DM i's reachable list from state s by a mild unilateral improvement;
(iii) $R_i^{U}(s) = \{q \in S : (s, q) \in A_i$ and $q\, U_i\, s\}$ denotes DM i's reachable list from state s by an uncertain move;
(iv) $R_i^{=}(s) = \{q \in S : (s, q) \in A_i$ and $q \sim_i s\}$ denotes DM i's reachable list from state s by an equally unilateral move;
(v) $R_i^{-}(s) = \{q \in S : (s, q) \in A_i$ and $s >_i q\}$ denotes DM i's reachable list from state s by a mild unilateral disimprovement; and
(vi) $R_i^{--}(s) = \{q \in S : (s, q) \in A_i$ and $s \gg_i q\}$ is DM i's reachable list from state s by a strong unilateral disimprovement.

DM i's reachable list from state s, $R_i(s)$, represents DM i's unilateral moves (UMs). Let $\cup$ denote the union operation. Hence, $R_i(s)$ is partitioned according to hybrid preference structure as

$$R_i(s) = R_i^{++}(s) \cup R_i^{+}(s) \cup R_i^{U}(s) \cup R_i^{=}(s) \cup R_i^{-}(s) \cup R_i^{--}(s).$$

The relationships among the subsets of S and the reachable lists of DM i from state s in the graph model having hybrid preference are shown in Fig. 7.1.

For ease of use, the notation with respect to UMs and subsets of the state set S for hybrid preference is presented as follows:

- $R_i^{+,++,U}(s) = R_i^{+}(s) \cup R_i^{++}(s) \cup R_i^{U}(s)$ stands for mild unilateral improvements, strong unilateral improvements, or unilateral uncertain moves called mild or strong unilateral improvements or uncertain moves (MSUIUMs) from state s for DM i;
- $\Phi_i^{--,U}(s) = \Phi_i^{--}(s) \cup \Phi_i^{U}(s)$; and
- $\Phi_i^{--,-,=,U}(s) = \Phi_i^{--}(s) \cup \Phi_i^{-}(s) \cup \Phi_i^{=}(s) \cup \Phi_i^{U}(s)$.

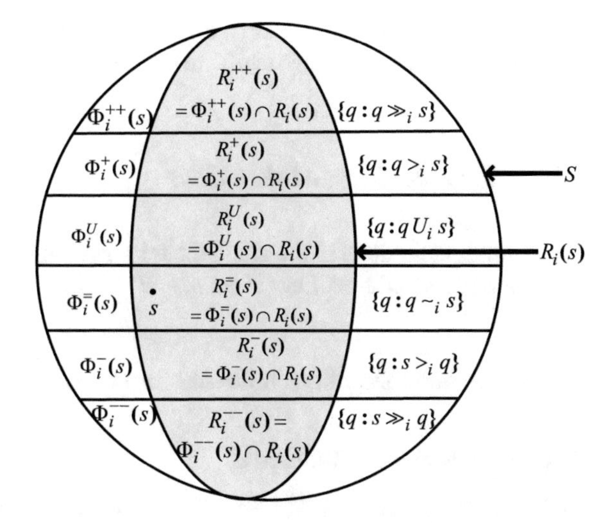

Fig. 7.1 Relations among subsets of S and reachable lists including hybrid preference

7.2 Logical Representation of Stability Definitions Under Hybrid Preference

The four stability definitions given in this section recognize two cases in which the degree of strength in hybrid preference is distinguished. Firstly, general stabilities are defined, and then the two subclasses, strong and weak, are determined. Stabilities of the first kind are referred to as general because they are in essence the same as the stability definitions using simple preference defined in Chap. 4. Stability definitions are called strong or weak stabilities in order to reflect the additional preference information contained in the strength of the preference relation. These more sophisticated definitions furnish expanded strategic insights into a conflict model that handles strength of preference.

In order to represent the following stabilities, appropriate notation is now defined. Four types of stability definitions in logical forms are proposed, with index a, b, c, and d, according to whether the DM would move to a state of uncertain preference and whether the DM would be sanctioned by a responding move to a state of uncertain preference, relative to the status quo (Li et al. 2004). Let l denote one of the four extensions indexed a, b, c, and d, i.e., $l = a, b, c$, or d. In the following theorems, the symbol GS denotes a graph model stability, GMR, SMR, or SEQ. Then GS_l refers to the GS solution concept, GMR, SMR, or SEQ, indexed l in the graph model for unknown preference. The symbols GGS, SGS, and WGS respectively denote a general stability, GGMR, GSMR, or GSEQ, the strong stability, SGMR, SSMR, or SSEQ, and the weak stability, WGMR, WSMR, or WSEQ, under strength of preference. Then GGS_l refers to the GGS solution concept indexed l, SGS_l refers to the strong solution concept SGS indexed l, and WGS_l refers to the weak solution concept WGS indexed l (defined below) in the graph model for hybrid preference. The symbol $s \in S_i^{GGS_l}$ denotes that $s \in S$ is stable for DM i according to general stability GGS indexed l. Similarly, $s \in S_i^{SGS_l}$ denotes that $s \in S$ is strongly stable for DM i according to strong stability SGS indexed l.

7.2.1 Two Decision Maker Case

The stability definitions in the graph model for two-DM conflicts with hybrid preference are presented first. Let $l \in \{a, b, c, d\}$.

7.2.1.1 General Stabilities Indexed l for Hybrid Preference

(1) General Stabilities Indexed a

For stabilities indexed a, DM i is willing to move to states that are mildly preferred or strongly preferred, as well as states having uncertain preference relative to the status quo but does not wish to be sanctioned by a strongly less preferred, mildly less preferred, or equally preferred state relative to the status quo. The definitions given below assume that $s \in S$ and $i \in N$.

Definition 7.1 State s is general $Nash_a$ for DM i, denoted by $s \in S_i^{Nash_a}$, iff $R_i^{+,++,U}(s) = \emptyset$.

Nash stability does not involve sanctions so only general Nash stability is defined within hybrid preference. One uses Nash to denote GNash in the following definitions.

Definition 7.2 State s is general GMR_a for DM i, denoted by $s \in S_i^{GGMR_a}$, iff for every $s_1 \in R_i^{+,++,U}(s)$ there exists at least one $s_2 \in R_j(s_1)$ with $s_2 \in \Phi_i^{--,-,=}(s)$.

Definition 7.3 State s is general SMR_a for DM i, denoted by $s \in S_i^{GSMR_a}$, iff for every $s_1 \in R_i^{+,++,U}(s)$ there exists at least one $s_2 \in R_j(s_1)$, such that $s_2 \in \Phi_i^{--,-,=}(s)$ and $s_3 \in \Phi_i^{--,-,=}(s)$ for any $s_3 \in R_i(s_2)$.

Definition 7.4 State s is general SEQ_a for DM i, denoted by $s \in S_i^{GSEQ_a}$, iff for every $s_1 \in R_i^{+,++,U}(s)$ there exists at least one $s_2 \in R_j^{+,++,U}(s_1)$ with $s_2 \in \Phi_i^{--,-,=}(s)$.

(2) General Stabilities Indexed b

For stabilities indexed b, DM i will move only to mildly or strongly preferred states from a status quo, but does not want to be sanctioned by a strongly less preferred, mildly less preferred, or equally preferred state relative to the status quo.

Definition 7.5 State s is general $Nash_b$ for DM i, denoted by $s \in S_i^{Nash_b}$, iff $R_i^{+,++}(s) = \emptyset$.

Definition 7.6 State s is general GMR_b for DM i, denoted by $s \in S_i^{GGMR_b}$, iff for every $s_1 \in R_i^{+,++}(s)$ there exists at least one $s_2 \in R_j(s_1)$ with $s_2 \in \Phi_i^{--,-,=}(s)$.

Definition 7.7 State s is general SMR_b for DM i, denoted by $s \in S_i^{GSMR_b}$, iff for every $s_1 \in R_i^{+,++}(s)$ there exists at least one $s_2 \in R_j(s_1)$, such that $s_2 \in \Phi_i^{--,-,=}(s)$ and $s_3 \in \Phi_i^{--,-,=}(s)$ for any $s_3 \in R_i(s_2)$.

Definition 7.8 State s is general SEQ_b for DM i, denoted by $s \in S_i^{GSEQ_b}$, iff for every $s_1 \in R_i^{+,++}(s)$ there exists at least one $s_2 \in R_j^{+,++,U}(s_1)$ with $s_2 \in \Phi_i^{--,-,=}(s)$.

The above definitions indexed b which exclude uncertainty in preference are different from those discussed in Sect. 6.3.1.1, since current definitions are utilized to analyze conflict models under combining preference uncertainty and strength of preference.

(3) General Stabilities Indexed c

For definitions indexed c, DM i can move to mildly preferred, strongly preferred states, as well as states having uncertain preference relative to the starting state. With respect to sanctioning, DM i does not want to end up at states that are mildly less preferred, strongly less preferred, or equally preferred, as well as states having uncertain preference relative to state s.

Definition 7.9 State s is general $Nash_c$ for DM i, denoted by $s \in S_i^{Nash_c}$, iff $R_i^{+,++,U}(s) = \emptyset$.

Definition 7.10 State s is general GMR_c for DM i, denoted by $s \in S_i^{GGMR_c}$, iff for every $s_1 \in R_i^{+,++,U}(s)$ there exists at least one $s_2 \in R_j(s_1)$ with $s_2 \in \Phi_i^{--,-,=,U}(s)$.

Definition 7.11 State s is general SMR_c for DM i, denoted by $s \in S_i^{GSMR_c}$, iff for every $s_1 \in R_i^{+,++,U}(s)$ there exists at least one $s_2 \in R_j(s_1)$, such that $s_2 \in \Phi_i^{--,-,=,U}(s)$ and $s_3 \in \Phi_i^{--,-,=,U}(s)$ for any $s_3 \in R_i(s_2)$.

Definition 7.12 State s is general SEQ_c for DM i, denoted by $s \in S_i^{GSEQ_c}$, iff for every $s_1 \in R_i^{+,++,U}(s)$ there exists at least one $s_2 \in R_j^{+,++,U}(s_1)$ with $s_2 \in \Phi_i^{--,-,=,U}(s)$.

(4) General Stabilities Indexed d

For the last set of stabilities, indexed d, a DM is not willing to move to a state with uncertain preference relative to the status quo, but is deterred by sanctions to states that have uncertain preference relative to the status quo.

Definition 7.13 State s is general $Nash_d$ for DM i, denoted by $s \in S_i^{Nash_d}$, iff $R_i^{+,++}(s) = \emptyset$.

Definition 7.14 State s is general GMR_d for DM i, denoted by $s \in S_i^{GGMR_d}$, iff for every $s_1 \in R_i^{+,++}(s)$ there exists at least one $s_2 \in R_j(s_1)$ with $s_2 \in \Phi_i^{--,-,=,U}(s)$.

Definition 7.15 State s is general SMR_d for DM i, denoted by $s \in S_i^{GSMR_d}$, iff for every $s_1 \in R_i^{+,++}(s)$ there exists at least one $s_2 \in R_j(s_1)$, such that $s_2 \in \Phi_i^{--,-,=,U}(s)$ and $s_3 \in \Phi_i^{--,-,=,U}(s)$ for any $s_3 \in R_i(s_2)$.

Definition 7.16 State s is general SEQ_d for DM i, denoted by $s \in S_i^{GSEQ_d}$, iff for every $s_1 \in R_i^{+,++}(s)$ there exists at least one $s_2 \in R_j^{+,++,U}(s_1)$ with $s_2 \in \Phi_i^{--,-,=,U}(s)$.

If the binary relation $\succ$ denotes $>$ or $\gg$ in this chapter, i.e., $s \succ q$ iff either $s > q$ or $s \gg q$, then Definitions 7.1–7.16 are identical to Definitions 5.2–5.17 in Chap. 5. On the other hand, when each DM does not consider including uncertain preference in stability analysis, the above definitions reduce to the general stability definitions from Definitions 6.1–6.4 in Chap. 6.

7.2.1.2 Strong Stabilities Indexed *l* for Hybrid Preference

With the hybrid preference framework introduced into the graph model, stable states can be classified into strongly stable or weakly stable according to strength of the possible sanctions and indexed a, b, c, or d by a DM's attitudes toward the risk associated with uncertain preferences. Strong and weak stabilities include only GMR, SMR, and SEQ because Nash stability does not involve sanctions.

(1) Strong Stabilities Indexed a

Definition 7.17 State s is strongly general metarational ($SGMR_a$) for DM i, denoted by $s \in S_i^{SGMR_a}$, iff for every $s_1 \in R_i^{+,++,U}(s)$ there exists at least one $s_2 \in R_j(s_1)$ such that $s_2 \in \Phi_i^{--}(s)$.

Definition 7.18 State s is strongly symmetric metarational ($SSMR_a$) for DM i, denoted by $s \in S_i^{SSMR_a}$, iff for every $s_1 \in R_i^{+,++,U}(s)$ there exists at least one $s_2 \in R_j(s_1)$, such that $s_2 \in \Phi_i^{--}(s)$ and $s_3 \in \Phi_i^{--}(s)$ for all $s_3 \in R_i(s_2)$.

Definition 7.19 State s is strongly sequentially stable ($SSEQ_a$) for DM i, denoted by $s \in S_i^{SSEQ_a}$, iff for every $s_1 \in R_i^{+,++,U}(s)$ there exists at least one $s_2 \in R_j^{+,++,U}(s)$ such that $s_2 \in \Phi_i^{--}(s)$.

With the above definitions indexed a, DM i would like to accept the risk associated with the uncertain status when considering whether to move away from the status quo state, but DM i will not consider states with uncertain preferences, when assessing sanctions.

(2) Strong Stabilities Indexed b

For the following definitions indexed b, DM i would move only to mildly or strongly preferred states and be deterred by sanctions to strongly less preferred states relative to the status quo.

Definition 7.20 State s is strongly general metarational ($SGMR_b$) for DM i, denoted by $s \in S_i^{SGMR_b}$, iff for every $s_1 \in R_i^{+,++}(s)$ there exists at least one $s_2 \in R_j(s_1)$ such that $s_2 \in \Phi_i^{--}(s)$.

Definition 7.21 State s is strongly symmetric metarational ($SSMR_b$) for DM i, denoted by $s \in S_i^{SSMR_b}$, iff for every $s_1 \in R_i^{+,++}(s)$ there exists at least one $s_2 \in R_j(s_1)$, such that $s_2 \in \Phi_i^{--}(s)$ and $s_3 \in \Phi_i^{--}(s)$ for all $s_3 \in R_i(s_2)$.

Definition 7.22 State s is strongly sequentially stable ($SSEQ_b$) for DM i, denoted by $s \in S_i^{SSEQ_b}$, iff for every $s_1 \in R_i^{+,++}(s)$ there exists at least one $s_2 \in R_j^{+,++,U}(s)$ such that $s_2 \in \Phi_i^{--}(s)$.

(3) Strong Stabilities Indexed c

The definitions indexed c refer to a DM's mixed attitudes toward the risk associated with uncertain preferences. Specifically, DM i is aggressive in deciding whether to move from the status quo, but is conservative when evaluating possible moves, because DM i is deterred by sanctions to states that are strongly less preferred and states that have uncertain preference relative to the status quo.

Definition 7.23 State s is strongly general metarational ($SGMR_c$) for DM i, denoted by $s \in S_i^{SGMR_c}$, iff for every $s_1 \in R_i^{+,++,U}(s)$ there exists at least one $s_2 \in R_j(s_1)$ such that $s_2 \in \Phi_i^{--,U}(s)$.

Definition 7.24 State s is strongly symmetric metarational ($SSMR_c$) for DM i, denoted by $s \in S_i^{SSMR_c}$, iff for every $s_1 \in R_i^{+,++,U}(s)$ there exists at least one $s_2 \in R_j(s_1)$, such that $s_2 \in \Phi_i^{--,U}(s)$ and $s_3 \in \Phi_i^{--,U}(s)$ for all $s_3 \in R_i(s_2)$.

Definition 7.25 State s is strongly sequentially stable ($SSEQ_c$) for DM i, denoted by $s \in S_i^{SSEQ_c}$, iff for every $s_1 \in R_i^{+,++,U}(s)$ there exists at least one $s_2 \in R_j^{+,++,U}(s)$ such that $s_2 \in \Phi_i^{--,U}(s)$.

(4) Strong Stabilities Indexed d

Definition 7.26 State s is strongly general metarational ($SGMR_d$) for DM i, denoted by $s \in S_i^{SGMR_d}$, iff for every $s_1 \in R_i^{+,++}(s)$ there exists at least one $s_2 \in R_j(s_1)$ such that $s_2 \in \Phi_i^{--,U}(s)$.

Definition 7.27 State s is strongly symmetric metarational ($SSMR_d$) for DM i, denoted by $s \in S_i^{SSMR_d}$, iff for every $s_1 \in R_i^{+,++}(s)$ there exists at least one $s_2 \in R_j(s_1)$, such that $s_2 \in \Phi_i^{--,U}(s)$ and $s_3 \in \Phi_i^{--,U}(s)$ for all $s_3 \in R_i(s_2)$.

Definition 7.28 State s is strongly sequentially stable ($SSEQ_d$) for DM i, denoted by $s \in S_i^{SSEQ_d}$, iff for every $s_1 \in R_i^{+,++}(s)$ there exists at least one $s_2 \in R_j^{+,++,U}(s)$ such that $s_2 \in \Phi_i^{--,U}(s)$.

The above definitions indexed d indicate that DM i would move only to mildly or strongly preferred states, but is deterred by sanctions that could move i to strongly less preferred states and states that have uncertain preference relative to the status quo. Therefore, definitions indexed d represent strong stabilities for the most conservative DMs.

7.2.1.3 Weak Stabilities Indexed l for Hybrid Preference

Let $l \in \{a, b, c, d\}$. A state is weakly stable if it is general stable GMR_l, SMR_l, or SEQ_l but not strongly stable. Specifically, weak stability concepts are defined next.

Definition 7.29 State s is weakly stable indexed as l, $WGMR_l$, $WSMR_l$, or $WSEQ_l$ for DM i, denoted by $s \in S_i^{WGMR_l}$, $s \in S_i^{WSMR_l}$, or $s \in S_i^{WSEQ_l}$, iff s is general stable but not strongly stable indexed as l.

7.2.2 Reachable List of a Coalition of Decision Makers Under Hybrid Preference

Analysis of a graph model involves searching paths in a graph but an important restriction of a graph model is that no DM can move twice in succession along any path. As explained earlier in Sects. 4.2.2 and 6.4.2, this means the GMCR approach takes into account intransitive as well as transitive moves. Any nonempty subset H of DMs, $H \subseteq N$ and $H \neq \emptyset$, is called a coalition. A legal sequence of unilateral moves (UMs) for a coalition of DMs is a sequence of states linked by unilateral moves by members of the coalition, in which a DM may move more than once, but not twice consecutively.

Within an n-DM model ($n \geq 2$), DM i's opponents, $N \setminus \{i\}$, where $\setminus$ refers to "set subtraction", consist of a group of one or more DMs. In order to analyze the stability of a state for DM $i \in N$, it is necessary to take into account possible responses by all other DMs $j \in N \setminus \{i\}$. The essential inputs of stability analysis are reachable lists of coalition $N \setminus \{i\}$ from state s, $R_{N\setminus\{i\}}(s)$ and $R_{N\setminus\{i\}}^{+,++,U}(s)$, for hybrid preference. Let the coalition $H \subseteq N$ satisfy $|H| \geq 2$ and let the status quo state be $s \in S$. $R_H(s) \subseteq S$, the reachable list of coalition H from state s by a legal sequence of UMs, is defined now.

Definition 7.30 A unilateral move by H is a member of $R_H(s) \subseteq S$ defined inductively by

(1) assuming $\Omega_H(s, s_1) = \emptyset$ for all $s_1 \in S$;
(2) if $j \in H$ and $s_1 \in R_j(s)$, then $s_1 \in R_H(s)$ and $\Omega_H(s, s_1) = \Omega_H(s, s_1) \cup \{j\}$;
(3) if $s_1 \in R_H(s), j \in H$, and $s_2 \in R_j(s_1)$, then, provided $\Omega_H(s, s_1) \neq \{j\}$, $s_2 \in R_H(s)$ and $\Omega_H(s, s_2) = \Omega_H(s, s_2) \cup \{j\}$.

Note that this definition is inductive: first, using (2), the states reachable from s are identified and added to $R_H(s)$; then, using (3), all states reachable from those states are identified and added to $R_H(s)$; afterwards the process is repeated until no further states are added to $R_H(s)$ by repeating (3). Because $R_H(s) \subseteq S$, and S is finite, this limit must be reached in finitely many steps.

To extend the definitions of the reachable lists for a coalition to take hybrid preference into account, a legal sequence of coalitional mild or strong unilateral improvements or uncertain moves (MSUIUMs) must be defined first. A legal sequence of

MSUIUMs is a sequence of allowable mild unilateral improvements, strong unilateral improvements, or uncertain moves by a coalition, with the same restriction that any member in the coalition may move more than once, but not twice consecutively. The formal definition for reachable lists of coalition H by the legal sequence of MSUIUMs is given as follows:

Definition 7.31 Let $s \in S$, $H \subseteq N$, and $H \neq \emptyset$. A mild or strong unilateral improvement or uncertain move (MSUIUM) by H is a member of $R_H^{+,++,U}(s) \subseteq S$, defined inductively by

(1) assuming $\Omega_H^{+,++,U}(s, s_1) = \emptyset$ for all $s_1 \in S$;
(2) if $j \in H$ and $s_1 \in R_j^{+,++,U}(s)$, then $s_1 \in R_H^{+,++,U}(s)$ and $\Omega_H^{+,++,U}(s, s_1) = \Omega_H^{+,++,U}(s, s_1) \cup \{j\}$;
(3) if $s_1 \in R_H^{+,++,U}(s)$, $j \in H$, and $s_2 \in R_j^{+,++,U}(s_1)$, then, provided $\Omega_H^{+,++,U}(s, s_1) \neq \{j\}$, $s_2 \in R_H^{+,++,U}(s)$ and $\Omega_H^{+,++,U}(s, s_2) = \Omega_H^{+,++,U}(s, s_2) \cup \{j\}$.

Like Definition 7.30, Definition 7.31 is an inductive definition. The roles and interpretations of $R_H^{+,++,U}(s)$ and $\Omega_H^{+,++,U}(s, s_1)$ are likewise analogous. To interpret Definition 7.31, note that if $s_1 \in R_H^{+,++,U}(s)$, then $\Omega_H^{+,++,U}(s, s_1) \subseteq H$ is the set of all last DMs in legal sequences from s to s_1. (If $s_1 \notin R_H^{+,++,U}(s)$, it can be assumed that $\Omega_H^{+,++,U}(s, s_1) = \emptyset$.) Suppose that $\Omega_H^{+,++,U}(s, s_1)$ contains only one DM, say $j \in N$. Then any move from s_1 to a subsequent state, say s_2, must be made by a member of H other than j; otherwise DM j would have to move twice in succession. On the other hand, if $|\Omega_H^{+,++,U}(s, s_1)| \geq 2$, any member of H who has a unilateral move from s_1 to s_2 may exercise it.

7.2.3 n-Decision Maker Case

The stability definitions in the graph model for two-DM conflicts with hybrid preference are special cases of the definitions for n-DM case ($n \geq 2$) presented in this subsection.

7.2.3.1 General Stabilities Indexed l for Hybrid Preference

(1) General Stabilities Indexed a

For stabilities indexed a, DM i is willing to move to states that are mildly preferred or strongly preferred, as well as states having uncertain preference relative to the status quo but does not wish to be sanctioned by a strongly less preferred, mildly less preferred, or equally preferred state relative to the status quo. The definitions given below assume that $s \in S$ and $i \in N$.

Definition 7.32 State s is general $Nash_a$ for DM i, denoted by $s \in S_i^{Nash_a}$, iff $R_i^{+,++,U}(s) = \emptyset$.

Nash stability does not involve sanctions so n-DM case is identical with two-DM case.

Definition 7.33 State s is general GMR_a for DM i, denoted by $s \in S_i^{GGMR_a}$, iff for every $s_1 \in R_i^{+,++,U}(s)$ there exists at least one $s_2 \in R_{N\setminus\{i\}}(s_1)$ with $s_2 \in \Phi_i^{--,-,=}(s)$.

Definition 7.34 State s is general SMR_a for DM i, denoted by $s \in S_i^{GSMR_a}$, iff for every $s_1 \in R_i^{+,++,U}(s)$ there exists at least one $s_2 \in R_{N\setminus\{i\}}(s_1)$, such that $s_2 \in \Phi_i^{--,-,=}(s)$ and $s_3 \in \Phi_i^{--,-,=}(s)$ for any $s_3 \in R_i(s_2)$.

Definition 7.35 State s is general SEQ_a for DM i, denoted by $s \in S_i^{GSEQ_a}$, iff for every $s_1 \in R_i^{+,++,U}(s)$ there exists at least one $s_2 \in R_{N\setminus\{i\}}^{+,++,U}(s_1)$ with $s_2 \in \Phi_i^{--,-,=}(s)$.

It should be pointed out that the same notation for stabilities indexed a for preference with uncertainty presented in Sect. 5.2.3 is used for hybrid preference. However, they have different meanings, since current definitions can analyze conflict models including hybrid preference. The following definitions are still presented using the same notation as those including preference uncertainty.

(2) General Stabilities Indexed b

For stabilities indexed b, DM i will move only to mildly or strongly preferred states from a status quo, but does not want to be sanctioned by a strongly less preferred, mildly less preferred, or equally preferred state relative to the status quo.

Definition 7.36 State s is general $Nash_b$ for DM i, denoted by $s \in S_i^{Nash_b}$, iff $R_i^{+,++}(s) = \emptyset$.

Definition 7.37 State s is general GMR_b for DM i, denoted by $s \in S_i^{GGMR_b}$, iff for every $s_1 \in R_i^{+,++}(s)$ there exists at least one $s_2 \in R_{N\setminus\{i\}}(s_1)$ with $s_2 \in \Phi_i^{--,-,=}(s)$.

Definition 7.38 State s is general SMR_b for DM i, denoted by $s \in S_i^{GSMR_b}$, iff for every $s_1 \in R_i^{+,++}(s)$ there exists at least one $s_2 \in R_{N\setminus\{i\}}(s_1)$, such that $s_2 \in \Phi_i^{--,-,=}(s)$ and $s_3 \in \Phi_i^{--,-,=}(s)$ for any $s_3 \in R_i(s_2)$.

Definition 7.39 State s is general SEQ_b for DM i, denoted by $s \in S_i^{GSEQ_b}$, iff for every $s_1 \in R_i^{+,++}(s)$ there exists at least one $s_2 \in R_{N\setminus\{i\}}^{+,++,U}(s_1)$ with $s_2 \in \Phi_i^{--,-,=}(s)$.

The above definitions indexed b which exclude uncertainty in preference are different from those discussed by Hamouda et al. (2006), since current definitions are utilized to analyze conflict models under combining preference uncertainty and strength of preference.

(3) General Stabilities Indexed c

For definitions indexed c, DM i can move to mildly preferred, strongly preferred states, as well as states having uncertain preference relative to the starting state. With respect to sanctioning, DM i does not want to end up at states that are mildly less preferred, strongly less preferred, or equally preferred, as well as states having uncertain preference relative to state s.

Definition 7.40 State s is general $Nash_c$ for DM i, denoted by $s \in S_i^{Nash_c}$, iff $R_i^{+,++,U}(s) = \emptyset$.

Definition 7.41 State s is general GMR_c for DM i, denoted by $s \in S_i^{GGMR_c}$, iff for every $s_1 \in R_i^{+,++,U}(s)$ there exists at least one $s_2 \in R_{N\setminus\{i\}}(s_1)$ with $s_2 \in \Phi_i^{--,-,=,U}(s)$.

Definition 7.42 State s is general SMR_c for DM i, denoted by $s \in S_i^{GSMR_c}$, iff for every $s_1 \in R_i^{+,++,U}(s)$ there exists at least one $s_2 \in R_{N\setminus\{i\}}(s_1)$, such that $s_2 \in \Phi_i^{--,-,=,U}(s)$ and $s_3 \in \Phi_i^{--,-,=,U}(s)$ for any $s_3 \in R_i(s_2)$.

Definition 7.43 State s is general SEQ_c for DM i, denoted by $s \in S_i^{GSEQ_c}$, iff for every $s_1 \in R_i^{+,++,U}(s)$ there exists at least one $s_2 \in R_{N\setminus\{i\}}^{+,++,U}(s_1)$ with $s_2 \in \Phi_i^{--,-,=,U}(s)$.

(4) General Stabilities Indexed d

For the last set of stabilities, indexed d, a DM is not willing to move to a state with uncertain preference relative to the status quo, but is deterred by sanctions to states that have uncertain preference relative to the status quo.

Definition 7.44 State s is general $Nash_d$ for DM i, denoted by $s \in S_i^{Nash_d}$, iff $R_i^{+,++}(s) = \emptyset$.

Definition 7.45 State s is general GMR_d for DM i, denoted by $s \in S_i^{GGMR_d}$, iff for every $s_1 \in R_i^{+,++}(s)$ there exists at least one $s_2 \in R_{N\setminus\{i\}}(s_1)$ with $s_2 \in \Phi_i^{--,-,=,U}(s)$.

Definition 7.46 State s is general SMR_d for DM i, denoted by $s \in S_i^{GSMR_d}$, iff for every $s_1 \in R_i^{+,++}(s)$ there exists at least one $s_2 \in R_{N\setminus\{i\}}(s_1)$, such that $s_2 \in \Phi_i^{--,-,=,U}(s)$ and $s_3 \in \Phi_i^{--,-,=,U}(s)$ for any $s_3 \in R_i(s_2)$.

Definition 7.47 State s is general SEQ_d for DM i, denoted by $s \in S_i^{GSEQ_d}$, iff for every $s_1 \in R_i^{+,++}(s)$ there exists at least one $s_2 \in R_{N\setminus\{i\}}^{+,++,U}(s_1)$ with $s_2 \in \Phi_i^{--,-,=,U}(s)$.

When $n = 2$, the DM set N becomes $\{i, j\}$ in Definitions 7.32–7.47, and the reachable lists for $H = N \setminus \{i\}$ by legal sequences of UMs and MSUIUMs from s_1, $R_{N\setminus\{i\}}(s_1)$ and $R_{N\setminus\{i\}}^{+,++,U}(s_1)$, respectively, degenerate to $R_j(s_1)$ and $R_j^{+,++,U}(s_1)$, DM j's corresponding reachable lists from s_1.

If the binary relation $\succ$ denotes $>$ or $\gg$ in this chapter, i.e., $s \succ q$ iff either $s > q$ or $s \gg q$, then Definitions 7.32–7.47 are identical to Definitions 5.19–5.34 in Chap. 5 proposed by Li et al. (2004). On the other hand, when each DM does not consider including uncertain preference in stability analysis, the above definitions reduce to the general stability definitions from Definitions 6.38–6.40 in Chap. 6 developed by Hamouda et al. (2006).

7.2.3.2 Strong Stabilities Indexed *l* for Hybrid Preference

(1) Strong Stabilities Indexed *a*

Definition 7.48 State s is strongly GMR_a ($SGMR_a$) for DM i, denoted by $s \in S_i^{SGMR_a}$, iff for every $s_1 \in R_i^{+,++,U}(s)$ there exists at least one $s_2 \in R_{N\backslash\{i\}}(s_1)$ such that $s_2 \in \Phi_i^{--}(s)$.

Definition 7.49 State s is strongly SMR_a ($SSMR_a$) for DM i, denoted by $s \in S_i^{SSMR_a}$, iff for every $s_1 \in R_i^{+,++,U}(s)$ there exists at least one $s_2 \in R_{N\backslash\{i\}}(s_1)$, such that $s_2 \in \Phi_i^{--}(s)$ and $s_3 \in \Phi_i^{--}(s)$ for all $s_3 \in R_i(s_2)$.

Definition 7.50 State s is strongly SEQ_a ($SSEQ_a$) for DM i, denoted by $s \in S_i^{SSEQ_a}$, iff for every $s_1 \in R_i^{+,++,U}(s)$ there exists at least one $s_2 \in R_{N\backslash\{i\}}^{+,++,U}(s_1)$ such that $s_2 \in \Phi_i^{--}(s)$.

(2) Strong Stabilities Indexed *b*

Definition 7.51 State s is strongly GMR_b ($SGMR_b$) for DM i, denoted by $s \in S_i^{SGMR_b}$, iff for every $s_1 \in R_i^{+,++}(s)$ there exists at least one $s_2 \in R_{N\backslash\{i\}}(s_1)$ such that $s_2 \in \Phi_i^{--}(s)$.

Definition 7.52 State s is strongly SMR_b ($SSMR_b$) for DM i, denoted by $s \in S_i^{SSMR_b}$, iff for every $s_1 \in R_i^{+,++}(s)$ there exists at least one $s_2 \in R_{N\backslash\{i\}}(s_1)$, such that $s_2 \in \Phi_i^{--}(s)$ and $s_3 \in \Phi_i^{--}(s)$ for all $s_3 \in R_i(s_2)$.

Definition 7.53 State s is strongly SEQ_b ($SSEQ_b$) for DM i, denoted by $s \in S_i^{SSEQ_b}$, iff for every $s_1 \in R_i^{+,++}(s)$ there exists at least one $s_2 \in R_{N\backslash\{i\}}^{+,++,U}(s_1)$ such that $s_2 \in \Phi_i^{--}(s)$.

(3) Strong Stabilities Indexed *c*

Definition 7.54 State s is strongly GMR_c ($SGMR_c$) for DM i, denoted by $s \in S_i^{SGMR_c}$, iff for every $s_1 \in R_i^{+,++,U}(s)$ there exists at least one $s_2 \in R_{N\backslash\{i\}}(s_1)$ such that $s_2 \in \Phi_i^{--,U}(s)$.

Definition 7.55 State s is strongly SMR_c ($SSMR_c$) for DM i, denoted by $s \in S_i^{SSMR_c}$, iff for every $s_1 \in R_i^{+,++,U}(s)$ there exists at least one $s_2 \in R_{N\backslash\{i\}}(s_1)$, such that $s_2 \in \Phi_i^{--,U}(s)$ and $s_3 \in \Phi_i^{--,U}(s)$ for all $s_3 \in R_i(s_2)$.

Definition 7.56 State s is strongly SEQ_c ($SSEQ_c$) for DM i, denoted by $s \in S_i^{SSEQ_c}$, iff for every $s_1 \in R_i^{+,++,U}(s)$ there exists at least one $s_2 \in R_{N\setminus\{i\}}^{+,++,U}(s_1)$ such that $s_2 \in \Phi_i^{--,U}(s)$.

(4) Strong Stabilities Indexed d

Definition 7.57 State s is strongly GMR_d ($SGMR_d$) for DM i, denoted by $s \in S_i^{SGMR_d}$, iff for every $s_1 \in R_i^{+,++}(s)$ there exists at least one $s_2 \in R_{N\setminus\{i\}}(s_1)$ such that $s_2 \in \Phi_i^{--,U}(s)$.

Definition 7.58 State s is strongly SMR_d ($SSMR_d$) for DM i, denoted by $s \in S_i^{SSMR_d}$, iff for every $s_1 \in R_i^{+,++}(s)$ there exists at least one $s_2 \in R_{N\setminus\{i\}}(s_1)$, such that $s_2 \in \Phi_i^{--,U}(s)$ and $s_3 \in \Phi_i^{--,U}(s)$ for all $s_3 \in R_i(s_2)$.

Definition 7.59 State s is strongly SEQ_d ($SSEQ_d$) for DM i, denoted by $s \in S_i^{SSEQ_d}$, iff for every $s_1 \in R_i^{+,++}(s)$ there exists at least one $s_2 \in R_{N\setminus\{i\}}^{+,++,U}(s_1)$ such that $s_2 \in \Phi_i^{--,U}(s)$.

Note that Definitions 7.48–7.59 will reduce to Definitions 7.17–7.28 in Sect. 7.2.1 if $n = 2$.

7.2.3.3 Weak Stabilities Indexed l for Hybrid Preference

Definition 7.60 Let $s \in S$ and $i \in N$. State s is weakly stable WGS_l for DM i according to stability WGS indexed l, denoted by $s \in S_i^{WGS_l}$, iff $s \in S_i^{GGS_l}$ and $s \notin S_i^{SGS_l}$.

7.2.4 Interrelationships Among Stabilities Under Hybrid Preference

Fang et al. (1993) established general relationships among Nash, GMR, SMR, and SEQ solution concepts. The following interrelationships are similar to those developed by Fang et al. (1993) among the solution concepts with simple preference. Let $l \in \{a, b, c, d\}$. The inclusion relationships among the four solution concepts indexed as l with preference uncertainty are shown in Fig. 5.5. Within the new structure of preference, the interrelationships of three types of stabilities, which are with preferences of general strength and uncertainty, with preferences of strong strength and uncertainty, and with preferences of weak strength and uncertainty, are presented as follows:

$$S_i^{GGMR_l} = S_i^{SGMR_l} \cup S_i^{WGMR_l},$$

$$S_i^{GSMR_l} = S_i^{SSMR_l} \cup S_i^{WSMR_l},$$

and

$$S_i^{GSEQ_l} = S_i^{SSEQ_l} \cup S_i^{WSEQ_l}.$$

Based on the above definitions, the interrelationships among the solution concepts with the new preference structure are concluded as follows:

Theorem 7.1 *The interrelationships among the four solution concepts are*

$$S_i^{Nash_l} \subseteq S_i^{SSMR_l} \subseteq S_i^{GSMR_l} \subseteq S_i^{GGMR_l}$$

and

$$S_i^{Nash_l} \subseteq S_i^{SSEQ_l} \subseteq S_i^{GSEQ_l} \subseteq S_i^{GGMR_l}.$$

Theorem 7.2 *The interrelationships among the solution concepts are*

$$S_i^{Nash_l} \subseteq S_i^{LSMR_l} \subseteq S_i^{LGMR_l}$$

and

$$S_i^{Nash_l} \subseteq S_i^{LSEQ_l} \subseteq S_i^{LGMR_l},$$

where L denotes the two levels of general strength and strong strength.

The proof of Theorems 7.1 and 7.2 can easily follow from the above definitions. Note that there is no necessary inclusion relationship between $S_i^{LSMR_l}$ and $S_i^{LSEQ_l}$, i.e., it may or may not be true that $S_i^{LSMR_l} \supseteq S_i^{LSEQ_l}$, or that $S_i^{LSMR_l} \subseteq S_i^{LSEQ_l}$.

Theorem 7.3 *The interrelationships among Nash stabilities for the two levels of general strength and strong strength are*

$$S_i^{Nash_a} = S_i^{Nash_c},$$

$$S_i^{Nash_b} = S_i^{Nash_d},$$

and

$$S_i^{Nash_a} \subseteq S_i^{Nash_b}.$$

This result holds obviously from the above stability definitions.

Theorem 7.4 *The interrelationships among the solution concepts with preferences of general strength and uncertainty are*

$$S_i^{LGMR_a} \subseteq S_i^{LGMR_b} \subseteq S_i^{LGMR_d},$$

$$S_i^{LGMR_a} \subseteq S_i^{LGMR_c} \subseteq S_i^{LGMR_d},$$

$$S_i^{LSMR_a} \subseteq S_i^{LSMR_b} \subseteq S_i^{LSMR_d},$$

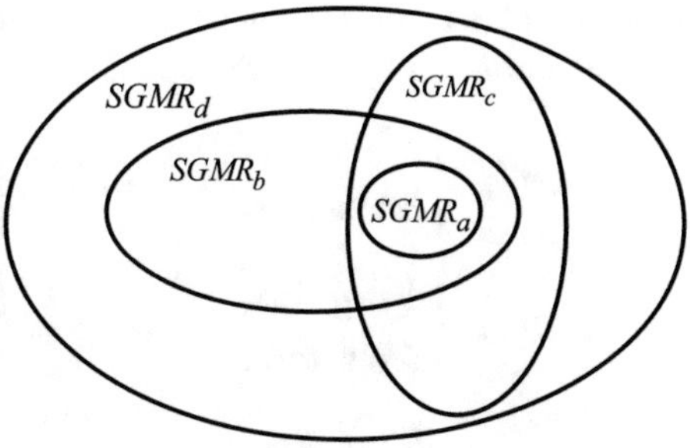

Fig. 7.2 Interrelationships among strong GMR stabilities indexed as a, b, c and d

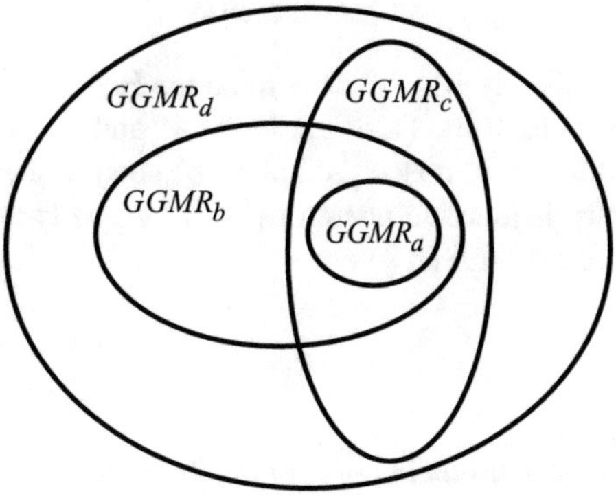

Fig. 7.3 Interrelationships among general GMR stabilities indexed as a, b, c and d

$$S_i^{LSMR_a} \subseteq S_i^{LSMR_c} \subseteq S_i^{LSMR_d},$$

$$S_i^{LSEQ_a} \subseteq S_i^{LSEQ_b} \subseteq S_i^{LSEQ_d}$$

and

$$S_i^{LSEQ_a} \subseteq S_i^{LSEQ_c} \subseteq S_i^{LSEQ_d}.$$

The proof of this theorem is similar to that presented by Li et al. (2005a, b). Figures 7.2 and 7.3 show the interrelationships among strong GMR stabilities indexed as a, b, c and d and among general GMR stabilities indexed as a, b, c and d, respectively.

7.3 Some Important Matrices Under Hybrid Preference

7.3.1 Preference Matrices Including Uncertainty and Strength

Let m denote the number of states. DM i's preference matrices including uncertainty and strength in the hybrid system are defined as follows.

Definition 7.61 For a graph model G under hybrid preference, the appropriate preference matrices for DM i are a set of $m \times m$ matrices with (s, q) entries

$$P_i^U(s,q) = \begin{cases} 1 \text{ if } s\ U_i\ q, \\ 0 \text{ otherwise,} \end{cases} \quad P_i^{=}(s,q) = \begin{cases} 1 \text{ if } s \sim_i q \text{ and } s \neq q, \\ 0 \text{ otherwise,} \end{cases}$$

$$P_i^{+}(s,q) = \begin{cases} 1 \text{ if } q >_i s, \\ 0 \text{ otherwise,} \end{cases} \quad P_i^{-}(s,q) = \begin{cases} 1 \text{ if } s >_i q, \\ 0 \text{ otherwise,} \end{cases}$$

$$P_i^{++}(s,q) = \begin{cases} 1 \text{ if } q \gg_i s, \\ 0 \text{ otherwise,} \end{cases} \quad P_i^{--}(s,q) = \begin{cases} 1 \text{ if } s \gg_i q, \\ 0 \text{ otherwise.} \end{cases}$$

Define $W = M \circ G$ to be the Hadamard product of two $m \times m$ matrices M and G, i.e., if the (s, q) entries of M and G are $M(s, q)$ and $G(s, q)$, respectively, then the $m \times m$ matrix W has (s, q) entry $W(s, q) = M(s, q) \cdot G(s, q)$. As well, define the disjunction operator ("or") "$\vee$" on two matrices: $B = M \vee G$ to be the $m \times m$ matrix with (s, q) entry

$$B(s,q) = \begin{cases} 1 \text{ if } M(s,q) + G(s,q) \neq 0, \\ 0 \text{ otherwise.} \end{cases}$$

The following $m \times m$ matrices are important in stability definitions under hybrid preference. Let E denote the $m \times m$ matrix with each entry 1 and let I be the $m \times m$ unit matrix. Then,

$$P_i^{--,=} = P_i^{--} \vee P_i^{=}, P_i^{--,U} = P_i^{--} \vee P_i^{U}, P_i^{--,-,=} = P_i^{--,=} \vee P_i^{-},$$

$$P_i^{+,++} = P_i^{+} \vee P_i^{++}, P_i^{+,++,U} = P_i^{+,++} \vee P_i^{U}, P_i^{--,-,=,U} = E - I - P_i^{+,++}. \quad (7.1)$$

It is well-known that matrices can efficiently describe adjacency of vertices, and incidence of arcs and vertices, in a graph, thereby permitting tracking of paths between any two vertices (Godsil and Royle 2001). Matrices possess useful algebraic properties that can be exploited to produce improved algorithms for solving graph problems. For instance, extensive research has been conducted to design effective algorithms and efficient search procedures using relationships between matrices and paths (Hoffman and Schiebe 2001). Here, the adjacency matrix is extended to a graph model with hybrid preference.

Definition 7.62 For a graph model G, DM i's adjacency matrix is the $m \times m$ matrix J_i with (s, q) entries

$$J_i(s,q) = \begin{cases} 1 \text{ if } (s,q) \in A_i, \\ 0 \text{ otherwise.} \end{cases}$$

It follows the set of matrices that play important roles in matrix representation of stabilities under hybrid preference. Let $i \in N$, then

$$J_i^{+,++} = J_i \circ P_i^{+,++} \text{and } J_i^{+,++,U} = J_i \circ P_i^{+,++,U}. \tag{7.2}$$

Note that the corresponding preference matrices in Eq. 7.2 are defined in Eq. 7.1.

7.3.2 *Reachability Matrices Under Hybrid Preference*

Preference matrices and reachability matrices are two important components of the matrix representation of the graph model. Under hybrid preference, preference matrices are defined in Sect. 7.3.1 while reachability matrices are defined now.

Definition 7.63 For the graph model G, the UM reachability matrix and the MSUIUM reachability matrix for H are the $m \times m$ matrices M_H and $M_H^{+,++,U}$ with (s, q) entries

$$M_H(s,q) = \begin{cases} 1 \text{ if } q \in R_H(s), \\ 0 \text{ otherwise,} \end{cases}$$

and

$$M_H^{+,++,U}(s,q) = \begin{cases} 1 \text{ if } q \in R_H^{+,++,U}(s), \\ 0 \text{ otherwise,} \end{cases}$$

respectively.

It is clear that $R_H(s) = \{q : M_H(s,q) = 1\}$ and $R_H^{+,++,U}(s) = \{q : M_H^{+,++,U}(s, q) = 1\}$. If $R_H(s)$ and $R_H^{+,++,U}(s)$ are written as 0–1 row vectors, then

$$R_H(s) = e_s^T \cdot M_H \text{ and } R_H^{+,++,U}(s) = e_s^T \cdot M_H^{+,++,U},$$

where e_s^T denotes the transpose of the sth standard basis vector of the m-dimensional Euclidean space. Therefore, the reachability matrices for coalition H, M_H and $M_H^{+,++,U}$, can be used to construct the reachable lists of H from state s, $R_H(s)$ and $R_H^{+,++,U}(s)$.

Fix $H \subseteq N$ such that $|H| \geq 2$, and let $s \in S$. One now demonstrates how to find matrices M_H and $M_H^{+,++,U}$ corresponding to $R_H(s)$ and $R_H^{+,++,U}(s)$ which are the reachable list of H from s by legal sequences of UMs and the reachable list of H from s by legal sequences of MSUIUMs, respectively.

Definition 7.64 For $i \in N$, $H \subseteq N$, and $t = 1, 2, 3, \cdots$, define the $m \times m$ matrices $M_i^{(t)}$ and $M_i^{(t,+,++,U)}$ with (s, q) entries as follows:

$$M_i^{(t)}(s,q) = \begin{cases} 1 \text{ if } q \in S \text{ is reachable from } s \in S \text{ in exactly} \\ \quad t \text{ legal UMs by } H \text{ with last mover DM } i, \\ 0 \text{ otherwise,} \end{cases}$$

and

$$M_i^{(t,+,++,U)}(s,q) = \begin{cases} 1 \text{ if } q \in S \text{ is reachable from } s \in S \text{ in exactly} \\ \quad t \text{ legal MSUIUMs by } H \text{ with last mover DM } i, \\ 0 \text{ otherwise.} \end{cases}$$

Based on Definition 7.64, one has:

Lemma 7.1 *For $i \in N$ and $H \subseteq N$, the two matrices $M_i^{(t)}$ and $M_i^{(t,+,++,U)}$ satisfy*

$$M_i^{(1)} = J_i \text{ and, for } t = 2, 3, \ldots, M_i^{(t)} = sign\left[\left(\bigvee_{j \in H \setminus \{i\}} M_j^{(t-1)}\right) \cdot J_i\right], \quad (7.3)$$

$$M_i^{(1,+,++,U)} = J_i^{+,++,U} \text{ and, for } t = 2, 3, \ldots,$$

$$M_i^{(t,+,++,U)} = sign\left[\left(\bigvee_{j \in H \setminus \{i\}} M_j^{(t-1,+,++,U)}\right) \cdot J_i^{+,++,U}\right]. \quad (7.4)$$

In Lemma 7.1, the sign function maps an $m \times m$ matrix with (s, q) entry $M(s, q)$ to the $m \times m$ matrix

$$sign[M(s,q)] = \begin{cases} 1 & M(s,q) > 0, \\ 0 & M(s,q) = 0, \\ -1 & M(s,q) < 0. \end{cases}$$

Generally, if there is no new appropriate arc produced, then the corresponding unilateral moves will stop. Therefore, the following Lemma 7.2 is easy to prove. Let $L_1 = |\bigcup_{i \in H} A_i|$ denote the number of UM arcs and $L_2 = |\bigcup_{i \in H} A_i^{+,++,U}|$ be the number of MSUIUM arcs in the following lemma and theorem.

Lemma 7.2 *For the graph model G, let $H \subseteq N$. $R_H(s)$ and $R_H^{+,++,U}(s)$ are the reachable lists of H by the legal sequences of UMs and MSUIUMs from s. δ_1 and δ_2 are the numbers of iteration steps required to find $R_H(s)$ and $R_H^{+,++,U}(s)$, respectively. Then*

$$\delta_1 \leq L_1 \text{ and } \delta_2 \leq L_2.$$

The following theorem can be derived using Lemmas 7.1 and 7.2.

Theorem 7.5 *Let $s \in S$, $H \subseteq N$, and $H \neq \emptyset$. The reachability matrices M_H and $M_H^{+,++,U}$ by H can be respectively expressed by*

$$M_H = \bigvee_{t=1}^{L_1} \bigvee_{i \in H} M_i^{(t)} \quad (7.5)$$

and

$$M_H^{+,++,U} = \bigvee_{t=1}^{L_2} \bigvee_{i \in H} M_i^{(t,+,++,U)}. \tag{7.6}$$

Proof The proofs of Eqs. 7.5 and 7.6 are similar. To prove Eq. 7.6, assume that $Q = \bigvee_{t=1}^{L_2} \bigvee_{i \in H} M_i^{(t,+,++,U)}$. Based on Definition 7.63, $M_H^{+,++,U}(s,q) = 1$ iff $q \in R_H^{+,++,U}(s)$. Since $L_2 = |\bigcup_{i \in H} A_i^{+,++,U}|$, then, using Lemma 7.2, $L_2 \geq \delta_2$. Therefore, by Definition 7.64, $q \in R_H^{+,++,U}(s)$ implies that there exists $1 \leq t_0 \leq \delta_2$ and $i_0 \in H$ such that $M_{i_0}^{(t_0,+,++,U)}(s,q) = 1$. This implies that matrix Q has (s,q) entry 1. Therefore, $M_H^{+,++,U}(s,q) = 1$ iff $Q(s,q) = 1$. Since $M_H^{+,++,U}$ and Q are 0–1 matrices, it follows that $M_H^{+,++,U} = Q = \bigvee_{t=1}^{L_2} \bigvee_{i \in H} M_i^{(t,+,++,U)}$. □

7.4 Matrix Representation of Stabilities Under Hybrid Preference

7.4.1 Matrix Representation of General Stabilities

In this section, let $m = |S|$ be the number of the states in S, and let $i \in N$ and $s \in S$. The algebraic representation of general graph model stabilities (GGSs) under hybrid preference is incorporated into the set of $m \times m$ matrices given in Table 7.1, $M_i^{GGS_l}$ for $l \in D = \{a, b, c, d\}$, which captures GGS_l stabilities for DM $i \in N$, where GGS_l represents $GGMR_l$, $GSMR_l$, or $GSEQ_l$ stability. Here, DMs' preferences may be hybrid. For example, the $m \times m$ matrix $M_i^{GGMR_a}$ representing $GGMR_a$ stability matrix shown in Table 7.1 is

$$M_i^{GGMR_a} = J_i^{+,++,U} \cdot [E - sign\left(M_{N\setminus\{i\}} \cdot (P_i^{--,-,=})^T\right)],$$

where E denotes the $m \times m$ matrix with each entry 1 and $M_{N\setminus\{i\}}$ is calculated using Eq. 7.5 when $H = N \setminus \{i\}$.

Define DM i's $m \times m$ $Nash_a$ and $Nash_c$ stability matrices as

$$M_i^{Nash_a} = M_i^{Nash_c} = J_i^{+,++,U} \cdot E.$$

The following theorem establishes the algebraic method to assess whether state s is $Nash_a$ or $Nash_c$ stable for a DM.

Theorem 7.6 *State $s \in S$ is $Nash_a$ or $Nash_c$ stable for DM i iff $M_i^{Nash_a}(s,s) = 0$.*

Note that $Nash_a$ is equivalent to $Nash_c$, which is proven by Xu et al. (2010a). Similarly, define DM i's $m \times m$ $Nash_b$ and $Nash_d$ stability matrices as

Table 7.1 General stability matrices under hybrid preference

Preference	Definition set	Stability matrices
Including strength and uncertainty	a	$M_i^{Nash_a} = J_i^{+,++,U} \cdot E$
		$M_i^{GGMR_a} = J_i^{+,++,U} \cdot [E - sign\left(M_{N\setminus\{i\}} \cdot (P_i^{--,-,=})^T\right)]$
		$M_i^{GSMR_a} = J_i^{+,++,U} \cdot [E - sign(M_{N\setminus\{i\}} \cdot Q)]$, with $Q = (P_i^{--,-,=})^T \circ [E - sign(J_i \cdot (P_i^{+,++,U})^T)]$
		$M_i^{GSEQ_a} = J_i^{+,++,U} \cdot [E - sign\left(M_{N\setminus\{i\}}^{+,++,U} \cdot (P_i^{--,-,=})^T\right)]$
	b	$M_i^{Nash_b} = J_i^{+,++} \cdot E$
		$M_i^{GGMR_b} = J_i^{+,++} \cdot [E - sign\left(M_{N\setminus\{i\}} \cdot (P_i^{--,-,=})^T\right)]$
		$M_i^{GSMR_b} = J_i^{+,++} \cdot [E - sign(M_{N\setminus\{i\}} \cdot Q)]$, with $Q = (P_i^{--,-,=})^T \circ [E - sign(J_i \cdot (P_i^{+,++,U})^T)]$
		$M_i^{GSEQ_b} = J_i^{+,++} \cdot [E - sign\left(M_{N\setminus\{i\}}^{+,++,U} \cdot (P_i^{--,-,=})^T\right)]$
	c	$M_i^{Nash_c} = J_i^{+,++,U} \cdot E$
		$M_i^{GGMR_c} = J_i^{+,++,U} \cdot [E - sign\left(M_{N\setminus\{i\}} \cdot (P_i^{--,-,=,U})^T\right)]$
		$M_i^{GSMR_c} = J_i^{+,++,U} \cdot [E - sign(M_{N\setminus\{i\}} \cdot Q)]$, with $Q = (P_i^{--,-,=,U})^T \circ [E - sign\left(J_i \cdot (P_i^{+,++})^T\right)]$
		$M_i^{GSEQ_c} = J_i^{+,++,U} \cdot [E - sign\left(M_{N\setminus\{i\}}^{+,++,U} \cdot (P_i^{--,-,=,U})^T\right)]$
	d	$M_i^{Nash_d} = J_i^{+,++} \cdot E$
		$M_i^{GGMR_d} = J_i^{+,++} \cdot [E - sign\left(M_{N\setminus\{i\}} \cdot (P_i^{--,-,=,U})^T\right)]$
		$M_i^{GSMR_d} = J_i^{+,++} \cdot [E - sign(M_{N\setminus\{i\}} \cdot Q)]$, with $Q = (P_i^{--,-,=,U})^T \circ [E - sign\left(J_i \cdot (P_i^{+,++})^T\right)]$
		$M_i^{GSEQ_d} = J_i^{+,++} \cdot [E - sign\left(M_{N\setminus\{i\}}^{+,++,U} \cdot (P_i^{--,-,=,U})^T\right)]$

$$M_i^{Nash_b} = M_i^{Nash_d} = J_i^{+,++} \cdot E,$$

then the following theorem holds.

Theorem 7.7 *State $s \in S$ is $Nash_b$ or $Nash_d$ stable for DM i iff $M_i^{Nash_b}(s, s) = 0$.*

Based on the stability matrix $M_i^{GGS_l}$ for $l \in D = \{a, b, c, d\}$, one has:

Theorem 7.8 *State $s \in S$ is GGS_l stable for DM i iff $M_i^{GGS_l}(s, s) = 0$.*

Theorem 7.8 contains twelve matrix representations of general stabilities that are GGMR, GSMR, and GSEQ with index a, b, c and d, respectively. The proofs of these results are similar, so one of the matrix representations is proven as follows. Note that the following proof applies to the case GGS = GSEQ, and $l = a$. In all other cases, the proof is analogous. Define DM i's $m \times m$ $GSEQ_a$ stability matrix as

$$M_i^{GSEQ_a} = J_i^{+,++,U} \cdot [E - sign\left(M_{N\setminus\{i\}}^{+,++,U} \cdot (P_i^{--,-,=})^T\right)].$$

Corollary 7.1 *State $s \in S$ is $GSEQ_a$ stable for DM i iff*

$$M_i^{GSEQ_a}(s, s) = 0. \tag{7.7}$$

Proof Equation 7.7 is equivalent to

$$(e_s^T \cdot J_i^{+,++,U}) \cdot [\left(E - sign\left(M_{N\setminus\{i\}}^{+,++,U} \cdot (P_i^{--,-,=})^T\right)\right) \cdot e_s] = 0.$$

Since $(e_s^T \cdot J_i^{+,++,U}) \cdot [\left(E - sign\left(M_{N\setminus\{i\}}^{+,++,U} \cdot (P_i^{--,-,=})^T\right)\right) \cdot e_s]$

$$= \sum_{s_1=1}^{m} J_i^{+,++,U}(s, s_1)[1 - sign\left((e_{s_1}^T \cdot M_{N\setminus\{i\}}^{+,++,U}) \cdot (e_s^T \cdot P_i^{--,-,=})^T\right)],$$

then Eq. 7.7 holds iff

$$J_i^{+,++,U}(s, s_1)[1 - sign\left((e_{s_1}^T \cdot M_{N\setminus\{i\}}^{+,++,U}) \cdot (e_s^T \cdot P_i^{--,-,=})^T\right)] = 0, \forall s_1 \in S. \tag{7.8}$$

It is clear that Eq. 7.8 is equivalent to

$$(e_{s_1}^T \cdot M_{N\setminus\{i\}}^{+,++,U}) \cdot (e_s^T \cdot P_i^{--,-,=})^T \neq 0, \forall s_1 \in R_i^{+,++,U}(s). \tag{7.9}$$

Based on Definitions 7.61 and 7.63, Eq. 7.9 implies that for any $s_1 \in R_i^{+,++,U}(s)$, there exists at least one $s_2 \in R_{N\setminus\{i\}}^{+,++,U}(s_1)$ with $s_2 \ll_i s$, $s_2 <_i s$, or $s_2 \sim_i s$, which is exactly the logical definition of general SEQ stability with index a presented in Definition 7.35 and by Xu et al. (2010a).

Hence, state s is $GSEQ_a$ stable for DM i iff $M_i^{GSEQ_a}(s, s) = 0$. □

If the binary relation $\succ$ denotes $>$ or $\gg$ in this chapter, i.e., $s \succ q$ iff either $s > q$ or $s \gg q$, then Theorems 7.6–7.8 are identical with the matrix representation of stabilities with preference uncertainty (Xu et al. 2010a) and equivalent to the results presented using logical representation under unknown preference (Li et al. 2004). On the other hand, when each DM does not consider including uncertain preference in stability analysis, Theorems 7.6–7.8 reduce to the general stabilities described using matrix representation (Xu et al. 2010a) and logical representation (Hamouda et al. 2006) under strength of preference. If a graph model contains neither strength nor uncertainty of preference, Theorems 7.6–7.8 reduce to the stabilities in matrix form (Xu et al. 2010a) and in logical form (Fang et al. 1993) under simple preference.

With the hybrid preference structure introduced into the graph model, general stable states can be classified into strongly stable or weakly stable according to strength of the possible sanctions and indexed a, b, c, or d by a DM's attitudes toward the risk associated with uncertain preference. Strong and weak stabilities include only GMR, SMR, and SEQ because Nash stability does not involve sanctions.

7.4.2 Matrix Representation of Strong and Weak Stabilities

If a particular state s is general stable, then s is either strongly stable or weakly stable. The algebraic system to represent matrix representation of strong graph model stabilities (SGSs) under hybrid preference is incorporated into the set of $m \times m$ matrices shown in Table 7.2. The set of matrices $M_i^{SGS_l}$ for $l \in D = \{a, b, c, d\}$ captures SGS_l stabilities for DM $i \in N$, where SGS_l represents $SGMR_l$, $SSMR_l$, or $SSEQ_l$ stability, and DMs' preference may include hybrid preference.

Based on the stability matrix $M_i^{SGS_l}$ for $l \in D = \{a, b, c, d\}$, one has:

Theorem 7.9 *State $s \in S$ is SGS_l stable for DM i iff $M_i^{SGS_l}(s, s) = 0$.*

Theorem 7.9 contains twelve matrix representations of strong stabilities that are SGMR, SSMR, and SSEQ with index a, b, c and d, respectively. Note that the following corollary applies to the case SGS=SSMR and $l = c$. In all other cases, the representation and the corresponding proof are analogous. Define DM i's $m \times m$ $SSMR_c$ stability matrix

$$M_i^{SSMR_c} = J_i^{+,++,U} \cdot [E - sign(M_{N\setminus\{i\}} \cdot Q)], \ with$$

$$Q = (P_i^{--,U})^T \circ [E - sign\left(J_i \cdot (E - P_i^{++,U})\right)].$$

Corollary 7.2 *State $s \in S$ is $SSMR_c$ stable for DM i iff*

$$M_i^{SSMR_c}(s, s) = 0. \tag{7.10}$$

Table 7.2 Strong stability matrices under hybrid preference

Preference	Definition set	Stability matrices
Including strength and uncertainty	a	$M_i^{Nash_a} = J_i^{+,++,U} \cdot E$
		$M_i^{SGMR_a} = J_i^{+,++,U} \cdot [E - sign\left(M_{N\setminus\{i\}} \cdot (P_i^{--})^T\right)]$
		$M_i^{SSMR_a} = J_i^{+,++,U} \cdot [E - sign(M_{N\setminus\{i\}} \cdot Q)]$, with $Q = (P_i^{--})^T \circ [E - sign\big(J_i \cdot (E - P_i^{++})\big)]$
		$M_i^{SSEQ_a} = J_i^{+,++,U} \cdot [E - sign\left(M_{N\setminus\{i\}}^{+,++,U} \cdot (P_i^{--})^T\right)]$
	b	$M_i^{Nash_b} = J_i^{+,++} \cdot E$
		$M_i^{SGMR_b} = J_i^{+,++} \cdot [E - sign\left(M_{N\setminus\{i\}} \cdot (P_i^{--})^T\right)]$
		$M_i^{SSMR_b} = J_i^{+,++} \cdot [E - sign(M_{N\setminus\{i\}} \cdot Q)]$, with $Q = (P_i^{--})^T \circ [E - sign\big(J_i \cdot (E - P_i^{++})\big)]$
		$M_i^{SSEQ_b} = J_i^{+,++} \cdot [E - sign\left(M_{N\setminus\{i\}}^{+,++,U} \cdot (P_i^{--})^T\right)]$
	c	$M_i^{Nash_c} = J_i^{+,++,U} \cdot E$
		$M_i^{SGMR_c} = J_i^{+,++,U} \cdot [E - sign\left(M_{N\setminus\{i\}} \cdot (P_i^{--,U})^T\right)]$
		$M_i^{SSMR_c} = J_i^{+,++,U} \cdot [E - sign(M_{N\setminus\{i\}} \cdot Q)]$, with $Q = (P_i^{--,U})^T \circ [E - sign\left(J_i \cdot (E - P_i^{++,U})\right)]$
		$M_i^{SSEQ_c} = J_i^{+,++,U} \cdot [E - sign\left(M_{N\setminus\{i\}}^{+,++,U} \cdot (P_i^{--,U})^T\right)]$
	d	$M_i^{Nash_d} = J_i^{+,++} \cdot E$
		$M_i^{SGMR_d} = J_i^{+,++} \cdot [E - sign\left(M_{N\setminus\{i\}} \cdot (P_i^{--,U})^T\right)]$
		$M_i^{SSMR_d} = J_i^{+,++} \cdot [E - sign(M_{N\setminus\{i\}} \cdot Q)]$, with $Q = (P_i^{--,U})^T \circ [E - sign\left(J_i \cdot (E - P_i^{++,U})\right)]$
		$M_i^{SSEQ_d} = J_i^{+,++} \cdot [E - sign\left(M_{N\setminus\{i\}}^{+,++,U} \cdot (P_i^{--,U})^T\right)]$

Proof Since the diagonal element of matrix $M_i^{SSMR_c}$

$$M_i^{SSMR_c}(s,s) = (e_s^T \cdot J_i^{+,++,U}) \cdot [(E - sign(M_{N\backslash\{i\}} \cdot Q)) \cdot e_s]$$

$$= \sum_{s_1=1}^{m} J_i^{+,++,U}(s,s_1)[1 - sign\left((M_{N\backslash\{i\}} \cdot Q)(s_1,s)\right)]$$

with

$$(M_{N\backslash\{i\}} \cdot Q)(s_1,s) = \sum_{s_2=1}^{m} M_{N\backslash\{i\}}(s_1,s_2) \cdot P_i^{--,U}(s,s_2)[1 - sign\left(\sum_{s_3=1}^{m}\left(J_i(s_2,s_3)(1 - P_i^{++,U}(s_3,s))\right)\right)],$$

then, $M_i^{SSMR_c}(s,s) = 0$ iff $(M_{N\backslash\{i\}} \cdot Q)(s_1,s) \neq 0$ for any $s_1 \in R_i^{+,++,U}(s)$, which is equivalent to the equation that, for any $s_1 \in R_i^{+,++,U}(s)$ there exists $s_2 \in R_{N\backslash\{i\}}(s_1)$ such that

$$P_i^{--,U}(s,s_2) \neq 0, \quad \text{and} \quad \sum_{s_3=1}^{m}\left(J_i(s_2,s_3) \cdot (1 - P_i^{++,U}(s_3,s)\right) = 0. \tag{7.11}$$

Obviously, for any $s_1 \in R_i^{+,++,U}(s)$ there exists $s_2 \in R_{N\backslash\{i\}}(s_1)$ such that Eq. 7.11 holds iff for every $s_1 \in R_i^{+,++,U}(s)$ there exists $s_2 \in R_{N\backslash\{i\}}(s_1)$ with $s \gg_i s_2$ or $s\ U_i\ s_2$ and $s \gg_i s_3$ or $s\ U_i\ s_3$ for all $s_3 \in R_i(s_2)$, which is exactly the logical definition for strong SMR stability with index c presented in Xu et al. (2010a).

Therefore, state s is $SSMR_c$ stable for DM i iff $M_i^{SSMR_c}(s,s) = 0$. □

Under hybrid preference, the interrelationships of general stabilities, strong stabilities, and weak stabilities are as follows:

Theorem 7.10 *Let $l = a, b, c,$ or d and $i \in N$. The interrelationships among general stability (GGS), strong stability (SGS), and weak stability (WGS) with index l for DM i are*

$$S_i^{WGS_l} = S_i^{GGS_l} - S_i^{SGS_l}.$$

For $n = 2$, the above theorems degenerate to those theorems shown in Xu et al. (2011).

From the matrix representation of general, strong, and weak stabilities, each of which is indexed by a, b, c, and d, presented above, it can be seen that stability (strong or weak) with index a constitutes strong or weak stability for the most aggressive DMs. First, the DM is aggressive in deciding whether to move from the status quo, in that he or she is willing to accept the risk associated with moves to states of unknown preference. In addition, when evaluating possible moves, the DM is deterred only by sanctions to states that are determinately less preferred than the status quo, and does not consider states of uncertain preference (relative to the status quo) to be sanctions. For the definitions indexed b, uncertainty in preferences is not considered by a DM.

The definitions indexed c incorporate a mixed attitude toward the risk associated with states of uncertain preference. Specifically, the DM is aggressive in deciding whether to move from the status quo, but is conservative when evaluating possible moves, being deterred by sanctions to states that are less preferred or have uncertain preference relative to the status quo. Finally, the definition indexed d represents stability for the most conservative DMs, who would move only to determinately preferred states from a status quo, but would be deterred by responses that result in states of uncertain preference.

7.5 Application

In this section, the algebraic approach developed in this chapter is applied to a practical problem—the Lake Gisborne conflict. As explained in Sect. 5.4, in 1995, a company called Canada Wet Incorporated in Newfoundland proposed a project to export bulk water from Lake Gisborne. The **Provincial Government** of Newfoundland and Labrador approved this project because of its potential economic benefits. Nonetheless, because of the risk of harmful impacts on local environment, a wide range of lobby groups opposed the proposal. The **Federal Government of Canada** supported the opposing groups and prohibited water exports. In view of the impoverished state, however, **several groups** supported the project, arguing to continue it. Thus, the Lake Gisborne conflict arose among the Federal Government of Canada, Provincial Government of Newfoundland and Labrador, and Support Groups. (See details in Fang et al. (2002) and Li et al. (2004)).

This conflict is modeled using three DMs and a total of three options that are shown as follows:

- Federal government of Canada (**Federal**): its only option is to continue a Canada wide accord on the prohibition of bulk water exports **(Continue)**, or not,
- Provincial government of Newfoundland and Labrador (**Provincial**): its only option is to lift the ban on bulk water exports **(Lift)**, or not, and
- Support groups (**Support**): its only option is to appeal for continuation of the Gisborne project **(Appeal)**, or not.

Because each option can either be selected (Y for yes) or not taken (N for no), there is a total of 2^3 possible states, $s_1, s_2, \cdots, s_8$, in the Lake Gisborne conflict. The results are presented in Table 7.3. One advantage of the graph model is its innate capability to systematically keep track of state transitions. State transition is the process by which a conflict moves from one state to another. If a DM can cause a state transition on his or her own, then this transition is called a unilateral move (UM) for that DM. Hence, the graph model of this conflict is depicted based on the eight feasible states in Fig. 7.4, in which a label on an arc indicates which DM controls the moves between the two states connected by the arc.

Since several groups supported the project, an economical-oriented provincial government might have considered supporting it because of the urgent need for cash.

Table 7.3 Feasible states for the Lake Gisborne model

Federal								
1. Continue	N	Y	N	Y	N	Y	N	Y
Provincial								
2. Lift	N	N	Y	Y	N	N	Y	Y
Support								
3. Appeal	N	N	N	N	Y	Y	Y	Y
State number	s_1	s_2	s_3	s_4	s_5	s_6	s_7	s_8

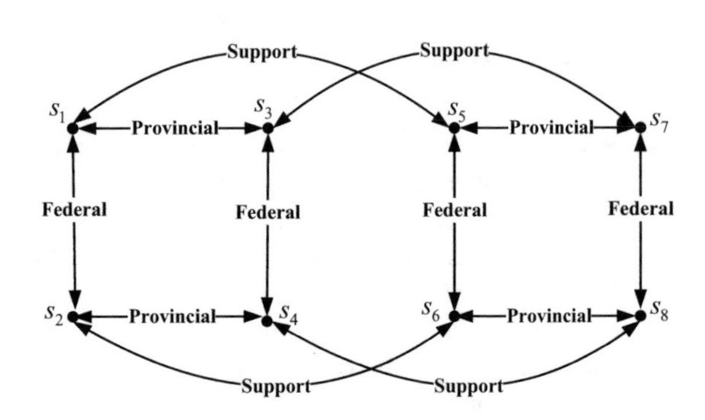

Fig. 7.4 Graph model for the Gisborne conflict

However, an environmental-oriented provincial government might have opposed it because of the possibility of devastating environmental consequences. In 1999, it was unclear which of these two different attitudes described the provincial government's thinking, resulting in uncertainty in preferences in the Gisborne conflict. The details can be found in Li et al. (2004). The graph model introduced by Li et al. (2004) is extended to include hybrid preferences of uncertainty and strength in the Gisborne dispute. The preference information for this conflict over the feasible states is given in Table 7.4. One assumes that state s_7 is strongly less preferred to all other states by the Federal Government, the Support Groups consider state s_2 to be strongly less preferred relative to all other states, and the Provincial Government strongly prefers state s_2 to state s_6. Note that DM **Provincial** only knows that it mildly prefers state s_3 to s_7, state s_4 to s_8, state s_1 to s_5, and strongly prefers state s_2 to s_6. It is obvious that DM **Provincial**'s preference information includes combining uncertainty and strength. Additionally, this representation of preference information presented in Table 7.4

Table 7.4 Certain preference information for the Gisborne model

DMs	Certain preferences
Federal	$s_2 > s_6 > s_4 > s_8 > s_1 > s_5 > s_3 \gg s_7$
Provincial	$s_3 > s_7, s_4 > s_8, s_1 > s_5, s_2 \gg s_6$, only
Support	$s_3 > s_4 > s_7 > s_8 > s_5 > s_6 > s_1 \gg s_2$

Table 7.5 Stability results of the Gisborne conflict with hybrid preferences

State		Nash				GGMR				GSMR				GSEQ				SGMR				SSMR				SSEQ				WGMR				WSMR				WSEQ			
		1	2	3	E	1	2	3	E	1	2	3	E	1	2	3	E	1	2	3	E	1	2	3	E	1	2	3	E	1	2	3	E	1	2	3	E	1	2	3	E
s_1	a																																								
	b		√				√				√				√				√				√				√														
	c						√				√				√				√				√				√														
	d		√				√				√				√				√				√				√														
s_2	a	√				√				√				√				√				√				√															
	b	√	√			√	√			√	√			√	√			√	√			√	√			√	√														
	c	√				√	√			√	√			√				√	√			√	√			√															
	d	√	√			√	√			√	√			√	√			√	√			√	√			√	√														
s_3	a			√				√				√				√				√				√				√													
	b		√	√			√	√			√	√			√	√			√	√			√	√			√	√													
	c			√			√	√			√	√			√	√			√	√			√	√			√	√													
	d		√	√			√	√			√	√			√	√			√	√			√	√			√	√													
s_4	a	√		√		√		√		√		√		√		√		√		√		√		√		√		√													
	b	√	√	√	√	√	√	√	√	√	√	√	√	√	√	√	√	√	√	√	√	√	√	√	√	√	√	√	√												
	c	√		√		√	√	√	√	√	√	√	√	√	√	√	√	√	√	√	√	√	√	√	√	√	√	√	√												
	d	√	√	√	√	√	√	√	√	√	√	√	√	√	√	√	√	√	√	√	√	√	√	√	√	√	√	√	√												
s_5	a			√				√				√				√				√				√				√													
	b		√	√			√	√			√	√			√	√			√	√			√	√			√	√													
	c			√			√	√			√	√			√	√			√	√			√	√			√	√													
	d		√	√			√	√			√	√			√	√			√	√			√	√			√	√													
s_6	a	√		√		√		√		√		√		√		√		√		√		√		√		√		√													
	b	√	√	√	√	√	√	√	√	√	√	√	√	√	√	√	√	√	√	√	√	√	√	√	√	√	√	√	√												
	c	√		√		√	√	√	√	√	√	√	√	√	√	√	√	√	√	√	√	√	√	√	√	√	√	√	√												
	d	√	√	√	√	√	√	√	√	√	√	√	√	√	√	√	√	√	√	√	√	√	√	√	√	√	√	√	√												
s_7	a							√				√				√				√								√								√					
	b		√				√	√			√	√			√	√			√	√			√				√	√								√					
	c						√	√			√	√			√	√			√	√			√				√	√								√					
	d		√				√	√			√	√			√	√			√	√			√				√	√								√					
s_8	a	√				√		√		√		√		√		√		√		√		√				√										√				√	
	b	√	√			√	√	√	√	√	√	√	√	√	√	√	√	√	√	√	√	√	√			√	√									√				√	
	c	√				√	√	√	√	√	√	√	√	√		√		√	√	√	√	√	√			√										√				√	
	d	√	√			√	√	√	√	√	√	√	√	√	√	√	√	√	√	√	√	√	√			√	√									√				√	

implies that the preferred relations, $>$ and $\gg$, are transitive. For instance, since $s_5 > s_3$ and $s_3 \gg s_7$, then $s_5 \gg s_7$, for DM **Federal**. However, in general, the preference structure presented in this book does not require the transitivity of preference relations and, hence, the developed results can be used to handle intransitive preferences.

In multiple decision maker graph models, the reachability matrices, M_H and $M_H^{+,++,U}$, are essential components of the matrix method for stability analysis under hybrid preference. Here, using the Gisborne model as an example, the procedures of constructing the reachability matrices are shown next.

Let $N = \{1, 2, 3\}$ and $H = N \setminus \{i\}$, $i = 1, 2, 3$.

- Construct preference matrices, $P_i^{+,++}$, $P_i^{++,U}$, $P_i^{+,++,U}$, P_i^{--}, and $P_i^{--,U}$, for $i = 1, 2,$ and 3, using information provided in Table 7.4, as well as $P^{--,-,=} = E - I - P_i^{+,++,U}$ and $P^{--,-,=,U} = E - I - P_i^{+,++}$;
- Construct DM i's adjacency matrix J_i and calculate matrices $J_i^{+,++}$ and $J_i^{+,++,U}$ for $i = 1, 2,$ and 3;
- Calculate matrices $M_i^{(t)}$ and $M_i^{(t,+,++,U)}$ using Lemma 7.1;
- Calculate the UM reachability matrix and the MSUIUM reachability matrix by H,

$$M_H = \bigvee_{t=1}^{L_1} \bigvee_{i \in H} M_i^{(t)}$$

and

$$M_H^{+,++,U} = \bigvee_{t=1}^{L_2} \bigvee_{i \in H} M_i^{(t,+,++,U)},$$

where $L_1 = |\bigcup_{i \in N} A_i| = 24$ and $L_2 = |\bigcup_{i \in N} A_i^{+,++,U}| = 16$;

- Construct the general stability and strong stability matrices using Tables 7.1 and 7.2;
- Analyze the general stabilities, $Nash_l$, $GGMR_l$, $GSMR_l$, and $GSEQ_l$ by Theorems 7.6–7.8, and calculate the strong stabilities, $SGMR_l$, $SSMR_l$, and $SSEQ_l$ using Theorem 7.9, as well as calculate the weak stabilities using Theorem 7.10;
- Present the stability results for the Gisborne conflict in Table 7.5.

7.6 Important Ideas

The fundamental design of GMCR permits the methodology to handle a rich range of preference structures for both transitive and intransitive preferences. In this chapter, both unknown (Chap. 5) and degree or strength of preference (Chap. 6) up to three degrees are combined as a type of the hybrid preference within the graph model paradigm. In this way, GMCR can simultaneously account for preference uncertainty and strength or degree of preference. More specifically, the main properties of the preference structure and reachable lists are introduced in Sect. 7.1 in this chapter. Because DMs make moves and countermoves when interacting with one another under conflict, reachable lists are used to keep track of the possible unilateral movements in one step from a given state for a particular DM with respect to hybrid preference. When considering stability definitions for more than two DMs, coalition moves are defined since two or more DMs can participate in blocking a unilateral improvement by another DM. Subsequently, four stability definitions consisting of Nash stability, general metarationality (GMR), symmetric metarationality (SMR), and sequential stability (SEQ), are defined for the graph model with this extended preference structure and relationships among them are investigated. Additionally, in this chapter, matrix representations of the four stabilities are presented for graph models having a preference structure of up to three degrees.

7.7 Problems

7.7.1 Describe a real-world application in which you believe one should simultaneously account for both unknown and degree or strength of preference.

7.7.2 A variety of approaches are available for modeling uncertain preferences within GMCR in different but complementary fashions such as unknown (Chap. 5), fuzzy (Hipel et al. 2011, Bashar et al. 2012, 2015, 2016, 2018), grey (Kuang et al. 2015a, b, Zhao and Xu 2017), and probabilistic preferences (Rego and dos Santos 2015). Qualitatively describe the type of uncertainty that each of these four approaches to preference uncertainty capture. Based on this, suggest meaningful combinations of uncertainty approaches for more comprehensively modeling uncertain preferences with GMCR. Discuss difficulties that may arise when considering other kinds of hybrid or combined preferences. Which combinations of uncertain preferences do you think could be meaningfully combined with degree of preference?

7.7.3 The normal form of Prisoner's Dilemma is given in Problem 3.5.1. Suggest a model of this generic conflict which requires both unknown preference plus degree of preference at level 3. Carry out a stability analysis using this hybrid preference based on the stability definitions for Nash and SEQ stability following a logical interpretation. Comment on any strategic insights that you detect in your stability results.

7.7.4 Analyze the Prisoner's Dilemma Problem specified in Problem 7.7.3 using the matrix interpretation of hybrid preferences.

7.7.5 The game of Chicken is presented in normal form in Problem 3.5.4. Create a version of this game which contains both unknown preference and three degrees of preference. Model and analyze this conflict using the hybrid preference approach furnished in this chapter using the logical definitions of stability for Nash and SEQ stability.

7.7.6 Model and analyze the game of Chicken mentioned in Problem 7.7.5 for your version of hybrid preference as it appears in Chicken using the matrix formulations given in this chapter.

7.7.7 The Elmira groundwater contamination conflict is described in Sects. 1.2.2 and 3.2, as well as elsewhere in the book. Develop what you think is a reasonable version of this conflict which contains hybrid preference consisting of unknown preference and three degrees of preference. Use either the logical or matrix methods of this chapter to carry out a full modeling and stability analysis. Discuss what you learned strategically about this conflict when using hybrid preferences.

7.7.8 A large-scale environmental problem over potential pollution caused by a proposed irrigation scheme called the Garrison Diversion Unit (GDU) is described in Sect. 6.6. Suggest a sensible hybrid version of this dispute in which both unknown preference and three degrees of preference are present. Use appropriate methods from this chapter to carry out a full conflict modeling and analysis of this dispute for the case of Nash and SEQ stability. Discuss interesting facts about your strategic findings.

7.7.9 A potential nuclear war between the two superpowers consisting of the USA and the USSR (Union of the Soviet Socialist Republics) is given in Problem 3.5.10. Develop a reasonable model in which the hybrid preference of this chapter is contained in the model. Carry out a strategic study using the ideas furnished in this chapter for the case of Nash and SEQ stability. Summarize and put into perspective your key strategic findings.

7.7.10 Find a current conflict which is of interest to you and which you think can be realistically modeled and analyzed using the hybrid preference ideas for modeling and analysis given in this chapter. After summarizing the background to this conflict, carry out a complete modeling and analysis of the dispute using the procedures from this chapter. Discuss your strategic findings especially with respect to interesting strategic insights.

References

Bashar, M. A., Kilgour, D. M., & Hipel, K. W. (2012). Fuzzy preferences in the graph model for conflict resolution. *IEEE Transactions on Fuzzy Systems*, *20*(4), 760–770. https://doi.org/10.1109/tfuzz.2012.2183603.

Bashar, M. A., Hipel, K. W., Kilgour, D. M., & Obeidi, A. (2015). Coalition fuzzy stability analysis in the graph model for conflict resolution. *Journal of Intelligent and Fuzzy Systems*, *29*(2), 593–607. https://doi.org/10.3233/ifs-141336.

Bashar, M. A., Obeidi, A., Kilgour, D. M., & Hipel, K. W. (2016). Modeling fuzzy and interval fuzzy preferences within a graph model framework. *IEEE Transactions on Fuzzy Systems*, *24*(4), 765–778. https://doi.org/10.1109/tfuzz.2015.2446536.

Bashar, M. A., Hipel, K. W., Kilgour, D. M., & Obeidi, A. (2018). Interval fuzzy preferences in the graph model for conflict resolution. *Fuzzy Optimization and Decision Making*. https://doi.org/10.1007/s10700-017-9279-7.

Fang, L., Hipel, K. W., & Kilgour, D. M. (1993). *Interactive decision making: The graph model for conflict resolution*. New York: Wiley. https://doi.org/10.2307/2583940.

Fang, L., Hipel, K. W., & Wang, L. (2002). Gisborne water export conflict study. In G. H. Schmiz (Ed.), *Proceedings of the Third International Conference on Water Resources and Environment Research* (Vol. 1, pp. 432–436). Dresden, Germany.

Godsil, C., & Royle, G. (2001). *Algebraic graph theory*. New York: Springer. https://doi.org/10.1007/978-1-4613-0163-9.

Hamouda, L., Kilgour, D. M., & Hipel, K. W. (2006). Strength of preference in graph models for multiple decision-maker conflicts. *Applied Mathematics and Computation*, *179*(1), 314–327. https://doi.org/10.1016/j.amc.2005.11.109.

Hipel, K. W., Kilgour, D. M., & Bashar, M. A. (2011). Fuzzy preferences in multiple participant decision making. *Scientia Iranica, Transactions D: Computer Science and Engineering and Electrical Engineering*, *18*(3), 627–638. https://doi.org/10.1016/j.scient.2011.04.016.

Hoffman, A. J., & Schiebe, B. (2001). The edge versus path incidence matrix of series-parallel graphs and greedy packing. *Discrete Applied Mathematics*, *113*, 275–284. https://doi.org/10.1016/s0166-218x(00)00294-8.

Kuang, H., Bashar, M. A., Hipel, K. W., & Kilgour, D. M. (2015a). Grey-based preference in a graph model for conflict resolution with multiple decision makers. *IEEE Transactions on Systems, Man and Cybernetics: Systems*, *45*(9), 1254–1267. https://doi.org/10.1109/tsmc.2014.2387096.

Kuang, H., Bashar, M. A., Kilgour, D. M., & Hipel, K. W. (2015b). Strategic analysis of a brownfield revitalization conflict using the grey-based graph model for conflict resolution. *EURO Journal on Decision Processes*, *3*(3), 219–248. https://doi.org/10.1007/s40070-015-0042-4.

Li, K. W., Hipel, K. W., Kilgour, D. M., & Fang, L. (2004). Preference uncertainty in the graph model for conflict resolution. *IEEE Transactions on Systems, Man, and Cybernetics, Part A, Systems and Humans*, *34*(4), 507–520. https://doi.org/10.1109/tsmca.2004.826282.

Li, K. W., Hipel, K. W., Kilgour, D. M., & Noakes, D. J. (2005a). Integrating uncertain preferences into status quo analysis with application to an environmental conflict. *Group Decision and Negotiation*, *14*(6), 461–479. https://doi.org/10.1007/s10726-005-9003-9.

Li, K. W., Kilgour, D. M., & Hipel, K. W. (2005b). Status quo analysis in the graph model for conflict resolution. *Journal of the Operational Research Society*, *56*, 699–707. https://doi.org/10.1057/palgrave.jors.2601870.

Rego, L. C., & dos Santos, A. M. (2015). Probabilistic preferences in the graph model for conflict resolution. *IEEE Transactions on Systems, Man, and Cybernetics: Systems*, *45*(4), 595–608. https://doi.org/10.1109/tsmc.2014.2379626.

Xu, H., Hipel, K. W., & Kilgour, D. M. (2008). Preference strength and uncertainty in the graph model for conflict resolution for two decision-makers. In *Proceedings of the 2008 IEEE International Conference on Systems, Man, and Cybernetics* (pp. 2907–2912). Singapore. https://doi.org/10.1109/icsmc.2008.4811739.

Xu, H., Hipel, K. W., Kilgour, D. M., & Chen, Y. (2010a). Combining strength and uncertainty for preferences in the graph model for conflict resolution with multiple decision makers. *Theory and Decision*, *69*(4), 497–521. https://doi.org/10.1007/s11238-009-9134-6.

Xu, H., Kilgour, D. M., & Hipel, K. W. (2010b). An integrated algebraic approach to conflict resolution with three-level preference. *Applied Mathematics and Computation*, *216*(3), 693–707. https://doi.org/10.1016/j.amc.2010.01.054.

Xu, H., Kilgour, D. M., & Hipel, K. W. (2010c). Matrix representation and extension of coalition analysis in group decision support. *Computers and Mathematics with Applications*, *60*(5), 1164–1176. https://doi.org/10.1016/j.camwa.2010.05.040.

Xu, H., Kilgour, D. M., & Hipel, K. W. (2011). Matrix representation of conflict resolution in multiple-decision-maker graph models with preference uncertainty. *Group Decision and Negotiation*, *20*(6), 755–779. https://doi.org/10.1007/s10726-010-9188-4.

Xu, H., Kilgour, D. M., Hipel, K. W., & McBean, E. A. (2013). Theory and application of conflict resolution with hybrid preference in colored graphs. *Applied Mathematical Modelling*, *37*(3), 989–1003. https://doi.org/10.1016/j.apm.2012.03.009.
Zhao, S., & Xu, H. (2017). Grey option prioritization for the graph model for conflict resolution. *Journal of Grey System*, *29*(3), 14–25.

Chapter 8
Coalitional Stabilities

Stability definitions for simple preference, unknown preference, degrees of preference, and hybrid preference (unknown combined with degree of preference) are presented in Chaps. 4–7, respectively. A typical stability analysis is built upon a noncooperative framework, with the underlying assumption being that each DM acts independently in its own self interest, after calculating moves and countermoves by its opponents. On the other hand, a coalitional analysis takes place in a cooperative framework, and assesses whether individual DMs can jointly improve their positions by forming a coalition (Kilgour et al. 2001, Inohara and Hipel 2008a, b, Xu et al. 2010, 2011, 2014). In fact, as emphasized in this book, after determining how well a DM can fare on his or her own by carrying out individual stability analyses, one should ascertain if a DM can do even better by cooperating with others via executing coalitional stability analyses, which is the focus of this chapter. Outside of Chap. 8, discussions regarding the importance of coalition investigations are put forward in Sect. 1.2.3 and portrayed in Fig. 1.5. Moreover, coalition modeling and analysis should be embedded as a key function of a decision support system for GMCR as explained in Sect. 10.2 and depicted in Figs. 10.2 and 10.4.

Coalition formation and stability analysis have long been active research areas in game theory (Aumann and Hart 1994, van Deeman 1997). The coalitional analysis considered in this book is confined to the Graph Model for Conflict Resolution (GMCR) paradigm. It assesses whether a subset of self-interested and independent DMs can gain by forming a coalition and coordinating their choices. The rationale is that a nonequilibrium state is not sustainable, because at least one DM can deviate from it in its own interest. An equilibrium, on the other hand, is expected to be sustainable, as no DM is motivated to depart from it. However, when a subset of DMs forms a coalition, an equilibrium may be upset via a sequence of joint moves by the coalition. In this case, the target state must also be an equilibrium, as any nonequilibrium state is transient. In Kilgour et al. (2001), this process is referred to

H. Xu et al., *Conflict Resolution Using the Graph Model: Strategic Interactions in Competition and Cooperation*, Studies in Systems, Decision and Control 153,
https://doi.org/10.1007/978-3-319-77670-5_8

as an "equilibrium jump". Understandably the target state of an equilibrium jump should make all members in the coalition better off and cannot be achieved by any DM acting individually. Coalition analysis, therefore, aims to alert the analyst that such a coalition exists and, if so, which equilibria are vulnerable to equilibrium jumps and how these jumps can be achieved by coalitional joint moves.

Coalition movements under various preference structures are introduced in Sect. 8.1. Subsequently, the logical representations of the four coalitional stability definitions, coalitional Nash stability, coalitional general metarationality, coalitional symmetric metarationality, and coalitional sequential stability, are defined in Sects. 8.2–8.5 under simple preference, unknown preference, three-level preference, and hybrid preference, respectively. Additionally, in this chapter, matrix representations of coalitional stabilities are presented in Sects. 8.6–8.9 for the four types of preference structures.

8.1 Coalition Movement Definitions

To define coalitional stabilities, concepts of *coalitional improvement* under various preferences must be introduced.

Definition 8.1 For a status quo state s and a nonempty coalition $H \subseteq N$, a state $s_1 \in R_H(s)$ is **a coalitional improvement** for H under simple preference from s, denoted by $s_1 \in CR_H^+(s)$, iff $s_1 \succ_i s$ for every $i \in H$.

It is worth noting that $CR_H^+(s) \neq R_H^+(s)$, as $R_H^+(s)$ denotes all states that are attainable by coalition H via legal sequences of UIs from s (see Definition 4.7). Although each individual move is a UI for the mover, there is no guarantee that the terminal state is preferred to s by any DM in H. On the contrary, $CR_H^+(s)$ is the subset of the terminal states preferred to s by all DMs in the coalition, although any individual move in the sequence may not be a UI for the mover.

Xu et al. (2010) extend the definition of coalitional improvement to weak coalitional improvement by including uncertain preference in the definition. A weak coalitional improvement for a coalition is a state that is the result of a sequence of moves from the status quo by members of the coalition, where each move is a *coalition improvement or uncertain move (CIUM)*, defined as follows.

Definition 8.2 For a status quo state s and a nonempty coalition $H \subseteq N$, a state $s_1 \in R_H(s)$ is **a coalition improvement or uncertain move** for H from s, denoted by $s_1 \in CR_H^{+,U}(s)$, iff $s_1 \succ_i s$ or $s_1 \; U_i \; s$ for every $i \in H$.

Here, $CR_H^{+,U}(s)$ differs from $R_H^{+,U}(s)$ in Definition 5.18 in that $R_H^{+,U}(s)$ reflects the steps of the process without taking into account the final result, while $CR_H^{+,U}(s)$ is the final result, instead of the process. In other words, $R_H^{+,U}(s)$ requires each move in a legal sequence to be a UIUM for the mover, but the relative preference of the final state and the status quo is not a concern. On the contrary, $CR_H^{+,U}(s)$ ensures that all coalition members prefer the terminal state to the status quo, or are uncertain about their preference between these two states, without examining the relative preference for each individual move along the legal sequence.

Similarly, a coalitional improvement can be extended to include strength of preference.

Definition 8.3 For a status quo state s and a nonempty coalition $H \subseteq N$, a state $s_1 \in R_H(s)$ is **a mild or strong coalitional improvement** for H from s under the three-degree preference, denoted by $s_1 \in CR_H^{+,++}(s)$, iff $s_1 >_i s$ or $s_1 \gg_i s$ for every $i \in H$.

This means that, under a model with three degrees of preference, a coalitional improvement is a state mildly preferred or strongly preferred to s by any DM in H and is reachable by the coalition H. As before, note that $CR_H^{+,++}(s) \neq R_H^{+,++}(s)$, because $R_H^{+,++}(s)$ (Definition 6.9) denotes the states attainable by coalition H via legal sequences of mild or strong unilateral improvements (MSUIs) from s. But there is no guarantee that every DM in H prefers the terminal state to state s. On the other hand, $CR_H^{+,++}(s)$ ensures that the terminal state is always mildly or strongly preferred to s by all DMs in H though any individual move in the sequence may not be an MSUI for the mover. The following definition of coalitional movement is for the combination of unknown preference with three degrees of preference.

Definition 8.4 For a status quo state s and a nonempty coalition $H \subseteq N$, a state $s_1 \in R_H(s)$ is **a mild or strong or uncertain coalitional improvement** for H from s under hybrid preference, denoted by $s_1 \in CR_H^{+,++,U}(s)$, iff $s_1 >_i s$, $s_1 \gg_i s$, or $s_1 \; U_i \; s$ for every $i \in H$.

Now that the important concept of coalitional improvement or coalitional uncertainty has been defined for various preference structures, the logical and matrix representations of coalitional stabilities can be presented as follows.

8.2 Logical Representation of Coalitional Stabilities Under Simple Preference

The logical representations of individual stabilities in the graph model for simple preference, unknown preference, three degrees of preference and hybrid preference are presented in Sects. 4.2, 5.2, 6.3 and 7.2, respectively. In this section, logical representations of coalitional stabilities are defined for the four kinds of preference structure.

Firstly, coalitional stabilities under Nash, GMR, SMR, and SEQ with simple preference are furnished.

Definition 8.5 Let $H \subseteq N$ be a nonempty coalition. State $s \in S$ is coalitional Nash stable for H, denoted by $s \in S_H^{CNash}$, iff $CR_H^+(s) = \emptyset$.

From Definition 8.1, $CR_H^+(s)$ honors the rule of no-successive-moves by the same DM and, hence, this definition is applicable to both transitive and intransitive graph models. As mentioned earlier, an empty coalition has no meaning, so it is assumed hereafter that $|H| > 0$. If $|H| = 1$, then $H = \{i\}$ and $CR_H^+(s) = R_i^+(s)$. In this special case, Definition 8.5 reduces to individual Nash stability defined in Chap. 4. However, for a nontrivial coalition $H \subseteq N, |H| \geq 2$, coalitional Nash stability depends on the coalitional improvement list $CR_H^+(s)$, rather than coalition members' individual UI lists, $R_i^+(s)$, for $i \in H$.

If state $s \in S$ is Nash stable for every nonempty coalition $H \subseteq N$, it is called universally coalitional Nash stable. The formal definition is described as follows.

Definition 8.6 State $s \in S$ is **universally coalitional Nash stable**, denoted by $s \in S^{UCNash}$, iff s is coalitional stable for every nonempty coalition $H \subseteq N$.

Note that S_H^{CNash} in Definition 8.5 is different from S^{UCNash}. S_H^{CNash} is the set of coalitional Nash stable states for some coalition H, whereas S^{UCNash} contains all coalitional Nash stable states.

For notational convenience, the notation to represent a preference relation in coalition H is defined as follows.

Definition 8.7 For the graph model G, let $H \subseteq N$ be a coalition. $\Phi_H^{\preceq}(s) = \{t \in S : s \succeq_i t \text{ for at least one } i \in H\}$ in which $s \succeq_i t$ denotes $s \succ_i t$ or $s \sim_i t$.

It is apparent that $\Phi_H^{\preceq}(s)$ considers only preference relative to state s without regard to reachability from s.

Definition 8.8 Let $H \subseteq N$ be a nonempty coalition. State $s \in S$ is coalitional general metarational (CGMR) for H, denoted by $s \in S_H^{CGMR}$, iff for every $s_1 \in CR_H^+(s)$, there exists $s_2 \in R_{N-H}(s_1)$ such that $s_2 \in \Phi_H^{\preceq}(s)$.

If $H = \{i\}$, this definition reduces to individual GMR, defined in Sect. 4.2.3. If a state is coalitional GMR for every coalition, it is called universally coalitional GMR stable, formally defined as follows.

Definition 8.9 State $s \in S$ is **universally coalitional GMR stable**, denoted by $s \in S^{UCGMR}$, iff s is coalitional GMR stable for every nonempty coalition $H \subseteq N$.

Definition 8.10 Let $H \subseteq N$ be a nonempty coalition. State $s \in S$ is coalitional symmetric metarational (CSMR) for H, denoted by $s \in S_H^{CSMR}$, iff for every $s_1 \in CR_H^+(s)$, there exists $s_2 \in R_{N-H}(s_1)$ such that $s_2 \in \Phi_H^{\preceq}(s)$ and $s_3 \in \Phi_H^{\preceq}(s)$ for all $s_3 \in R_H(s_2)$.

As usual, if $H = \{i\}$, Definition 8.10 reduces to individual SMR, defined in Sect. 4.2.3. If a state is coalitional SMR for every coalition, it is called universally coalitional SMR stable, defined as follows.

Definition 8.11 State $s \in S$ is **universally coalitional SMR stable** , denoted by $s \in S^{UCSMR}$, iff s is coalitional SMR stable for every nonempty coalition $H \subseteq N$.

Normally, coalition H's opponents $N - H$ may be treated as a coalition or as individual DMs in the next two definitions.

Definition 8.12 Let $H \subseteq N$ be a nonempty coalition. State $s \in S$ is coalitional sequentially stable ($CSEQ_1$) for H, denoted by $s \in S_H^{CSEQ_1}$, iff for every $s_1 \in CR_H^+(s)$, there exists $s_2 \in CR_{N-H}^+(s_1)$ such that $s_2 \in \Phi_H^{\precsim}(s)$.

Definition 8.13 Let $H \subseteq N$ be a nonempty coalition. State $s \in S$ is coalitional sequentially stable ($CSEQ_2$) for H, denoted by $s \in S_H^{CSEQ_2}$, iff for every $s_1 \in CR_H^+(s)$, there exists $s_2 \in R_{N-H}^+(s_1)$ such that $s_2 \in \Phi_H^{\precsim}(s)$.

Remark: By employing the subclass improvement list concept, the SEQ stability definition for coalition H introduced by Inohara and Hipel (2008a, b) considers credible sanctions by subcoalitions of opponents. But their result assumes that the rule of no consecutive moves by the same DM has been lifted for the sake of tractability. The implication is that the definition is applicable only to transitive graph models, so in this book, one retains this restriction for coalitional stabilities. Because the number of subcoalitions increases exponentially with the number of DMs in the opponents, making the calculation of subclass improvement lists prohibitively difficult, H's opponents $N - H$ are treated here as a coalition or individual DMs, as shown in Definitions 8.12 and 8.13, respectively.

As usual, when $H = \{i\}$, coalitional SEQ would be reduced to individual SEQ stability. Similarly, if state is coalitional SEQ for every coalition, it is called universally coalitional SEQ stable. Specifically,

Definition 8.14 State $s \in S$ is **universally coalitional** SEQ_1 **stable**, denoted by $s \in S^{UCSEQ_1}$, iff s is coalitional SEQ_1 stable for every nonempty coalition $H \subseteq N$.

Definition 8.15 State $s \in S$ is **universally coalitional** SEQ_2 **stable**, denoted by $s \in S^{UCSEQ_2}$, iff s is coalitional SEQ_2 stable for every nonempty coalition $H \subseteq N$.

From the discussions above, it is clear that coalitional stability analysis extends individual stabilities under simple preference. Next, the coalitional stabilities are extended to preference with uncertainty.

8.3 Logical Representation of Coalitional Stabilities Under Unknown Preference

DMs may exhibit different attitudes toward preference uncertainty when making choices. For instance, an optimistic DM tends to view uncertainty as a potential opportunity, while a pessimistic DM may regard an uncertain outcome as a risk. In addition, a DM's attitude towards uncertainty may change with the status quo state: a DM who has little to lose is more likely to take an aggressive attitude towards uncertainty and treat it as a potential gain. On the contrary, a DM who has little to gain is highly likely to regard uncertain outcomes as a risk and adopt a conservative stance.

To accommodate different attitudes toward preference uncertainty, Li et al. (2004) define individual Nash, GMR, SMR, and SEQ stabilities with preference uncertainty under four forms, a, b, c, and d (see Chap. 5). The purpose of these four extensions is to characterize a focal DM with diverse attitudes toward preference uncertainty, ranging from aggressive to mixed to conservative. When coalitional GMR, SMR, and SEQ stability definitions are extended from graph models with simple preference, as presented in Sect. 8.2, to those with unknown preference, these four extensions apply, depending on the focal coalition's attitude towards preference uncertainty.

First, the coalitional Nash, GMR, SMR, and SEQ stabilities with indices a, b, c, and d for unknown preference are described as follows. Let $l \in \{a, b, c, d\}$.

8.3.1 *Logical Representation of Coalitional Stabilities Indexed l*

(1) Logical Representation of Coalitional Stabilities Indexed a

Definition 8.16 State $s \in S$ is coalitional $Nash_a$ stable for $H \subseteq N$, denoted by $s \in S_H^{CNash_a}$, iff $CR_H^{+,U}(s) = \emptyset$.

Definition 8.17 State $s \in S$ is coalitional GMR_a for $H \subseteq N$, denoted by $s \in S_H^{CGMR_a}$, iff for every $s_1 \in CR_H^{+,U}(s)$, there exists $s_2 \in R_{N-H}(s_1)$ such that $s_2 \in \Phi_H^{\preceq}(s)$.

Definition 8.18 State $s \in S$ is coalitional SMR_a for $H \subseteq N$, denoted by $s \in S_H^{CSMR_a}$, iff for every $s_1 \in CR_H^{+,U}(s)$, there exists $s_2 \in R_{N-H}(s_1)$ such that $s_2 \in \Phi_H^{\preceq}(s)$ and $s_3 \in \Phi_H^{\preceq}(s)$ for all $s_3 \in R_H(s_2)$.

Definition 8.19 State $s \in S$ is coalitional SEQ_a for $H \subseteq N$, denoted by $s \in S_H^{CSEQ_a}$, iff for every $s_1 \in CR_H^{+,U}(s)$, there exists $s_2 \in R_{N-H}^{+,U}(s_1)$ such that $s_2 \in \Phi_H^{\preceq}(s)$.

In extension a, the focal coalition members are conceived to be aggressive. They are willing to deviate from the status quo state for uncertain outcomes in that uncertainty is allowed at the incentive end for the focal coalition. Therefore, $s \in S_H^{CNash_a}$ is also said to be Nash stable for aggressive DMs in H. While assessing sanctions by opponents, at least one coalition member must end up in a no-better-off position in order to successfully block the focal coalition. Thus, uncertainty is not allowed at the sanction end for the focal coalition.

(2) Logical Representation of Coalitional Stabilities Indexed b

Definition 8.20 State $s \in S$ is coalitional $Nash_b$ stable for $H \subseteq N$, denoted by $s \in S_H^{CNash_b}$, iff $CR_H^{+}(s) = \emptyset$.

Definition 8.21 State $s \in S$ is coalitional GMR_b for $H \subseteq N$, denoted by $s \in S_H^{CGMR_b}$, iff for every $s_1 \in CR_H^{+}(s)$, there exists $s_2 \in R_{N-H}(s_1)$ such that $s_2 \in \Phi_H^{\preceq}(s)$.

Definition 8.22 State $s \in S$ is coalitional SMR_b for $H \subseteq N$, denoted by $s \in S_H^{CSMR_b}$, iff for every $s_1 \in CR_H^+(s)$, there exists $s_2 \in R_{N-H}(s_1)$ such that $s_2 \in \Phi_H^{\precsim}(s)$ and $s_3 \in \Phi_H^{\precsim}(s)$ for all $s_3 \in R_H(s_2)$.

Definition 8.23 State $s \in S$ is coalitional SEQ_b for $H \subseteq N$, denoted by $s \in S_H^{CSEQ_b}$, iff for every $s_1 \in CR_H^+(s)$, there exists $s_2 \in R_{N-H}^{+,U}(s_1)$ such that $s_2 \in \Phi_H^{\precsim}(s)$.

Compared to the stability definitions for a coalition in extension a, this extension does not treat uncertain moves as sufficient incentive for the focal coalition to deviate from the status quo. The focal coalition under this extension presumably exhibits a mixed attitude towards preference uncertainty, conservative at the incentive end but aggressive at the sanction end (Li et al. 2004). Although Definitions 8.20–8.23, respectively, look the same as Definitions 8.5, 8.8, 8.10, and 8.13, they are in fact different in the sense that Definitions 8.20–8.23 assume preference uncertainty but uncertain moves are neither strong enough motivation for the focal coalition to deviate from the status quo nor allowed as valid sanctions to deter the focal coalition. On the other hand, Definitions 8.5, 8.8, 8.10, and 8.13 assume graph models with simple preference.

(3) Logical Representation of Coalitional Stabilities Indexed c

For convenience, let $\Phi_H^{\precsim,U}(s) = \{t \in S : s \succsim_i t \text{ or } s\ U_i\ t \text{ for at least one } i \in H\}$. As Nash stability does not examine countermoves by the opponents, similar to the individual stability case in Chap. 5, $S_H^{CNash_c} = S_H^{CNash_a}$.

Definition 8.24 State $s \in S$ is coalitional GMR_c for $H \subseteq N$, denoted by $s \in S_H^{CGMR_c}$, iff for every $s_1 \in CR_H^{+,U}(s)$, there exists $s_2 \in R_{N-H}(s_1)$ such that $s_2 \in \Phi_H^{\precsim,U}(s)$.

Definition 8.25 State $s \in S$ is coalitional SMR_c for $H \subseteq N$, denoted by $s \in S_H^{CSMR_c}$, iff for every $s_1 \in CR_H^{+,U}(s)$, there exists $s_2 \in R_{N-H}(s_1)$ such that $s_2 \in \Phi_H^{\precsim,U}(s)$ and $s_3 \in \Phi_H^{\precsim,U}(s)$ for all $s_3 \in R_H(s_2)$.

Definition 8.26 State $s \in S$ is coalitional SEQ_c for $H \subseteq N$, denoted by $s \in S_H^{CSEQ_c}$, iff for every $s_1 \in CR_H^{+,U}(s)$, there exists $s_2 \in R_{N-H}^{+,U}(s_1)$ such that $s_2 \in \Phi_H^{\precsim,U}(s)$.

Extension c assumes that uncertain moves are allowed as sufficient incentives and sanctions for the focal coalition and is designed to characterize focal coalition members with mixed attitude towards preference uncertainty: aggressive at the incentive end but conservative at the sanction end.

(4) Logical Representation of Coalitional Stabilities Indexed d

Similar to the individual stability case, $S_H^{CNash_d} = S_H^{CNash_b}$.

Definition 8.27 State $s \in S$ is coalitional GMR_d for $H \subseteq N$, denoted by $s \in S_H^{CGMR_d}$, iff for every $s_1 \in CR_H^+(s)$, there exists $s_2 \in R_{N-H}(s_1)$ such that $s_2 \in \Phi_H^{\precsim,U}(s)$.

Definition 8.28 State $s \in S$ is coalitional SMR_d for $H \subseteq N$, denoted by $s \in S_H^{CSMR_d}$, iff for every $s_1 \in CR_H^{+}(s)$, there exists $s_2 \in R_{N-H}(s_1)$ such that $s_2 \in \Phi_H^{\precsim,U}(s)$ and $s_3 \in \Phi_H^{\precsim,U}(s)$ for all $s_3 \in R_H(s_2)$.

Definition 8.29 State $s \in S$ is coalitional SEQ_d for $H \subseteq N$, denoted by $s \in S_H^{CSEQ_d}$, iff for every $s_1 \in CR_H^{+}(s)$, there exists $s_2 \in R_{N-H}^{+,U}(s_1)$ such that $s_2 \in \Phi_H^{\precsim,U}(s)$.

Coalitional stability definitions in extension d are devised for conservative focal coalitions: When contemplating incentives, they do not envision uncertain moves as opportunities (preference uncertainty is not allowed as incentives); while assessing sanctions, these DMs would view uncertain moves as potential harm (preference uncertainty is allowed as valid sanctions).

Let $l \in \{a, b, c, d\}$. As usual, if state $s \in S$ is coalitional Nash, GMR, SMR, or SEQ stable for each coalition $H \subseteq N$ under a particular extension l, it is called universally coalitional Nash, GMR, SMR, or SEQ stable indexed l, and denoted by $s \in S^{UCNash_l}$, $s \in S^{UCGMR_l}$, $s \in S^{UCSMR_l}$, or $s \in S^{UCSEQ_l}$. It is obvious that $S^{UCGMR_l} = \cap_{H \subseteq N} S_H^{CGMR_l}$, $S^{UCSMR_l} = \cap_{H \subseteq N} S_H^{CSMR_l}$, and $S^{UCSEQ_l} = \cap_{H \subseteq N} S_H^{CSEQ_l}$.

The logical representations of the coalitional stabilities for simple preference and unknown preference have been described in Sects. 8.2 and 8.3. The logical representation of coalitional stabilities when there are three degrees of preference are presented next.

8.4 Logical Representation of Coalitional Stabilities Under Three Degrees of Preference

Two-degree preference (simple preference) is often inadequate for modeling the complex strategic conflicts that arise in practical applications, so it is natural to explore how to expand coalitional stability from two-degree preference, presented in Sect. 8.2, to the three-degree version. The coalitional stability definitions given below for three degrees of preference recognize three distinct categories of stability that are general coalitional stability, strong coalitional stability, and weak coalitional stability. Coalitional stability definitions are called strong or weak to reflect the additional preference information contained in the strength of preference relation. General coalitional stabilities are defined first.

8.4.1 General Coalitional Stabilities

In order to analyze the coalitional stability of a state for a coalition $H \subseteq N$, it is necessary to take into account possible responses from the opponents of H, $j \in N - H$.

The reachable lists of coalition H from state s, $R_H(s)$ and $R_H^{+,++}(s)$, defined in Sects. 4.2.2 and 6.3.2, respectively, are used in this subsection for coalitional stability definitions for three degrees of preference . A mild or strong coalitional improvement from s for H, $CR_H^{+,++}(s)$, is presented in Definition 8.3. General coalitional stabilities are defined next.

Definition 8.30 For $H \subseteq N$, state $s \in S$ is general coalitional Nash stable for coalition H, denoted by $s \in S_H^{GCNash}$, iff $CR_H^{+,++}(s) = \emptyset$.

State s is general coalitional Nash stable for coalition H iff H has no coalitional improvements from state s. Nash stability takes no account of possible responses by the opponents of H for any move by H away from s.

To develop the coalitional versions of GMR, SMR, and SEQ, it is necessary to identify coalition H's UMs, $R_H(s)$, MSUIs, $R_H^{+,++}(s)$, and coalitional improvements, $CR_H^{+,++}(s)$, from state s.

Definition 8.31 For $H \subseteq N$, state s is general coalitional GMR (GCGMR) for coalition H, denoted by $s \in S_H^{GCGMR}$, iff for every $s_1 \in CR_H^{+,++}(s)$ there exists at least one $s_2 \in R_{N-H}(s_1)$ such that $s \gg_i s_2$, $s >_i s_2$, or $s \sim_i s_2$ for some DM $i \in H$.

Definition 8.32 State s is general coalitional SMR (GCSMR) stable for coalition H, denoted by $s \in S_H^{GCSMR}$, iff for every $s_1 \in CR_H^{+,++}(s)$ there exists $s_2 \in R_{N-H}(s_1)$, such that $s \gg_i s_2$, $s >_i s_2$, or $s \sim_i s_2$ for at least one $i \in H$ and $s \gg_i s_3$, $s >_i s_3$, or $s \sim_i s_3$ for all $s_3 \in R_H(s_2)$.

State s is general coalitional SMR stable for H iff, for every s_1 that H can attain from s, and that is mildly or strongly preferred to s by everyone in H, there exists s_2 that $N - H$ can reach from s_1 that someone in H finds no more preferable than s, and, moreover, every s_3 that H can attain from s_2 is no more preferable than s for some member of H. If the sanction imposed by the opponents on H's improvement cannot be mitigated by coalition H's counterresponse, then coalition H is better off staying at the original state. Coalitional SMR presumes one step more foresight than coalitional GMR.

Coalitional SEQ stability examines the credibility of sanctions of coalition H's improvements by its opponents. The legality of sequences of improvements by subcoalitions of $N - H$ is another issue. Similar to Sect. 8.2, H's opponents $N - H$ may be treated as a coalition or as individual DMs in the next two definitions.

Definition 8.33 For $H \subseteq N$, state s is general coalitional SEQ_1 ($GCSEQ_1$) for coalition H, denoted by $s \in S_H^{GCSEQ_1}$, iff for every $s_1 \in CR_H^{+,++}(s)$ there exists at least one $s_2 \in CR_{N-H}^{+,++}(s_1)$ such that $s \gg_i s_2$, $s >_i s_2$, or $s \sim_i s_2$ for some $i \in H$.

The state $s \in S$ is general coalitional SEQ_1 stable for H iff, for every s_1 that H can reach from s which everyone in H mildly or strongly prefers to s, there exists s_2 that $N - H$ can reach from s_1 such that everyone in $N - H$ mildly or strongly prefers s_2 to s_1 and someone in H finds s_2 no more preferable than s. (Note that s_2 may be reachable from s_1 by unilateral moves rather than unilateral improvements.)

This is the same as saying that, for every coalitional unilateral improvement by H from s, there is a response that can be achieved by $N-H$ such that at least one person in H finds the coalitional improvement sanctioned. In this case, at least one person in H would rather be at s than at s_2. This person therefore refuses to contribute to the move from s to s_1. (Of course, if this person is not essential to making the move from s to s_1 in the first place, then he or she could be dropped from the coalition.)

Alternatively, H's opponents can be treated as individual DMs, producing the general coalitional SEQ_2 stability, defined as follows:

Definition 8.34 For $H \subseteq N$, state s is general coalitional SEQ_2 ($GCSEQ_2$) for coalition H, denoted by $s \in S_H^{GCSEQ_2}$, iff for every $s_1 \in CR_H^{+,++}(s)$ there exists at least one $s_2 \in R_{N-H}^{+,++}(s_1)$ such that $s \gg_i s_2$, $s >_i s_2$, or $s \sim_i s_2$ for some $i \in H$.

8.4.2 *Strong or Weak Coalitional Stabilities*

When degree of preference is introduced into the graph model, general coalitional stability definitions can be strong or weak, according to the degree of sanctioning. For a risk-averse coalition H, if all of coalition H's improvements from a particular state are strongly sanctioned, then the status quo state possesses an extra degree of stability, called strong stability. A coalitional improvement of a focal H is sanctioned strongly if it could result in a greatly less preferred state relative to the initial state, and this sanction cannot be avoided by an appropriate counterresponse.

Definition 8.35 For $H \subseteq N$, state s is strong coalitional GMR (SCGMR) for coalition H, denoted by $s \in S_H^{SCGMR}$, iff for every $s_1 \in CR_H^{+,++}(s)$ there exists at least one $s_2 \in R_{N-H}(s_1)$ such that $s \gg_i s_2$ for some DM $i \in H$.

Under strong coalitional GMR stability, all H's coalitional improvements can be strongly sanctioned by the opponents.

Definition 8.36 State s is strong coalitional SMR (SCSMR) stable for coalition H, denoted by $s \in S_H^{SCSMR}$, iff for every $s_1 \in CR_H^{+,++}(s)$ there exists $s_2 \in R_{N-H}(s_1)$, such that $s \gg_i s_2$ for at least one $i \in H$ and $s \gg_i s_3$ for all $s_3 \in R_H(s_2)$.

If the strong sanction imposed by the opponents on H's improvements cannot be mitigated by coalition H's counterresponse, then at least one member of the coalition H is better off staying at the original state. Two following definitions are analogous to Definitions 8.33 and 8.34.

Definition 8.37 For $H \subseteq N$, state s is strong coalitional SEQ_1 ($SCSEQ_1$) for coalition H, denoted by $s \in S_H^{SCSEQ_1}$, iff for every $s_1 \in CR_H^{+,++}(s)$ there exists at least one $s_2 \in CR_{N-H}^{+,++}(s_1)$ such that $s \gg_i s_2$ for at least one $i \in H$.

Definition 8.38 For $H \subseteq N$, state s is strong coalitional SEQ_2 ($SCSEQ_2$) for coalition H, denoted by $s \in S_H^{SCSEQ_2}$, iff for every $s_1 \in CR_H^{+,++}(s)$ there exists at least one $s_2 \in R_{N-H}^{+,++}(s_1)$ such that $s \gg_i s_2$ for at least one $i \in H$.

For three-degree preference, general coalitional stabilities are classified as strong and weak according to the strength of the possible sanctions. Let GCGS and SCGS denote general coalitional graph model stability, GCNash, GCGMR, GCSMR, $GCSEQ_1$, or $GCSEQ_2$, and strong coalitional graph model stability, SCGMR, SCSMR, $SCSEQ_1$, or $SCSEQ_2$, respectively. Strong coalitional Nash stability is excluded because CNash stability does not involve sanctions. The symbol WCGS denotes weak coalitional graph model stability , WCGMR, WCSMR, or WCSEQ, under three-degree preference. Weak coalitional stability is defined as follows:

Definition 8.39 For $H \subseteq N$, state s is weak coalitional stable for coalition H, denoted by $s \in S_H^{WCGS}$, iff $s \in S_H^{GCGS}$ but $s \notin S_H^{SCGS}$.

A weak coalitional stable state means that it is general coalitional stable for some stability, but not strong coalitional stable for the corresponding stability. Hence, if a particular state s is general coalitional stable, then s is either strong coalitional stable or weak coalitional stable.

8.5 Logical Representation of Coalitional Stability with Hybrid Preference

The logical representations of coalitional stabilities under unknown preference and three-level preference have been defined in Sects. 8.3 and 8.4, respectively. The two types of preference are combined into the hybrid preference structure. The coalitional stabilities under the hybrid preference are discussed in this section.

8.5.1 General Coalitional Stabilities with Hybrid Preference

The hybrid preference is to combine three-level preference and unknown preference together. Therefore, general coalitional stabilities within hybrid preference expand the general coalitional stabilities under simple preference, unknown preference, and three-degree preference. Let $l \in \{a, b, c, d\}$.

8.5.1.1 General Coalitional Stabilities Indexed l

(1) General Coalitional Stabilities Indexed a

For coalitional stabilities indexed a, coalition H is willing to move to states that are mildly preferred or strongly preferred, as well as states having uncertain preference relative to the status quo but does not wish to be sanctioned by a strongly less preferred, mildly less preferred, or equally preferred state relative to the status quo. The definitions given below assume that $s \in S$ and $i \in N$.

Definition 8.40 For the graph model G, let $H \subseteq N$ be a coalition. Define $\Phi_H^{\ll,<,\sim}(s) = \{t \in S : s \gg_i t, s >_i t, \text{ or } s \sim_i \text{ for at least one } i \in H\}$ and $\Phi_H^{\ll,<,\sim,U}(s) = \{t \in S : s \gg_i t, s >_i t, s \sim_i t, \text{ or } s\ U_i\ t \text{ for at least one } i \in H\}$.

Note that $\Phi_H^{\ll,<,\sim}(s)$ and $\Phi_H^{\ll,<,\sim,U}(s)$ do not consider the reachability from s.

Definition 8.41 State $s \in S$ is general coalitional $Nash_a$ stable for coalition $H \subseteq N$, denoted by $s \in S_H^{GCNash_a}$, iff $CR_H^{+,++,U}(s) = \emptyset$.

Definition 8.42 State $s \in S$ is general coalitional GMR_a for coalition $H \subseteq N$, denoted by $s \in S_H^{GCGMR_a}$, iff for every $s_1 \in CR_H^{+,++,U}(s)$, there exists $s_2 \in R_{N-H}(s_1)$ such that $s_2 \in \Phi_H^{\ll,<,\sim}(s)$.

Definition 8.43 State $s \in S$ is general coalitional SMR_a for coalition $H \subseteq N$, denoted by $s \in S_H^{GCSMR_a}$, iff for every $s_1 \in CR_H^{+,++,U}(s)$, there exists $s_2 \in R_{N-H}(s_1)$ such that $s_2 \in \Phi_H^{\ll,<,\sim}(s)$ and $s_3 \in \Phi_H^{\ll,<,\sim}(s)$ for all $s_3 \in R_H(s_2)$.

Definition 8.44 State $s \in S$ is general coalitional SEQ_a for coalition $H \subseteq N$, denoted by $s \in S_H^{GCSEQ_a}$, iff for every $s_1 \in CR_H^{+,++,U}(s)$, there exists $s_2 \in R_{N-H}^{+,++,U}(s_1)$ such that $s_2 \in \Phi_H^{\ll,<,\sim}(s)$.

(2) General Coalitional Stabilities Indexed b

Definition 8.45 State $s \in S$ is general $Nash_b$ stable for coalition $H \subseteq N$, denoted by $s \in S_H^{GCNash_b}$, iff $CR_H^{+,++}(s) = \emptyset$.

Definition 8.46 State $s \in S$ is general coalitional GMR_b for coalition $H \subseteq N$, denoted by $s \in S_H^{GCGMR_b}$, iff for every $s_1 \in CR_H^{+,++}(s)$, there exists $s_2 \in R_{N-H}(s_1)$ such that $s_2 \in \Phi_H^{\ll,<,\sim}(s)$.

Definition 8.47 State $s \in S$ is general coalitional SMR_b for coalition $H \subseteq N$, denoted by $s \in S_H^{GCSMR_b}$, iff for every $s_1 \in CR_H^{+,++}(s)$, there exists $s_2 \in R_{N-H}(s_1)$ such that $s_2 \in \Phi_H^{\ll,<,\sim}(s)$ and $s_3 \in \Phi_H^{\ll,<,\sim}(s)$ for all $s_3 \in R_H(s_2)$.

Definition 8.48 State $s \in S$ is general coalitional SEQ_b for coalition $H \subseteq N$, denoted by $s \in S_H^{GCSEQ_b}$, iff for every $s_1 \in CR_H^{+,++}(s)$, there exists $s_2 \in R_{N-H}^{+,++,U}(s_1)$ such that $s_2 \in \Phi_H^{\ll,<,\sim}(s)$.

(3) General Coalitional Stabilities Indexed c

Definition 8.49 State $s \in S$ is general coalitional $Nash_c$ stable for coalition $H \subseteq N$, denoted by $s \in S_H^{GCNash_c}$, iff $CR_H^{+,++,U}(s) = \emptyset$.

Definition 8.50 State $s \in S$ is general coalitional GMR_c for coalition $H \subseteq N$, denoted by $s \in S_H^{GCGMR_c}$, iff for every $s_1 \in CR_H^{+,++,U}(s)$, there exists $s_2 \in R_{N-H}(s_1)$ such that $s_2 \in \Phi_H^{\ll,<,\sim,U}(s)$.

Definition 8.51 State $s \in S$ is general coalitional SMR_c for coalition $H \subseteq N$, denoted by $s \in S_H^{GCSMR_c}$, iff for every $s_1 \in CR_H^{+,++,U}(s)$, there exists $s_2 \in R_{N-H}(s_1)$ such that $s_2 \in \Phi_H^{\ll,<,\sim,U}(s)$ and $s_3 \in \Phi_H^{\ll,<,\sim,U}(s)$ for all $s_3 \in R_H(s_2)$.

Definition 8.52 State $s \in S$ is general coalitional SEQ_c for coalition $H \subseteq N$, denoted by $s \in S_H^{GCSEQ_c}$, iff for every $s_1 \in CR_H^{+,++,U}(s)$, there exists $s_2 \in R_{N-H}^{+,++,U}(s_1)$ such that $s_2 \in \Phi_H^{\ll,<,\sim,U}(s)$.

(4) General Coalitional Stabilities Indexed d

Definition 8.53 State $s \in S$ is general coalitional $Nash_d$ stable for coalition $H \subseteq N$, denoted by $s \in S_H^{GCNash_d}$, iff $CR_H^{+,++}(s) = \emptyset$.

Definition 8.54 State $s \in S$ is general coalitional GMR_d for coalition $H \subseteq N$, denoted by $s \in S_H^{GCGMR_d}$, iff for every $s_1 \in CR_H^{+,++}(s)$, there exists $s_2 \in R_{N-H}(s_1)$ such that $s_2 \in \Phi_H^{\ll,<,\sim,U}(s)$.

Definition 8.55 State $s \in S$ is general coalitional SMR_d for coalition $H \subseteq N$, denoted by $s \in S_H^{GCSMR_d}$, iff for every $s_1 \in CR_H^{+,++}(s)$, there exists $s_2 \in R_{N-H}(s_1)$ such that $s_2 \in \Phi_H^{\ll,<,\sim,U}(s)$ and $s_3 \in \Phi_H^{\ll,<,\sim,U}(s)$ for all $s_3 \in R_H(s_2)$.

Definition 8.56 State $s \in S$ is general coalitional SEQ_d for coalition $H \subseteq N$, denoted by $s \in S_H^{GCSEQ_d}$, iff for every $s_1 \in CR_H^{+,++}(s)$, there exists $s_2 \in R_{N-H}^{+,++,U}(s_1)$ such that $s_2 \in \Phi_H^{\ll,<,\sim,U}(s)$.

8.5.2 Strong Coalitional Stabilities with Hybrid Preference

The notation related to strong preference is defined within the hybrid preference framework.

Definition 8.57 For the graph model G, let $H \subseteq N$ be a coalition. $\Phi_H^{\ll}(s) = \{t \in S : s \gg_i t \text{ for at least one } i \in H\}$.

Definition 8.58 Let $l \in \{a, b, c, d\}$. Strong coalitional $Nash_l$ stable for coalition $H \subseteq N$ is identical with general coalitional $Nash_l$ stable for coalition $H \subseteq N$. In other words, $S_H^{SCNash_l} = S_H^{GCNash_l}$.

For example, when $l = a$, then $S_H^{SCNash_a} = S_H^{GCNash_a}$.

8.5.2.1 Strong Coalitional Stabilities Indexed *l*

(1) Strong Coalitional Stabilities Indexed a

Definition 8.59 State $s \in S$ is strong coalitional GMR_a for coalition $H \subseteq N$, denoted by $s \in S_H^{SCGMR_a}$, iff for every $s_1 \in CR_H^{+,++,U}(s)$, there exists $s_2 \in R_{N-H}(s_1)$ such that $s_2 \in \Phi_H^{\ll}(s)$.

Definition 8.60 State $s \in S$ is strong coalitional SMR_a for coalition $H \subseteq N$, denoted by $s \in S_H^{SCSMR_a}$, iff for every $s_1 \in CR_H^{+,++,U}(s)$, there exists $s_2 \in R_{N-H}(s_1)$ such that $s_2 \in \Phi_H^{\ll}(s)$ and $s_3 \in \Phi_H^{\ll}(s)$ for all $s_3 \in R_H(s_2)$.

Definition 8.61 State $s \in S$ is strong coalitional SEQ_a for coalition $H \subseteq N$, denoted by $s \in S_H^{SCSEQ_a}$, iff for every $s_1 \in CR_H^{+,++,U}(s)$, there exists $s_2 \in R_{N-H}^{+,++,U}(s_1)$ such that $s_2 \in \Phi_H^{\ll}(s)$.

(2) Strong Coalitional Stabilities Indexed b

Definition 8.62 State $s \in S$ is strong coalitional GMR_b for coalition $H \subseteq N$, denoted by $s \in S_H^{SCGMR_b}$, iff for every $s_1 \in CR_H^{+,++}(s)$, there exists $s_2 \in R_{N-H}(s_1)$ such that $s_2 \in \Phi_H^{\ll}(s)$.

Definition 8.63 State $s \in S$ is strong coalitional SMR_b for coalition $H \subseteq N$, denoted by $s \in S_H^{SCSMR_b}$, iff for every $s_1 \in CR_H^{+,++}(s)$, there exists $s_2 \in R_{N-H}(s_1)$ such that $s_2 \in \Phi_H^{\ll}(s)$ and $s_3 \in \Phi_H^{\ll}(s)$ for all $s_3 \in R_H(s_2)$.

Definition 8.64 State $s \in S$ is strong coalitional SEQ_b for coalition $H \subseteq N$, denoted by $s \in S_H^{SCSEQ_b}$, iff for every $s_1 \in CR_H^{+,++}(s)$, there exists $s_2 \in R_{N-H}^{+,++,U}(s_1)$ such that $s_2 \in \Phi_H^{\ll}(s)$.

(3) Strong Coalitional Stabilities Indexed c

Definition 8.65 State $s \in S$ is strong coalitional GMR_c for coalition $H \subseteq N$, denoted by $s \in S_H^{SCGMR_c}$, iff for every $s_1 \in CR_H^{+,++,U}(s)$, there exists $s_2 \in R_{N-H}(s_1)$ such that $s_2 \in \Phi_H^{\ll,U}(s)$.

Definition 8.66 State $s \in S$ is strong coalitional SMR_c for coalition $H \subseteq N$, denoted by $s \in S_H^{SCSMR_c}$, iff for every $s_1 \in CR_H^{+,++,U}(s)$, there exists $s_2 \in R_{N-H}(s_1)$ such that $s_2 \in \Phi_H^{\ll,U}(s)$, and $s_3 \in \Phi_H^{\ll,U}(s)$ for all $s_3 \in R_H(s_2)$.

Definition 8.67 State $s \in S$ is strong coalitional SEQ_c for coalition $H \subseteq N$, denoted by $s \in S_H^{SCSEQ_c}$, iff for every $s_1 \in CR_H^{+,++,U}(s)$, there exists $s_2 \in R_{N-H}^{+,++,U}(s_1)$ such that $s_2 \in \Phi_H^{\ll,U}(s)$.

(4) Strong Coalitional Stabilities Indexed d

Definition 8.68 State $s \in S$ is strong coalitional GMR_d for coalition $H \subseteq N$, denoted by $s \in S_H^{SCGMR_d}$, iff for every $s_1 \in CR_H^{+,++}(s)$, there exists $s_2 \in R_{N-H}(s_1)$ such that $s_2 \in \Phi_H^{\ll,U}(s)$.

Definition 8.69 State $s \in S$ is strong coalitional SMR_d for coalition $H \subseteq N$, denoted by $s \in S_H^{SCSMR_d}$, iff for every $s_1 \in CR_H^{+,++}(s)$, there exists $s_2 \in R_{N-H}(s_1)$ such that $s_2 \in \Phi_H^{\ll,U}(s)$, and $s_3 \in \Phi_H^{\ll,U}(s)$ for all $s_3 \in R_H(s_2)$.

Definition 8.70 State $s \in S$ is strong coalitional SEQ_d for coalition $H \subseteq N$, denoted by $s \in S_H^{SCSEQ_d}$, iff for every $s_1 \in CR_H^{+,++}(s)$, there exists $s_2 \in R_{N-H}^{+,++,U}(s_1)$ such that $s_2 \in \Phi_H^{\ll,U}(s)$.

8.6 Matrix Representation of Coalitional Stability Under Simple Preference

Although the four basic coalitional stabilities are defined for simple preference in Sect. 8.2, unknown preference in Sect. 8.3, three degree-preference in Sect. 8.4 and hybrid preference in Sect. 8.5, they are represented logically, which make coding difficult. In order to develop algorithms to implement these coalitional stabilities more easily, matrix representation of coalitional stabilities under various preference structures is introduced in the following sections. The matrix version of coalitional stability under simple preference is presented first (Xu et al. 2014).

8.6.1 *Coalitional Improvement Matrix*

Let $m = |S|$ denote the number of states, E be the $m \times m$ matrix with each entry equal to 1, and e_s denote the sth standard basis vector of the m-dimensional Euclidean space, $\mathbb{R}^S$. Recall that the UM reachability matrix M_H is constructed using two approaches that are based on the incidence matrix B and the adjacency matrix J presented in Chaps. 4 and 5, respectively.

A matrix approach is proposed in this section to construct the coalitional improvements from state s, $CR_H^+(s)$, given in Definition 8.1 in logical form.

Definition 8.71 For the graph model G, let $H \subseteq N$ be a nonempty coalition. The coalitional improvement matrix for H is defined as the $m \times m$ matrix CM_H^+ with (s, q) entry

$$CM_H^+(s, q) = \begin{cases} 1 \text{ if } q \in CR_H^+(s), \\ 0 \text{ otherwise.} \end{cases}$$

It is clear that $CR_H^+(s) = \{q : CM_H^+(s, q) = 1\}$. Then

$$CR_H^+(s) = e_s^T \cdot CM_H^+,$$

if $CR_H^+(s)$ is written as 0–1 row vectors, where a "1" at the jth element indicates coalition H has a coalitional improvement from s to s_j. Note that e_s^T denotes the transpose of e_s, the sth standard basis vector of m-dimensional Euclidean space. Therefore, the coalitional improvement matrix for coalition H, CM_H^+, can be used to construct the coalitional improvements of H from state s, $CR_H^+(s)$.

Using Definition 8.1, the coalitional improvement matrix of H can be constructed by the following theorem. Recall that $P_H^{-,=} = \bigvee_{i \in H} P_i^{-,=}$ ("$\bigvee$" denotes the disjunction operator described in Definition 3.16).

Theorem 8.1 *For the graph model G, let $H \subseteq N$ be a nonempty coalition. The coalitional improvement matrix for H is expressed as*

$$CM_H^{+} = M_H \circ (E - P_H^{-,=}). \tag{8.1}$$

Proof To prove Eq. 8.1, assume that $C = M_H \circ (E - P_H^{-,=})$. Using the definition for matrix M_H given in Chaps. 4 and 5, $C(s, q) = 1$ iff $M_H(s, q) = 1$ and $P_H^{-,=}(s, q) = 0$, which together imply that there is $q \in R_H(s)$ such that $P_i^{-,=}(s, q) = 0$ for every DM $i \in H$. Therefore, $C(s, q) = 1$ iff there is $q \in R_H(s)$ with $q \succ_i s$ for every $i \in H$, so that $q \in CR_H^{+}(s)$, according to Definition 8.1. Thus, $CM_H^{+}(s, q) = 1$ using Definition 8.71. Hence, $CM_H^{+}(s, q) = 1$ iff $C(s, q) = 1$. Since CM_H^{+} and C are 0–1 matrices, it follows that $CM_H^{+} = M_H \circ (E - P_H^{-,=})$. □

8.6.2 Matrix Representation of Coalitional Stabilities

For a fixed state $s \in S$, let e_s be an m-dimensional vector with 1 as its sth element and 0 everywhere else and e be an m-dimensional vector with every entry 1. Let $(\overrightarrow{0})^T$ denote the transpose of $\overrightarrow{0}$.

Theorem 8.2 *For the graph model G, let $H \subseteq N$ be a nonempty coalition. State $s \in S$ is coalitional Nash stable for H, denoted by $s \in S_H^{CNash}$, iff $e_s^T \cdot CM_H^{+} \cdot e = 0$.*

Proof Since $e_s^T \cdot CM_H^{+} \cdot e = 0$ iff $e_s^T \cdot CM_H^{+} = (\overrightarrow{0})^T$, then $CR_H^{+}(s) = \emptyset$ using Definition 8.71. Consequently, the proof of this theorem follows by Definition 8.5. □

Coalitional Nash stability extends individual Nash stability. For example, If $|H| = 1$, Theorem 8.2 reduces to the matrix representation of individual Nash stability presented in Theorem 4.3. Specifically,

Corollary 8.1 *For the graph model G, let $i \in N$. If $e_s^T \cdot CM_{\{i\}}^{+} \cdot e = 0$, then s is Nash stable for DM i.*

From Corollary 8.1, coalitional Nash stability is a generalization of individual Nash stability.

Theorem 8.3 *For the graph model G, state $s \in S$ is universally coalitional Nash stable for every $H \subseteq N$, denoted by $s \in S^{UCNash}$, iff $\sum_{\forall H \subseteq N} e_s^T \cdot CM_H^{+} \cdot e = 0$.*

Proof Since $\sum_{\forall H \subseteq N} e_s^T \cdot CM_H^{+} \cdot e = 0$ iff for any $H \subseteq N$, $e_s^T \cdot CM_H^{+} \cdot e = 0$. By Theorem 8.2, $e_s^T \cdot CM_H^{+} \cdot e = 0$ iff $s \in S$ is coalitional Nash stable for H. Consequently, $\sum_{H \subseteq N} e_s^T \cdot CM_H^{+} \cdot e = 0$ iff $s \in S$ is coalitional Nash stable for every coalition $H \subseteq N$. The proof is completed by Definition 8.6. □

Theorem 8.3 shows the matrix representation of universally coalitional Nash stability equivalent to logical representation stated in Definition 8.6.

Similar to the individual GMR stability, define coalitional GMR stability matrix as

$$M_H^{CGMR} = CM_H^+ \cdot [E - sign\left(M_{N-H} \cdot (P_H^{-,=})^T\right)], \tag{8.2}$$

where $H \subseteq N$. The following theorem establishes the matrix method to assess whether state s is coalitional GMR stable for H.

Theorem 8.4 *For the graph model G, let $H \subseteq N$ be a nonempty coalition. State $s \in S$ is coalitional GMR stable for H, denoted by $s \in S_H^{CGMR}$, iff $M_H^{CGMR}(s, s) = 0$.*

Proof Since

$$M_H^{CGMR}(s, s) = (e_s^T \cdot CM_H^+) \cdot [(E - sign\left(M_{N-H} \cdot (P_H^{-,=})^T\right)) \cdot e_s]$$

$$= \sum_{s_1=1}^{m} CM_H^+(s, s_1) \cdot [1 - sign\left((e_{s_1}^T \cdot M_{N-H}) \cdot (e_s^T \cdot P_H^{-,=})^T\right)],$$

then $M_H^{CGMR}(s, s) = 0$ holds iff

$$CM_H^+(s, s_1) \cdot [1 - sign\left((e_{s_1}^T \cdot M_{N-H}) \cdot (e_s^T \cdot P_H^{-,=})^T\right)] = 0, \tag{8.3}$$

for every $s_1 \in S - \{s\}$. It is clear that Eq. 8.3 is equivalent to

$$(e_{s_1}^T \cdot M_{N-H}) \cdot (e_s^T \cdot P_H^{-,=})^T \neq 0,$$

for every $s_1 \in CR_H^+(s)$. Therefore, for a coalitional improvement from s, $s_1 \in CR_H^+(s)$, there exists at least one $s_2 \in R_{N-H}(s_1)$ with $P_H^{-,=}(s, s_2) = 1$ that is equivalent to $s \succeq_i s_2$ for some DM $i \in H$. According to Definition 8.8, $M_H^{CGMR}(s, s) = 0$ implies that s is coalitional GMR stable for H. □

Theorem 8.4 shows that this matrix method, called matrix representation of coalitional GMR stability, is equivalent to the logical version of the same stability given in Definition 8.8. To analyze the coalitional GMR stability at s for coalition H, one only needs to identify whether the diagonal entry $M_H^{CGMR}(s, s)$ of the coalitional GMR matrix is zero. If so, s is coalitional GMR stable for H; otherwise, s is coalitional GMR unstable for H. Similar to individual GMR stability, all information about coalitional GMR stability is contained in the diagonal entries of the coalitional GMR stability matrix.

If $|H| = 1$, Theorem 8.4 reduces to the matrix representation of individual GMR stability presented in Theorem 4.10. Specifically,

Corollary 8.2 *For the graph model G, let $i \in N$. if $M_{\{i\}}^{CGMR}(s, s) = 0$, then s is GMR stable for DM i.*

Coalitional SMR is similar to coalitional GMR except that coalition H expects to have a chance to counterrespond to its opponent $(N-H)$'s response to H's original move. Define the coalitional SMR stability matrix as

$$M_H^{CSMR} = CM_H^+ \cdot [E - sign(F)]$$

in which

$$F = M_{N-H} \cdot [(P_H^{-,=})^T \circ \left(E - sign\left(M_H \cdot (E - P_H^{-,=})^T\right)\right)],$$

for $H \subseteq N$. The following theorem establishes the matrix method to determine whether state s is coalitional SMR stable for H.

Theorem 8.5 *For the graph model G, let $H \subseteq N$ be a coalition. State $s \in S$ is coalitional SMR for H, denoted by $s \in S_H^{CSMR}$, iff $M_H^{CSMR}(s,s) = 0$.*

Proof Since

$$M_H^{CSMR}(s,s) = (e_s^T \cdot CM_H^+) \cdot [(E - sign(F)) \cdot e_s]$$

$$= \sum_{s_1=1}^{m} CM_H^+(s, s_1)[1 - sign\,(F(s_1, s))]$$

with

$$F(s_1, s) = \sum_{s_2=1}^{m} M_{N-H}(s_1, s_2) \cdot W(s_2, s),$$

and

$$W(s_2, s) = P_H^{-,=}(s, s_2) \cdot \left[1 - sign\left(\sum_{s_3=1}^{m}\left(M_H(s_2, s_3) \cdot (1 - P_H^{-,=}(s, s_3))\right)\right)\right],$$

then $M_H^{CSMR}(s,s) = 0$ holds iff $F(s_1, s) \neq 0$, for every $s_1 \in CR_H^+(s)$, which is equivalent to the statement that, for every $s_1 \in CR_H^+(s)$, there exists $s_2 \in R_{N-H}(s_1)$ such that

$$P_H^{-,=}(s, s_2) \neq 0, \tag{8.4}$$

and

$$\sum_{s_3=1}^{m} M_H(s_2, s_3) \cdot (1 - P_H^{-,=}(s, s_3)) = 0. \tag{8.5}$$

Equation 8.4 means that $s \succeq_i s_2$ for at least one DM $i \in H$, i.e., $s_2 \in \Phi_H^{\preceq}(s)$ that is given in Definition 8.7. Equation 8.5 is equivalent to

$$P_H^{-,=}(s, s_3) = 1 \text{ for any } s_3 \in R_H(s_2). \tag{8.6}$$

Obviously, for every $s_1 \in CR_H^+(s)$, there exists $s_2 \in R_{N-H}(s_1)$ such that Eqs. 8.4 and 8.5 hold iff for every $s_1 \in CR_H^+(s)$ there exists $s_2 \in R_{N-H}(s_1)$ such that $s_2 \in \Phi_H^{\preceq}(s)$ and $s_3 \in \Phi_H^{\preceq}(s)$ for all $s_3 \in R_H(s_2)$. Therefore, the proof of this theorem follows using Definition 8.10. □

Theorem 8.5 displays this matrix method, called matrix representation of coalitional SMR stability, which is equivalent to the logical version given in Definition 8.10. To calculate coalitional SMR stability at s for H, one only needs to assess whether the diagonal entry $M_H^{CSMR}(s, s)$ of coalitional SMR stability matrix is zero. If so, s is coalitional SMR stable for H; otherwise, s is coalitional SMR unstable for H.

Corollary 8.3 *For the graph model G, let $i \in N$. if $M_{\{i\}}^{CSMR}(s, s) = 0$, then s is SMR stable for DM i.*

Coalitional sequential stability is similar to coalitional GMR stability, but includes only those sanctions that are "credible". If H's opponents are treated as a coalition, the coalitional SEQ_1 stability matrix $M_H^{CSEQ_1}$ is defined as

$$M_H^{CSEQ_1} = CM_H^+ \cdot [E - sign\left(CM_{N-H}^+ \cdot (P_H^{-,=})^T\right)].$$

The following theorem provides the matrix method to analyze whether state s is coalitional SEQ_1 stable for H when H's opponents, $N - H$, are in a coalition.

Theorem 8.6 *For the graph model G, let $H \subseteq N$ be a coalition. State $s \in S$ is coalitional SEQ_1 stable for H, denoted by $s \in S_H^{CSEQ_1}$, iff $M_H^{CSEQ_1}(s, s) = 0$.*

Proof Since

$$M_H^{CSEQ_1}(s, s) = (e_s^T CM_H^+) \cdot [\left(E - sign(CM_{N-H}^+ \cdot (P_H^{-,=})^T)\right) e_s]$$

$$= \sum_{s_1=1}^{|S|} CM_H^+(s, s_1)[1 - sign\left((e_{s_1}^T CM_{N-H}^+) \cdot (e_s^T P_H^{-,=})^T\right)],$$

then $M_H^{CSEQ_1}(s, s) = 0$ holds iff

$$CM_H^+(s, s_1)[1 - sign\left((e_{s_1}^T CM_{N-H}^+) \cdot (e_s^T P_H^{-,=})^T\right)] = 0, \forall s_1 \in S. \tag{8.7}$$

It is clear that Eq. 8.7 is equivalent to

$$(e_{s_1}^T CM_{N-H}^+) \cdot (e_s^T P_H^{-,=})^T \neq 0 \text{ for any } s_1 \in CR_H^+(s).$$

This implies that for any $s_1 \in CR_H^+(s)$, there exists at least one $s_2 \in CR_{N-H}^+(s_1)$ with $s \succeq_i s_2$ for some DM $i \in H$ that satisfies $s_2 \in \Phi_H^{\preceq}(s)$. The proof of this theorem follows using Definition 8.12. □

Note that the coalitional SEQ_1 stability matrix is identical to the coalitional GMR stability matrix except that the UM reachability matrix for H's opponents, M_{N-H}, is replaced by the coalitional improvement matrix CM^+_{N-H}.

Similar to the previous two theorems, the matrix representation of coalitional SEQ stability is equivalent to the logical version given in Definition 8.12. Once, when the diagonal entry at (s, s) is zero, the state s under consideration is coalitional SEQ_1 stable for H. The following theorem is equivalent to the coalitional SEQ_2 stability given in Definition 8.13. Define the coalitional SEQ_2 stability matrix $M_H^{CSEQ_2}$ is defined as

$$M_H^{CSEQ_2} = CM_H^+ \cdot [E - sign\left(M^+_{N-H} \cdot (P_H^{-,=})^T\right)].$$

Theorem 8.7 *For the graph model G, let $H \subseteq N$ be a coalition. State $s \in S$ is coalitional SEQ_2 stable for H, denoted by $s \in S_H^{CSEQ_2}$, iff $M_H^{CSEQ_2}(s, s) = 0$.*

Corollary 8.4 *For the graph model G, let $i \in N$. Then, (1) $M_{\{i\}}^{CSEQ_1} = M_{\{i\}}^{CSEQ_2}$; (2) If $M_{\{i\}}^{CSEQ_1}(s, s) = 0$ or $M_{\{i\}}^{CSEQ_2}(s, s) = 0$, then s is SEQ stable for DM i.*

8.7 Matrix Representation of Coalitional Stabilities Under Unknown Preference

8.7.1 Matrix Representation of Coalitional Improvement or Uncertain Move

Let $m = |S|$ denote the number of states, E be the $m \times m$ matrix with each entry equal to 1, and e_s denote the sth standard basis vector of the m-dimensional Euclidean space, $\mathbb{R}^S$. Recall that the UM reachability matrix M_H is constructed using Theorem 4.9.

A matrix approach is presented in this section to construct the coalitional improvements and coalitional improvements or uncertain moves from state s, $CR_H^+(s)$ and $CR_H^{+,U}(s)$, given in Definitions 8.1 and 8.2, respectively, in logical form.

Definition 8.72 For the graph model G, let $H \subseteq N$ be a nonempty coalition. The coalitional improvement matrix for H is defined as the $m \times m$ matrix CM_H^+ with (s, q) entry

$$CM_H^+(s, q) = \begin{cases} 1 \text{ if } q \in CR_H^+(s), \\ 0 \text{ otherwise.} \end{cases}$$

Moreover, the coalitional improvement or uncertain move matrix for H is defined as the $m \times m$ matrix $CM_H^{+,U}$ with (s, q) entry

$$CM_H^{+,U}(s, q) = \begin{cases} 1 \text{ if } q \in CR_H^{+,U}(s), \\ 0 \text{ otherwise.} \end{cases}$$

It is clear that $CR_H^+(s) = \{q : CM_H^+(s,q) = 1\}$ and $CR_H^{+,U}(s) = \{q : CM_H^{+,U}(s,q) = 1\}$. Then

$$CR_H^+(s) = e_s^T \cdot CM_H^+ \text{ and } CR_H^{+,U}(s) = e_s^T \cdot CM_H^{+,U},$$

if $CR_H^+(s)$ and $CR_H^{+,U}(s)$ are written as 0–1 row vectors, where a "1" at the jth element indicates coalition H has a coalitional improvement from s to s_j and coalition H has a coalitional improvement or uncertain move from s to s_j, respectively. Note that e_s^T denotes the transpose of e_s, the sth standard basis vector of m-dimensional Euclidean space. Therefore, the coalitional improvement and coalitional improvement or uncertain move matrices for coalition H, CM_H^+ and $CM_H^{+,U}$, can be used to construct the coalitional improvements and the coalitional improvements or uncertain moves of H from state s, $CR_H^+(s)$ and $CR_H^{+,U}(s)$, respectively.

Theorem 8.8 *For the graph model G, let $H \subseteq N$ be a nonempty coalition. The coalitional improvement matrix for H is expressed as*

$$CM_H^+ = M_H \circ (E - P_H^{-,=,U}). \tag{8.8}$$

Proof To prove Eq. 8.8, assume that $C = M_H \circ (E - P_H^{-,=,U})$. Using the definition for matrix M_H presented in Chap. 4, $C(s,q) = 1$ iff $M_H(s,q) = 1$ and $P_H^{-,=,U}(s,q) = 0$, which together imply that there is $q \in R_H(s)$ such that $P_i^{-,=,U}(s,q) = 0$ for every DM $i \in H$. Therefore, $C(s,q) = 1$ iff there is $q \in R_H(s)$ with $q \succ_i s$ for every $i \in H$, so that $q \in CR_H^+(s)$, according to Definition 8.1. Thus, $CM_H^+(s,q) = 1$ using Definition 8.72. Hence, $CM_H^+(s,q) = 1$ iff $C(s,q) = 1$. Since CM_H^+ and C are 0–1 matrices, it follows that $CM_H^+ = M_H \circ (E - P_H^{-,=,U})$. □

Note that $CM_H^+ \neq M_H \circ P_H^+$. Recall that matrix $P_H^+ = \bigvee_{i \in H} P_i^+$ ("$\bigvee$" denotes the disjunction operator described in Definition 3.16). $(M_H \circ P_H^+)(s,q) = 1$ iff $M_H(s,q) = 1$ and $P_H^+(s,q) = 1$, which means that there is $q \in R_H(s)$ such that $P_i^+(s,q) = 1$ for some DM $i \in H$. This is not consistent with the definition of CM_H^+.

It is worth to note that matrix CM_H^+ defined in Theorem 8.8 is different from the matrix specified in Theorem 8.1 that cannot be used to analyze conflict models with preference uncertainty. The matrix defined in Theorem 8.8 contains information about uncertain preference. Using Definition 8.2, the coalitional improvement or uncertain move matrix of H can be constructed by the following theorem. Recall that $P_H^{-,=} = \bigvee_{i \in H} P_i^{-,=}$.

Theorem 8.9 *For the graph model G, let $H \subseteq N$ be a nonempty coalition. The coalitional improvement or uncertain move matrix for H is expressed as*

$$CM_H^{+,U} = M_H \circ (E - P_H^{-,=}). \tag{8.9}$$

Proof To prove Eq. 8.9, assume that $C = M_H \circ (E - P_H^{-,=})$. Using the definition for matrix M_H presented in Chap. 4, $C(s, q) = 1$ iff $M_H(s, q) = 1$ and $P_H^{-,=}(s, q) = 0$, which together imply that there is $q \in R_H(s)$ such that $P_i^{-,=}(s, q) = 0$ for every DM $i \in H$. Therefore, $C(s, q) = 1$ iff there is $q \in R_H(s)$ with $q \succ_i s$ or $q\ U_i\ s$ for every $i \in H$, so that $q \in CR_H^{+,U}(s)$, according to Definition 8.2. Thus, $CM_H^{+,U}(s, q) = 1$ using Definition 8.72. Hence, $CM_H^{+,U}(s, q) = 1$ iff $C(s, q) = 1$. Since $CM_H^{+,U}$ and C are 0–1 matrices, it follows that $CM_H^{+,U} = M_H \circ (E - P_H^{-,=})$. □

Theorems 8.8 and 8.9 provide a matrix approach to construct the coalitional improvements from state s by H, $CR_H^{+}(s)$, and coalitional improvements or uncertain moves for state s by H, $CR_H^{+,U}(s)$. After obtaining the two important components of coalitional stability definitions with unknown preference, the matrix representation of coalitional stabilities can be constructed as follows. Let $l \in \{a, b, c, d\}$.

8.7.2 *Matrix Representation of Coalitional Stabilities Indexed l*

(1) Matrix Representation of Coalitional Stabilities Indexed a

For a fixed state $s \in S$, let e_s be an m-dimensional vector with 1 as its sth element and 0 everywhere else and e be an m-dimensional vector with every entry 1. Let $(\overrightarrow{0})^T$ denote the transpose of $\overrightarrow{0}$.

Theorem 8.10 *For the graph model G, let $H \subseteq N$ be a nonempty coalition. State $s \in S$ is coalitional $Nash_a$ stable for H, denoted by $s \in S_H^{CNash_a}$, iff $e_s^T \cdot CM_H^{+,U} \cdot e$=0.*

Proof Since $e_s^T \cdot CM_H^{+,U} \cdot e = 0$ iff $e_s^T \cdot CM_H^{+,U} = (\overrightarrow{0})^T$, then $CR_H^{+,U}(s) = \emptyset$ using Definition 8.72. Consequently, the proof of the theorem follows by Definition 8.16. □

Define coalitional $CGMR_a$ stability matrix for coalition H as

$$M_H^{CGMR_a} = CM_H^{+,U} \cdot [E - sign\left(M_{N-H} \cdot (P_H^{-,=})^T\right)], \tag{8.10}$$

where $H \subseteq N$. The following theorem establishes the matrix method to assess whether state s is coalitional GMR_a stable for H.

Theorem 8.11 *For the graph model G, let $H \subseteq N$ be a nonempty coalition. State $s \in S$ is coalitional GMR_a stable for H, denoted by $s \in S_H^{CGMR_a}$, iff $M_H^{CGMR_a}$ $(s, s) = 0$.*

Proof Since

$$M_H^{CGMR_a}(s, s) = (e_s^T \cdot CM_H^{+,U}) \cdot [\left(E - sign\left(M_{N-H} \cdot (P_H^{-,=})^T\right)\right) \cdot e_s]$$

$$= \sum_{s_1=1}^{m} CM_H^{+,U}(s, s_1) \cdot [1 - sign\left((e_{s_1}^T \cdot M_{N-H}) \cdot (e_s^T \cdot P_H^{-,=})^T\right)],$$

then $M_H^{CGMR_a}(s, s) = 0$ holds iff

$$CM_H^{+,U}(s, s_1) \cdot [1 - sign\left((e_{s_1}^T \cdot M_{N-H}) \cdot (e_s^T \cdot P_H^{-,=})^T\right)] = 0, \tag{8.11}$$

for every $s_1 \in S - \{s\}$. It is clear that Eq. 8.11 is equivalent to

$$(e_{s_1}^T \cdot M_{N-H}) \cdot (e_s^T \cdot P_H^{-,=})^T \neq 0,$$

for every $s_1 \in CR_H^{+,U}(s)$. Therefore, for a coalitional improvement or uncertain move from s, $s_1 \in CR_H^{+,U}(s)$, there exists at least one $s_2 \in R_{N-H}(s_1)$ with $P_H^{-,=}(s, s_2) = 1$ that is equivalent to $s \succeq_i s_2$ for some DM $i \in H$. According to Definition 8.17, $M_H^{CGMR_a}(s, s) = 0$ implies that s is coalitional GMR_a stable for H. □

Theorem 8.11 shows that this matrix method, called matrix representation of coalitional GMR_a stability, is equivalent to the logical version of the same stability given in Definition 8.17.

Coalitional SMR_a is similar to coalitional GMR_a except that coalition H expects to have a chance to counterrespond to its opponent $(N-H)$'s response to H's original move. Define the coalitional SMR_a stability matrix as

$$M_H^{CSMR_a} = CM_H^{+,U} \cdot [E - sign(F)]$$

in which

$$F = M_{N-H} \cdot [(P_H^{-,=})^T \circ \left(E - sign\left(M_H \cdot (E - P_H^{-,=})^T\right)\right)],$$

for $H \subseteq N$. The following theorem establishes a matrix method to determine whether state s is coalitional SMR stable for H.

Theorem 8.12 *For the graph model G, let $H \subseteq N$ be a coalition. State $s \in S$ is coalitional SMR_a for H, denoted by $s \in S_H^{CSMR_a}$, iff $M_H^{CSMR_a}(s, s) = 0$.*

Proof Since

$$M_H^{CSMR_a}(s, s) = (e_s^T \cdot CM_H^{+,U}) \cdot [(E - sign(F)) \cdot e_s]$$

$$= \sum_{s_1=1}^{m} CM_H^{+,U}(s, s_1)[1 - sign\,(F(s_1, s))]$$

with

$$F(s_1, s) = \sum_{s_2=1}^{m} M_{N-H}(s_1, s_2) \cdot W(s_2, s),$$

and

$$W(s_2, s) = P_H^{-,=}(s, s_2) \cdot \left[1 - sign\left(\sum_{s_3=1}^{m} \left(M_H(s_2, s_3) \cdot (1 - P_H^{-,=}(s, s_3)\right)\right)\right],$$

then $M_H^{CSMR_a}(s, s) = 0$ holds iff $F(s_1, s) \neq 0$, for every $s_1 \in CR_H^{+,U}(s)$, which is equivalent to the statement that, for every $s_1 \in CR_H^{+,U}(s)$, there exists $s_2 \in R_{N-H}(s_1)$ such that

$$P_H^{-,=}(s, s_2) \neq 0, \tag{8.12}$$

and

$$\sum_{s_3=1}^{m} M_H(s_2, s_3) \cdot (1 - P_H^{-,=})(s, s_3) = 0. \tag{8.13}$$

Equation 8.12 means that $s \succeq_i s_2$ for at least one DM $i \in H$. Equation 8.13 is equivalent to

$$P_H^{-,=}(s, s_3) \neq 0 \text{ for any } s_3 \in R_H(s_2). \tag{8.14}$$

Obviously, for every $s_1 \in CR_H^{+,U}(s)$, there exists $s_2 \in R_{N-H}(s_1)$ such that $s \succeq_i s_2$ and Eq. 8.13 hold iff for every $s_1 \in CR_H^{+,U}(s)$ there exists $s_2 \in R_{N-H}(s_1)$ such that $s_2 \in \Phi_H^{\preceq}(s)$ and $s_3 \in \Phi_H^{\preceq}(s)$ for all $s_3 \in R_H(s_2)$. Therefore, the proof of this theorem follows using Definition 8.18. □

Theorem 8.12 displays that this matrix method, called matrix representation of coalitional SMR stability, is equivalent to the logical version given in Definition 8.18. To calculate coalitional SMR_a stability at s for H, one only needs to assess whether the diagonal entry $M_H^{CSMR_a}(s, s)$ of coalitional SMR_a stability matrix is zero. If so, s is coalitional SMR_a stable for H; otherwise, s is coalitional SMR_a unstable for H.

Coalitional sequential stability is similar to coalitional GMR stability, but includes only those sanctions that are "credible". The coalitional SEQ_a stability matrix $M_H^{CSEQ_a}$ is defined as

$$M_H^{CSEQ_a} = CM_H^{+,U} \cdot [E - sign\left(M_{N-H}^{+,U} \cdot (P_H^{-,=})^T\right)].$$

The following theorem provides the matrix method to analyze whether state s is coalitional SEQ_a stable for H when H's opponents, $N - H$, are in a coalition.

Theorem 8.13 *For the graph model G, let $H \subseteq N$ be a coalition. State $s \in S$ is coalitional SEQ_a stable for H, denoted by $s \in S_H^{CSEQ_a}$, iff $M_H^{CSEQ_a}(s, s) = 0$.*

Proof Since

$$M_H^{CSEQ_a}(s,s) = (e_s^T CM_H^{+,U}) \cdot [\left(E - sign(M_{N-H}^{+,U} \cdot (P_H^{-,=})^T)\right) e_s]$$
$$= \sum_{s_1=1}^{|S|} CM_H^{+,U}(s,s_1)[1 - sign\left((e_{s_1}^T M_{N-H}^{+,U}) \cdot (e_s^T P_H^{-,=})^T\right)],$$

then $M_H^{SEQ_a}(s,s) = 0$ holds iff

$$CM_H^{+,U}(s,s_1)[1 - sign\left((e_{s_1}^T M_{N-H}^{+,U}) \cdot (e_s^T P_H^{-,=})^T\right)] = 0, \forall s_1 \in S. \tag{8.15}$$

It is clear that Eq. 8.15 is equivalent to

$$(e_{s_1}^T M_{N-H}^{+,U}) \cdot (e_s^T P_H^{-,=})^T \neq 0 \text{ for any } s_1 \in CR_H^{+,U}(s).$$

It implies that for any $s_1 \in CR_H^{+,U}(s)$, there exists at least one $s_2 \in R_{N-H}^{+,U}(s_1)$ with $s \succeq_i s_2$ for some DM $i \in H$ that satisfies $s_2 \in \Phi_H^{\preceq}(s)$. The proof of this theorem follows using Definition 8.19. □

Note that the coalitional SEQ_a stability matrix is identical to the coalitional GMR_a stability matrix except that the UM reachability matrix for H's opponents, M_{N-H}, is replaced by the coalitional improvement or uncertain move matrix $CM_{N-H}^{+,U}$.

(2) Matrix Representation of Coalitional Stabilities Indexed b

The following theorems establish relationships between logical and matrix representations for coalitional stabilities indexed b under unknown preference. The extension indexed b excludes uncertainty in preferences when the focal coalition H considers incentives to leave a state and evaluates sanctions from its opponents. However, the following coalitional definitions are different from the coalitional stability definitions without preference uncertainty as discussed in Sect. 8.6, because the previous definitions cannot be used to analyze coalitional stabilities with uncertain preference.

Theorem 8.14 *For the graph model G, let $H \subseteq N$ be a nonempty coalition. State $s \in S$ is coalitional $Nash_b$ stable for H, denoted by $s \in S_H^{CNash_b}$, iff $e_s^T \cdot CM_H^{+} \cdot e = 0$.*

Define coalitional $CGMR_b$ stability matrix for coalition H as

$$M_H^{CGMR_b} = CM_H^{+} \cdot [E - sign\left(M_{N-H} \cdot (P_H^{-,=})^T\right)], \tag{8.16}$$

where $H \subseteq N$. The following theorem establishes the matrix method to assess whether state s is coalitional GMR_b stable for H.

Theorem 8.15 *For the graph model G, let $H \subseteq N$ be a nonempty coalition. State $s \in S$ is coalitional GMR_b stable for H, denoted by $s \in S_H^{CGMR_b}$, iff $M_H^{CGMR_b}(s,s) = 0$.*

Theorem 8.15 shows that this matrix method, called matrix representation of coalitional GMR_b stability, is equivalent to the logical version of the same coalitional GMR_b stability given in Definition 8.21.

Define the coalitional SMR_b stability matrix as

$$M_H^{CSMR_b} = CM_H^+ \cdot [E - sign(Q)]$$

in which

$$Q = M_{N-H} \cdot [(P_H^{-,=})^T \circ \left(E - sign\left(M_H \cdot (E - P_H^{-,=})^T\right)\right)],$$

for $H \subseteq N$. The following theorem establishes the matrix method to determine whether state s is coalitional SMR_b stable for H.

Theorem 8.16 *For the graph model G, let $H \subseteq N$ be a coalition. State $s \in S$ is coalitional SMR_b for H, denoted by $s \in S_H^{CSMR_b}$, iff $M_H^{CSMR_b}(s, s) = 0$.*

The coalitional SEQ_b stability matrix $M_H^{CSEQ_b}$ is defined as

$$M_H^{CSEQ_b} = CM_H^+ \cdot [E - sign\left(M_{N-H}^{+,U} \cdot (P_H^{-,=})^T\right)].$$

The following theorem provides the matrix method to analyze whether state s is coalitional SEQ_b stable for H when H's opponents, $N - H$, are in a coalition.

Theorem 8.17 *For the graph model G, let $H \subseteq N$ be a coalition. State $s \in S$ is coalitional SEQ_b stable for H, denoted by $s \in S_H^{CSEQ_b}$, iff $M_H^{CSEQ_b}(s, s) = 0$.*

The proofs of the above theorems on coalitional stabilities indexed b are similar to the proofs for the matrix representation of coalitional stabilities indexed a. Therefore, these proofs are left as exercises.

(3) Matrix Representation of Coalitional Stabilities Indexed c

Coalitional Nash stability similar to the individual stability case in Chap. 5 does not examine countermoves by the opponents, so $S_H^{Nash_c} = S_H^{Nash_a}$.

Define coalitional $CGMR_c$ stability matrix for coalition H as

$$M_H^{CGMR_c} = CM_H^{+,U} \cdot [E - sign\left(M_{N-H} \cdot (P_H^{-,=,U})^T\right)], \tag{8.17}$$

where $H \subseteq N$. The following theorem establishes the matrix method to assess whether state s is coalitional GMR_c stable for H.

Theorem 8.18 *For the graph model G, let $H \subseteq N$ be a nonempty coalition. State $s \in S$ is coalitional GMR_c stable for H, denoted by $s \in S_H^{CGMR_c}$, iff $M_H^{CGMR_c}(s, s) = 0$.*

Theorem 8.18 shows that this matrix method, called matrix representation of coalitional GMR_c stability, is equivalent to the logical version of the same coalitional GMR_c stability given in Definition 8.24.

Define the coalitional SMR_c stability matrix as

$$M_H^{CSMR_c} = CM_H^{+,U} \cdot [E - sign(Q)]$$

in which

$$Q = M_{N-H} \cdot [(P_H^{-,=,U})^T \circ \left(E - sign\left(M_H \cdot (E - P_H^{-,=,U})^T\right)\right)],$$

for $H \subseteq N$. The following theorem establishes the matrix method to determine whether state s is coalitional SMR_c stable for H.

Theorem 8.19 *For the graph model G, let $H \subseteq N$ be a coalition. State $s \in S$ is coalitional SMR_c for H, denoted by $s \in S_H^{CSMR_c}$, iff $M_H^{CSMR_c}(s, s) = 0$.*

The coalitional SEQ_c stability matrix $M_H^{CSEQ_c}$ is defined as

$$M_H^{CSEQ_c} = CM_H^{+,U} \cdot [E - sign\left(M_{N-H}^{+,U} \cdot (P_H^{-,=,U})^T\right)].$$

The following theorem provides the matrix method to analyze whether state s is coalitional SEQ_c stable for H when H's opponents, $N - H$, are in a coalition.

Theorem 8.20 *For the graph model G, let $H \subseteq N$ be a coalition. State $s \in S$ is coalitional SEQ_c stable for H, denoted by $s \in S_H^{CSEQ_c}$, iff $M_H^{CSEQ_c}(s, s) = 0$.*

The proofs of the above theorems on coalitional stabilities indexed c are left as exercises.

(4) Matrix Representation of Coalitional Stabilities Indexed d

As mentioned before, similar to the individual stability case in Chap. 5 coalitional Nash stability does not examine countermoves by the opponents, so $S_H^{Nash_d} = S_H^{Nash_b}$.

Define coalitional $CGMR_d$ stability matrix for coalition H as

$$M_H^{CGMR_d} = CM_H^{+} \cdot [E - sign\left(M_{N-H} \cdot (P_H^{-,=,U})^T\right)], \tag{8.18}$$

where $H \subseteq N$. The following theorem establishes the matrix method to assess whether state s is coalitional GMR_d stable for H.

Theorem 8.21 *For the graph model G, let $H \subseteq N$ be a nonempty coalition. State $s \in S$ is coalitional GMR_d stable for H, denoted by $s \in S_H^{CGMR_d}$, iff $M_H^{CGMR_d}(s, s) = 0$.*

Proof Since

$$M_H^{CGMR_d}(s, s) = (e_s^T \cdot CM_H^{+}) \cdot [\left(E - sign\left(M_{N-H} \cdot (P_H^{-,=,U})^T\right)\right) \cdot e_s]$$

$$= \sum_{s_1=1}^{m} CM_H^{+}(s, s_1) \cdot [1 - sign\left((e_{s_1}^T \cdot M_{N-H}) \cdot (e_s^T \cdot P_H^{-,=,U})^T\right)],$$

then $M_H^{CGMR_d}(s, s) = 0$ holds iff

$$CM_H^{+}(s, s_1) \cdot [1 - sign\left((e_{s_1}^T \cdot M_{N-H}) \cdot (e_s^T \cdot P_H^{-,=,U})^T\right)] = 0, \tag{8.19}$$

for every $s_1 \in S - \{s\}$. It is clear that Eq. 8.19 is equivalent to

$$(e_{s_1}^T \cdot M_{N-H}) \cdot (e_s^T \cdot P_H^{-,=,U})^T \neq 0,$$

for every $s_1 \in CR_H^{+}(s)$. Therefore, for a coalitional improvement from s, $s_1 \in CR_H^{+}(s)$, there exists at least one $s_2 \in R_{N-H}(s_1)$ with $P_H^{-,=,U}(s, s_2) = 1$ that is equivalent to $s \succeq_i s_2$ or $s\ U_i\ s_2$ for some DM $i \in H$. According to Definition 8.27, $M_H^{CGMR_d}(s, s) = 0$ implies that s is coalitional GMR_d stable for H. □

Theorem 8.21 shows that this matrix method, called matrix representation of coalitional GMR_d stability, is equivalent to the logical version of the same coalitional GMR_d stability given in Definition 8.27.

Define the coalitional SMR_d stability matrix as

$$M_H^{CSMR_d} = CM_H^{+} \cdot [E - sign(F)]$$

in which

$$F = M_{N-H} \cdot [(P_H^{-,=,U})^T \circ \left(E - sign\left(M_H \cdot (E - P_H^{-,=,U})^T\right)\right)],$$

for $H \subseteq N$. The following theorem establishes the matrix method to determine whether state s is coalitional SMR_d stable for H.

Theorem 8.22 *For the graph model G, let $H \subseteq N$ be a coalition. State $s \in S$ is coalitional SMR_d for H, denoted by $s \in S_H^{CSMR_d}$, iff $M_H^{CSMR_d}(s, s) = 0$.*

Proof Since

$$M_H^{CSMR_d}(s, s) = (e_s^T \cdot CM_H^{+}) \cdot [(E - sign(F)) \cdot e_s]$$

$$= \sum_{s_1=1}^{m} CM_H^{+}(s, s_1)[1 - sign\left(F(s_1, s)\right)]$$

with

$$F(s_1, s) = \sum_{s_2=1}^{m} M_{N-H}(s_1, s_2) \cdot W(s_2, s),$$

and

$$W(s_2, s) = P_H^{-,=,U}(s, s_2) \cdot \left[1 - sign\left(\sum_{s_3=1}^{m}\left(M_H(s_2, s_3) \cdot (1 - P_H^{-,=,U}(s, s_3))\right)\right)\right],$$

then $M_H^{CSMR_d}(s, s) = 0$ holds iff $F(s_1, s) \neq 0$, for every $s_1 \in CR_H^+(s)$, which is equivalent to the statement that, for every $s_1 \in CR_H^+(s)$, there exists $s_2 \in R_{N-H}(s_1)$ such that

$$P_H^{-,=,U}(s, s_2) \neq 0, \tag{8.20}$$

and

$$\sum_{s_3=1}^{m} M_H(s_2, s_3) \cdot (1 - P_H^{-,=,U})(s, s_3) = 0. \tag{8.21}$$

Equation 8.20 means that $s \succeq_i s_2$ or $s\ U_i\ s_2$ for at least one DM $i \in H$. Equation 8.21 is equivalent to

$$P_H^{-,=,U}(s, s_3) \neq 0 \text{ for any } s_3 \in R_H(s_2). \tag{8.22}$$

Obviously, for every $s_1 \in CR_H^+(s)$, there exists $s_2 \in R_{N-H}(s_1)$ such that $s \succeq_i s_2$ or $s\ U_i\ s_2$ and Eq. 8.22 hold iff for every $s_1 \in CR_H^+(s)$ there exists $s_2 \in R_{N-H}(s_1)$ such that $s_2 \in \Phi_H^{\preceq,U}(s)$ and $s_3 \in \Phi_H^{\preceq,U}(s)$ for all $s_3 \in R_H(s_2)$. Therefore, the proof of this theorem follows using Definition 8.28. □

The coalitional SEQ_d stability matrix $M_H^{CSEQ_d}$ is defined as

$$M_H^{CSEQ_d} = CM_H^+ \cdot [E - sign\left(M_{N-H}^{+,U} \cdot (P_H^{-,=,U})^T\right)].$$

The following theorem provides the matrix method to analyze whether state s is coalitional SEQ_d stable for H when H's opponents, $N - H$, are in a coalition.

Theorem 8.23 *For the graph model G, let $H \subseteq N$ be a coalition. State $s \in S$ is coalitional SEQ_d stable for H, denoted by $s \in S_H^{CSEQ_d}$, iff $M_H^{CSEQ_d}(s, s) = 0$.*

Proof Since

$$M_H^{CSEQ_d}(s, s) = (e_s^T CM_H^+) \cdot [\left(E - sign(M_{N-H}^{+,U} \cdot (P_H^{-,=,U})^T)\right) e_s]$$
$$= \sum_{s_1=1}^{|S|} CM_H^+(s, s_1)[1 - sign\left((e_{s_1}^T M_{N-H}^{+,U}) \cdot (e_s^T P_H^{-,=,U})^T\right)],$$

then $M_H^{SEQ_d}(s, s) = 0$ holds iff

$$CM_H^+(s, s_1)[1 - sign\left((e_{s_1}^T M_{N-H}^{+,U}) \cdot (e_s^T P_H^{-,=,U})^T\right)] = 0, \forall s_1 \in S. \tag{8.23}$$

It is clear that Eq. 8.23 is equivalent to

$$(e_{s_1}^T M_{N-H}^{+,U}) \cdot (e_s^T P_H^{-,=,U})^T \neq 0 \text{ for any } s_1 \in CR_H^+(s).$$

It implies that for any $s_1 \in CR_H^+(s)$, there exists at least one $s_2 \in R_{N-H}^{+,U}(s_1)$ with $s \succeq_i s_2$ or $s\ U_i\ s_2$ for some DM $i \in H$ that satisfies $s_2 \in \Phi_H^{\preceq,U}(s)$. The proof of this theorem follows using Definition 8.29. □

8.8 Matrix Representation of Coalitional Stability with Three Degrees of Preference

The logical representation of coalitional stabilities under three-degree preference is discussed in Sect. 8.4. The matrix form of these coalitional stabilities is introduced as follows.

8.8.1 Matrix Representation of Mild or Strong Coalitional Improvement

Definition 8.73 For the graph model G, the mild or strong coalitional improvement matrix for coalition H is an $m \times m$ matrix $CM_H^{+,++}$ with (s, q) entry

$$CM_H^{+,++}(s,q) = \begin{cases} 1 & \text{if } q \in CR_H^{+,++}(s), \\ 0 & \text{otherwise.} \end{cases}$$

The mild or strong coalitional improvement matrix is equivalent to the coalitional reachable list, $CR_H^{+,++}(s)$, defined in Sect. 8.1. The matrix $CM_H^{+,++}$ can be constructed as follows.

To carry out coalitional stability analysis, recall a set of matrices corresponding to three-level preference defined in Chap. 6.

$$P_i^{++}(s,q) = \begin{cases} 1 \text{ if } q \gg_i s, \\ 0 \text{ otherwise,} \end{cases}$$

$$P_i^{--}(s,q) = \begin{cases} 1 \text{ if } s \gg_i q, \\ 0 \text{ otherwise,} \end{cases}$$

$$P_i^{+,++}(s,q) = \begin{cases} 1 \text{ if } q >_i s \text{ or } q \gg_i s, \\ 0 \text{ otherwise,} \end{cases}$$

and

$$P_i^{--,-,=}(s,q) = \begin{cases} 1 \text{ if } s >_i q, s \gg_i q, \text{ or } (s \sim_i q \text{ and } s \neq q), \\ 0 \text{ otherwise.} \end{cases}$$

Based on the above definitions, the UM adjacency matrix J_i, mild or strong unilateral improvement adjacency matrix $J_i^{+,++}$, and preference matrix $P_i^{+,++}$ for DM i have the relationship among them:

$$J_i^{+,++} = J_i \circ P_i^{+,++}.$$

Theorem 8.24 *For the graph model G, let $H \subseteq N$ be a nonempty coalition. The mild or strong coalitional improvement matrix for H is expressed as*

$$CM_H^{+,++} = M_H \circ (E - P_H^{-,--,=}). \tag{8.24}$$

Proof To prove Eq. 8.24, assume that $C = M_H \circ (E - P_H^{-,--,=})$. Using the definition for matrix M_H presented in Chap. 4, $C(s,q) = 1$ iff $M_H(s,q) = 1$ and $P_H^{-,--,=}(s,q) = 0$, which together imply that there is $q \in R_H(s)$ such that $P_i^{-,--,=}(s,q) = 0$ for every DM $i \in H$. Therefore, $C(s,q) = 1$ iff there is $q \in R_H(s)$ with $q >_i s$ or $q \gg_i s$ for every $i \in H$, so that $q \in CR_H^{+,++}(s)$, according to Definition 8.3. Hence, $CM_H^{+,++}(s,q) = 1$ iff $C(s,q) = 1$. Since CM_H^{+} and C are 0–1 matrices, it follows that $CM_H^{+,++} = M_H \circ (E - P_H^{-,--,=})$. □

Note that $CM_H^{+,++} \neq M_H \circ P_H^{+,++}$. Recall that matrix $P_H^{+,++} = \bigvee_{i \in H} P_i^{+,++}$ ("$\bigvee$" denotes the disjunction operator described in Definition 3.16). $(M_H \circ P_H^{+,++})(s,q) = 1$ iff $M_H(s,q) = 1$ and $P_H^{+,++}(s,q) = 1$, which means that there is $q \in R_H(s)$ such that $P_i^{+,++}(s,q) = 1$ for some DM $i \in H$. This is not consistent with the definition of $CM_H^{+,++}$.

8.8.2 *Matrix Representation of General Coalitional Stabilities*

Let $m = |S|$ denote the number of states and E be the $m \times m$ matrix with each entry equal to 1. For a fixed state $s \in S$, let e_s be an m-dimensional vector with 1 as its sth element and 0 everywhere else and e be an m-dimensional vector with every entry 1. Let $(\overrightarrow{0})^T$ denote the transpose of $\overrightarrow{0}$. Recall that the UM reachability matrix M_H is constructed using Theorem 4.9. General coalitional stabilities are presented using matrix approach next.

Theorem 8.25 *For the graph model G, let $H \subseteq N$ be a nonempty coalition. State $s \in S$ is general coalitional Nash stable for H, denoted by $s \in S_H^{GCNash}$, iff $e_s^T \cdot CM_H^{+,++} \cdot e = 0$.*

Proof Since $e_s^T \cdot CM_H^{+,++} \cdot e = 0$ iff $e_s^T \cdot CM_H^{+,++} = (\vec{0})^T$, then $CR_H^{+,++}(s) = \emptyset$. Consequently, the proof of the theorem follows by Definition 8.30. □

Similar to the individual general GMR stability for the three-degree preference, define general coalitional GMR stability matrix as

$$M_H^{GCGMR} = CM_H^{+,++} \cdot [E - sign\left(M_{N-H} \cdot (P_H^{-,--,=})^T\right)], \tag{8.25}$$

where $H \subseteq N$. The following theorem establishes the matrix method to assess whether state s is general coalitional GMR stable for H.

Theorem 8.26 *For the graph model G, let $H \subseteq N$ be a nonempty coalition. State $s \in S$ is general coalitional GMR stable for H, denoted by $s \in S_H^{GCGMR}$, iff $M_H^{GCGMR}(s, s) = 0$.*

Proof Since

$$M_H^{GCGMR}(s, s) = (e_s^T \cdot CM_H^{+,++}) \cdot [\left(E - sign\left(M_{N-H} \cdot (P_H^{-,--,=})^T\right)\right) \cdot e_s]$$

$$= \sum_{s_1=1}^{m} CM_H^{+,++}(s, s_1) \cdot [1 - sign\left((e_{s_1}^T \cdot M_{N-H}) \cdot (e_s^T \cdot P_H^{-,--,=})^T\right)],$$

then $M_H^{GCGMR}(s, s) = 0$ holds iff

$$CM_H^{+,++}(s, s_1) \cdot [1 - sign\left((e_{s_1}^T \cdot M_{N-H}) \cdot (e_s^T \cdot P_H^{-,--,=})^T\right)] = 0, \tag{8.26}$$

for every $s_1 \in S - s$. It is clear that Eq. 8.26 is equivalent to

$$(e_{s_1}^T \cdot M_{N-H}) \cdot (e_s^T \cdot P_H^{-,--,=})^T \neq 0,$$

for every $s_1 \in CR_H^{+,++}(s)$. Therefore, for a coalitional mild or strong improvement from s, $s_1 \in CR_H^{+,++}(s)$, there exists at least one $s_2 \in R_{N-H}(s_1)$ with $P_H^{-,--,=}(s, s_2) = 1$ that is equivalent to $s >_i s_2$, $s \gg_i s_2$ or $s \sim_i s_2$ for some DM $i \in H$. According to Definition 8.31, $M_H^{GCGMR}(s, s) = 0$ implies that s is general coalitional GMR stable for H. □

Theorem 8.26 shows that this matrix method, called matrix representation of general coalitional GMR stability, is equivalent to the logical version of the same stability given in Definition 8.31. To analyze the general coalitional GMR stability at s for coalition H, one only needs to identify whether the diagonal entry $M_H^{GCGMR}(s, s)$ is zero. If so, s is general coalitional GMR stable for H; otherwise, s is general coalitional GMR unstable for H.

General coalitional SMR stability is similar to general coalitional GMR except that coalition H expects to have a chance to counterrespond to its opponents' $(N-H)$ response to H's original move. Define the general coalitional SMR stability matrix as

$$M_H^{GCSMR} = CM_H^{+,++} \cdot [E - sign(F)]$$

in which

$$F = M_{N-H} \cdot [(P_H^{-,--,=})^T \circ \left(E - sign\left(M_H \cdot (E - P_H^{-,--,=})^T\right)\right)],$$

for $H \subseteq N$. The following theorem establishes the matrix method to determine whether state s is general coalitional SMR stable for H.

Theorem 8.27 *For the graph model G, let $H \subseteq N$ be a coalition. State $s \in S$ is general coalitional SMR for H, denoted by $s \in S_H^{GCSMR}$, iff $M_H^{GCSMR}(s, s) = 0$.*

Proof Since

$$M_H^{GCSMR}(s, s) = (e_s^T \cdot CM_H^{+,++}) \cdot [(E - sign(F)) \cdot e_s]$$

$$= \sum_{s_1=1}^{m} CM_H^{+,++}(s, s_1)[1 - sign\,(F(s_1, s))]$$

with

$$F(s_1, s) = \sum_{s_2=1}^{m} M_{N-H}(s_1, s_2) \cdot W(s_2, s),$$

and

$$W(s_2, s) = P_H^{-,--,=}(s, s_2) \cdot \left[1 - sign\left(\sum_{s_3=1}^{m} \left(M_H(s_2, s_3) \cdot (1 - P_H^{+,++}(s, s_3))\right)\right)\right],$$

then $M_H^{GCSMR}(s, s) = 0$ holds iff $F(s_1, s) \neq 0$, for every $s_1 \in CR_H^{+,++}(s)$, which is equivalent to the statement that, for every $s_1 \in CR_H^{+,++}(s)$, there exists $s_2 \in R_{N-H}(s_1)$ such that

$$P_H^{-,--,=}(s, s_2) \neq 0, \tag{8.27}$$

and

$$\sum_{s_3=1}^{m} M_H(s_2, s_3) \cdot (1 - P_H^{-,--,=}(s, s_3)) = 0. \tag{8.28}$$

Equation 8.27 means that $s >_i s_2$, $s \gg_i s_2$, or $s \sim_i s_2$ for at least one DM $i \in H$. Equation 8.28 is equivalent to

$$P_H^{-,--,=}(s, s_3) = 1 \text{ for any } s_3 \in R_H(s_2). \tag{8.29}$$

Obviously, for every $s_1 \in CR_H^{+,++}(s)$, there exists $s_2 \in R_{N-H}(s_1)$ such that Eqs. 8.27 and 8.28 hold iff for every $s_1 \in CR_H^{+,++}(s)$ there exists $s_2 \in R_{N-H}(s_1)$ such that $s >_i s_2$, $s \gg_i s_2$, or $s \sim_i s_2$ and $s >_i s_3$, $s \gg_i s_3$, or $s \sim_i s_3$ for some DM i with all $s_3 \in R_H(s_2)$. Therefore, the proof of this theorem follows using Definition 8.32. □

Theorem 8.27 displays that this matrix method, called matrix representation of general coalitional SMR stability, is equivalent to the logical version given in Definition 8.32. To calculate general coalitional SMR stability at s for H, one only needs to assess whether the diagonal entry $M_H^{GCSMR}(s, s)$ is zero. If so, s is general coalitional SMR stable for H; otherwise, s is general coalitional SMR unstable for H.

General coalitional sequential stability is similar to general coalitional GMR stability, but includes only those sanctions that are "credible". If H's opponents are treated as a coalition, the general coalitional SEQ_1 stability matrix $M_H^{GCSEQ_1}$ is defined as

$$M_H^{GCSEQ_1} = CM_H^{+,++} \cdot [E - sign\left(CM_{N-H}^{+,++} \cdot (P_H^{-,--,=})^T\right)].$$

The following theorem provides the matrix method to analyze whether state s is general coalitional SEQ_1 stable for H when H's opponents, $N-H$, are in a coalition.

Theorem 8.28 *For the graph model G, let $H \subseteq N$ be a coalition. State $s \in S$ is general coalitional SEQ_1 stable for H, denoted by $s \in S_H^{GCSEQ_1}$, iff $M_H^{GCSEQ_1}(s, s) = 0$.*

Proof Since

$$M_H^{GCSEQ_1}(s, s) = (e_s^T \cdot CM_H^{+,++}) \cdot [\left(E - sign(CM_{N-H}^{+,++} \cdot (P_H^{-,--,=})^T)\right) e_s]$$

$$= \sum_{s_1=1}^{|S|} CM_H^{+,++}(s, s_1)[1 - sign\left((e_{s_1}^T CM_{N-H}^{+,++}) \cdot (e_s^T P_H^{-,--,=})^T\right)],$$

then $M_H^{GCSEQ_1}(s, s) = 0$ holds iff

$$CM_H^{+,++}(s, s_1)[1 - sign\left((e_{s_1}^T \cdot CM_{N-H}^{+,++}) \cdot (e_s^T \cdot P_H^{-,--,=})^T\right)] = 0, \forall s_1 \in S. \tag{8.30}$$

It is clear that Eq. 8.30 is equivalent to

$$(e_{s_1}^T \cdot CM_{N-H}^{+,++}) \cdot (e_s^T \cdot P_H^{-,--,=})^T \neq 0 \text{ for any } s_1 \in CR_H^{+,++}(s).$$

It implies that for any $s_1 \in CR_H^{+,++}(s)$, there exists at least one $s_2 \in CR_{N-H}^{+,++}(s_1)$ with $s >_i s_2$, $s \gg_i s_2$ or $s \sim_i s_2$ for some DM $i \in H$. The proof of this theorem follows using Definition 8.33. □

Note that the general coalitional SEQ_1 stability matrix is identical to the general coalitional GMR stability matrix except that the UM reachability matrix for H's opponents, M_{N-H}, is replaced by the mild or strong coalitional improvement matrix $CM^{+,++}_{N-H}$.

Similar to the previous two theorems, the matrix representation of general coalitional SEQ_1 stability is equivalent to the logical version given in Definition 8.33. When the diagonal entry at (s, s) is zero, the state s under consideration is general coalitional SEQ_1 stable for H. The following theorem is equivalent to the coalitional SEQ_2 stability presented in Definition 8.34. Define the coalitional SEQ_2 stability matrix $M_H^{GCSEQ_2}$ as

$$M_H^{GCSEQ_2} = CM_H^{+,++} \cdot [E - sign\left(M_{N-H}^{+,++} \cdot (P_H^{-,--,=})^T\right)].$$

Theorem 8.29 *For the graph model G, let $H \subseteq N$ be a coalition. State $s \in S$ is general coalitional SEQ_2 stable for H, denoted by $s \in S_H^{GCSEQ_2}$, iff $M_H^{GCSEQ_2}(s, s) = 0$.*

The proof of this theorem is similar to the proof of Theorem 8.28.

8.8.3 Matrix Representation of Strong Coalitional Stabilities

When three degrees of preference is introduced into the graph model, general coalitional stability definitions may be strong or weak, according to the strength of sanctioning. The following matrix representations of strong or weak coalitional stabilities are equivalent to the logical forms presented in Sect. 8.4.2.

Define the strong coalitional GMR stability matrix as

$$M_H^{SCGMR} = CM_H^{+,++} \cdot [E - sign\left(M_{N-H} \cdot (P_H^{--})^T\right)], \tag{8.31}$$

where $H \subseteq N$. The following theorem establishes the matrix method to assess whether state s is strong coalitional GMR stable for H.

Theorem 8.30 *For the graph model G, let $H \subseteq N$ be a nonempty coalition. State $s \in S$ is strong coalitional GMR stable for H, denoted by $s \in S_H^{SCGMR}$, iff $M_H^{SCGMR}(s, s) = 0$.*

Proof Since

$$M_H^{SCGMR}(s, s) = (e_s^T \cdot CM_H^{+,++}) \cdot [\left(E - sign\left(M_{N-H} \cdot (P_H^{--})^T\right)\right) \cdot e_s]$$

$$= \sum_{s_1=1}^{m} CM_H^{+,++}(s, s_1) \cdot [1 - sign\left((e_{s_1}^T \cdot M_{N-H}) \cdot (e_s^T \cdot P_H^{--})^T\right)],$$

then $M_H^{SCGMR}(s,s)=0$ holds iff

$$CM_H^{+,++}(s,s_1)\cdot[1-sign\left((e_{s_1}^T\cdot M_{N-H})\cdot(e_s^T\cdot P_H^{--})^T\right)]=0, \tag{8.32}$$

for every $s_1\in S-\{s\}$. It is clear that Eq. 8.32 is equivalent to

$$(e_{s_1}^T\cdot M_{N-H})\cdot(e_s^T\cdot P_H^{--})^T\neq 0,$$

for every $s_1\in CR_H^{+,++}(s)$. Therefore, for a mild or strong coalitional improvement from s, $s_1\in CR_H^{+,++}(s)$, there exists at least one $s_2\in R_{N-H}(s_1)$ with $P_H^{--}(s,s_2)=1$ that is equivalent to $s\gg_i s_2$ for some DM $i\in H$. According to Definition 8.35, $M_H^{SCGMR}(s,s)=0$ implies that s is strong coalitional GMR stable for H. □

Theorem 8.30 shows that this matrix method, called matrix representation of strong coalitional GMR stability, is equivalent to the logical version of the same stability given in Definition 8.35. To analyze the strong coalitional GMR stability at s for coalition H, the diagonal entry (s,s) of matrix M_H^{SCGMR} is identified whether it is zero. If so, s is strong coalitional GMR stable for H.

Define the strong coalitional SMR stability matrix as

$$M_H^{SCSMR}=CM_H^{+,++}\cdot[E-sign(F)]$$

in which

$$F=M_{N-H}\cdot[(P_H^{--})^T\circ\left(E-sign\left(M_H\cdot(E-P_H^{--})^T\right)\right)],$$

for $H\subseteq N$. The following theorem establishes the matrix method to determine whether state s is strong coalitional SMR stable for H.

Theorem 8.31 *For the graph model G, let $H\subseteq N$ be a coalition. State $s\in S$ is strong coalitional SMR for H, denoted by $s\in S_H^{SCSMR}$, iff $M_H^{SCSMR}(s,s)=0$.*

Proof Since

$$M_H^{SCSMR}(s,s)=(e_s^T\cdot CM_H^{+,++})\cdot[(E-sign(F))\cdot e_s]$$

$$=\sum_{s_1=1}^{m}CM_H^{+,++}(s,s_1)[1-sign\left(F(s_1,s)\right)]$$

with

$$F(s_1,s)=\sum_{s_2=1}^{m}M_{N-H}(s_1,s_2)\cdot W(s_2,s),$$

and

$$W(s_2, s) = P_H^{--}(s, s_2) \cdot \left[1 - sign\left(\sum_{s_3=1}^{m} \left(M_H(s_2, s_3) \cdot (1 - P_H^{--}(s, s_3))\right)\right)\right],$$

then $M_H^{SCSMR}(s, s) = 0$ holds iff $F(s_1, s) \neq 0$, for every $s_1 \in CR_H^{+,++}(s)$, which is equivalent to the statement that, for every $s_1 \in CR_H^{+,++}(s)$, there exists $s_2 \in R_{N-H}(s_1)$ such that

$$P_H^{--}(s, s_2) \neq 0, \tag{8.33}$$

and

$$\sum_{s_3=1}^{m} M_H(s_2, s_3) \cdot (1 - P_H^{--}(s, s_3)) = 0. \tag{8.34}$$

Equation 8.33 means that $s \gg_i s_2$ for at least one DM $i \in H$. Equation 8.34 is equivalent to

$$P_H^{--}(s, s_3) = 1 \text{ for any } s_3 \in R_H(s_2). \tag{8.35}$$

Obviously, for every $s_1 \in CR_H^{+,++}(s)$, there exists $s_2 \in R_{N-H}(s_1)$ such that Eqs. 8.33 and 8.34 hold iff for every $s_1 \in CR_H^{+,++}(s)$ there exists $s_2 \in R_{N-H}(s_1)$ such that $s \gg_i s_2$ and $s \gg_i s_3$ for some DM i with all $s_3 \in R_H(s_2)$. Therefore, the proof of this theorem follows using Definition 8.36. □

Theorem 8.31 provides a matrix method, called matrix representation of strong coalitional SMR stability, which is equivalent to the logical version given in Definition 8.36. The following theorem displays the matrix method to identify whether state s is strong coalitional SEQ_1 stable for H when H's opponents, $N - H$, are in a coalition. Let the strong coalitional SEQ_1 stability matrix $M_H^{SCSEQ_1}$ be defined as

$$M_H^{SCSEQ_1} = CM_H^{+,++} \cdot [E - sign\left(CM_{N-H}^{+,++} \cdot (P_H^{--})^T\right)].$$

Theorem 8.32 *For the graph model G, let $H \subseteq N$ be a coalition. State $s \in S$ is strong coalitional SEQ_1 stable for H, denoted by $s \in S_H^{SCSEQ_1}$, iff $M_H^{SCSEQ_1}(s, s) = 0$.*

Proof Since

$$M_H^{SCSEQ_1}(s, s) = (e_s^T \cdot CM_H^{+,++}) \cdot [\left(E - sign(CM_{N-H}^{+,++} \cdot (P_H^{--})^T)\right) e_s]$$

$$= \sum_{s_1=1}^{|S|} CM_H^{+,++}(s, s_1)[1 - sign\left((e_{s_1}^T CM_{N-H}^{+,++}) \cdot (e_s^T P_H^{--})^T\right)],$$

then $M_H^{SCSEQ_1}(s, s) = 0$ holds iff

$$CM_H^{+,++}(s, s_1)[1 - sign\left((e_{s_1}^T \cdot CM_{N-H}^{+,++}) \cdot (e_s^T \cdot P_H^{--})^T\right)] = 0, \forall s_1 \in S. \quad (8.36)$$

It is clear that Eq. 8.36 is equivalent to

$$(e_{s_1}^T \cdot CM_{N-H}^{+,++}) \cdot (e_s^T \cdot P_H^{--})^T \neq 0 \text{ for any } s_1 \in CR_H^{+,++}(s).$$

It implies that for any $s_1 \in CR_H^{+,++}(s)$, there exists at least one $s_2 \in CR_{N-H}^{+,++}(s_1)$ with $s \gg_i s_2$ for some DM $i \in H$. The proof of this theorem follows using Definition 8.37. □

The following theorem is equivalent to the strong coalitional SEQ_2 stability presented in Definition 8.38. The strong coalitional SEQ_2 stability matrix $M_H^{SCSEQ_2}$ is defined as

$$M_H^{SCSEQ_2} = CM_H^{+,++} \cdot [E - sign\left(M_{N-H}^{+,++} \cdot (P_H^{--})^T\right)].$$

Theorem 8.33 *For the graph model G, let $H \subseteq N$ be a coalition. State $s \in S$ is strong coalitional SEQ_2 stable for H, denoted by $s \in S_H^{SCSEQ_2}$, iff $M_H^{SCSEQ_2}(s, s) = 0$.*

The proof of this theorem is left as an exercise.

8.9 Matrix Representation of Coalitional Stability with Hybrid Preference

After discussing matrix representations of coalitional stabilities with unknown preference and with three degrees of preference, respectively, it is nature to construct the matrix form of the coalitional stabilities under hybrid preference.

8.9.1 Matrix Representation of Coalitional Improvement Under Hybrid Preference

Definition 8.74 For the graph model G, the mild or strong or uncertain coalitional improvement matrix for coalition H is an $m \times m$ matrix $CM_H^{+,++,U}$ with (s, q) entry

$$CM_H^{+,++,U}(s, q) = \begin{cases} 1 & \text{if } q \in CR_H^{+,++,U}(s), \\ 0 & \text{otherwise.} \end{cases}$$

The mild or strong or uncertain coalitional improvement matrix is equivalent to the coalitional reachable list $CR_H^{+,++,U}(s)$ given in Definition 8.4. The matrix $CM_H^{+,++,U}$ can be constructed as follows. To carry out coalitional stability analysis, recall a set of matrices corresponding to hybrid preference defined in Chap. 7.

The following $m \times m$ matrices are important in stability definitions under hybrid preference. Let E denote the $m \times m$ matrix with each entry 1 and let I be the $m \times m$ unit matrix. Then, $m \times m$ preference matrix $P_i^{+,++,U}$ is defined as

$$P_i^{+,++,U}(s,q) = \begin{cases} 1 \text{ if } q >_i s, \; q \gg_i s, \text{ or } q \; U_i \; s, \\ 0 \text{ otherwise.} \end{cases}$$

For hybrid preference, $P_i^{--,-,=} = E - I - P_i^{+,++,U}$.

Theorem 8.34 *For the graph model G, let $H \subseteq N$ be a nonempty coalition. The mild or strong coalitional improvement matrix, $CM_H^{+,++}$, and mild or strong or uncertain coalitional improvement matrix, $CM_H^{+,++,U}$, for H are expressed as*

$$CM_H^{+,++} = M_H \circ (E - P_H^{-,--,=,U}), \tag{8.37}$$

$$CM_H^{+,++,U} = M_H \circ (E - P_H^{-,--,=}), \tag{8.38}$$

respectively.

Proof Equation 8.37 is left as an exercise. To prove Eq. 8.38, assume that $C = M_H \circ (E - P_H^{-,--,=})$. Using the definition for matrix M_H given in Chap. 4, $C(s,q) = 1$ iff $M_H(s,q) = 1$ and $P_H^{-,--,=}(s,q) = 0$, which together imply that there is $q \in R_H(s)$ such that $P_i^{+,++,U}(s,q) = 1$ for every DM $i \in H$. Therefore, $C(s,q) = 1$ iff there is $q \in R_H(s)$ with $q >_i s$, $q \gg_i s$ or $q \; U_i \; s$ for every $i \in H$, so that $q \in CR_H^{+,++,U}(s)$, according to Definition 8.74. Hence, $CM_H^{+,++,U}(s,q) = 1$ iff $C(s,q) = 1$. Since $CM_H^{+,++,U}$ and C are 0–1 matrices, it follows that $CM_H^{+,++,U} = M_H \circ (E - P_H^{-,--,=})$. □

Note that $CM_H^{+,++}$ here is different from the matrix in Theorem 8.24 which cannot be used to analyze situations with uncertain preference. Furthermore, $CM_H^{+,++,U} \neq M_H \circ P_H^{+,++,U}$. Recall that matrix $P_H^{+,++,U} = \bigvee_{i \in H} P_i^{+,++,U}$ ("$\bigvee$" denotes the disjunction operator described in Definition 3.16). $(M_H \circ P_H^{+,++,U})(s,q) = 1$ iff $M_H(s,q) = 1$ and $P_H^{+,++,U}(s,q) = 1$, which means that there is $q \in R_H(s)$ such that $P_i^{+,++,U}(s,q) = 1$ for some DM $i \in H$. This is not consistent with the definition of $CM_H^{+,++,U}$.

8.9.2 Matrix Representation of General Coalitional Stabilities with Hybrid Preference

8.9.2.1 Matrix Representation of General Coalitional Stabilities Indexed *l*

(1) Matrix Representation of General Coalitional Stabilities Indexed *a*

Let $m = |S|$ denote the number of states and E be the $m \times m$ matrix with each entry equal to 1. For a fixed state $s \in S$, let e_s be an m-dimensional vector with 1 as its sth element and 0 everywhere else and e be an m-dimensional vector with every entry 1. Let $(\overrightarrow{0})^T$ denote the transpose of $\overrightarrow{0}$. Recall that the UM reachability matrix M_H is constructed using Theorem 4.9. General coalitional stabilities are presented using matrix approach next.

Theorem 8.35 *For the graph model G, let $H \subseteq N$ be a nonempty coalition. State $s \in S$ is general coalitional $Nash_a$ stable for H, denoted by $s \in S_H^{GCNash_a}$, iff $e_s^T \cdot CM_H^{+,++,U} \cdot e = 0$.*

Proof Since $e_s^T \cdot CM_H^{+,++,U} \cdot e = 0$ iff $e_s^T \cdot CM_H^{+,++,U} = (\overrightarrow{0})^T$, then $CR_H^{+,++,U}(s) = \emptyset$ using Definition 8.74. Consequently, the proof of the theorem follows by Definition 8.41. □

Similar to the individual general GMR_a stability for the hybrid preference, define general coalitional GMR_a stability matrix as

$$M_H^{GCGMR_a} = CM_H^{+,++,U} \cdot [E - sign\left(M_{N-H} \cdot (P_H^{-,--,=})^T\right)], \tag{8.39}$$

where $H \subseteq N$. The following theorem establishes the matrix method to assess whether state s is general coalitional GMR_a stable for H.

Theorem 8.36 *For the graph model G, let $H \subseteq N$ be a nonempty coalition. State $s \in S$ is general coalitional GMR_a stable for H, denoted by $s \in S_H^{GCGMR_a}$, iff $M_H^{GCGMR_a}(s, s) = 0$.*

Proof Since

$$M_H^{GCGMR_a}(s, s) = (e_s^T \cdot CM_H^{+,++,U}) \cdot [\left(E - sign\left(M_{N-H} \cdot (P_H^{-,--,=})^T\right)\right) \cdot e_s]$$

$$= \sum_{s_1=1}^{m} CM_H^{+,++,U}(s, s_1) \cdot [1 - sign\left((e_{s_1}^T \cdot M_{N-H}) \cdot (e_s^T \cdot P_H^{-,--,=})^T\right)],$$

then $M_H^{GCGMR_a}(s, s) = 0$ holds iff

$$CM_H^{+,++,U}(s, s_1) \cdot [1 - sign\left((e_{s_1}^T \cdot M_{N-H}) \cdot (e_s^T \cdot P_H^{-,--,=})^T\right)] = 0, \tag{8.40}$$

for every $s_1 \in S - \{s\}$. It is clear that Eq. 8.40 is equivalent to

$$(e_{s_1}^T \cdot M_{N-H}) \cdot (e_s^T \cdot P_H^{-,--,=})^T \neq 0,$$

for every $s_1 \in CR_H^{+,++,U}(s)$. Therefore, for a coalitional mild, strong, or uncertain improvement from s, $s_1 \in CR_H^{+,++,U}(s)$, there exists at least one $s_2 \in R_{N-H}(s_1)$ with $P_H^{-,--,=}(s, s_2) = 1$ that is equivalent to $s_2 \in \Phi_H^{\ll,<,\sim}(s)$, i.e., $s >_i s_2$, $s \gg_i s_2$, or $s \sim_i s_2$ for some DM $i \in H$. According to Definition 8.42, $M_H^{GCGMR_a}(s, s) = 0$ implies that s is general coalitional GMR_a stable for H. □

Theorem 8.36 shows that the matrix representation of general coalitional GMR_a stability is equivalent to the logical version of the same stability given in Definition 8.42. To analyze the general coalitional GMR_a stability at s for coalition H, one only needs to identify whether the diagonal entry $M_H^{GCGMR_a}(s, s)$ is zero.

General coalitional SMR_a stability is similar to general coalitional GMR_a except that coalition H expects to have a chance to counterrespond to its opponents' $(N - H)$ response to H's original move. Define the general coalitional SMR_a stability matrix as

$$M_H^{GCSMR_a} = CM_H^{+,++,U} \cdot [E - sign(F)]$$

in which

$$F = M_{N-H} \cdot [(P_H^{-,--,=})^T \circ \left(E - sign\left(M_H \cdot (E - P_H^{-,--,=})^T\right)\right)],$$

for $H \subseteq N$. The following theorem establishes the matrix method to determine whether state s is general coalitional SMR_a stable for H.

Theorem 8.37 *For the graph model G, let $H \subseteq N$ be a coalition. State $s \in S$ is general coalitional SMR_a for H, denoted by $s \in S_H^{GCSMR_a}$, iff $M_H^{GCSMR_a}(s, s) = 0$.*

Proof Since

$$M_H^{GCSMR_a}(s, s) = (e_s^T \cdot CM_H^{+,++,U}) \cdot [(E - sign(F)) \cdot e_s]$$

$$= \sum_{s_1=1}^{m} CM_H^{+,++,U}(s, s_1)[1 - sign\,(F(s_1, s))]$$

with

$$F(s_1, s) = \sum_{s_2=1}^{m} M_{N-H}(s_1, s_2) \cdot W(s_2, s),$$

and

$$W(s_2, s) = P_H^{-,--,=}(s, s_2) \cdot \left[1 - sign\left(\sum_{s_3=1}^{m} \left(M_H(s_2, s_3) \cdot (1 - P_H^{-,--,=}(s, s_3))\right)\right)\right],$$

then $M_H^{GCSMR_a}(s,s) = 0$ holds iff $F(s_1, s) \neq 0$, for every $s_1 \in CR_H^{+,++,U}(s)$, which is equivalent to the statement that, for every $s_1 \in CR_H^{+,++,U}(s)$, there exists $s_2 \in R_{N-H}(s_1)$ such that

$$P_H^{-,--,=}(s, s_2) \neq 0, \tag{8.41}$$

and

$$\sum_{s_3=1}^{m} M_H(s_2, s_3) \cdot (1 - P_H^{-,--,=}(s, s_3)) = 0. \tag{8.42}$$

Equation 8.41 means that $s >_i s_2$, $s \gg_i s_2$, or $s \sim_i s_2$ for at least one DM $i \in H$. Equation 8.42 is equivalent to

$$P_H^{-,--,=}(s, s_3) = 1 \text{ for any } s_3 \in R_H(s_2). \tag{8.43}$$

Obviously, for every $s_1 \in CR_H^{+,++,U}(s)$, there exists $s_2 \in R_{N-H}(s_1)$ such that Eqs. 8.41 and 8.42 hold iff for every $s_1 \in CR_H^{+,++,U}(s)$ there exists $s_2 \in R_{N-H}(s_1)$ such that $s_2 \in \Phi_H^{\ll,<,\sim}(s)$ and $s_3 \in \Phi_H^{\ll,<,\sim}(s)$ with all $s_3 \in R_H(s_2)$. Therefore, the proof of this theorem follows using Definition 8.43. □

Theorem 8.37 displays the matrix representation of general coalitional SMR_a stability, which is equivalent to the logical version given in Definition 8.43. To calculate general coalitional SMR_a stability at s for H, one only needs to assess whether the diagonal entry $M_H^{GCSMR_a}(s,s)$ is zero.

General coalitional sequential stability is similar to general coalitional GMR stability, but includes only those sanctions that are "credible". The logical representation of two types of coalitional SEQ stability under hybrid preference was discussed, the matrix form is provided here for $CSEQ_2$ only. If H's opponents are treated as a coalition, the general coalitional SEQ_a stability matrix $M_H^{GCSEQ_a}$ is defined as

$$M_H^{GCSEQ_a} = CM_H^{+,++,U} \cdot [E - sign\left(M_{N-H}^{+,++,U} \cdot (P_H^{-,--,=})^T\right)].$$

The following theorem provides the matrix method to analyze whether state s is general coalitional SEQ_a stable for H when H's opponents, $N-H$, are in a coalition.

Theorem 8.38 *For the graph model G, let $H \subseteq N$ be a coalition. State $s \in S$ is general coalitional SEQ_a stable for H, denoted by $s \in S_H^{GCSEQ_a}$, iff $M_H^{GCSEQ_a}(s,s) = 0$.*

Proof Since

$$M_H^{GCSEQ_a}(s,s) = (e_s^T \cdot CM_H^{+,++,U}) \cdot [\left(E - sign(M_{N-H}^{+,++,U} \cdot (P_H^{-,--,=})^T)\right) e_s]$$

$$= \sum_{s_1=1}^{|S|} CM_H^{+,++,U}(s, s_1)[1 - sign\left((e_{s_1}^T M_{N-H}^{+,++,U}) \cdot (e_s^T P_H^{-,--,=})^T\right)],$$

then $M_H^{GCSEQ_a}(s, s) = 0$ holds iff

$$CM_H^{+,++,U}(s, s_1)[1 - sign\left((e_{s_1}^T \cdot M_{N-H}^{+,++,U}) \cdot (e_s^T \cdot P_H^{-,--,=})^T\right)] = 0, \forall s_1 \in S. \tag{8.44}$$

It is clear that Eq. 8.44 is equivalent to

$$(e_{s_1}^T \cdot M_{N-H}^{+,++,U}) \cdot (e_s^T \cdot P_H^{-,--,=})^T \neq 0 \text{ for any } s_1 \in CR_H^{+,++,U}(s).$$

It implies that for any $s_1 \in CR_H^{+,++,U}(s)$, there exists at least one $s_2 \in R_{N-H}^{+,++,U}(s_1)$ with $s_2 \in \Phi_H^{\ll,<,\sim}(s)$. The proof of this theorem follows using Definition 8.44. □

Note that the general coalitional SEQ_a stability matrix is identical to the general coalitional GMR_a stability matrix except that the UM reachability matrix for H's opponents, M_{N-H}, is replaced by the mild, strong or uncertain reachability improvement matrix $M_{N-H}^{+,++,U}$.

(2) Matrix Representation of General Coalitional Stabilities Indexed b

Theorem 8.39 *For the graph model G, let $H \subseteq N$ be a nonempty coalition. State $s \in S$ is general coalitional $Nash_b$ stable for H, denoted by $s \in S_H^{GCNash_b}$, iff $e_s^T \cdot CM_H^{+,++} \cdot e = 0$.*

This theorem is different from Theorem 8.25 presented in Sect. 8.8.2 though their representations are identical. Theorem 8.39 can analyze Nash stability with hybrid preference.

Define the general coalitional GMR_b stability matrix as

$$M_H^{GCGMR_b} = CM_H^{+,++} \cdot [E - sign\left(M_{N-H} \cdot (P_H^{-,--,=})^T\right)], \tag{8.45}$$

where $H \subseteq N$. The following theorem establishes the matrix method to assess whether state s is general coalitional GMR_b stable for H.

Theorem 8.40 *For the graph model G, let $H \subseteq N$ be a nonempty coalition. State $s \in S$ is general coalitional GMR_b stable for H, denoted by $s \in S_H^{GCGMR_b}$, iff $M_H^{GCGMR_b}(s, s) = 0$.*

General coalitional SMR_b stability is similar to general coalitional GMR_b except that coalition H expects to have a chance to counterrespond to its opponents' $(N - H)$ response to H's original move. Define the general coalitional SMR_b stability matrix as

$$M_H^{GCSMR_b} = CM_H^{+,++} \cdot [E - sign(Q)]$$

in which

$$Q = M_{N-H} \cdot [(P_H^{-,--,=})^T \circ (E - sign(M_H \cdot (E - P_H^{-,--,=})^T))],$$

for $H \subseteq N$. The following theorem establishes the matrix method to determine whether state s is general coalitional SMR_b stable for H.

Theorem 8.41 *For the graph model G, let $H \subseteq N$ be a coalition. State $s \in S$ is general coalitional SMR_b for H, denoted by $s \in S_H^{GCSMR_b}$, iff $M_H^{GCSMR_b}(s,s) = 0$.*

Although matrix representations of $GCNash_b$, $GCGMR_b$ and $GCSMR_b$ do not include uncertain preference, they may be used to analyze situations with preference uncertainty. If H's opponents are treated as a coalition, the general coalitional SEQ_b stability matrix $M_H^{GCSEQ_b}$ is defined as

$$M_H^{GCSEQ_b} = CM_H^{+,++} \cdot [E - sign\left(M_{N-H}^{+,++,U} \cdot (P_H^{-,--,=})^T\right)].$$

The following theorem provides the matrix method to analyze whether state s is general coalitional SEQ_b stable for H when H's opponents, $N-H$, are in a coalition.

Theorem 8.42 *For the graph model G, let $H \subseteq N$ be a coalition. State $s \in S$ is general coalitional SEQ_b stable for H, denoted by $s \in S_H^{GCSEQ_b}$, iff $M_H^{GCSEQ_b}(s,s) = 0$.*

The proofs of the general coalitional stabilities indexed b are similar to the general coalitional stabilities indexed a. The proofs are left for readers.

(3) Matrix Representation of General Coalitional Stabilities Indexed c

Theorem 8.43 *For the graph model G, let $H \subseteq N$ be a nonempty coalition. $S_H^{GCNash_c} = S_H^{GCNash_a}$.*

Let $M_H^{GCGMR_c}$ denote the general coalitional GMR_c matrix. It is defined by

$$M_H^{GCGMR_c} = CM_H^{+,++,U} \cdot [E - sign\left(M_{N-H} \cdot (P_H^{-,--,=,U})^T\right)], \tag{8.46}$$

where $H \subseteq N$. The following theorem establishes the matrix method to assess whether state s is general coalitional GMR_c stable for H.

Theorem 8.44 *For the graph model G, let $H \subseteq N$ be a nonempty coalition. State $s \in S$ is general coalitional GMR_c stable for H, denoted by $s \in S_H^{GCGMR_c}$, iff $M_H^{GCGMR_c}(s,s) = 0$.*

Proof Since

$$M_H^{GCGMR_c}(s,s) = (e_s^T \cdot CM_H^{+,++,U}) \cdot [\left(E - sign\left(M_{N-H} \cdot (P_H^{-,--,=,U})^T\right)\right) \cdot e_s]$$

$$= \sum_{s_1=1}^{m} CM_H^{+,++,U}(s,s_1) \cdot [1 - sign\left((e_{s_1}^T \cdot M_{N-H}) \cdot (e_s^T \cdot P_H^{-,--,=,U})^T\right)],$$

then $M_H^{GCGMR_c}(s,s) = 0$ holds iff

$$CM_H^{+,++,U}(s,s_1) \cdot [1 - sign\left((e_{s_1}^T \cdot M_{N-H}) \cdot (e_s^T \cdot P_H^{-,--,=,U})^T\right)] = 0, \quad (8.47)$$

for every $s_1 \in S - \{s\}$. It is clear that Eq. 8.47 is equivalent to

$$(e_{s_1}^T \cdot M_{N-H}) \cdot (e_s^T \cdot P_H^{-,--,=,U})^T \neq 0,$$

for every $s_1 \in CR_H^{+,++,U}(s)$. Therefore, for a coalitional mild, strong, or uncertain improvement from s, $s_1 \in CR_H^{+,++,U}(s)$, there exists at least one $s_2 \in R_{N-H}(s_1)$ with $P_H^{-,--,=,U}(s,s_2) = 1$ that is equivalent to $s_2 \in \Phi_H^{\ll,<,\sim,U}(s)$, i.e., $s >_i s_2$, $s \gg_i s_2$, $s \sim_i s_2$, or $s\ U_i\ s_2$ for some DM $i \in H$. According to Definition 8.50, $M_H^{GCGMR_c}(s,s) = 0$ implies that s is general coalitional GMR_c stable for H. □

General coalitional SMR_c stability is similar to general coalitional GMR_c except that coalition H expects to have a chance to counterrespond to its opponents' $(N-H)$ response to H's original move. Define the general coalitional SMR_c stability matrix as

$$M_H^{GCSMR_c} = CM_H^{+,++,U} \cdot [E - sign(F)]$$

in which

$$F = M_{N-H} \cdot [(P_H^{-,--,=,U})^T \circ \left(E - sign\left(M_H \cdot (E - P_H^{-,--,=,U})^T\right)\right)],$$

for $H \subseteq N$. The following theorem establishes the matrix method to determine whether state s is general coalitional SMR_c stable for H.

Theorem 8.45 *For the graph model G, let $H \subseteq N$ be a coalition. State $s \in S$ is general coalitional SMR_c for H, denoted by $s \in S_H^{GCSMR_c}$, iff $M_H^{GCSMR_c}(s,s) = 0$.*

Proof Since

$$M_H^{GCSMR_c}(s,s) = (e_s^T \cdot CM_H^{+,++,U}) \cdot [(E - sign(F)) \cdot e_s]$$

$$= \sum_{s_1=1}^{m} CM_H^{+,++,U}(s,s_1)[1 - sign\,(F(s_1,s))]$$

with

$$F(s_1,s) = \sum_{s_2=1}^{m} M_{N-H}(s_1,s_2) \cdot W(s_2,s),$$

and

$$W(s_2, s) = P_H^{-,--,=,U}(s, s_2) \cdot \left[1 - sign\left(\sum_{s_3=1}^{m}\left(M_H(s_2, s_3) \cdot (1 - P_H^{-,--,=,U}(s, s_3))\right)\right)\right],$$

then $M_H^{GCSMR_c}(s, s) = 0$ holds iff $F(s_1, s) \neq 0$, for every $s_1 \in CR_H^{+,++,U}(s)$, which is equivalent to the statement that, for every $s_1 \in CR_H^{+,++,U}(s)$, there exists $s_2 \in R_{N-H}(s_1)$ such that

$$P_H^{-,--,=,U}(s, s_2) \neq 0, \tag{8.48}$$

and

$$\sum_{s_3=1}^{m} M_H(s_2, s_3) \cdot (1 - P_H^{-,--,=,U}(s, s_3)) = 0. \tag{8.49}$$

Equation 8.48 means that $s >_i s_2$, $s \gg_i s_2$, $s \sim_i s_2$, or $s\ U_i\ s_2$ for at least one DM $i \in H$. Equation 8.49 is equivalent to

$$P_H^{-,--,=,U}(s, s_3) = 1 \text{ for any } s_3 \in R_H(s_2). \tag{8.50}$$

Obviously, for every $s_1 \in CR_H^{+,++,U}(s)$, there exists $s_2 \in R_{N-H}(s_1)$ such that Eqs. 8.48 and 8.49 hold iff for every $s_1 \in CR_H^{+,++,U}(s)$ there exists $s_2 \in R_{N-H}(s_1)$ such that $s_2 \in \Phi_H^{\ll,<,\sim,U}(s)$ and $s_3 \in \Phi_H^{\ll,<,\sim,U}(s)$ with all $s_3 \in R_H(s_2)$. Therefore, the proof of this theorem follows using Definition 8.51. □

The general coalitional SEQ_c stability matrix $M_H^{GCSEQ_c}$ is defined as

$$M_H^{GCSEQ_c} = CM_H^{+,++,U} \cdot [E - sign\left(M_{N-H}^{+,++,U} \cdot (P_H^{-,--,=,U})^T\right)].$$

The following theorem provides the matrix method to analyze whether state s is general coalitional SEQ_c stable for H when H's opponents, $N-H$, are in a coalition.

Theorem 8.46 *For the graph model G, let $H \subseteq N$ be a coalition. State $s \in S$ is general coalitional SEQ_c stable for H, denoted by $s \in S_H^{GCSEQ_c}$, iff $M_H^{GCSEQ_c}(s, s) = 0$.*

Proof Since

$$M_H^{GCSEQ_c}(s, s) = (e_s^T \cdot CM_H^{+,++,U}) \cdot [\left(E - sign(M_{N-H}^{+,++,U} \cdot (P_H^{-,--,=,U})^T)\right) e_s]$$

$$= \sum_{s_1=1}^{|S|} CM_H^{+,++,U}(s, s_1)[1 - sign\left((e_{s_1}^T M_{N-H}^{+,++,U}) \cdot (e_s^T P_H^{-,--,=,U})^T\right)],$$

then $M_H^{GCSEQ_c}(s, s) = 0$ holds iff

$$CM_H^{+,++,U}(s,s_1)[1 - sign\left((e_{s_1}^T \cdot M_{N-H}^{+,++,U}) \cdot (e_s^T \cdot P_H^{-,--,=,U})^T\right)] = 0, \forall s_1 \in S. \tag{8.51}$$

It is clear that Eq. 8.51 is equivalent to

$$(e_{s_1}^T \cdot M_{N-H}^{+,++,U}) \cdot (e_s^T \cdot P_H^{-,--,=,U})^T \neq 0 \text{ for any } s_1 \in CR_H^{+,++,U}(s).$$

It implies that for any $s_1 \in CR_H^{+,++,U}(s)$, there exists at least one $s_2 \in R_{N-H}^{+,++,U}(s_1)$ with $s_2 \in \Phi_H^{\ll,<,\sim,U}(s)$. The proof of this theorem follows using Definition 8.52. □

(4) Matrix Representation of General Coalitional Stabilities Indexed d

Theorem 8.47 *For the graph model G, let $H \subseteq N$ be a nonempty coalition.* $S_H^{GCNash_d} = S_H^{GCNash_b}$.

Define the general coalitional GMR_d stability matrix as

$$M_H^{GCGMR_d} = CM_H^{+,++} \cdot [E - sign\left(M_{N-H} \cdot (P_H^{-,--,=,U})^T\right)], \tag{8.52}$$

where $H \subseteq N$. The following theorem establishes the matrix method to assess whether state s is general coalitional GMR_d stable for H.

Theorem 8.48 *For the graph model G, let $H \subseteq N$ be a nonempty coalition. State $s \in S$ is general coalitional GMR_d stable for H, denoted by $s \in S_H^{GCGMR_d}$, iff $M_H^{GCGMR_d}(s,s) = 0$.*

General coalitional SMR_d stability is similar to general coalitional GMR_d except that coalition H expects to have a chance to counterrespond to its opponents' $(N-H)$ response to H's original move. Define the general coalitional SMR_d stability matrix as

$$M_H^{GCSMR_d} = CM_H^{+,++} \cdot [E - sign(Q)]$$

in which

$$Q = M_{N-H} \cdot [(P_H^{-,--,=,U})^T \circ \left(E - sign\left(M_H \cdot (E - P_H^{-,--,=,U})^T\right)\right)],$$

for $H \subseteq N$. The following theorem establishes the matrix method to determine whether state s is general coalitional SMR_d stable for H.

Theorem 8.49 *For the graph model G, let $H \subseteq N$ be a coalition. State $s \in S$ is general coalitional SMR_d for H, denoted by $s \in S_H^{GCSMR_d}$, iff $M_H^{GCSMR_d}(s,s) = 0$.*

The general coalitional SEQ_d stability matrix $M_H^{GCSEQ_d}$ is defined as

$$M_H^{GCSEQ_d} = CM_H^{+,++} \cdot [E - sign\left(M_{N-H}^{+,++,U} \cdot (P_H^{-,--,=,U})^T\right)].$$

The following theorem provides the matrix method to analyze whether state s is general coalitional SEQ_d stable for H when H's opponents, $N-H$, are in a coalition.

Theorem 8.50 *For the graph model G, let $H \subseteq N$ be a coalition. State $s \in S$ is general coalitional SEQ_d stable for H, denoted by $s \in S_H^{GCSEQ_d}$, iff $M_H^{GCSEQ_d}(s,s)=0$.*

The proofs of the general coalitional stabilities indexed d are similar to the general coalitional stabilities indexed a. The proofs are left for readers.

8.9.3 Matrix Representation of Strong Coalitional Stabilities with Hybrid Preference

When hybrid preference is introduced into the graph model, general coalitional stability definitions indexed a, b, c or d may be strong or weak coalitional stability definitions indexed a, b, c or d, according to the degree of sanctioning. The following matrix representations of strong coalitional stabilities under hybrid preference are equivalent to the logical forms presented in Sect. 8.5.2.

Theorem 8.51 *For the graph model G, let $H \subseteq N$ be a nonempty coalition and $l \in \{a, b, c, d\}$. State $s \in S$ is general or strong coalitional $Nash_l$ stable for H, denoted by $s \in S_H^{GNash_l}$ or $s \in S_H^{SNash_l}$, respectively. Then $S_H^{SNash_l} = S_H^{GNash_l}$.*

8.9.3.1 Matrix Representation of Strong Coalitional Stabilities Indexed l

(1) Matrix Representation of Strong Coalitional Stabilities Indexed a

The strong coalitional GMR_a stability matrix is defined as

$$M_H^{SCGMR_a} = CM_H^{+,++,U} \cdot [E - sign\left(M_{N-H} \cdot (P_H^{--})^T\right)], \tag{8.53}$$

where $H \subseteq N$. The following theorem establishes the matrix method to assess whether state s is strong coalitional GMR_a stable for H.

Theorem 8.52 *For the graph model G, let $H \subseteq N$ be a nonempty coalition. State $s \in S$ is strong coalitional GMR_a stable for H, denoted by $s \in S_H^{SCGMR_a}$, iff $M_H^{SCGMR_a}(s,s) = 0$.*

Theorem 8.52 shows that the matrix representation of strong coalitional GMR_a stability is equivalent to the logical version of the same stability given in Definition 8.59. The diagonal entry (s, s) of matrix $M_H^{SCGMR_a}$ is identified whether it is zero. If so, s is strong coalitional GMR_a stable for H.

Define the strong coalitional SMR_a stability matrix as

$$M_H^{SCSMR_a} = CM_H^{+,++,U} \cdot [E - sign(Q)]$$

in which

$$Q = M_{N-H} \cdot [(P_H^{--})^T \circ \left(E - sign\left(M_H \cdot (E - P_H^{--})^T\right)\right)],$$

for $H \subseteq N$. The following theorem establishes the matrix method to determine whether state s is strong coalitional SMR_a stable for H.

Theorem 8.53 *For the graph model G, let $H \subseteq N$ be a coalition. State $s \in S$ is strong coalitional SMR_a for H, denoted by $s \in S_H^{SCSMR_a}$, iff $M_H^{SCSMR_a}(s, s) = 0$.*

Theorem 8.53 provides a matrix method, which is equivalent to the logical version given in Definition 8.60. The following theorem displays the matrix method to identify whether state s is strong coalitional SEQ_a stable. Let the strong coalitional SEQ_a stability matrix $M_H^{SCSEQ_a}$ be defined as

$$M_H^{SCSEQ_a} = CM_H^{+,++,U} \cdot [E - sign\left(M_{N-H}^{+,++,U} \cdot (P_H^{--})^T\right)].$$

Theorem 8.54 *For the graph model G, let $H \subseteq N$ be a coalition. State $s \in S$ is strong coalitional SEQ_a stable for H, denoted by $s \in S_H^{SCSEQ_a}$, iff $M_H^{SCSEQ_a}(s, s) = 0$.*

(2) Matrix Representation of Strong Coalitional Stabilities Indexed b

Define the strong coalitional GMR_b stability matrix as

$$M_H^{SCGMR_b} = CM_H^{+,++} \cdot [E - sign\left(M_{N-H} \cdot (P_H^{--})^T\right)], \tag{8.54}$$

where $H \subseteq N$. The following theorem establishes the matrix method to assess whether state s is strong coalitional GMR_b stable for H.

Theorem 8.55 *For the graph model G, let $H \subseteq N$ be a nonempty coalition. State $s \in S$ is strong coalitional GMR_b stable for H, denoted by $s \in S_H^{SCGMR_b}$, iff $M_H^{SCGMR_b}(s, s) = 0$.*

Define the strong coalitional SMR_b stability matrix as

$$M_H^{SCSMR_b} = CM_H^{+,++} \cdot [E - sign(Q)]$$

in which

$$Q = M_{N-H} \cdot [(P_H^{--})^T \circ \left(E - sign\left(M_H \cdot (E - P_H^{--})^T\right)\right)],$$

for $H \subseteq N$. The following theorem establishes the matrix method to determine whether state s is strong coalitional SMR_b stable for H.

Theorem 8.56 *For the graph model G, let $H \subseteq N$ be a coalition. State $s \in S$ is strong coalitional SMR_b for H, denoted by $s \in S_H^{SCSMR_b}$, iff $M_H^{SCSMR_b}(s, s) = 0$.*

The following theorem displays the matrix method to identify whether state s is strong coalitional SEQ_b stable. Let the strong coalitional SEQ_b stability matrix $M_H^{SCSEQ_b}$ be defined as

$$M_H^{SCSEQ_b} = CM_H^{+,++} \cdot [E - sign\left(M_{N-H}^{+,++,U} \cdot (P_H^{--})^T\right)].$$

Theorem 8.57 *For the graph model G, let $H \subseteq N$ be a coalition. State $s \in S$ is strong coalitional SEQ_b stable for H, denoted by $s \in S_H^{SCSEQ_b}$, iff $M_H^{SCSEQ_b}(s, s) = 0$.*

(3) Matrix Representation of Strong Coalitional Stabilities Indexed c

The strong coalitional GMR_c stability matrix is defined as

$$M_H^{SCGMR_c} = CM_H^{+,++,U} \cdot [E - sign\left(M_{N-H} \cdot (P_H^{--,U})^T\right)], \tag{8.55}$$

where $H \subseteq N$. The following theorem establishes the matrix method to assess whether state s is strong coalitional GMR_c stable for H.

Theorem 8.58 *For the graph model G, let $H \subseteq N$ be a nonempty coalition. State $s \in S$ is strong coalitional GMR_c stable for H, denoted by $s \in S_H^{SCGMR_c}$, iff $M_H^{SCGMR_c}(s, s) = 0$.*

Theorem 8.58 shows that the matrix representation of strong coalitional GMR_c stability is equivalent to the logical version of the same stability given in Definition 8.65. The diagonal entry (s, s) of matrix $M_H^{SCGMR_c}$ is identified whether it is zero. If so, s is strong coalitional GMR_c stable for H.

Define the strong coalitional SMR_c stability matrix as

$$M_H^{SCSMR_c} = CM_H^{+,++,U} \cdot [E - sign(Q)]$$

in which

$$Q = M_{N-H} \cdot [(P_H^{--,U})^T \circ \left(E - sign\left(M_H \cdot (E - P_H^{--,U})^T\right)\right)],$$

for $H \subseteq N$. The following theorem establishes the matrix method to determine whether state s is strong coalitional SMR_c stable for H.

Theorem 8.59 *For the graph model G, let $H \subseteq N$ be a coalition. State $s \in S$ is strong coalitional SMR_c for H, denoted by $s \in S_H^{SCSMR_c}$, iff $M_H^{SCSMR_c}(s, s) = 0$.*

Theorem 8.59 provides a matrix method, which is equivalent to the logical version given in Definition 8.66. The following theorem displays the matrix method to identify whether state s is strong coalitional SEQ_c stable. Let the strong coalitional SEQ_c stability matrix $M_H^{SCSEQ_c}$ be defined as

$$M_H^{SCSEQ_c} = CM_H^{+,++,U} \cdot [E - sign\left(M_{N-H}^{+,++,U} \cdot (P_H^{--,U})^T\right)].$$

Theorem 8.60 *For the graph model G, let $H \subseteq N$ be a coalition. State $s \in S$ is strong coalitional SEQ_c stable for H, denoted by $s \in S_H^{SCSEQ_c}$, iff $M_H^{SCSEQ_c}(s, s) = 0$.*

(4) Matrix Representation of Strong Coalitional Stabilities Indexed d

Define the strong coalitional GMR_d stability matrix as

$$M_H^{SCGMR_d} = CM_H^{+,++} \cdot [E - sign\left(M_{N-H} \cdot (P_H^{--,U})^T\right)], \tag{8.56}$$

where $H \subseteq N$. The following theorem establishes the matrix method to assess whether state s is strong coalitional GMR_d stable for H.

Theorem 8.61 *For the graph model G, let $H \subseteq N$ be a nonempty coalition. State $s \in S$ is strong coalitional GMR_d stable for H, denoted by $s \in S_H^{SCGMR_d}$, iff $M_H^{SCGMR_d}(s, s) = 0$.*

Theorem 8.61 shows that the matrix representation of strong coalitional GMR_d stability is equivalent to the logical version of the same stability given in Definition 8.68. The diagonal entry (s, s) of matrix $M_H^{SCGMR_d}$ is identified whether it is zero. If so, s is strong coalitional GMR_d stable for H.

Define the strong coalitional SMR_d stability matrix as

$$M_H^{SCSMR_d} = CM_H^{+,++} \cdot [E - sign(Q)]$$

in which

$$Q = M_{N-H} \cdot [(P_H^{--,U})^T \circ \left(E - sign\left(M_H \cdot (E - P_H^{--,U})^T\right)\right)],$$

for $H \subseteq N$. The following theorem establishes the matrix method to determine whether state s is strong coalitional SMR_d stable for H.

Theorem 8.62 *For the graph model G, let $H \subseteq N$ be a coalition. State $s \in S$ is strong coalitional SMR_d for H, denoted by $s \in S_H^{SCSMR_d}$, iff $M_H^{SCSMR_d}(s, s) = 0$.*

Theorem 8.62 provides a matrix method, which is equivalent to the logical version given in Definition 8.69. The following theorem displays the matrix method to identify whether state s is strong coalitional SEQ_d stable. Define the strong coalitional SEQ_d stability matrix $M_H^{SCSEQ_d}$ as

$$M_H^{SCSEQ_d} = CM_H^{+,++} \cdot [E - sign\left(M_{N-H}^{+,++,U} \cdot (P_H^{--,U})^T\right)].$$

Table 8.1 Options and feasible states for the Lake Gisborne conflict

Federal								
1. Continue	N	Y	N	Y	N	Y	N	Y
Provincial								
2. Lift	N	N	Y	Y	N	N	Y	Y
Support								
3. Appeal	N	N	N	N	Y	Y	Y	Y
States	s_1	s_2	s_3	s_4	s_5	s_6	s_7	s_8

Theorem 8.63 *For the graph model G, let $H \subseteq N$ be a coalition. State $s \in S$ is strong coalitional SEQ_d stable for H, denoted by $s \in S_H^{SCSEQ_d}$, iff $M_H^{SCSEQ_d}(s, s) = 0$.*

The matrix representation of coalitional stabilities under hybrid preference presented in this section is identical with the logical form discussed in Sect. 8.5. However, the matrix form is more efficient for calculating coalitional stabilities than logical representation.

8.10 Application: Coalition Analysis for Lake Gisborne Conflict with Simple Preference

In this section, the matrix approach is used to analyze the coalitional stability for the Lake Gisborne conflict with simple preference. Recall from Sects. 5.4 and 7.5 that the graph model for the Lake Gisborne Conflict has the following DMs and options:

- Federal Government of Canada (**Federal**): its option is to continue a Canada-wide accord on the prohibition of bulk water export **(Continue)** or not,
- Provincial Government of Newfoundland and Labrador (**Provincial**): its option is to lift the ban on bulk water exports **(Lift)** or not, and
- Support groups (**Support**): their option is to appeal for continuing the Lake Gisborne project **(Appeal)** or not.

The three DMs and the options they control are listed on the left in Table 8.1. Together, the three options create eight possible states as listed on the right in Table 8.1, where a "Y" indicates that an option is selected by the DM controlling it and an "N" means that the option is not chosen. Each state, shown as a column of Ys and Ns in Table 8.1, represents a possible scenario as to what could occur. For instance, s_4 means that the Federal Government will continue prohibiting bulk water exports, the Provincial Government will lift the ban on bulk water exports, and the Support Groups will not appeal for implementing this project. The graph model capturing the possible moves by the three DMs in the Lake Gisborne conflict

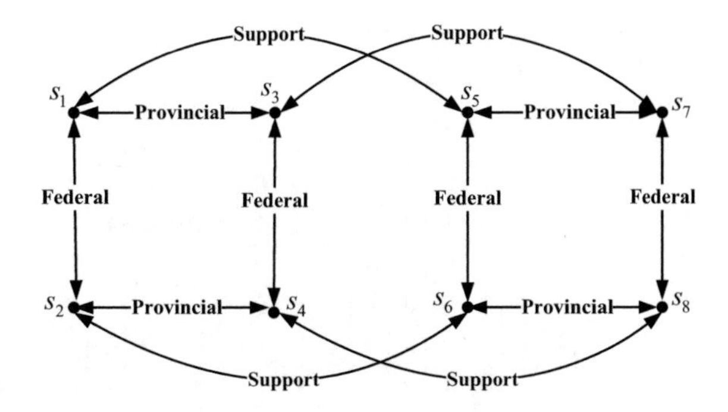

Fig. 8.1 Graph model of moves for the Lake Gisborne conflict

Table 8.2 Preference information for the Lake Gisborne model with low water price

DMs	Certain preferences
Federal	$s_2 \succ s_6 \succ s_4 \succ s_8 \succ s_1 \succ s_5 \succ s_3 \succ s_7$
Provincial	$s_2 \succ s_6 \succ s_1 \succ s_5 \succ s_4 \succ s_8 \succ s_3 \succ s_7$
Support	$s_3 \succ s_4 \succ s_7 \succ s_8 \succ s_5 \succ s_6 \succ s_1 \succ s_2$

Table 8.3 Preference information for the Lake Gisborne model with high water price

DMs	Certain preferences
Federal	$s_2 \succ s_6 \succ s_4 \succ s_8 \succ s_1 \succ s_5 \succ s_3 \succ s_7$
Provincial	$s_3 \succ s_7 \succ s_4 \succ s_8 \succ s_1 \succ s_5 \succ s_2 \succ s_6$
Support	$s_3 \succ s_4 \succ s_7 \succ s_8 \succ s_5 \succ s_6 \succ s_1 \succ s_2$

is shown in Fig. 8.1, where the labels on the arcs identify the DMs who control the relevant moves.

Besides the DMs, states, and potential moves, the other key component of a graph model is the relative preferences for each DM. Tables 8.2 and 8.3 provide the preferences for the situations in which the price of water is low and high, respectively. Notice that only the preferences for the Provincial Government are different for these two conflicts. In these tables, the symbol given by $\succ$ means more preferred. When the price of water is low, the Provincial most prefers state s_2 from Table 8.2. State s_2 indicates that the Provincial sides with the Federal for protecting the environment. With the increasing price of water, Table 8.3 shows that state s_3 is most preferred by the Provincial. It means that the economical-oriented provincial government will lift the ban on bulk water exports. Two attitudes of the Provincial will result in different coalitions and different coalitional stability resolutions. Due to the methods to analyze the two models for the Lake Gisborne conflict are similar, the following discussions will be based on the second case in which the Provincial sides with the Support Groups. The reachability matrices for the Lake Gisborne model is constructed using the algebraic approach next.

Table 8.4 UM reachability matrices by $N-\{i\}$ for $i=1,2,$ and 3 for the Lake Gisborne model with high water price

Matrix	$M_{N-\{1\}}$								$M_{N-\{2\}}$								$M_{N-\{3\}}$							
State	1	2	3	4	5	6	7	8	1	2	3	4	5	6	7	8	1	2	3	4	5	6	7	8
1	0	0	1	0	1	0	1	0	0	1	0	0	1	1	0	0	0	1	1	1	0	0	0	0
2	0	0	0	1	0	1	0	1	1	0	0	0	1	1	0	0	1	0	1	1	0	0	0	0
3	1	0	0	0	1	0	1	0	0	0	0	1	0	0	1	1	1	1	0	1	0	0	0	0
4	0	1	0	0	0	1	0	1	0	0	1	0	0	0	1	1	1	1	1	0	0	0	0	0
5	1	0	1	0	0	0	1	0	1	1	0	0	0	1	0	0	0	0	0	0	0	1	1	1
6	0	1	0	1	0	0	0	1	1	1	0	0	1	0	0	0	0	0	0	0	1	0	1	1
7	1	0	1	0	1	0	0	0	0	0	1	1	0	0	0	1	0	0	0	0	1	1	0	1
8	0	1	0	1	0	1	0	0	0	0	1	1	0	0	1	0	0	0	0	0	1	1	1	0

Table 8.5 UI reachability matrices by $N-\{i\}$ for $i=1,2,$ and 3 for the Lake Gisborne model with high water price

Matrix	$M^{+}_{N-\{1\}}$								$M^{+}_{N-\{2\}}$								$M^{+}_{N-\{3\}}$							
State	1	2	3	4	5	6	7	8	1	2	3	4	5	6	7	8	1	2	3	4	5	6	7	8
1	0	0	1	0	0	0	1	0	0	1	0	0	1	1	0	0	0	1	1	1	0	0	0	0
2	0	0	0	1	0	1	0	1	0	0	0	0	0	1	0	0	0	0	0	1	0	0	0	0
3	0	0	0	0	0	0	0	0	0	0	0	1	0	0	0	0	0	0	0	1	0	0	0	0
4	0	0	0	0	0	0	0	0	0	0	0	0	0	0	0	0	0	0	0	0	0	0	0	0
5	0	0	1	0	0	0	1	0	0	0	0	0	0	1	0	0	0	0	0	0	0	1	1	1
6	0	0	0	1	0	0	0	1	0	0	0	0	0	0	0	0	0	0	0	0	0	0	0	1
7	0	0	1	0	0	0	0	0	0	0	1	1	0	0	0	0	0	0	0	0	0	0	0	0
8	0	0	0	1	0	0	0	0	0	0	0	1	0	0	0	0	0	0	0	0	0	0	0	0

8.10.1 Reachability Matrices in the Lake Gisborne Model

$N=\{1,2,3\}$=$\{Federal, Provincial, Support\}$ is the set of three DMs. Use the Lake Gisborne model as an example to demonstrate how the algebraic approach works for building UM, UI, and CI (Coalitional Improvement) reachability matrices (Xu et al. 2014). One can adhere to the following steps:

- Construct matrices, J_i, J_i^{+}, P_i^{+}, and $P_i^{-,=}$, for $i=1,2,$ and 3, using information provided in Fig. 8.1 and Table 8.3;
- Calculate the UM, UI, and CI reachability matrices, M_H, M_H^{+}, and CM_H^{+} by $H=N-\{i\}$ for $i=1,2,$ and 3, respectively;
- Three reachability matrices are shown in Tables 8.4, 8.5, and 8.6.

For example, using Table 8.5, one has:

Table 8.6 CI reachability matrices by $N - \{i\}$ for $i = 1, 2,$ and 3 for the Lake Gisborne model with high water price

Matrix	$CM^+_{N-\{1\}}$								$CM^+_{N-\{2\}}$								$CM^+_{N-\{3\}}$							
State	1	2	3	4	5	6	7	8	1	2	3	4	5	6	7	8	1	2	3	4	5	6	7	8
1	0	0	1	0	0	0	1	0	0	0	0	0	0	1	0	0	0	0	0	1	0	0	0	0
2	0	0	0	1	0	0	0	1	0	0	0	0	0	0	0	0	0	0	0	0	0	0	0	0
3	1	0	0	0	0	0	0	0	0	0	0	0	0	0	0	0	0	0	0	0	0	0	0	0
4	0	0	0	0	0	0	0	0	0	0	0	0	0	0	0	0	0	0	0	0	0	0	0	0
5	0	0	1	0	0	0	1	0	0	0	0	0	0	0	0	0	0	0	0	0	0	0	0	1
6	0	0	0	1	0	0	0	1	0	0	0	0	0	0	0	0	0	0	0	0	0	0	0	0
7	0	0	1	0	0	0	0	0	0	0	1	1	0	0	0	0	0	0	0	0	0	0	0	0
8	0	0	0	1	0	0	0	0	0	0	0	1	0	0	0	0	0	0	0	0	0	0	0	0

$$e_2^T \cdot M^+_{N-\{1\}} = (0, 0, 0, 1, 0, 1, 0, 1),$$

which means that the reachable list of $H = N - \{1\}$ by the legal UIs from state s_2, $R^+_H(s_2) = \{s_4, s_6, s_8\}$, i.e., states s_4, s_6, and s_8 can be reached by any legal UI sequence, by coalition $H = \{2, 3\}$, from the status quo $s = s_2$. However, from Table 8.6,

$$e_2^T \cdot CM^+_{N-\{1\}} = (0, 0, 0, 1, 0, 0, 0, 1),$$

which indicates that the coalitional improvements from s_2 by coalition $H = \{2, 3\}$ are $CR^+_H = (0, 0, 0, 1, 0, 0, 0, 1)$. As mentioned after Definition 8.1, normally, $R^+_H(s) \neq CR^+_H(s)$. It is clear from this example that $R^+_H(s_2) \neq CR^+_H(s_2)$ for $H = \{2, 3\}$. In fact, although $s_6 \in R^+_H(s_2)$, $s_6 \notin CR^+_H(s_2)$ for $H = \{2, 3\}$, since $s_2 \succ_2 s_6$.

8.10.2 Coalitional Stability Results in the Lake Gisborne Model

After obtaining three important components, UM, UI, and CI reachability matrices (M_H, M^+_H, and CM^+_H, respectively), coalitional stabilities, CNash, CGMR, CSMR, $CSEQ_1$, and $CSEQ_2$, can be calculated using Theorems 8.2 and 8.4–8.7 and are shown in Table 8.7.

Both of the foregoing water export conflicts were thoroughly analyzed using the algebraic methodology for coalitional analysis provided in this chapter. In the first dispute for which the price of water is low, the only equilibrium according to both noncooperative stability calculations and coalitional stability when the Federal and Provincial Governments form a coalition ($H = \{1, 2\}$) is s_6. From Table 8.1, state

Table 8.7 Coalitional stabilities of the Lake Gisborne model for various coalitions with high water price

State	Stability	{1}	{2}	{3}	{1, 2}	{1, 3}	{2, 3}
s_1	CNash						
	CGMR						
	CSMR						
	$CSEQ_1$						
	$CSEQ_2$						
s_2	CNash	√			√	√	
	CGMR	√			√	√	
	CSMR	√			√	√	
	$CSEQ_1$	√			√	√	
	$CSEQ_2$	√			√	√	
s_3	CNash		√	√	√	√	√
	CGMR		√	√	√	√	√
	CSMR		√	√	√	√	√
	$CSEQ_1$		√	√	√	√	√
	$CSEQ_2$		√	√	√	√	√
s_4	CNash	√	√	√	√	√	√
	CGMR	√	√	√	√	√	√
	CSMR	√	√	√	√	√	√
	$CSEQ_1$	√	√	√	√	√	√
	$CSEQ_2$	√	√	√	√	√	√
s_5	CNash			√		√	
	CGMR			√		√	
	CSMR			√		√	
	$CSEQ_1$			√		√	
	$CSEQ_2$			√		√	
s_6	CNash	√		√	√	√	
	CGMR	√		√	√	√	
	CSMR	√		√	√	√	
	$CSEQ_1$	√		√	√	√	
	$CSEQ_2$	√		√	√	√	
s_7	CNash		√		√		
	CGMR		√	√	√	√	√
	CSMR		√	√	√	√	√
	$CSEQ_1$		√		√		√
	$CSEQ_2$		√		√		√
s_8	CNash	√	√				
	CGMR	√	√	√	√	√	
	CSMR	√	√	√	√	√	
	$CSEQ_1$	√	√		√		
	$CSEQ_2$	√	√		√		

s_6 is the situation in which the Federal Government continues to promote a ban, the Provincial Government does not lift the ban, and the Support Groups appeal. In this case, the Provincial Government is environmentally oriented. For the second conflict, in which the price of water is high, the noncooperative stability results are listed in the 3, 4, and 5th columns of Table 8.7. Obviously, state s_4 is an equilibrium for all individual noncooperative stability definitions consisting of Nash, GMR, SMR, and SEQ; s_8 is also an equilibrium for GMR and SMR individual stabilities. However, notice from Table 8.7 that when the Provincial Government and the Support Groups form a coalition ($H = \{2, 3\}$), s_8 is coalitionally unstable for CGMR and CSMR. As can be seen from Table 8.1, at s_8, the Support Groups are appealing, which is not necessary because the Provincial Government and the Support Groups are cooperating. Therefore, state s_8 is not long-term stable. For nontrivial coalitions, the cooperative stabilities are listed in the three columns on the right of Table 8.7. Observe that s_4 is universally CNash, CGMR, CSMR, $CSEQ_1$, and $CSEQ_2$ stable, which means that at state s_4, the Federal Government continues with the ban, the Provincial Government lifts the ban and the Support Groups do not appeal. State s_4 is a resolution of the conflict when the price of water is high. In this case, the water export project will proceed.

8.11 Important Ideas

Coalition analysis should form a key component of every formal conflict resolution investigation. After determining what a given DM can accomplish on his or her own and in his own self-interest, one should determine if the DM can do even better by cooperating with others. The coalition ideas presented in this chapter provide a solid mathematical foundation for coalition modeling and analysis, which can be programmed into a decision support system (DSS) for GMCR, as explained in Sect. 10.2. Hence, an encompassing coalition approach to formal conflict studies can be fully operationalized for employment by researchers, teachers, students, and practitioners working in many fields. The logical representation of coalitional stability analyses for four key solution concepts are presented in this chapter for the four types of preference structures given in Chaps. 4–7. Moreover, the matrix representation of coalitional analysis under a range of preference framework given later in this chapter means that coalitional analysis can be readily incorporated into the construction of the engine for a DSS for GMCR, as explained in Sect. 10.2. Accordingly, coalitional analysis is now a fully mature decision technology within the paradigm of GMCR, which can be readily utilized as evidenced by the water export conflict application presented in Sect. 8.10.

8.12 Problems

8.12.1 Select a current conflict, such as an international trading dispute or negotiating a climate change agreement, which is of direct interest to you. Explain why you think coalition modeling and analysis may or may not be an important tool for better resolving this conflict.

8.12.2 In a coalition improvement given in Definition 8.1, a state is a coalition improvement for the members of a coalition with respect to another state if and only if the state to which the DMs are jointly moving is more preferred by all of the members of the coalition. For a conflict of your choice, provide an example of a coalition improvement. Explain how this move could be carried out in practice via appropriate communication among the coalition members.

8.12.3 The game of Prisoner's Dilemma is presented in Problem 3.5.1. If both DMs were to move together from state s_4 to state s_1 in this conflict, this constitutes an example of a coalition improvement. Write a short discussion about interpreting Prisoner's Dilemma as some type of typical or generic real-world dispute, such as a trading or environmental dispute. Explain sensible steps that could be taken in practice to ensure that both DMs move together from state s_4 to s_1 and, hence, no DM defects during this process. Why is the move from state s_4 to s_1 called an equilibrium jump?

8.12.4 In the game of Chicken in Problem 3.5.4, both DMs or drivers moving together from state s_1 to state s_4 is an example of a coalition improvement as presented in Definition 8.1. Furnish an example of a real-world interpretation of the game of Chicken. Explain how the two DMs could improve together from state s_1 to state s_4 via taking appropriate measures.

8.12.5 If a conflict consists of only two DMs, these two DMs can still participate in a coalition improvement as presented in Definition 8.1. However, when there are only two DMs in a conflict, there are no other DMs left in the conflict to block possible coalition improvements. Explain why the definitions for coalitional stabilities given in Sect. 8.2 for simple preference work when there are only two DMs. Why are these coalitional stability definitions identical to the stability definitions given in Chap. 4 for simple preference with no coalitions having two or more DMs?

8.12.6 The Lake Gisborne conflict over the proposed exportation of water is presented in Sect. 8.10. Apply the logical form of the coalitional stability definitions given in Sect. 8.2 to the Lake Gisborne example to show by hand how you calculate various coalitional stabilities. Be sure to present the special situation for which there are no coalitions for each stability definition.

8.12.7 For the Elmira groundwater contamination dispute first presented in Sect. 1.2.2 in the book, calculate by hand the coalitional stabilities for simple preference using the matrix formulation given in Sect. 8.6. Be sure to include sample calculations and the stability results for the special situation in which there are no coalitions.

8.12.8 For the coalition investigation approach presented in this chapter, it is assumed that DMs will form a coalition during a conflict when it is in their interest to do so, as reflected by the way a coalition improvement is defined in Definition 8.1. However, other ways to study coalitions exist. In particular, in some situations, such as a military alliance among nations during warfare, a coalition may last throughout the duration of the dispute. Accordingly, researchers developed a procedure for determining the preference of a coalition based on the preferences of the individual coalition members (Kuhn et al. 1983, Hipel and Fraser 1991, Meister et al. 1992, Hipel and Meister 1994). By referring to the research of these authors, outline how coalition preferences are ascertained. Explain how these authors identify possible coalition formation and how coalitional stability analyses are executed. Describe a specific actual dispute for which you think this approach could be useful for obtaining strategic insights.

8.12.9 Logical definitions of coalitional stabilities under unknown preference are presented in Sect. 8.3. By employing a real-world application of your choice, explain a situation in which you think this kind of coalitional analysis could prove to be informative.

8.12.10 For the case of three degrees of preference, logical definitions for coalitional stability are provided in Sect. 8.4. Describe an actual situation in which you think this kind of coalitional stability analysis could provide insightful strategic findings.

8.12.11 Hybrid preference coalitional stability definitions, in which both unknown preference and three degrees of preference are simultaneously taken into account, are presented in Sect. 8.5. Based on an actual dispute which is of direct interest to you, describe why you think this hybrid coalitional stability approach could provide insightful strategic findings.

8.12.12 The matrix representations of coalitional stability under simple, unknown, three degree, and hybrid preference are presented in Sects. 8.6–8.9, respectively. As explained in Sect. 10.2, these matrix representations are needed for designing and programming a flexible decision support system (DSS) especially for the analysis engine. Using diagrams, outline how you would design a DSS which can handle coalitional analyses.

References

Aumann, R. J., & Hart, S. (1994). *Handbook of game theory with economic applications*. Amsterdam: Elsevier.

Hipel, K. W., & Fraser, N. M. (1991). Cooperation in conflict analysis. *Applied Mathematics and Computation*, *43*, 181–206. https://doi.org/10.1016/0096-3003(91)90033-j.

Hipel, K. W., & Meister, D. B. G. (1994). Conflict analysis methodology for modelling coalition in multilateral negotiations. *Information and Decision Technologies*, *19*(2), 85–103.

Inohara, T., & Hipel, K. W. (2008a). Coalition analysis in the graph model for conflict resolution. *Systems Engineering*, *11*, 343–359. https://doi.org/10.1002/sys.20104.

Inohara, T., & Hipel, K. W. (2008b). Interrelationships among noncooperative and coalition stability concepts. *Journal of Systems Science and Systems Engineering*, *17*(1), 1–29. https://doi.org/10.1007/s11518-008-5070-1.

Kilgour, D. M., Hipel, K. W., Fang, L., & Peng, X. (2001). Coalition analysis in group decision support. *Group Decision and Negotiation*, *10*(2), 159–175.

Kuhn, J. R. D., Hipel, K. W., & Fraser, N. M. (1983). A coalition analysis algorithm with application to the Zimbabwe conflict. *IEEE Transactions on Systems, Man and Cybernetics*, *13*(3), 338–352. https://doi.org/10.1109/tsmc.1983.6313166.

Li, K. W., Hipel, K. W., Kilgour, D. M., & Fang, L. (2004). Preference uncertainty in the graph model for conflict resolution. *IEEE Transactions on Systems, Man, and Cybernetics, Part A, Systems and Humans*, *34*(4), 507–520. https://doi.org/10.1109/tsmca.2004.826282.

Meister, D. B. G., Hipel, K. W., & De, M. (1992). Coalition formation. *Journal of Scientific and Industrial Research*, *51*(8–9), 612–625.

Van Deeman, M. A. (1997). *Coalition formation and social choice*. Dordrecht: Kluwer Academic. https://doi.org/10.1007/978-1-4757-2578-0.

Xu, H., Kilgour, D. M., & Hipel, K. W. (2010). Matrix representation and extension of coalition analysis in group decision support. *Computers and Mathematics with Applications*, *60*(5), 1164–1176. https://doi.org/10.1016/j.camwa.2010.05.040.

Xu, H., Kilgour, D. M., & Hipel, K. W. (2011). Matrix representation of conflict resolution in multiple-decision-maker graph models with preference uncertainty. *Group Decision and Negotiation*, *20*(6), 755–779. https://doi.org/10.1007/s10726-010-9188-4.

Xu, H., Kilgour, D. M., Hipel, K. W., & McBean, E. A. (2014). Theory and implementation of coalitional analysis in cooperative decision making. *Theory and Decision*, *76*(2), 147–171. https://doi.org/10.1007/s11238-013-9363-6.

Chapter 9
Follow-Up Analysis: Conflict Evolution

When executing a comprehensive conflict resolution investigation, one first ascertains how well each decision maker (DM) can fare on his or her own by utilizing the modeling concepts in Chap. 3 along with the stability definitions provided in Chaps. 4–7 under simple, unknown, degrees of preference, and hybrid preference, respectively. As emphasized right at the beginning of the book in Fig. 1.1 in Sect. 1.2.1, as well as in the design for a decision support system (DSS) in Sect. 10.2 in the last chapter, one should also determine if one or more of the DMs can do even better via forming coalitions, which is the focus of Chap. 8. The analysis system of GMCR consists of stability analysis and post-stability or follow-up analysis that includes coalition analysis given in Chap. 8 and status quo analysis in a graph model.

The purpose of this chapter is to present formal algorithms for tracing possible evolutions of a conflict from a given selected starting state, which is often the status quo situation, to any final desired state, such as an attractive win/win equilibrium. In the original graph model, individual and coalitional stability analyses were carried out within the easily understandable logical structures described in the earlier sections of Chaps. 4–8. Likewise, when algorithms for finding potential evolutionary path of a conflict, initially called status quo analyses, were created by Li et al. (2004b, 2005a, b), logical structures were retained and presented in pseudocode. Although the logical approach to tracing specified types of paths between any two chosen states is easy to comprehend, it is difficult to incorporate into an existing or new DSS, as explained in Sect. 10.2. Accordingly, to overcome this challenge, an innovative matrix system to determine the paths between two states within a graph model is presented in this chapter (Xu et al. 2009a, b, 2010a, b).

A legal path in a graph model satisfies the usual restriction that no DM may move twice consecutively. It is shown that the fundamental problem of finding paths is equivalent to a search of all colored paths from a given initial state to a desirable state within an edge-colored multidigraph.

H. Xu et al., *Conflict Resolution Using the Graph Model: Strategic Interactions in Competition and Cooperation*, Studies in Systems, Decision and Control 153,
https://doi.org/10.1007/978-3-319-77670-5_9

The key concepts underlying the tracing of the potential evolution paths of a conflict between any two states are presented in Sect. 9.1 for logical representations where the preferences can be simple, unknown, three-degree, and hybrid. Subsequently, two algebraic approaches based on the adjacency matrix and incidence matrix for status quo analysis for these kinds of preferences are presented in Sects. 9.2 and 9.3, respectively. The Elmira groundwater contamination, Gisborne water export, and Garrison Diversion Unit irrigation conflicts are utilized to demonstrate how these algorithms work in practice.

9.1 Logical Representation of Conflict Evolution

When a conflict is modeled as a graph model, a point in time must be selected first; the current (or initial) state of the conflict is then referred to as the status quo. Two fundamental steps are involved in analyzing a graph model: stability analysis and follow-up analysis (Kilgour and Hipel 2010). When the stability of a state is assessed at the first stage, it is not a concern whether this state is actually achievable from the status quo state. But as one form of a follow-up analysis, status quo analysis aims to determine whether a particular equilibrium is reachable from the status quo and, if so, how to reach it. Thus, in contrast to stability analysis, which identifies states that would be stable if attained, status quo analysis provides a dynamic and forward-looking perspective, identifying states that are attainable, and describing how to reach them. Pseudocodes for status quo analysis under simple preference are presented next (Li et al. 2004b, 2005a, b). As a follow-up analysis, status quo analysis is designed to trace the evolution of a conflict from the status quo state to any desirable outcome.

9.1.1 *Simple Preference*

Let $i \in N$ and $H \subseteq N$ and let $k \geq 1$ be an integer. New notation is required, as follows:

- SQ denotes the status quo state;
- The state sets, $S_i^{(k)}(s)$ and $S_i^{(k,+)}(s)$, denote the states reachable from $SQ = s$ in legal sequences of exactly k steps of unilateral moves (UMs) and unilateral improvements (UIs), respectively, with last mover i;
- The state sets, $V_i^{(k)}(s)$ and $V_i^{(k,+)}(s)$, denote the sets of states reachable from $SQ = s$ in legal sequences of at most k UMs and UIs, respectively, with last mover i;
- The state sets, $V_H^{(k)}(s)$ and $V_H^{(k,+)}(s)$, stand for the sets of states reachable from $SQ = s$ in legal sequences of at most k UMs and UIs, respectively, by H; (If $H = N$, then $V_H^{(k)}(s) = V^{(k)}(s)$ and $V_H^{(k,+)}(s) = V^{(k,+)}(s)$.)

- The arc sets, $A_i^{(k)}(s)$ and $A_i^{(k,+)}(s)$, controlled by DM i, contain the final arcs in legal sequences of at most k UMs and UIs, respectively, from $SQ = s$.

Recall that A_i is DM i's arc set in a graph model. Let A_i^+ denote i's UI arc set. For $s \in S$, let $A_i(s)$ and $A_i^+(s)$ stand for the respective subsets of these two arc sets with initial state s. Therefore, these arc sets are expressed by $A_i = \bigcup_{s \in S} A_i(s)$ and $A_i^+ = \bigcup_{s \in S} A_i^+(s)$. Note that for all $i \in N, s \in S$, and all positive integers k, $A_i^{(k)}(s) \subseteq A_i$ and $A_i^{(k,+)}(s) \subseteq A_i^+$. Similarly, $S_i^{(k,+)}(s) \subseteq S_i^{(k)}(s) \subseteq S$ and $V_i^{(k,+)}(s) \subseteq V_i^{(k)}(s) \subseteq S$.

The algorithm presented in Table 9.1 allows for UMs as opposed to UIs under simple preference. If the algorithm stops at step k, the graph defined by $(V_H^{(k)}(SQ), \bigcup_{i \in H} A_i^{(k)}(SQ))$, is called the status quo diagram of permitted UMs. An algorithm that permits only UIs can be found in Table 9.2 and lead to a similar status quo diagram for UIs. Two important components of stability analysis under simple preference, the reachable list of UMs from status quo s by coalition H, $R_H(s)$, and the reachable list of UIs from status quo s by coalition H, $R_H^+(s)$, can be constructed using Tables 9.1 and 9.2, respectively.

Employing a status quo diagram of the evolution of a conflict, useful information regarding the conflict under study can be garnered. In particular, if an equilibrium appears in the diagram, at least one path from the status quo to the equilibrium exists. By tracing this path in the status quo diagram, one can ascertain a viable evolution path from the status quo state to this resolution. On the other hand, if a predicted equilibrium does not appear in the status quo diagram, there is no way for the conflict to settle at this equilibrium, as it cannot be reached from the status quo (Li et al. 2005b).

The algorithms for status quo analysis in the graph model with simple preference can be extended to models with unknown preference.

9.1.2 *Unknown Preference*

First, some additional notation for unknown preference is required, as follows:

- The set, $S_i^{(k,+U)}(s) \subseteq S$, contains all states reachable from $SQ = s$ in legal sequences of exactly k unilateral improvements or uncertain moves (UIUMs) with last mover i;
- The set, $V_i^{(k,+U)}(s) \subseteq S$, contains all states reachable from $SQ = s$ in legal sequences of at most k UIUMs with last mover i;
- The set, $V_H^{(k,+U)}(s) \subseteq S$, is comprised of all states reachable from $SQ = s$ in legal sequences of at most k UIUMs by H; (if $H = N$, then $V_H^{(k,+U)}(s) = V^{(k,+U)}(s)$.)
- The arc set, $A_i^{(k,+U)}(s)$ controlled by DM i, contains the final arcs in legal sequences of at most k UIUMs from $SQ = s$.

Table 9.1 Pseudocode for constructing UM set from status quo state s

```
Initialize //initialize the necessary parameters
```

Initialize //initialize the necessary parameters

H: nonempty subset of DMs;

n: number of DMs in H;

m: the number of states;

s: status quo state;

δ_1: maximum number of loop repetitions;

$R_i(s)$: reachable list from state s for DM i, $i = 1, \cdots, n$;

$k = 1$

$S_i^{(k)}(s) = R_i(s), i = 1, \cdots, n$

$V_i^{(k)}(s) = S_i^{(k)}(s), i = 1, \cdots, n$

$A_i^{(k)}(s) = \bigcup_{q \in R_i(s)} (s, q), i = 1, \cdots, n$

loop 1

  $k = k + 1$

  loop 2 i from 1 to n // the last mover is DM i

    $S' = \bigcup_{j \in H \setminus \{i\}} S_j^{(k-1)}(s)$

    $S_i^{(k)}(s) = \bigcup_{s' \in S'} R_i(s')$

    $V_i^{(k)}(s) = V_i^{(k-1)}(s) \bigcup S_i^{(k)}(s)$

    $A_i^{(k)}(s) = A_i^{(k-1)}(s) \bigcup \{(s_1, s_2) : s_1 \in \bigcup_{j \in H \setminus \{i\}} S_j^{(k-1)}(s), \text{ and } s_2 \in R_i(s_1)\}$

  return to loop 2

  $V_H^{(k)}(s) = \bigcup_{i \in H} V_i^{(k)}(s)$

return to loop 1 if $\bigcup_{i \in H} A_i^{(k)}(s) \neq \bigcup_{i \in H} A_i^{(k-1)}(s)$

  $\delta_1 = k$

  $R_H(s) = V_H^{(\delta_1)}(s)$.

Recall that $R_i^{+,U}(s)$ stands for DM i's reachable list from s by UIUMs, which contains all states DM i can reach in one step from s. The algorithm presented in Table 9.3 permits only UIUMs for unknown preference. Similarly, the graph $(V_H^{(k,+U)}(SQ), \bigcup_{i \in H} A_i^{(k,+U)}(SQ))$ represents the status quo diagram, if only UIUMs are permitted, when the above algorithm stops at iteration step k.

The algorithm for status quo analysis in the graph model with unknown preference is described in Table 9.3. The important component of stability analysis under unknown preference, the reachable list of UIUMs from status quo s by coalition H, $R_H^{+,U}(s)$, may be obtained using the algorithm presented in Table 9.3. Preference with strength (three degrees of preference) is discussed in Chap. 6. The status quo analysis for graph models with this preference structure is presented in the next section.

Table 9.2 Pseudocode for constructing UI set from status quo state s

Initialize //initialize the necessary parameters
H: nonempty subset of DMs;
n: the number of DMs in H;
m: number of states;
s: status quo state;
δ_2: maximum number of loop repetitions;
$R_i^+(s)$: UIs from state s for DM i, $i = 1, \cdots, n$;
$k = 1$
$S_i^{(k,+)}(s) = R_i^+(s), i = 1, \cdots, n$
$V_i^{(k,+)}(s) = S_i^{(k,+)}(s), i = 1, \cdots, n$
$A_i^{(k,+)}(s) = \bigcup_{q \in R_i^+(s)} (s, q), i = 1, \cdots, n$

loop 1
  $k = k + 1$
  loop 2 i from 1 to n // the last mover is DM i
    $S' = \bigcup_{j \in H \setminus \{i\}} S_j^{(k-1,+)}(s)$
    $S_i^{(k,+)}(s) = \bigcup_{s' \in S'} R_i^+(s')$
    $V_i^{(k,+)}(s) = V_i^{(k-1,+)}(s) \bigcup S_i^{(k,+)}(s)$
    $A_i^{(k,+)}(s) = A_i^{(k-1,+)}(s) \bigcup \{(s_1, s_2) : s_1 \in \bigcup_{j \in H \setminus \{i\}} S_j^{(k-1,+)}(s)$, and $s_2 \in R_i^+(s_1)\}$
  return to loop 2
  $V_H^{(k,+)}(s) = \bigcup_{i \in H} V_i^{(k,+)}(s)$
return to loop 1 if $\bigcup_{i \in H} A_i^{(k,+)}(s) \neq \bigcup_{i \in H} A_i^{(k-1,+)}(s)$

  $\delta_2 = k$
  $R_H^+(s) = V_H^{(\delta_2,+)}(s)$.

9.1.3 *Three Degrees of Preference*

Some additional notation for the three types of preference (indifference "$\sim$", mild preference "$>$", and strong preference "$\gg$"), is required, as follows:

- $S_i^{(k,+,++)}(s)$ denotes the set of states reachable from $SQ = s$ in legal sequences of exactly k steps of mild or strong unilateral improvements (MSUIs) with last mover i;
- $V_i^{(k,+,++)}(s)$ denotes the set of all states reachable from $SQ = s$ in legal sequences of at most k MSUIs with last mover i;

Table 9.3 Pseudocode for constructing UIUM set from status quo state s

```
Initialize //initialize the necessary parameters
H: nonempty set of DMs;
n: number of DMs in H;
m: number of states;
s: status quo state;
δ3: maximum number of loop repetitions;
R_i^{+,U}(s): UIUMs from state s for DM i, i = 1, ..., n;
k = 1
S_i^{(k,+U)}(s) = R_i^{+,U}(s), i = 1, ..., n
V_i^{(k,+U)}(s) = S_i^{(k,+U)}(s), i = 1, ..., n
A_i^{(k,+U)}(s) = ∪_{q∈R_i^{+,U}(s)} (s, q), i = 1, ..., n

loop 1
  k = k + 1
  loop 2   i from 1 to n   // the last mover is DM i
    S' = ∪_{j∈H\{i}} S_j^{(k-1,+U)}(s)
    S_i^{(k,+U)}(s) = ∪_{s'∈S'} R_i^{+,U}(s')
    V_i^{(k,+U)}(s) = V_i^{(k-1,+U)}(s) ∪ S_i^{(k,+U)}(s)
    A_i^{(k,+U)}(s) = A_i^{(k-1,+U)}(s) ∪{(s1, s2) : s1 ∈ ∪_{j∈H\{i}} S_j^{(k-1,+U)}(s),
  and s2 ∈ R_i^{+,U}(s1)}
  return to loop 2
  V_H^{(k,+U)}(s) = ∪_{i∈H} V_i^{(k,+U)}(s)
return to loop 1 if ∪_{i∈H} A_i^{(k,+U)}(s) ≠ ∪_{i∈H} A_i^{(k-1,+U)}(s)

  δ3 = k
  R_H^{+,U}(s) = V_H^{(δ3,+U)}(s).
```

- $V_H^{(k,+,++)}(s)$ denotes the set of all states reachable from $SQ = s$ in legal sequences of at most k MSUIs by H; (If $H = N$, then $V_H^{(k,+,++)}(s) = V^{(k,+,++)}(s)$.)
- $A_i^{(k,+,++)}(s)$ is the set of all arcs controlled by DM i that contains the final arcs in legal sequences of at most k MSUIs from $SQ = s$.

Recall that $R_i^{+,++}(s)$ contains all states reachable for DM i by MSUIs in one step from s. The algorithm presented in Table 9.4 allows only MSUIs for the three degrees of preference. Similarly, the graph $(V_H^{(k,+,++)}(SQ), \bigcup_{i\in H} A_i^{(k,+,++)}(SQ))$ represents the status quo diagram, if only MSUIs are allowed, when the above algorithm stops at iteration step k. The important component of stability analysis under three degrees of

Table 9.4 Pseudocode for constructing MSUI set from status quo state s

```
Initialize //initialize the necessary parameters
H: nonempty subset of DMs;
n: number of DMs in H;
m: number of states;
s: status quo state;
δ4: maximum number of loop repetitions;
R_i^{+,++}(s): MSUIs from state s for DM i, i = 1, ..., n;
k = 1
S_i^{(k,+,++)}(s) = R_i^{+,++}(s), i = 1, ..., n
V_i^{(k,+,++)}(s) = S_i^{(k,+,++)}(s), i = 1, ..., n
A_i^{(k,+,++)}(s) = ∪_{q ∈ R_i^{+,++}(s)} (s, q), i = 1, ..., n

loop 1
  k = k + 1
  loop 2   i from 1 to n   // the last mover is DM i
    S' = ∪_{j ∈ H\{i}} S_j^{(k-1,+,++)}(s)
    S_i^{(k,+,++)}(s) = ∪_{s' ∈ S'} R_i^{+,++}(s')
    V_i^{(k,+,++)}(s) = V_i^{(k-1,+,++)}(s) ∪ S_i^{(k,+,++)}(s)
  A_i^{(k,+,++)}(s) = A_i^{(k-1,+,++)}(s) ∪ {(s1, s2) : s1 ∈ ∪_{j ∈ H\{i}} S_j^{(k-1,+,++)}(s), and s2 ∈
  R_i^{+,++}(s1)}
  return to loop 2
  V_H^{(k,+,++)}(s) = ∪_{i ∈ H} V_i^{(k,+,++)}(s)
return to loop 1 if ∪_{i ∈ H} A_i^{(k,+,++)}(s) ≠ ∪_{i ∈ H} A_i^{(k-1,+,++)}(s)

  δ4 = k
  R_H^{+,++}(s) = V_H^{(δ4,+,++)}(s).
```

preference, the reachable list of MSUIs by coalition H from status quo s, $R_H^{+,++}(s)$, may be obtained using the algorithm presented in Table 9.4. Status quo analysis can provide guidance for DMs and analysts by identifying how to attain reachable equilibria from a status quo state under appropriate preferences.

9.1.4 Hybrid Preference

The hybrid preference framework combines preference uncertainty and three degrees of preference together into the paradigm of the Graph Model for Conflict Resolution. The preference structure and stability analysis under hybrid preference are discussed in Chap. 7. The status quo analysis under hybrid preference is presented in this subsection.

First, some additional notation for the hybrid preference is presented here:

- $S_i^{(k,+,++,U)}(s) \subseteq S$ contains all states reachable from $SQ = s$ in legal sequences of exactly k mild or strong unilateral improvements or uncertain moves (MSUIUMs) with last mover i;
- $V_i^{(k,+,++,U)}(s) \subseteq S$ contains all states reachable from $SQ = s$ in at most k legal MSUIUMs with last mover i;
- $V_H^{(k,+,++,U)}(s) \subseteq S$ contains all states reachable from $SQ = s$ in at most k legal MSUIUMs by H;
- $A_i^{(k,+,++,U)}(s)$ is the set of all arcs controlled by DM i that denotes the final arcs in legal sequences of at most k MSUIUMs from $SQ = s$.

Note that A_i, $A_i^{+,++}$, and $A_i^{+,++,U} \subseteq S \times S$ are three oriented arc sets that denote that DM i can make one step UM, MSUI, and MSUIUM from the initial state of the arc to its terminal state, respectively. Let $A_i(s)$, $A_i^{+,++}(s)$, and $A_i^{+,++,U}(s)$ denote the sets of arcs associated with DM i in one step UM, MSUI, and MSUIUM from state s, respectively. Therefore, $A_i = \bigcup_{s \in S} A_i(s)$, $A_i^{+,++} = \bigcup_{s \in S} A_i^{+,++}(s)$, and $A_i^{+,++,U} = \bigcup_{s \in S} A_i^{+,++,U}(s)$.

The algorithm presented in Table 9.5 permits only mild or strong unilateral improvements or uncertain moves for the hybrid preference. Similarly, the graph $(V_H^{(k,+,++,U)}(SQ), \bigcup_{i \in H} A_i^{(k,+,++,U)}(SQ))$ represents the status quo diagram in which only MSUIUMs are permitted, when the above algorithm stops at iteration step k. The important component of stability analysis under hybrid preference, the reachable list of MSUIUMs by coalition H from status quo s, $R_H^{+,++,U}(s)$, may be constructed using the algorithm presented in Table 9.5.

Even though pseudocodes for status quo analyses have been presented for simple preference, unknown preference, three degrees of preference and hybrid preference, they are not yet implemented into a user-friendly decision support system for employment in practical applications except simple preference. To gain these additional insights, one has to rely on tedious manual computations. The approach based on matrix formulation described in the next section provides a flexible and easy-to-implement procedure for executing a status quo analysis. Analysts having basic matrix operation knowledge will be able to take advantage of the procedure to carry out a status quo analysis. An innovative matrix system to represent various preference structures and calculate corresponding stabilities in a graph model has been presented in Chaps. 3–7. The matrix representation effectively converts the stability analysis from a logical structure to an algebraic system. Because of the difficulty in integrating status quo analysis into the DSS GMCR II (Fang et al. 2003a, b) and the ease of implementing the matrix representation of stability analysis, one should utilize the matrix approach to perform status quo analysis.

Table 9.5 Pseudocode for constructing MSUIUM set from status quo state s

```
Initialize //initialize the necessary parameters
H: any set of DMs;
n: the number of DMs in H;
m: the number of states;
s: the start state;
δ5: the max step one wants to calculate;
R_i^{+,++,U}(s): MSUIUMs from state s by DM i, i = 1, ..., n;
k = 1
S_i^{(k,+,++,U)}(s) = R_i^{+,++,U}(s), i = 1, ..., n
V_i^{(k,+,++,U)}(s) = S_i^{(k,+,++,U)}(s), i = 1, ..., n
A_i^{(k,+,++,U)}(s) = ∪_{q∈R_i^{+,++,U}(s)} (s, q), (s, q) is a MSUIUM arc from state s to state q, i =
1, ..., n

loop 1
  k = k + 1
  loop 2   i from 1 to n   // the last mover is DM i
    S' = ∪_{j∈H\{i}} S_j^{(k-1,+,++,U)}(s)
    S_i^{(k,+,++,U)}(s) = ∪_{s'∈S'} R_i^{+,++,U}(s')
    V_i^{(k,+,++,U)}(s) = V_i^{(k-1,+,++,U)}(s) ∪ S_i^{(k,+,++,U)}(s)
    A_i^{(k,+,++,U)}(s) = A_i^{(k-1,+,++,U)}(s) ∪(s1, s2)
    (s1, s2) : s1 ∈ ∪_{j∈H\{i}} S_j^{(k-1,+,++,U)}(s), and s2 ∈ R_i^{+,++,U}(s1)
  return to loop 2
  V_H^{(k,+,++,U)}(s) = ∪_{i∈H} V_i^{(k,+,++,U)}(s)
return to loop 1 if ∪_{i∈H} A_i^{(k,+,++,U)}(s) ≠ ∪_{i∈H} A_i^{(k-1,+,++,U)}(s)

  δ5 = k
  R_H^{+,++,U}(s) = V_H^{(δ3,+,++,U)}(s).
```

9.2 Matrix Representation of Conflict Evolution Based on Adjacency Matrix

It is well-known that matrices can efficiently describe adjacency of vertices, and incidence of arcs and vertices, in a graph, thereby permitting tracking of paths between any two vertices (Godsil and Royle 2001). Matrices possess useful algebraic properties that can be exploited to produce improved algorithms for solving graph problems. For instance, extensive research has been conducted to design effective algorithms and efficient search procedures using relationships between matrices and paths (Gondran and Minoux 1979, Shiny and Pujari 1998, Hoffman and Schiebe 2001).

In a graph model of a conflict, status quo analysis is a form of follow-up analysis designed to trace the evolution of the conflict from a status quo state to any stable state. A legal path in the graph model has the usual restriction that any DM may move more than once, but not twice consecutively. The fundamental problem of status quo analysis is thus equivalent to a search of the colored paths from a given initial state to a desirable state within an edge-colored multidigraph. The new approach to using an adjacency matrix to track conflict evolution is presented next (Xu et al. 2009a).

9.2.1 t-Legal Unilateral Move Matrix Under Various Preference Structures

One now demonstrates how to find matrices to trace conflict evolution by the legal sequences of unilateral moves (UMs), unilateral improvements (UIs), unilateral improvements or uncertain moves (UIUMs), mild or strong unilateral improvements (MSUIs), or mild or strong unilateral improvements or uncertain moves (MSUIUMs) from a status quo s with the last mover i for simple preference, unknown preference, three degrees of preference and hybrid preference, respectively. First, recall t-legal unilateral move matrices under the four kinds of preference, which are $m \times m$ matrices $M_i^{(t)}$, $M_i^{(t,+)}$, $M_i^{(t,+,U)}$, $M_i^{(t,+,++)}$, and $M_i^{(t,+,++,U)}$ presented in Chaps. 5–7, respectively. The (s, q) entries of these matrices are summarized as follows:

Definition 9.1 In the graph model $G = (S, A)$, let $H \subseteq N$ and $H \neq \emptyset$. For $i \in H$ and $t = 1, 2, 3, \cdots$,

$$M_i^{(t)}(s,q) = \begin{cases} 1 & \text{if } q \in S \text{ is reachable by } H \text{ from } s \in S \text{ in exactly } t \text{ legal} \\ & \text{UMs with last mover } i, \\ 0 & \text{otherwise,} \end{cases}$$

$$M_i^{(t,+)}(s,q) = \begin{cases} 1 & \text{if } q \in S \text{ is reachable by } H \text{ from } s \in S \text{ in exactly } t \text{ legal} \\ & \text{UIs with last mover } i, \\ 0 & \text{otherwise,} \end{cases}$$

$$M_i^{(t,+,U)}(s,q) = \begin{cases} 1 & \text{if } q \in S \text{ is reachable by } H \text{ from } s \in S \text{ in exactly } t \text{ legal} \\ & \text{UIUMs with last mover } i, \\ 0 & \text{otherwise,} \end{cases}$$

$$M_i^{(t,+,++)}(s,q) = \begin{cases} 1 & \text{if } q \in S \text{ is reachable by } H \text{ from } s \in S \text{ in exactly } t \text{ legal} \\ & \text{MSUIs with last mover } i, \\ 0 & \text{otherwise,} \end{cases}$$

and

$$M_i^{(t,+,++,U)}(s,q) = \begin{cases} 1 & \text{if } q \in S \text{ is reachable by } H \text{ from } s \in S \text{ in exactly } t \text{ legal} \\ & \text{MSUIUMs with last mover } i, \\ 0 & \text{otherwise.} \end{cases}$$

In fact, all matrices of Definition 9.1 have been presented in Chaps. 5–7. Here, one only summarizes them for four types of preference. Finally, in Definition 3.10 G is referred to as the integrated graph model IG, but for simplicity here it is referred to simply as the graph model G.

Lemma 9.1 *In the graph model $G = (S, A)$, let $H \subseteq N$ and $H \neq \emptyset$. Then the set of $m \times m$ matrices under the respective preference structures are expressed inductively as*

$$M_i^{(1)}(s,q) = J_i(s,q) \text{ and, for } t = 2, 3, \ldots, M_i^{(t)} = sign\left[\left(\bigvee_{j \in H\backslash\{i\}} M_j^{(t-1)}\right) \cdot J_i\right], \tag{9.1}$$

$$M_i^{(1,+)}(s,q) = J_i^+(s,q) \text{ and, for } t = 2, 3, \ldots, M_i^{(t,+)} = sign\left[\left(\bigvee_{j \in H\backslash\{i\}} M_j^{(t-1,+)}\right) \cdot J_i^+\right], \tag{9.2}$$

$$M_i^{(1,+,U)}(s,q) = J_i^{+,U}(s,q) \text{ and, for } t = 2, 3, \ldots,$$

$$M_i^{(t,+,U)} = sign\left[\left(\bigvee_{j \in H\backslash\{i\}} M_j^{(t-1,+,U)}\right) \cdot J_i^{+,U}\right], \tag{9.3}$$

$$M_i^{(1,+,++)}(s,q) = J_i^{+,++}(s,q) \text{ and, for } t = 2, 3, \ldots,$$

$$M_i^{(t,+,++)} = sign\left[\left(\bigvee_{j \in H\backslash\{i\}} M_j^{(t-1,+,++)}\right) \cdot J_i^{+,++}\right], \tag{9.4}$$

$$M_i^{(1,+,++,U)}(s,q) = J_i^{+,++,U}(s,q) \text{ and, for } t = 2, 3, \ldots,$$

$$M_i^{(t,+,++,U)} = sign\left[\left(\bigvee_{j \in H\backslash\{i\}} M_j^{(t-1,+,++,U)}\right) \cdot J_i^{+,++,U}\right]. \tag{9.5}$$

Proof The verifications of Eqs. 9.1–9.5 are similar. Now Eq. 9.2 is verified. For $t = 2$, the definition of matrix multiplication shows that $G(s,q)$, the (s,q) entry of the matrix $G = (\bigvee_{j \in H\backslash\{i\}} J_j^+) \cdot J_i^+$, is nonzero iff state q is reachable from state s by H in exactly two UIs, with last mover DM i. The condition $j \in H\backslash\{i\}$ implies that DM i does not make two moves consecutively. Hence, $G(s,q) \neq 0$ iff state q is reachable by H from state s in exactly two legal UIs. Then

$$sign\left[\left(\bigvee_{j\in H\backslash\{i\}} J_j^+\right)\cdot J_i^+\right] = sign\left[\left(\bigvee_{j\in H\backslash\{i\}} M_j^{(1,+)}\right)\cdot J_i^+\right] = M_i^{(2,+)}.$$

Now suppose that $t > 2$. Since

$$M_j^{(t-1,+)}(s,q) = \begin{cases} 1 & \text{if } q \in S \text{ is reachable by } H \text{ from } s \in S \text{ in exactly } t-1 \text{ legal} \\ & \text{UIs with last mover } j, \\ 0 & \text{otherwise,} \end{cases}$$

the definition of matrix multiplication implies that the (s, q) entry of matrix

$$B = sign\left[\left(\bigvee_{j\in H\backslash\{i\}} M_j^{(t-1,+)}\right)\cdot J_i^+\right]$$

indicates

$$B(s,q) = \begin{cases} 1 & \text{if } q \in S \text{ is reachable by } H \text{ from } s \in S \text{ in exactly } t \text{ legal} \\ & \text{UIs with last mover } i, \\ 0 & \text{otherwise,} \end{cases}$$

which confirms Eq. 9.2. □

Based on Lemma 9.1, matrices are constructed to trace conflict evolution for various preference structures.

9.2.2 *Status Quo Matrices Under Various Preference Structures*

The status quo matrices $M_i^{SQ^{(t)}}$, $M_i^{SQ^{(t,+)}}$, $M_i^{SQ^{(t,+,U)}}$, $M_i^{SQ^{(t,+,++)}}$, and $M_i^{SQ^{(t,+,++,U)}}$ are defined for respective preference structures to trace conflict evolution from a status quo to any equilibrium by the legal sequences of UMs, UIs, UIUMs, MSUIs, and MSUIUMs, respectively.

Definition 9.2 In the graph model $G = (S, A)$, let $H \subseteq N$. For $i \in H$ and $t = 1, 2, 3, \cdots$, the UM, UI, UIUM, MSUI, and MSUIUM status quo matrices are $m \times m$ matrices with (s, q) entries

$$M_i^{SQ^{(t)}}(s,q) = \begin{cases} 1 & \text{if } q \in S \text{ is reachable by } H \text{ from } s \in S \text{ in at most } t \text{ legal UMs} \\ & \text{with last mover } i, \\ 0 & \text{otherwise,} \end{cases}$$

$$M_i^{SQ^{(t,+)}}(s,q) = \begin{cases} 1 & \text{if } q \in S \text{ is reachable by } H \text{ from } s \in S \text{ in at most } t \text{ legal UIs} \\ & \text{with last mover } i, \\ 0 & \text{otherwise,} \end{cases}$$

$$M_i^{SQ^{(t,+,U)}}(s,q) = \begin{cases} 1 & \text{if } q \in S \text{ is reachable by } H \text{ from } s \in S \text{ in at most } t \text{ legal} \\ & \text{UIUMs with last mover } i, \\ 0 & \text{otherwise,} \end{cases}$$

$$M_i^{SQ^{(t,+,++)}}(s,q) = \begin{cases} 1 & \text{if } q \in S \text{ is reachable by } H \text{ from } s \in S \text{ in at most } t \text{ legal} \\ & \text{MSUIs with last mover } i, \\ 0 & \text{otherwise,} \end{cases}$$

$$M_i^{SQ^{(t,+,++,U)}}(s,q) = \begin{cases} 1 & \text{if } q \in S \text{ is reachable by } H \text{ from } s \in S \text{ in at most } t \text{ legal} \\ & \text{MSUIUMs with last mover } i, \\ 0 & \text{otherwise.} \end{cases}$$

For example, $M_i^{SQ^{(t)}}(s,q) = 1$ and $M_i^{SQ^{(t,+)}}(s,q) = 1$ denote that state q is reachable from status quo state s in at most t legal UMs and legal UIs by H, respectively, with last mover i. Based on Definitions 9.1 and 9.2, Theorem 9.1 can be derived.

Theorem 9.1 *In the graph model $G = (S, A)$, let $H \subseteq N$, $i \in H$ and $k \geq 1$ be an integer. Then status quo matrices within four types of preference satisfy that*

$$M_i^{SQ^{(k)}} = \bigvee_{t=1}^{k} M_i^{(t)}, \tag{9.6}$$

$$M_i^{SQ^{(k,+)}} = \bigvee_{t=1}^{k} M_i^{(t,+)}, \tag{9.7}$$

$$M_i^{SQ^{(k,+,U)}} = \bigvee_{t=1}^{k} M_i^{(t,+,U)}, \tag{9.8}$$

$$M_i^{SQ^{(k,+,++)}} = \bigvee_{t=1}^{k} M_i^{(t,+,++)}, \tag{9.9}$$

and

$$M_i^{SQ^{(k,+,++,U)}} = \bigvee_{t=1}^{k} M_i^{(t,+,++,U)}. \tag{9.10}$$

Proof The proofs of Eqs. 9.6–9.10 are similar. Equation 9.10 is proved here. Let $M_i^{SQ^{(k,+,++,U)}}(s,q)$ denote the (s,q) entry of the matrix $M_i^{SQ^{(k,+,++,U)}}$. Based on Definition 9.2, $M_i^{SQ^{(k,+,++,U)}}(s,q) = 1$ iff q is reachable by H from $SQ = s$ in at most k legal sequences of mild improvements, strong improvements, or uncertain moves, with last mover $i \in H$.

Let $(\bigvee_{t=1}^{k} M_i^{(t,+,++,U)})(s,q)$ denote the (s,q) entry of the matrix $\bigvee_{t=1}^{k} M_i^{(t,+,++,U)}$.

By Definition 9.1, $(\bigvee_{t=1}^{k} M_i^{(t,+,++,U)})(s,q) = 1$ iff there exists $1 \leq t \leq k$, such that $M_i^{(t,+,++,U)}(s,q) = 1$. Hence, q is reachable by H from $SQ = s$ in exactly t legal MSUIUMs, with last mover i. It means that q is reachable from $SQ = s$ in at most k legal MSUIUMs, with last mover i. Consequently, $(\bigvee_{t=1}^{k} M_i^{(t,+,++,U)})(s,q) = 1$ iff $M_i^{SQ^{(k,+,++,U)}}(s,q) = 1$. Since $M_i^{SQ^{(k,+,++,U)}}$ and $\bigvee_{t=1}^{k} M_i^{(t,+,++,U)}$ are $m \times m$ 0-1 matrices, it follows that $M_i^{SQ^{(k,+,++,U)}} = \bigvee_{t=1}^{k} M_i^{(t,+,++,U)}$. □

Any nonzero entry (s,q) of these status quo matrices shows that the desired outcome state q is reachable from the status quo state s in at most t legal UMs, UIs, UIUMs, MSUIs, or MSUIUMs, respectively, with last mover i.

9.2.3 *Application: Status Quo Analysis for Elmira Conflict Under Simple Preference*

In this subsection, the matrix approach to status quo analysis is applied to the Elmira conflict to illustrate how the procedure works. The Elmira conflict model, described in Sect. 4.5, has three DMs: MoE, UR, and LG; and nine feasible states.

Let $N = \{1, 2, 3\}$ be the set of three DMs (1 = MoE, 2 = UR, and 3 = LG). To carry out status quo analysis for the Elmira model by using the matrix approach, the following steps are required:

- Construct matrices J_i and P_i^+ for $i = 1, 2,$ and 3, using information provided in Fig. 4.8;
- Calculate the UI adjacency matrices $J_i^+ = J_i \circ P_i^+$ for $i = 1, 2,$ and 3;
- Determine the matrices $M_i^{(t)}$ and $M_i^{(t,+)}$ for $i = 1, 2,$ and 3, using inductive formulations provided by Lemma 9.1; and
- Calculate the status quo analysis matrices $M_i^{SQ^{(k)}}$ and $M_i^{SQ^{(k,+)}}$ for $i = 1, 2,$ and 3, using Theorem 9.1.

Status quo analysis is mainly concerned with the attainability of predicted equilibria. Therefore, stability analysis, which identifies equilibria, is usually conducted first, often using the DSS GMCR II. To demonstrate the effectiveness of the matrix approach, stability analyses are carried out using the matrix method developed in Chap. 4 for the four basic solution concepts, Nash, GMR, SMR, and SEQ, under simple preference. The findings are summarized in Table 9.6, in which "√" for a given state under a DM indicates that this state is stable for the DM; and "$\surd$" for a state under "Eq" means that this state is an equilibrium according to the corresponding solution concept. It is easy to verify that the stability results for the four solution concepts are identical to the findings generated by GMCR II. Table 9.6 identifies three states, s_5, s_8, and s_9 as major equilibria because they are stable for all DMs and for all four solution concepts.

Matrix manipulations generate the status quo analysis matrices in Tables 9.7 (with all UMs) and 9.8 (with UIs only). As the status quo state is s_1, one can assess the attainability of any state from the status quo by examining its corresponding entry in the first row for each DM, where a value of 1 indicates that the associated state is reachable from s_1 and a value of 0 means that the corresponding state is not reachable. Given the three matrices in Table 9.7, it is obvious that the three major equilibria, s_5, s_8, and s_9, are all attainable. For instance, $M_{MoE}^{SQ^{(3)}}(s_1, s_8) = 1$, $M_{UR}^{SQ^{(3)}}(s_1, s_8) = 1$, and $M_{LG}^{SQ^{(3)}}(s_1, s_8) = 1$ demonstrate that the major equilibrium state s_8 is reachable from s_1 in at most three UMs, with the last mover being any of the three DMs. Of the three matrices, only $M_{LG}^{SQ^{(3)}}(s_1, s_5)$ has a nonzero $(1, 5)$ entry. Therefore, equilibrium s_5 can be reached from the status quo in at most three UMs, provided that LG is the last mover. Similarly, the major equilibrium s_9 is reachable from s_1 in at most three UMs, but the last mover must be UR.

When only UIs are allowed, as shown in Table 9.8, only the major equilibrium s_5 can be reached from state s_1 in at most three UIs. The last mover must be LG, because the unique nonzero entry in the first row of the three matrices is $M_{LG}^{SQ^{(3,+)}}(s_1, s_5)$.

If a state other than s_1 is chosen as the status quo state, a similar process can be used. The elements of the corresponding row in the relevant status quo analysis matrices can be investigated to evaluate the attainability from the new status quo of any state that is of interest.

By using the inductive formulations in Theorem 9.1, the status quo analysis result can also be presented in a tableau form as shown in Table 9.9 in which 1, 2, and 3 denote DM 1, DM 2, and DM 3, respectively. As well, $\Omega^{(k)}$ and $\Omega^{(k,+)}$ are the sets of all last DMs in legal sequences from some status quo of at most k UMs and UIs, respectively. Note that in Table 9.9, state s_1 ($\surd$), and state s_2 ($\surd$), are selected as the status quo by the legal sequence of UMs and UIs, respectively. It is easy to verify the equivalence of these results with those given by Li et al. (2005b), except for the different way of recording the last mover. This table offers a wealth of information, such as the specific DM(s) who can be last mover(s), and the shortest path(s) to reach a state. For example, the shortest path to the major equilibrium s_8 from s_1 requires three legal UMs, and any of the three DMs may be the last mover.

By carrying out status quo analysis, additional insights are revealed about the attainability of a potential resolution and, if it is attainable, the dynamics of the evolution from the status quo state. The results offered by Table 9.9 are identical to those obtained using the logical representation (leave as an exercise).

The novel matrix approach to status quo analysis discussed above is convenient for computer implementation and easy to employ, as illustrated by its application to a real-world case, the Elmira conflict. However, the proposed approach is based

Table 9.6 Stability results of the Elmira conflict

State number	Nash				GMR				SMR				SEQ			
	MoE	UR	LG	Eq	MoE	UR	LG	Eq	MoE	UR	LG	Eq	MoE	UR	LG	Eq
s_1	√	√			√	√	√	√	√	√	√	√	√	√		
s_2	√				√		√		√		√		√		√	
s_3	√				√		√		√		√		√		√	
s_4	√	√			√	√	√	√	√	√	√	√	√	√		
s_5	√	√	√	√	√	√	√	√	√	√	√	√	√	√	√	√
s_6	√		√		√		√		√		√		√		√	
s_7	√		√		√		√		√		√		√		√	
s_8	√	√	√	√	√	√	√	√	√	√	√	√	√	√	√	√
s_9	√	√	√	√	√	√	√	√	√	√	√	√	√	√	√	√

Table 9.7 UM status quo matrices for the Elmira conflict

Matrix	$M_{MoE}^{SQ(3)}$									$M_{UR}^{SQ(3)}$									$M_{LG}^{SQ(3)}$								
State	s_1	s_2	s_3	s_4	s_5	s_6	s_7	s_8	s_9	s_1	s_2	s_3	s_4	s_5	s_6	s_7	s_8	s_9	s_1	s_2	s_3	s_4	s_5	s_6	s_7	s_8	s_9
s_1	0	1	0	1	0	1	0	1	0	0	0	1	1	0	0	1	1	1	0	1	1	0	1	1	1	1	0
s_2	0	0	0	0	0	0	0	0	0	0	0	0	1	0	0	0	1	1	0	0	0	1	0	1	0	1	0
s_3	0	0	0	1	0	0	0	1	0	0	0	0	0	0	0	0	0	1	0	0	0	1	0	0	1	1	0
s_4	0	0	0	0	0	0	0	0	0	0	0	0	0	0	0	0	0	1	0	0	0	0	0	0	0	1	0
s_5	0	1	0	1	0	1	0	1	0	0	0	1	1	0	0	1	1	1	1	1	1	1	0	1	1	0	0
s_6	0	0	0	0	0	0	0	0	0	0	0	0	1	0	0	0	1	1	0	1	0	1	0	0	0	1	0
s_7	0	0	0	1	0	0	0	1	0	0	0	0	0	0	0	0	0	1	0	0	1	1	0	0	0	1	0
s_8	0	0	0	0	0	0	0	0	0	0	0	0	0	0	0	0	0	1	0	0	0	1	0	0	0	0	0
s_9	0	0	0	0	0	0	0	0	0	0	0	0	0	0	0	0	0	0	0	0	0	0	0	0	0	0	0

Table 9.8 UI status quo matrices for the Elmira conflict

Matrix	$M_{MoE}^{SQ(3,+)}$									$M_{UR}^{SQ(3,+)}$									$M_{LG}^{SQ(3,+)}$								
State	s_1	s_2	s_3	s_4	s_5	s_6	s_7	s_8	s_9	s_1	s_2	s_3	s_4	s_5	s_6	s_7	s_8	s_9	s_1	s_2	s_3	s_4	s_5	s_6	s_7	s_8	s_9
s_1	0	0	0	0	0	0	0	0	0	0	0	0	0	0	0	0	0	0	0	0	0	0	1	0	0	0	0
s_2	0	0	0	0	0	0	0	0	0	0	0	0	1	0	0	0	1	1	0	0	0	0	0	1	0	1	0
s_3	0	0	0	0	0	0	0	0	0	0	0	0	0	0	0	0	0	1	0	0	0	0	0	0	1	0	0
s_4	0	0	0	0	0	0	0	0	0	0	0	0	0	0	0	0	0	0	0	0	0	0	0	0	0	1	0
s_5	0	0	0	0	0	0	0	0	0	0	0	0	0	0	0	0	0	0	0	0	0	0	0	0	0	0	0
s_6	0	0	0	0	0	0	0	0	0	0	0	0	0	0	0	0	1	1	0	0	0	0	0	0	0	0	0
s_7	0	0	0	0	0	0	0	0	0	0	0	0	0	0	0	0	0	1	0	0	0	0	0	0	0	0	0
s_8	0	0	0	0	0	0	0	0	0	0	0	0	0	0	0	0	0	0	0	0	0	0	0	0	0	0	0
s_9	0	0	0	0	0	0	0	0	0	0	0	0	0	0	0	0	0	0	0	0	0	0	0	0	0	0	0

on the use of the adjacency matrix to search state-by-state paths. If a graph model contains different arcs between the same two states, controlled by different DMs, state-by-state paths will not track all aspects of the evolution of a conflict from the status quo state, and an expanded model is needed to search arc-by-arc paths. This expanded model is the subject of the next section.

9.3 Matrix Representation of Conflict Evolution Based on Edge Consecutive Matrix

Analysis of a graph model involves searching paths in a graph but an important restriction of a graph model is that no DM can move twice in succession along any path. Therefore, a graph model must be treated as an edge-weighted, colored multidigraph in which each arc represents a legal unilateral move and distinct colors refer to different DMs. The weight of an arc could represent some preference attribute. Tracing the evolution of a conflict in status quo analysis is converted to searching all

Table 9.9 The results of status quo analysis for the Elmira conflict

State	$\Omega^{(0)}$	$\Omega^{(1)}$	$\Omega^{(2)}$	$\Omega^{(3)}$	$\Omega^{(4)}$	State	$\Omega^{(0,+)}$	$\Omega^{(1,+)}$	$\Omega^{(2,+)}$	$\Omega^{(3,+)}$	$\Omega^{(4,+)}$
s_1	√					s_2	√				
s_2		1	1	1, 3	1, 3	s_4		2	2	2	2
s_3		2	2	2, 3	2, 3	s_6		3	3	3	3
s_5		3	3	3	3	s_9		2	2, 3	2, 3	2, 3
s_9		2	2	2	2	s_8			2	2	2
s_4			1, 2	1, 2	1, 2	s_1					
s_6			1, 3	1, 3	1, 3	s_3					
s_7			2, 3	2, 3	2, 3	s_5					
s_8				1, 2, 3	1, 2, 3	s_7					

colored paths from a status quo to a particular outcome in an edge-weighted, colored multidigraph.

From the discussions above, an adjacency matrix can determine a simple digraph and all state-by-state paths between any two vertices. However, if a graph model contains multiple arcs between the same two states controlled by different DMs, the adjacency matrix would be unable to track all aspects of conflict evolution from the status quo. To bridge the gap, a conversion function using the matrix representation is designed to transform the original problem of searching edge-weighted, colored paths in a colored multidigraph to a standard problem of finding paths in a simple digraph with no color constraints. As well, several unexpected and useful links among status quo analysis, stability analysis, and coalition analysis are revealed using the conversion function.

9.3.1 *Weighted Conversion Function for Finding Colored Paths*

9.3.1.1 Weighted Colored Multidigraph

The definitions of the colored multidigraph and the edge consecutive matrix are presented in Sect. 3.3.1. One will extend the definitions to weighted colored multidigraphs in this section.

Definition 9.3 For a colored multidigraph $G = (V, A, N, \psi, c)$, the **reduced line digraph** $L_r(G) = (A, LA_r)$ of G is a simple vertex-colored digraph with vertex set A and edge set $LA_r = \{d = (a, b) \in A \times A : a$ and b are consecutive (in the order ab) and $c(a) \neq c(b)\}$.

Recall that if $a \in A$ such that $\psi(a) = (u, v)$ and $c(a) = i$ for $i \in N$, then a can be written as $a = d_i(u, v)$. The line digraph of $G = (V, A, N, \psi, c)$, $L(G)$, is a simple digraph and each vertex in $L(G)$ corresponds to an edge in the multidigraph G. Hence, coloring edges in G is equivalent to assigning colors to vertices in $L(G)$.

Definition 9.4 A **weighted colored multidigraph** (V, A, N, ψ, c, w) is a colored multidigraph (V, A, N, ψ, c) together with a map $w : A \rightarrow \mathbb{R}_0^+$ (the set of non-negative real numbers).

Thus an arc $a \in A$, $a = d_i(u, v)$, carries a weight $w(a)$, representing some attribute of the move from node u to node v along the arc a, which is assigned color i. A network, for instance, is a multidigraph with weighted edges. Let $H \subseteq N$ be a subset of the color set N in the following definitions. An edge-weighted, colored path is defined as follows:

Definition 9.5 Let $H \subseteq N$. For a weighted colored multidigraph (V, A, N, ψ, c, w), an **edge-weighted, colored path by** H **from vertex** $u \in V$ **to vertex** $v \in V$, $PA_H^{(W)}(u, v)$, is a path from u to v in the multidigraph (V, A, ψ) in which any two consecutive edges have different colors and each edge a on the path carries a weight $w(a) \geq 0$ and $c(a) = i \in H$.

Definition 9.6 For a weighted colored multidigraph (V, A, N, ψ, c, w), the **shortest colored path between two vertices** is the colored path that minimizes the sum of the weights of its constituent edges.

Definition 9.7 Let $H \subseteq N$. For a weighted colored multidigraph (V, A, N, ψ, c, w), the **weighted arc set for** H denotes $A_H^{(W)} = \{a \in A : w(a) > 0 \text{ and } c(a) = i \in H.\}$.

Note that a colored multidigraph (V, A, N, ψ, c) is a unit weighted colored multidigraph if $w(u, v) = 1$ for any $a \in A$ such that $\psi(a) = (u, v)$.

Let $l = |A|$ denote the cardinality of A in G. The weight matrix of a weighted colored multidigraph (V, A, N, ψ, c, w) is defined as follows:

Definition 9.8 For a weighted colored multidigraph (V, A, N, ψ, c, w), let $H \subseteq N$ and w_k denote the weight of arc $a_k \in A$. The **weight matrix for** H is an $l \times l$ diagonal matrix W_H with (k, k) entry

$$W_H(k, k) = \begin{cases} w_k & \text{if } c(a_k) = i \in H, \\ 0 & \text{otherwise.} \end{cases}$$

It should be pointed out that if $H = N$, then W_N is expressed as W; if $H = \{i\}$, then $W_H = W_i$. A weighted line digraph $L^{(W)}(G) = (A, LA, w)$ is a set of vertices A together with a set of oriented edges LA, and a map $w : A \rightarrow \mathbb{R}_0^+$. In traditional graph coloring problems, such as vertex coloring and edge coloring, colors are assigned to vertices or edges such that adjacent vertices or consecutive edges have different colors, and the number of colors needed is minimized (Dieste 1997). In this chapter, the edge-weighted, colored graph problem is not concerned with coloring edges, but aims at searching edge-weighted, colored paths in a given weighted colored multidigraph.

Important matrices associated with a digraph include the adjacency matrix J and the incidence matrix B (Godsil and Royle 2001). J and B can be extended to the weighted adjacency and incidence matrices. Let $m = |V|$ denote the cardinality of V in G.

Definition 9.9 Let $H \subseteq N$. For a weighted colored multidigraph (V, A, N, ψ, c, w), the weighted adjacency matrix for H is the $m \times m$ matrix $J_H^{(W)}$ with (s, q) entry

$$J_H^{(W)}(s,q) = \begin{cases} 1 \text{ if there exists } a \in A_H^{(W)} \text{ such that } \psi(a) = (s,q) \text{ for } s, q \in V, \\ 0 \text{ otherwise.} \end{cases}$$

Definition 9.10 For a weighted colored multidigraph (V, A, N, ψ, c, w), w_a denotes the weight of arc $a \in A$. The **weighted incidence matrix for** H is the $m \times l$ matrix $B^{(W_H)}$ with (v, a) entry

$$B^{(W_H)}(v,a) = \begin{cases} -w_a & \text{if } a = (v,x) \text{ for some } x \in V \text{ and } c(a) = i \in H, \\ w_a & \text{if } a = (x,v) \text{ for some } x \in V \text{ and } c(a) = i \in H, \\ 0 & \text{otherwise,} \end{cases}$$

where $v \in V$.

According to the signed entries, the weighted incidence matrix can be separated into the weighted in-incidence matrix and the weighted out-incidence matrix.

Definition 9.11 For a weighted colored multidigraph (V, A, N, ψ, c, w), let $H \subseteq N$ and w_a denote the weight of arc $a \in A$. The **weighted in-incidence matrix for** H and the **weighted out-incidence matrix for** H are two $m \times l$ matrices $B_{in}^{(W_H)}$ and $B_{out}^{(W_H)}$ with (v, a) entries

$$B_{in}^{(W_H)}(v,a) = \begin{cases} w_a & \text{if } a = (x,v) \text{ for some } x \in V \text{ and } c(a) = i \in H, \\ 0 & \text{otherwise,} \end{cases}$$

and

$$B_{out}^{(W_H)}(v,a) = \begin{cases} w_a & \text{if } a = (v,x) \text{ for some } x \in V \text{ and } c(a) = i \in H, \\ 0 & \text{otherwise,} \end{cases}$$

where $v \in V$.

It is obvious that

$$B_{in}^{(W_H)} = (B^{(W_H)} + abs(B^{(W_H)}))/2 \text{ and } B_{out}^{(W_H)} = (abs(B^{(W_H)}) - B^{(W_H)})/2,$$

where $abs(B^{(W_H)})$ denotes the matrix in which each entry equals the absolute value of the corresponding entry of $B^{(W_H)}$. Let I denote the identity matrix. If $W_H = I$, then $B^{(W_H)} = B$, $B_{in}^{(W_H)} = B_{in}$, and $B_{out}^{(W_H)} = B_{out}$.

A weighted reachability matrix by H is used to describe the reachability by the weighted colored paths for H. Its formal definition is given as follows.

Definition 9.12 Let $H \subseteq N$. For a weighted colored multidigraph (V, A, N, ψ, c, w), the weighted reachability matrix by H is the $m \times m$ matrix $M_H^{(W)}$ with (s, q) entry

$$M_H^{(W)}(s,q) = \begin{cases} 1 \text{ if } q \text{ is reachable from vertex } s \text{ by a weighted} \\ \quad \text{colored path } PA_H^{(W)}(s,q), \text{ for } s, q \in V, \\ 0 \text{ otherwise.} \end{cases}$$

Let $l_H^{(W)} = |A_H^{(W)}|$ denote the number of arcs in $A_H^{(W)}$. Since all arcs are distinct on a path, the length of any path in $PA_H^{(W)}$ is less than $l_H^{(W)}$.

For a weighted colored multidigraph $G = (V, A, N, \psi, c, w)$, recall that the **adjacency matrix of the line graph** of G is the $l \times l$ matrix LJ with (a, b) entry

$$LJ(a,b) = \begin{cases} 1 & \text{if edges } a \text{ and } b \text{ are consecutive in order } ab \text{ in the graph } G, \\ 0 & \text{otherwise.} \end{cases}$$

In this section, LJ matrix is called an **edge consecutive matrix**.

Definition 9.13 For a weighted colored multidigraph $G = (V, A, N, \psi, c, w)$, let $H \subseteq N$ and w_a and w_b denote the weights of arcs $a, b \in A$. The **weighted edge consecutive matrix for** H is the $l \times l$ matrix $LJ^{(W_H)}$with (a, b) entry

$$LJ^{(W_H)}(a,b) = \begin{cases} w_a \cdot w_b & \text{if edges } a \text{ and } b \text{ are consecutive in order } ab \\ & \text{and } c(a) = i \text{ and } c(b) = j \text{ for } i, j \in H, \\ 0 & \text{otherwise.} \end{cases}$$

Definition 9.14 For a weighted colored multidigraph $G = (V, A, N, \psi, c, w)$, the **reduced weighted edge consecutive matrix for** H is the $l \times l$ matrix $LJ_r^{(W_H)}$ with (a, b) entry

$$LJ_r^{(W_H)}(a,b) = \begin{cases} w_a \cdot w_b & \text{if edges } a \text{ and } b \text{ are consecutive in order } ab \text{ and} \\ & c(a) = i \text{ and } c(b) = j \text{ such that } i, j \in H \text{ and } i \neq j, \\ 0 & \text{otherwise.} \end{cases}$$

Let c_i denote the cardinality of the arc set in color i. I_{c_i} is defined as a $c_i \times c_i$ identity matrix with each diagonal entry being set to 1 for $i = 1, 2, \cdots, n$. Let I_i denote an $l \times l$ diagonal matrix for which

$$I_i = \begin{pmatrix} 0 & & \cdots\cdots & & 0 \\ \vdots & \ddots & & & \vdots \\ 0 & \cdots & I_{c_i} & \cdots & 0 \\ \vdots & & & \ddots & \vdots \\ 0 & & \cdots\cdots & & 0 \end{pmatrix}.$$

For $H \subseteq N$, $H \neq \emptyset$, and $I_H = \bigvee_{i \in H} I_i$, $W_H = W \circ I_H$. ("$\circ$" denotes the Hadamard product.)

9.3.1.2 A Weighted Conversion Function

Lemma 9.2 *For a weighted colored multidigraph (V, A, N, ψ, c, w), the weighted incidence matrix $B^{(W_H)}$ for H and the incidence matrix B have the following relation*

$$B^{(W_H)} = B \cdot W_H = B \cdot (W \circ I_H).$$

Lemma 9.2 shows a conversion function to transform an original colored multidigraph in the color set N to a reduced weighted colored multidigraph in the color set $H \subseteq N$.

Now let W be a weight matrix and let $L^{(W)}(G)$ denote the weighted line digraph of G. The following theorem is obtained based on Definition 9.11, on the weighted in-incidence and out-incidence matrices $B_{in}^{(W)}$ and $B_{out}^{(W)}$, and Definition 9.13, on the weighted edge consecutive matrix $LJ^{(W)}$ of the digraph $L^{(W)}(G)$.

Theorem 9.2 *For a weighted colored multidigraph $G = (V, A, N, \psi, c, w)$, W is the weight matrix, $B_{in}^{(W)}$ is the weighted in-incidence matrix, and $B_{out}^{(W)}$ is the weighted out-incidence matrix of the graph G. Then, the weighted edge consecutive matrix $LJ^{(W)}$ satisfies $LJ^{(W)} = (B_{in}^{(W)})^T \cdot (B_{out}^{(W)})$.*

Obviously, when W is reduced to W_H, $LJ^{(W_H)} = (B_{in}^{(W_H)})^T \cdot (B_{out}^{(W_H)})$.

Let $T_1(B^{(W)}) = (B_{in}^{(W)})^T \cdot (B_{out}^{(W)}) = LJ^{(W)}$ denote a conversion function. The conversion function, $T_1(B^{(W)})$, maps the weighted incidence matrix $B^{(W)}$ to the weighted edge consecutive matrix $LJ^{(W)}$ of the graph G. It shows that this conversion function transforms the original edge-weighted, colored multidigraph G to a simple vertex-weighted-colored line digraph $L(G)$. When $W = I$, $LJ = (B_{in})^T \cdot (B_{out})$. This matrix captures the adjacency relation between pairs of consecutive edges without considering the color(s) of the consecutive edges. Another conversion function is thus presented next to transform the original problem of searching edge-colored paths in a colored multidigraph to the standard problem of finding paths in a simple digraph without color constraints.

Recall that c_i denotes the cardinality of the arc set in color i and let E_{c_i} denote a $c_i \times c_i$ matrix with each entry being set to 1 for $i = 1, 2, \cdots, n$. Then, D is defined in Chap. 3 as the following block diagonal matrix

$$D = \begin{pmatrix} E_{c_1} & 0 & \cdots & 0 \\ 0 & E_{c_2} & \cdots & 0 \\ \vdots & \vdots & \ddots & \vdots \\ 0 & 0 & \cdots & E_{c_n} \end{pmatrix}. \tag{9.11}$$

It is obvious that this matrix D encodes the color scheme in the graph G, where the dimension of each diagonal block E_{c_i} depends on the number of edges in color i. More specifically, recall that $\varepsilon_i = \sum_{j=1}^{i} c_j$ for $1 \le i \le n$. According to the Rule of Priority for labeling edges, for any $a_k \in A$ and $\varepsilon_{i-1} < k \le \varepsilon_i$, the edge a_k has color i. Hence, for any $a_k, a_h \in A$, if there exists $1 \le i \le n$ such that $k, h \in (\varepsilon_{i-1}, \varepsilon_i]$, then edges a_k and a_h have the same color i, and $D(k, h) = 1$. Also, $D(k, h) = 0$ iff edges a_k and a_h have different colors.

The conversion function can now be obtained in matrix form by the following theorem.

Theorem 9.3 *For the weighted colored multidigraph $G = (V, A, N, \psi, c, w)$, let E_l be the $l \times l$ matrix with each entry equal to 1. Then the reduced matrix $LJ_r^{(W)}$ satisfies $LJ_r^{(W)} = LJ^{(W)} \circ (E_l - D)$, where "$\circ$" denotes the Hadamard product.*

Obviously, when W is reduced to W_H, $LJ_r^{(W_H)} = LJ^{(W_H)} \circ (E_l - D)$ satisfies that

$$LJ_r^{(W_H)}(a, b) = \begin{cases} w_a \cdot w_b & \text{if edges } a \text{ and } b \text{ are consecutive in order } ab \text{ and} \\ & c(a) = i \text{ and } c(b) = j \text{ such that } i \neq j \text{ for } i, j \in H, \\ 0 & \text{otherwise.} \end{cases} \tag{9.12}$$

From Theorem 9.3, $T_2(LJ^{(W)}) = LJ^{(W)} \circ (E_l - D) = LJ_r^{(W)}$. The conversion function, $T_2(LJ^{(W)})$, maps the weighted edge consecutive matrix $LJ^{(W)}$ of the weighted line digraph $L^{(W)}(G)$ to its reduced matrix $LJ_r^{(W)}$. It reveals that this conversion function T_2 converts the simple vertex-weighted, colored line digraph $L^{(W)}(G)$ to its reduced subgraph $L_r^{(W)}(G)$, called reduced weighted line digraph, which is a simple digraph with no color constraints.

Theorems 9.2 and 9.3 together present a conversion function $F(B^{(W)})$ such that

$$F(B^{(W)}) = [(B_{in}^{(W)})^T \cdot B_{out}^{(W)}] \circ (E_l - D), \tag{9.13}$$

where $B_{in}^{(W)} = (B^{(W)} + abs(B^{(W)}))/2$ and $B_{out}^{(W)} = (abs(B^{(W)}) - B^{(W)})/2$. Therefore, $F(B^{(W)})$ transforms a problem of searching weighted colored paths in an edge-weighted, colored multidigraph to a standard problem of finding paths in a simple digraph with no color constraints. Note that the incident relations between vertices and edges of a graph can uniquely characterize the graph. Therefore, the incidence matrix is treated as the original graph and used for computer implementation.

Example 9.1 Figure 9.1 shows a colored multidigraph $G = (V, A, N, \psi, c)$. If G is associated with a map $w : A \to \mathbb{R}_0^+$, then $G = (V, A, N, \psi, c, w)$ is a weighted colored multidigraph. Construct conversion functions to determine the vertex labeled weighted line digraph $L^{(W)}(G)$ and its reduced line digraph $L_r^{(W)}(G)$.

The colored multidigraph is labeled using the Rule of Priority presented in Sect. 3.3.2. Obviously, the colored graph shown in Fig. 9.1 is labeled in Fig. 9.2a.

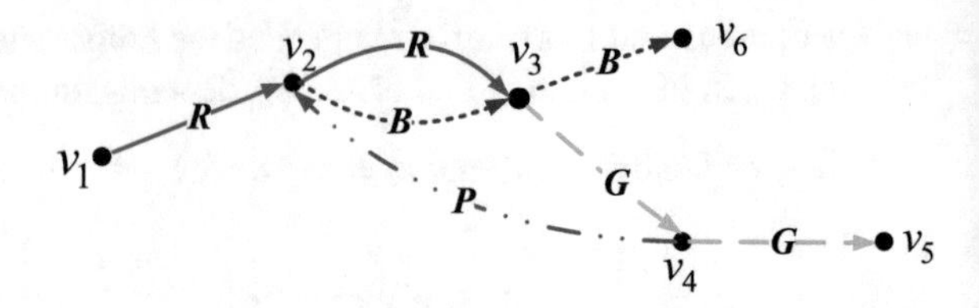

Fig. 9.1 The colored multidigraph G

It is easy to obtain incident relations between vertices and edges from the graph. Thus, matrices $B_{in}^{(W)}$ and $B_{out}^{(W)}$ are constructed by Definition 9.11 as follows:

$$B_{in}^{(W)} = \begin{pmatrix} 0 & 0 & 0 & 0 & 0 & 0 & 0 \\ w_1 & 0 & 0 & 0 & 0 & 0 & w_7 \\ 0 & w_2 & w_3 & 0 & 0 & 0 & 0 \\ 0 & 0 & 0 & 0 & w_5 & 0 & 0 \\ 0 & 0 & 0 & 0 & 0 & w_6 & 0 \\ 0 & 0 & 0 & w_4 & 0 & 0 & 0 \end{pmatrix},$$

and

$$B_{out}^{(W)} = \begin{pmatrix} w_1 & 0 & 0 & 0 & 0 & 0 & 0 \\ 0 & w_2 & w_3 & 0 & 0 & 0 & 0 \\ 0 & 0 & 0 & w_4 & w_5 & 0 & 0 \\ 0 & 0 & 0 & 0 & 0 & w_6 & w_7 \\ 0 & 0 & 0 & 0 & 0 & 0 & 0 \\ 0 & 0 & 0 & 0 & 0 & 0 & 0 \end{pmatrix}.$$

From Theorems 9.2 and 9.3, one obtains that

$$T_1(B^{(W)}) = \begin{pmatrix} 0 & w_1w_2 & w_1w_3 & 0 & 0 & 0 & 0 \\ 0 & 0 & 0 & w_2w_4 & w_2w_5 & 0 & 0 \\ 0 & 0 & 0 & w_3w_4 & w_3w_5 & 0 & 0 \\ 0 & 0 & 0 & 0 & 0 & 0 & 0 \\ 0 & 0 & 0 & 0 & 0 & w_5w_6 & w_5w_7 \\ 0 & 0 & 0 & 0 & 0 & 0 & 0 \\ 0 & w_7w_2 & w_7w_3 & 0 & 0 & 0 & 0 \end{pmatrix}$$

and

$$T_2(LJ^{(W)}) = \begin{pmatrix} 0 & 0 & w_1w_3 & 0 & 0 & 0 & 0 \\ 0 & 0 & 0 & w_2w_4 & w_2w_5 & 0 & 0 \\ 0 & 0 & 0 & 0 & w_3w_5 & 0 & 0 \\ 0 & 0 & 0 & 0 & 0 & 0 & 0 \\ 0 & 0 & 0 & 0 & 0 & 0 & w_5w_7 \\ 0 & 0 & 0 & 0 & 0 & 0 & 0 \\ 0 & w_7w_2 & w_7w_3 & 0 & 0 & 0 & 0 \end{pmatrix}.$$

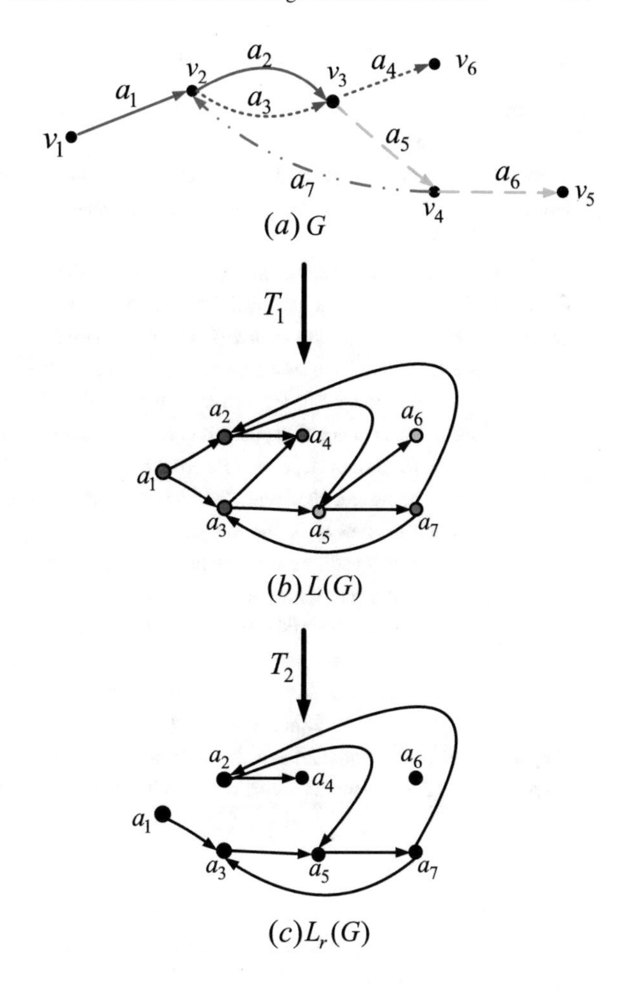

Fig. 9.2 Transformed graphs of G

The weight matrix designed here is convenient, since edge-weighted (0 or 1) can be used to flexibly control any move between any two vertices in G. For instance, if $w_4 = 0$, then the original graph will be reduced to a new graph with no edge a_4. If $W = I$, then the conversion function T_1 transforms the edge-labeled multidigraph G portrayed in Fig. 9.2a to the vertex-labeled line digraph $L(G)$ shown in Fig. 9.2b. Then, the reduced line digraph $L_r(G)$ presented in Fig. 9.2c for finding colored paths is obtained by using the conversion function T_2. The conversion process is illustrated in Fig. 9.2.

9.3.2 Computer Implementation

Many well-known algorithms have been developed to solve the shortest path problems in digraphs. Some other algorithms are available for searching for all paths

Table 9.10 Pseudocode for finding colored paths based on edge consecutive matrix

Step 0: Input the starting arc set A_S, the ending arc set A_E, and the reduced weighted edge consecutive matrix $LJ_r^{(W)}$.
Step 1: For each arc $a_s \in A_S$ and each arc $a_e \in A_E$, set a_s as the starting arc and a_e as the ending arc. For each pair of a_s and a_e, repeat the steps from Step 2 to Step 5.
Step 2: Put a_s into Path-Recorder as the last arc $a_l(1)$ of the first path.
Step 3: In Path-Recorder, for each path i, e.g., $PA^{(W)}(i)$, check its last arc $a_l(i)$. Obtain all the new arcs starting from $a_l(i)$ based on matrix $LJ_r^{(W)}$.
Case 1: If there is no arc starting from $a_l(i)$, path $PA^{(W)}(i)$ ends. Eliminate $PA^{(W)}(i)$ from Path-Recorder;
Case 2: If a new arc has appeared in the path, which means that the path forms a cycle, do not record the new path. If all the new arcs have appeared, eliminate $PA^{(W)}(i)$ from Path-Recorder;
Case 3: If the new arc is the end arc a_e, add a_e to the path $PA^{(W)}(i)$ to form a new path. Reserve the path into Path-Recorder and set an *end-mark* at the end of the path;
Otherwise: Add each new arc to path $PA^{(W)}(i)$, respectively, to form several new paths.
Reserve these paths into Path-Recorder, and eliminate the original path $PA^{(W)}(i)$ from Path-Recorder.
Step 4: Repeat Step 3 until all the paths in Path-Recorder have the *end-mark* at the end.
Step 5: Output Path-Recorder, which records all paths starting from a_s and ending at a_e.

in undirected graphs, such as the algorithm presented by Migliore et al. (1990). Although finding path problems in general graph classes has been extensively investigated, searching colored paths in weighted colored multidigraphs is still a novel topic.

Let $A_S = \{a \in A : B_{out}^{(W)}(s, a) \neq 0\}$ and $A_E = \{b \in A : B_{in}^{(W)}(q, b) \neq 0\}$ for $s, q \in V$. Here, matrices W, $B_{out}^{(W)}$, and $B_{in}^{(W)}$ have been introduced by Definitions 9.8 and 9.11. A_S is the set of arcs starting from vertex s and A_E is the arc set ending at vertex q. The matrix $LJ_r^{(W)}$ provided by Theorem 9.3 is used to search the edge-weighted, colored paths between any two arcs in a weighted colored multidigraph. Let $PA^{(W)}(a, b)$ for $a, b \in A$ denote the weighted colored paths between two edges a and b. The weighted colored paths between two vertices s and q for $s, q \in V$ are expressed as $PA^{(W)}(s, q)$. A vertex-by-vertex path between any two vertices in the graph G can be obtained by tracing arc-by-arc paths between two appropriate arcs in the line graph $L(G)$. Specifically, the paths between s and q can be expressed as $PA^{(W)}(s, q) = \{PA^{(W)}(a, b) : a \in A_S,\ b \in A_E\}$.

The algebraic method developed here is convenient for computer implementation. A pseudocode for the proposed algorithm is presented in Table 9.10.

Because the algebraic expressions are explicitly given, the developed method facilitates the development of improved algorithms to search colored paths and is easy to adapt to new path searching problems. For instance, a transportation network problem of finding the shortest path with specific constraints can be solved by using the conversion function $F(B^{(W)}) = [(B_{in}^{(W)})^T \cdot B_{out}^{(W)}] \circ M$, where $B^{(W)}$ denotes the original network and matrix M is designed to capture constraint requirements, to transform the original problem to a general shortest path searching problem without the constraints.

9.3.2.1 Weight Matrix for GMCR with Simple Preference

In the original information, the preference of DM i is coded by a pair of relations $\{\succ_i, \sim_i\}$ on S. This preference structure is called simple preference.

Definition 9.8 presents a weight matrix W_H for a weighted colored multidigraph $G = (V, A, N, \psi, c, w)$. In a graph model $G = (S, A)$, let $H \subseteq N$. By the Rule of Priority, the oriented arcs in the graph model are labeled according to the DM order; within each DM, according to the sequence of initial states; and within each DM and initial state, according to the sequence of terminal states. When an edge $a_k = d_i(u, v)$ for $u, v \in S$ and $i \in H \subseteq N$, then its weight w_k can be defined by

$$w_k = \begin{cases} P_w & \text{if } v \succ_i u \text{ and } i \in H, \\ E_w & \text{if } u \sim_i v \text{ and } i \in H, \\ N_w & \text{if } u \succ_i v \text{ and } i \in H, \\ 0 & \text{otherwise.} \end{cases} \tag{9.14}$$

The weight matrix W_H represents preference information of each edge in the graph model for simple preference. Recall that notation UMs and UIs denote unilateral moves and unilateral improvements, respectively. Based on Eq. 9.14, the UM weight matrix and the UI weight matrix for H are defined as follows.

Definition 9.15 For the graph model $G = (S, A)$, let $H \subseteq N$.

- When $P_w = E_w = N_w = 1$, the weight matrix W_H is called the UM weight matrix by H, denoted by $W_H^{(UM)}$;
- When $P_w = 1$ and $E_w = N_w = 0$, the weight matrix W_H is called the UI weight matrix by H, denoted by $W_H^{(UI)}$ or W_H^+.

Recall that each arc of A_i and A_i^+ denotes that DM i can make a UM and a UI (in one step) from the initial state to the terminal state of the arc, respectively. Therefore, $A_H = \bigcup_{i \in H} A_i$ and $A_H^+ = \bigcup_{i \in H} A_i^+$ denote the UM and the UI arcs associated with any DM in H. Based on Definition 9.7, on the weighted arc set for H, the following result relative to the UM arc set and the UI arc set is obvious for the graph model with simple preference.

Corollary 9.1 *For the graph model $G = (S, A)$, let $H \subseteq N$.*

- *If* $W_H = W_H^{(UM)}$*, then the arc set* $A_H^{(W)} = A_H$*;*
- *If* $W_H = W_H^+$*, the arc set* $A_H^{(W)} = A_H^+$*.*

Note that when $H = N$, A_H and A_H^+ are denoted by A and A^+, respectively.

In a weighted colored multidigraph, the edge-weighted, colored paths by H between two vertices u and v are described in Definition 9.5 which can represent conflict evolution by the legal UMs and the legal UIs in a graph model for simple preference.

Corollary 9.2 *For the graph model* $G = (S, A)$*, let* $u, v \in S$ *and* $H \subseteq N$*.*

- *If* $W_H = W_H^{(UM)}$*, the weighted colored paths between states* u *and* v*,* $PA_H^{(W)}(u, v)$*, give all paths from* u *to* v *where all legal UMs are allowed. Then* $PA_H^{(W)}(u, v)$ *are called legal UM paths from* u *to* v *by coalition* H*, denoted by* $PA_H(u, v)$*;*
- *If* $W_H = W_H^+$*, the weighted colored paths between states* u *and* v*,* $PA_H^{(W)}(u, v)$*, give all paths from* u *to* v *where only legal UIs are allowed. Then* $PA_H^{(W)}(u, v)$ *are called legal UI paths from* u *to* v *by coalition* H*, denoted by* $PA_H^+(u, v)$*.*

The weighted colored paths $PA_H^{(W)}$ can be used to trace conflict evolution of status quo analysis for simple preference. When u is selected as a status quo and v is an equilibrium for some stability in a graph model, $PA_H(u, v)$ and $PA_H^+(u, v)$ trace conflict evolution to confirm that the equilibrium is in fact reachable from the status quo and reveal how to reach it.

Definition 9.16 In the graph model $G = (S, A)$, the legal UM and the legal UI edge consecutive matrices are two $l \times l$ matrices $LJ_r^{(UM)}$ and LJ_r^+ with (a, b) entries

$$LJ_r^{(UM)}(a, b) = \begin{cases} 1 & \text{if edges } a \text{ and } b \text{ are consecutive in order } ab \text{ and} \\ & \text{are controlled by different DMs for } a, b \in A, \\ 0 & \text{otherwise,} \end{cases}$$

$$LJ_r^+(a, b) = \begin{cases} 1 & \text{if edges } a \text{ and } b \text{ are consecutive in order } ab \text{ and} \\ & \text{are controlled by different DMs for } a, b \in A^+, \\ 0 & \text{otherwise.} \end{cases}$$

Let LJ_{H_r} and $LJ_{H_r}^+$ denote the legal UM and the legal UI edge consecutive matrices in the graph model (S, A_H). Based on Definition 9.14, on the reduced weighted edge consecutive matrix by H, and Definition 9.16, the following result is obvious.

Corollary 9.3 *For the graph model* $G = (S, A)$*, let* $W^{(UM)}$ *and* W^+ *denote the UM and the UI weight matrices, and* $W_H^{(UM)}$ *and* W_H^+ *be the UM and the UI weight matrices for* H*. Then*

$$LJ_r^{(W^{(UM)})} = LJ_r^{(UM)} = LJ_r, \; LJ_r^{(W^+)} = LJ_r^+,$$

and

$$LJ_r^{(W_H^{(UM)})} = LJ_{H_r}, \; LJ_r^{(W_H^+)} = LJ_{H_r}^+.$$

As the algorithm given in Table 9.10 for searching weighted colored paths in a weighted colored multidigraph, the legal UM and UI edge consecutive matrices LJ_{H_r} and $LJ^+_{H_r}$ are applied to find paths PA_H and PA^+_H between any two states for status quo analysis in a graph model.

For simple preference, the key inputs of stability analysis, $R_H(s)$ and $R^+_H(s)$, are the reachable lists by coalition H from state $s \in S$ by the legal UMs and the legal UIs. This section provides an algebraic approach to construct $R_H(s)$ and $R^+_H(s)$ using the weighted reachability matrix $M^{(W)}_H$ shown by Definition 9.12.

9.3.2.2 Weight Matrix for GMCR with Unknown Preference

Preference information plays an important role in the decision analysis. To incorporate preference uncertainty into the graph model methodology, Li et al. (2004a) proposed a new preference structure in which DM i's preferences are expressed by a triple of relations $\{\succ_i, \sim_i, U_i\}$ on S, where $s \succ_i q$ indicates strict preference, $s \sim_i q$ indicates indifference, and $s\ U_i\ q$ means DM i may prefer state s to state q, may prefer q to s, or may be indifferent between s and q.

The weight matrix W_H can be employed to represent preference with uncertainty. When an edge $a_k = d_i(u, v)$ for $u, v \in S$ and $i \in H \subseteq N$, then its weight w_k can be defined by

$$w_k = \begin{cases} P_w & \text{if } v \succ_i u \text{ and } i \in H, \\ N_w & \text{if } u \succ_i v \text{ and } i \in H, \\ E_w & \text{if } u \sim_i v \text{ and } i \in H, \\ U_w & \text{if } u\ U_i\ v \text{ and } i \in H, \\ 0 & \text{otherwise.} \end{cases} \tag{9.15}$$

Recall that notation UIUMs denotes unilateral improvements or uncertain moves. Based on Eq. 9.15, the UIUM weight matrix for H is defined as follows.

Definition 9.17 For the graph model $G = (S, A)$, let $H \subseteq N$. When $P_w = U_w = 1$ and $E_w = N_w = 0$, the weight matrix W_H is called the UIUM weight matrix for H, denoted by $W_H^{(UIUM)}$ or $W_H^{+,U}$.

Each arc of arc set $A_i^{+,U}$ denotes that DM i can make a UIUM from the initial state to the terminal state of the arc. Therefore, $A_H^{+,U} = \bigcup_{i \in H} A_i^{+,U}$ indicates the UIUM arcs associated with any DM in H. By Definition 9.7 for the weighted arc set $A_H^{(W)}$, the UIUM arc set is obtained for a graph model with unknown preference by the following corollary.

Corollary 9.4 *For the graph model $G = (S, A)$, let $H \subseteq N$. If $W_H = W_H^{+,U}$, then the arc set $A_H^{(W)} = A_H^{+,U}$.*

Note that when $H = N$, $A_H^{+,U}$ is expressed by $A^{+,U}$.

The weighted colored paths $PA_H^{(W)}$ can be applied to trace conflict evolution by the legal UIUMs for the graph model with preference uncertainty.

Corollary 9.5 *For the graph model $G = (S, A)$, let $u, v \in S$ and $H \subseteq N$. If $W_H = W_H^{+,U}$, the weighted colored paths between states u and v, $PA_H^{(W)}(u, v)$, give all paths from u to v where only the legal UIUMs are allowed. Then $PA_H^{(W)}(u, v)$ are called the legal UIUM paths from u to v by coalition H, denoted by $PA_H^{+,U}(u, v)$.*

The conflict evolution by the legal UIUMs can be tracked using the reduced weighted edge consecutive matrix. The legal UIUM edge consecutive matrix is defined first.

Definition 9.18 In the graph model $G = (S, A)$, the legal UIUM edge consecutive matrix is an $l \times l$ matrix $LJ_r^{+,U}$ with (a, b) entry

$$LJ_r^{+,U}(a,b) = \begin{cases} 1 & \text{if edges } a \text{ and } b \text{ are consecutive in order } ab \text{ and} \\ & \text{are controlled by different DMs for } a, b \in A^{+,U}, \\ 0 & \text{otherwise.} \end{cases}$$

Let $LJ_{H_r}^{+,U}$ denote the legal UIUM edge consecutive matrix for the graph model (V, A_H). Based on Definitions 9.14 and 9.18, the following result is obtained.

Corollary 9.6 *For the graph model $G = (S, A)$, let $W^{+,U}$ denote the UIUM weight matrix and $W_H^{+,U}$ be the UIUM weight matrix for H. Then*

$$LJ_r^{(W^{+,U})} = LJ_r^{+,U},$$

and

$$LJ_r^{(W_H^{+,U})} = LJ_{H_r}^{+,U}.$$

The key input of stability analysis for the graph model with preference uncertainty is the reachable list $R_H^{+,U}(s)$ of coalition $H \subseteq N$ from state $s \in S$ by the legal UIUMs. The algebraic approach to searching weighted colored paths can also be used to construct $R_H^{+,U}(s)$.

9.3.2.3 Weight Matrix for GMCR with Three Degrees of Preference

Another triplet relation $\{\gg_i, >_i, \sim_i\}$ on S that expresses strength of preference (strong or mild preference) is presented in Chap. 6. For $s, q \in S$, $s \gg_i q$ denotes DM i strongly prefers s to q, $s >_i q$ means DM i mildly prefers s to q, and $s \sim_i q$ indicates that DM i is indifferent between states s and q. The weight matrix W_H can represent strength of preference. When an edge $a_k = d_i(u, v)$ for $u, v \in S$ and $i \in H \subseteq N$, then its weight w_k is defined by

$$w_k = \begin{cases} P_s & \text{if } v \gg_i u \text{ and } i \in H, \\ P_m & \text{if } v >_i u \text{ and } i \in H, \\ E_w & \text{if } u \sim_i v \text{ and } i \in H, \\ N_w & \text{if } u \gg_i v \text{ or } u >_i v \text{ and } i \in H, \\ 0 & \text{otherwise.} \end{cases} \tag{9.16}$$

Recall that notation MSUIs denotes mild or strong unilateral improvements. Based on Eq. 9.16, the MSUI weight matrix for H is defined as follows.

Definition 9.19 For the graph model $G = (S, A)$, let $H \subseteq N$. When $P_s = P_m = 1$ and $E_w = N_w = 0$, the weight matrix W_H is called the MSUI weight matrix for H, denoted for $W_H^{(MSUI)}$ or $W_H^{+,++}$.

Each arc of the arc set $A_i^{+,++}$ denotes that DM i can make a MSUI from the initial state to the terminal state of the arc. Therefore, $A_H^{+,++} = \bigcup_{i \in H} A_i^{+,++}$ denotes the MSUI arcs associated with any DM in H. By Definition 9.7 for the weighted arc set $A_H^{(W)}$, the MSUI arc set is obtained for a graph model with strength of preference by the following corollary.

Corollary 9.7 *For the graph model $G = (S, A)$, let $H \subseteq N$. If $W_H = W_H^{+,++}$, then the arc set $A_H^{(W)} = A_H^{+,++}$.*

Note that when $H = N$, $A_H^{+,++}$ is expressed by $A^{+,++}$.

The weighted colored paths $PA_H^{(W)}$ can be applied to trace conflict evolution by the legal MSUIs for the graph model with strength of preference.

Corollary 9.8 *For the graph model $G = (S, A)$, let $u, v \in S$ and $H \subseteq N$. If $W_H = W_H^{+,++}$, the weighted colored paths between states u and v, $PA_H^{(W)}(u, v)$, give all paths from u to v where only the legal MSUIs are allowed. Then $PA_H^{(W)}(u, v)$ are called the legal MSUI paths from u to v by coalition H, denoted by $PA_H^{+,++}(u, v)$.*

Definition 9.20 In the graph model $G = (S, A)$, the legal MSUI edge consecutive matrix is an $l \times l$ matrix $LJ_r^{+,++}$ with (a, b) entry

$$LJ_r^{+,++}(a, b) = \begin{cases} 1 & \text{if edges } a \text{ and } b \text{ are consecutive in order } ab \text{ and} \\ & \text{are controlled by different DMs for } a, b \in A^{+,++}, \\ 0 & \text{otherwise.} \end{cases}$$

Let $LJ_{H_r}^{+,++}$ denote the legal MSUI edge consecutive matrix for the graph model (V, A_H). Based on Definition 9.14, on the reduced weighted edge consecutive matrix by H, and Definition 9.20, the following result can be easily obtained.

Corollary 9.9 *For the graph model $G = (S, A)$, let $W^{+,++}$ denote the MSUI weight matrix and $W_H^{+,++}$ be the MSUI weight matrix for H. Then*

$$LJ_r^{(W^{+,++})} = LJ_r^{+,++},$$

and

$$LJ_r^{(W_H^{+,++})} = LJ_{H_r}^{+,++}.$$

The key input of stability analysis in the graph model with three degrees of preference is state set $R_H^{+,++}(s)$, the reachable list of coalition $H \subseteq N$ from state $s \in S$ by the legal MSUIs. The algebraic approach provides a new method to construct $R_H^{+,++}(s)$.

9.3.2.4 Weight Matrix for GMCR with Hybrid Preference

A hybrid preference framework is presented in Chap. 7 to combine preference uncertainty and strength of preference using a quadruple relation $\{\gg_i, >_i, \sim_i, U_i\}$ in a graph model for DM i. The weight matrix W_H can also represent the combination of preference uncertainty and strength of preference. When an edge $a_k = d_i(u, v)$ for $u, v \in S$ and $i \in H \subseteq N$, then its weight w_k is defined by

$$w_{a_k} = \begin{cases} P_s & \text{if } v \gg_i u \text{ and } i \in H, \\ P_m & \text{if } v >_i u \text{ and } i \in H, \\ E_w & \text{if } u \sim_i v \text{ and } i \in H, \\ U_w & \text{if } u\ U_i\ v \text{ and } i \in H, \\ N_w & \text{if } u \gg_i v \text{ or } u >_i v \text{ and } i \in H, \\ 0 & \text{otherwise.} \end{cases} \tag{9.17}$$

Recall that notation MSUIUMs denotes mild or strong unilateral improvements or uncertain moves. By Eq. 9.17, the MSUIUM weight matrix for H is defined as follows.

Definition 9.21 For the graph model $G = (S, A)$, let $H \subseteq N$. When $P_s = P_m = U_w = 1$ and $E_w = N_w = 0$, the weight matrix W_H is called the MSUIUM weight matrix for H, denoted by $W_H^{(MSUIUM)}$ or $W_H^{+,++,U}$.

Each arc of the arc set $A_i^{+,++,U}$ denotes that DM i can make a MSUIUM from the initial state to the terminal state of the arc. Therefore, $A_H^{+,++,U} = \bigcup_{i \in H} A_i^{+,++,U}$ denotes the MSUIUM arcs associated with any DM in H. By Definition 9.7 for the weighted arc set, the MSUIUM arc set is obtained for a graph model with hybrid preference by the following corollary.

Corollary 9.10 *For the graph model $G = (S, A)$, let $H \subseteq N$. If $W_H = W_H^{+,++,U}$, then the arc set $A_H^{(W)} = A_H^{+,++,U}$.*

Note that when $H = N$, $A_H^{+,++,U}$ is expressed by $A^{+,++,U}$.

The weighted colored paths $PA_H^{(W)}$ can be applied to trace conflict evolution by the legal MSUIUMs for the graph model with hybrid preference.

Corollary 9.11 *For the graph model $G = (S, A)$, let $u, v \in S$ and $H \subseteq N$. If $W_H = W_H^{+,++,U}$, the weighted colored paths between states u and v, $PA_H^{(W)}(u, v)$, give all paths from u to v where only the legal MSUIUMs are allowed. Then $PA_H^{(W)}(u, v)$ are called the legal MSUIUM paths from u to v by coalition H, denoted by $PA_H^{+,++,U}(u, v)$.*

Definition 9.22 In the graph model $G = (S, A)$, the legal MSUIUM edge consecutive matrix is an $l \times l$ matrix $LJ_r^{+,++,U}$ with (a, b) entry

$$LJ_r^{+,++,U}(a, b) = \begin{cases} 1 & \text{if edges } a \text{ and } b \text{ are consecutive in order } ab \text{ and} \\ & \text{are controlled by different DMs for } a, b \in A^{+,++,U}, \\ 0 & \text{otherwise.} \end{cases}$$

Let $LJ_{H_r}^{+,++,U}$ denote the legal MSUIUM edge consecutive matrix for the graph model (V, A_H). Based on Definition 9.14, on the reduced weighted edge consecutive matrix for H, and Definition 9.22, the following result is obtained.

Corollary 9.12 *For the graph model $G = (S, A)$, let $W^{+,++,U}$ denote the MSUIUM weight matrix and $W_H^{+,++,U}$ be the MSUIUM weight matrix for $H \subseteq N$. Then*

$$LJ_r^{(W^{+,++,U})} = LJ_r^{+,++,U},$$

and

$$LJ_r^{(W_H^{+,++,U})} = LJ_{H_r}^{+,++,U}.$$

9.3.3 Procedures of Employing the Algebraic Approach Based on Edge Consecutive Matrix

The algebraic approach developed uses the results of Graph Theory to assist in analyzing a graph model, and understanding evolution of a conflict, by carrying out the following steps:

- If the state set S is treated as a vertex set V and DM i's oriented arcs $A_i \subseteq A$ are coded in color $i \in N$, then a graph model (S, A) of a conflict is equivalent to a colored multidigraph (V, A, N, ψ, c) with induced preference relations on V, where ψ and c are functions with $\psi : A \to V \times V$ such that $\psi(a) = (u, v)$ for $a \in A$ and $u, v \in V$, and $c : A \to N$ such that $c(a) \in N$ is the color of $a \in A$;
- By the proposed *Rule of Priority*, the oriented arcs in the colored multidigraph are labeled according to the color order; within each color, according to the sequence of initial nodes; and within each color and initial node, according to the sequence of terminal nodes;
- The incidence matrix B represents the colored multidigraph after all edges are labeled;
- Based on preference structures such as simple preference, unknown preference, three degrees of preference and hybrid preference, a weight matrix W is designed to represent preference information for some preference framework;
- A graph model is thus conveniently treated as an edge-weighted, colored multidigraph (V, A, N, ψ, c, w) in which each arc represents a legal unilateral move, distinct colors refer to different DMs, and the weight along the arc identifies some preference attribute;
- Tracing the evolution of a conflict in status quo analysis is converted to searching all weighted colored paths between a status quo and a possible equilibrium for some preference structure;
- Let the weighted incidence matrix $B^{(W)}$ represent an original edge-weighted, colored multidigraph (V, A, N, ψ, c, w). Then the conversion function

$$F(B^{(W)}) = [(B_{in}^{(W)})^T \cdot B_{out}^{(W)}] \circ (E_l - D)$$

transforms the problem of searching edge-weighted, colored paths in a weighted colored multidigraph to a standard problem of finding paths in a simple digraph with no color constraints;

- Using existing algorithms or the proposed algorithm presented in Table 9.10, the paths between any two edges can be found in a simple digraph;
- If A_S and A_E are the two sets of arcs starting from vertex s and arcs ending at vertex q with

$$A_S = \{a \in A : B_{out}^{(W)}(s, a) \neq 0\} \text{ and } A_E = \{b \in A : B_{in}^{(W)}(q, b) \neq 0\},$$

then paths between any two vertices, $PA^{(W)}(s, q)$ for $s, q \in V$, can be obtained by the paths between two appropriate arcs by

$$PA^{(W)}(s, q) = \{PA^{(W)}(a, b) : a \in A_S,\ b \in A_E\}.$$

According to the procedures, the proposed algebraic approach based on edge consecutive matrix may be employed in practice.

9.3.4 *Applications: Analysis of Conflict Evolution Based on Edge Consecutive Matrix*

9.3.4.1 Analysis of Elmira Conflict Evolution with Simple Preference

The background of the Elmira conflict is introduced in Sect. 1.2.2 and the model is analyzed in Sect. 4.5. If the state set $S = \{s_1, s_2, \cdots, s_9\}$ is treated as a vertex set $V = \{v_1, v_2, \cdots, v_9\}$ and DM i's oriented arcs are coded in colors blue, red, and black for $i = 1, 2,$ and 3, respectively, then the graph model of the Elmira conflict shown in Figs. 3.5 and 4.8 with preference information is equivalent to a weighted colored multidigraph given in Fig. 9.3, in which $w_k(u, v)$ denotes the weight of arc $a_k = (u, v)$. Although no DM is explicitly shown in the labeled graph, the index number of an arc uniquely determines the DM who controls it when all arcs have been numbered according to the Rule of Priority. Recall that c_i denotes the cardinality of arc set assigned color i, i.e., $c_i = |A_i|$, where $A_i = \{x \in A : c(x) = i\}$ for each $i \in N$. Specifically, based on the number of arcs in i's graph G_i for $i = 1, 2,$ and 3, $c_1 = |A_1| = 4$, $c_2 = |A_2| = 12$, and $c_3 = |A_3| = 8$ provided by Figs. 3.5 and 4.8 for the graph model of the Elmira conflict, arcs a_1 to a_4 are controlled by DM 1 or MoE, arcs a_5 to a_{16} by DM 2 or UR, and arcs a_{17} to a_{24} by DM 3 or LG. The weight of each arc in Fig. 9.3 is assigned based on preference information

$$s_7 \succ_1 s_3 \succ_1 s_4 \succ_1 s_8 \succ_1 s_5 \succ_1 s_1 \succ_1 s_2 \succ_1 s_6 \succ_1 s_9;$$

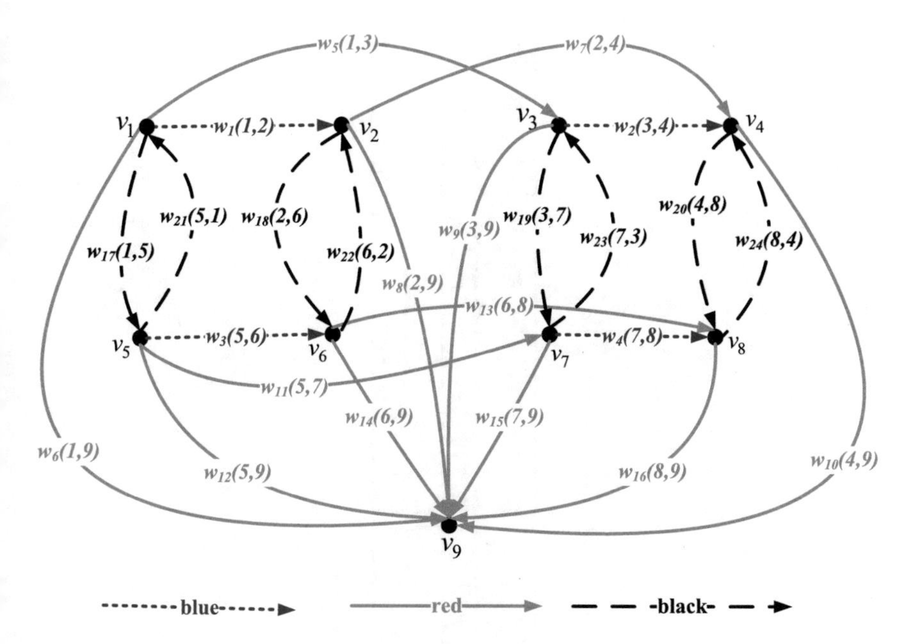

Fig. 9.3 The weighted colored graph for the Elmira conflict

$$s_1 \succ_2 s_4 \succ_2 s_8 \succ_2 s_5 \succ_2 s_9 \succ_2 s_3 \succ_2 s_7 \succ_2 s_2 \succ_2 s_6;$$

$$s_7 \succ_3 s_3 \succ_3 s_5 \succ_3 s_1 \succ_3 s_8 \succ_3 s_6 \succ_3 s_4 \succ_3 s_2 \succ_3 s_9.$$

Therefore, the diagonal weight matrix, the UM weight matrix, and the UI weight matrix of the Elmira conflict are constructed in Table 9.11.

Let

$$F(B^{(W^{(UM)})}) = [(B_{in}^{(W^{(UM)})})^T \cdot (B_{out}^{(W^{(UM)})})] \circ (E_l - D)$$

denote a conversion function. It transforms the labeled multidigraph by node-by-node to the reduced weighted line digraph by arc-by-arc that is a simple digraph with no color constraints to find all evolution paths of the Elmira conflict by allowing all UMs. The conversion process is depicted in Fig. 9.4 in which each hexagon denotes an arc. Status quo analysis is mainly concerned with the attainability of predicted equilibria. Therefore, stability analysis is usually conducted first. Table 9.6 provides states s_5, s_8, and s_9 are likely resolutions for the Elmira conflict. The three major equilibria are reachable from status quo $s = s_1$ by the legal UM paths $PA(s_1, s)$ for $s = s_5$, s_8, and s_9.

Let $B \Longrightarrow B^{(W^+)}$, then the labeled graph is converted to the reduced colored multidigraph as shown in Fig. 9.5a including UI arcs only. Let

$$F(B^{(W^+)}) = [(B_{in}^{(W^+)})^T \cdot (B_{out}^{(W^+)})] \circ (E_l - D).$$

Table 9.11 UM and UI weight matrices for the Elmira conflict

Arc number	a_1	a_2	a_3	a_4	a_5	a_6	a_7	a_8	a_9	a_{10}	a_{11}	a_{12}	a_{13}	a_{14}	a_{15}	a_{16}	a_{17}	a_{18}	a_{19}	a_{20}	a_{21}	a_{22}	a_{23}	a_{24}
W	N_w	N_w	N_w	N_w	N_w	N_w	P_w	P_w	P_w	N_w	N_w	N_w	P_w	P_w	P_w	N_w	P_w	P_w	P_w	P_w	N_w	N_w	N_w	N_w
$W^{(UM)}$	1	1	1	1	1	1	1	1	1	1	1	1	1	1	1	1	1	1	1	1	1	1	1	1
W^{+}	0	0	0	0	0	0	1	1	1	0	0	0	1	1	1	0	1	1	1	1	0	0	0	0

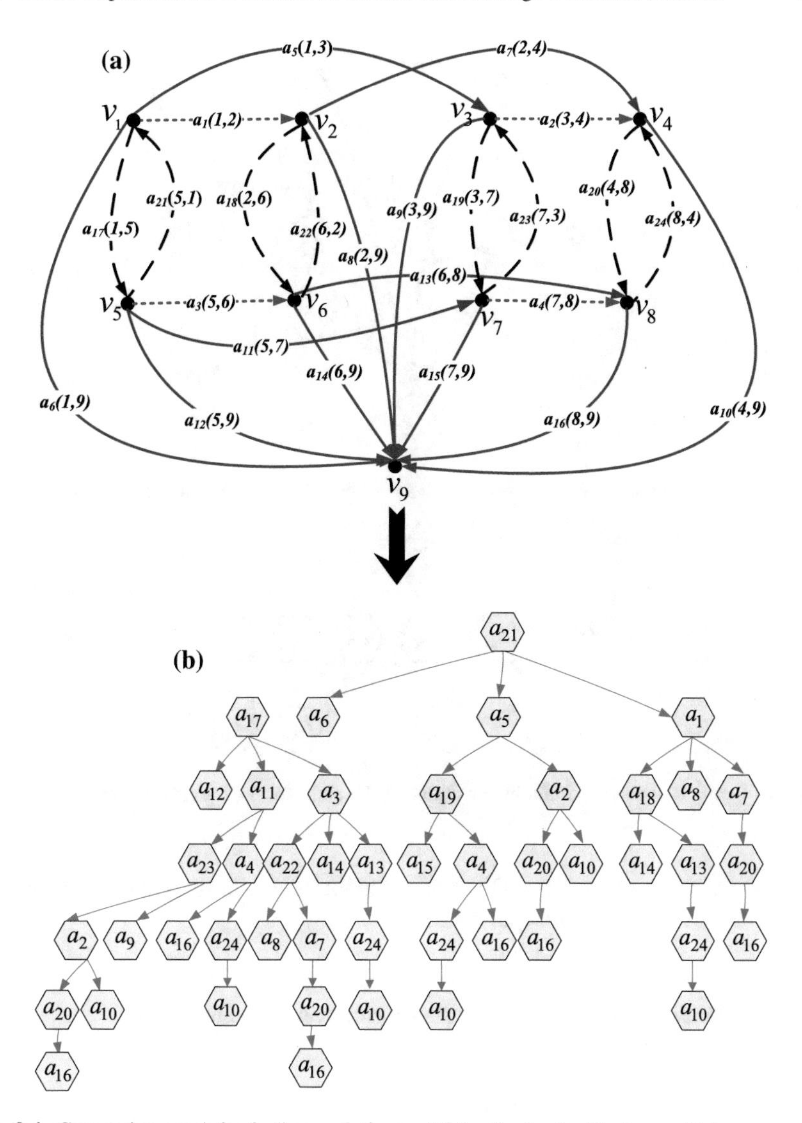

Fig. 9.4 Conversion graph for finding evolutionary UM paths for the Elmira conflict

The conversion function transforms the original problem of searching the legal UI paths in an edge-colored graph with no repeated colors to the standard problem of finding the UI paths on a graph with no color constraints (See Fig. 9.5b). For example, if status quo is selected as s_2, then Fig. 9.6a shows the UI conflict evolution by arc-by-arc from s_2 for the Elmira conflict. Note that the single arc a_8 does not appear in Fig. 9.6a though it is a UI arc and states are denoted by their indices to make figures clear. Figure 9.6b depicts all possible UI paths from state s_2 by state-by-state and

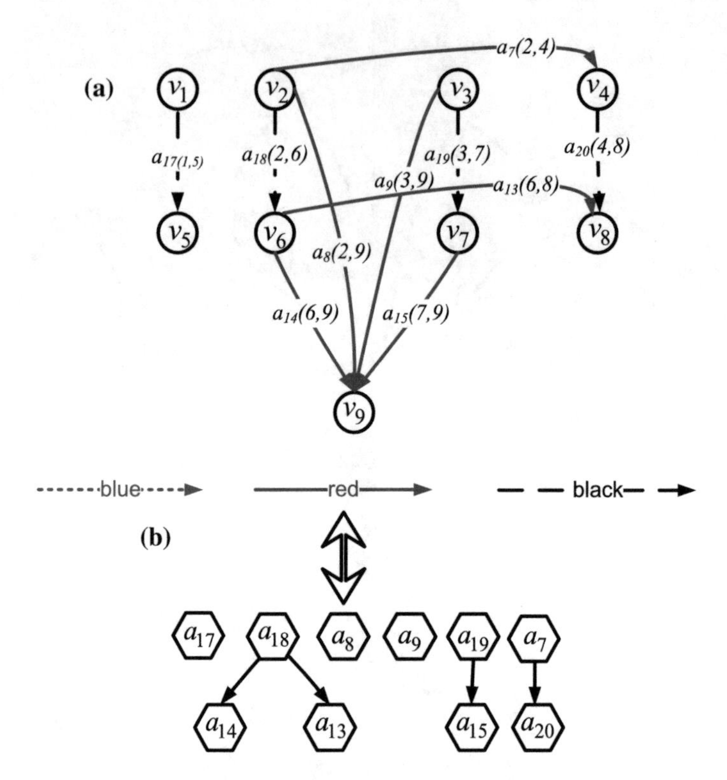

Fig. 9.5 Graph conversion for finding evolutionary UI paths for the Elmira conflict

includes the paths of length 1. Obviously, the major equilibrium state s_5 cannot be reachable by UIs from status quo state s_2.

9.3.4.2 Analysis of Gisborne Conflict Evolution with Unknown Preference

In this subsection, the matrix method developed in this section is applied to a case study — status quo analysis of the Gisborne conflict including preference uncertainty (Xu et al. 2010b). The background, modeling and analysis of the Gisborne conflict is presented in Sect. 5.4. The edge labeled multidigraph is portrayed in Fig. 9.7a equivalent to the graph model shown in Fig. 8.1. The weight of each arc in Fig. 9.7a is assigned based on preference information

$$s_2 \succ_1 s_6 \succ_1 s_4 \succ_1 s_8 \succ_1 s_1 \succ_1 s_5 \succ_1 s_3 \succ_1 s_7;$$

$$s_3 \succ_2 s_7, s_4 \succ_2 s_8, s_1 \succ_2 s_5, s_2 \succ_2 s_6, \text{ only};$$

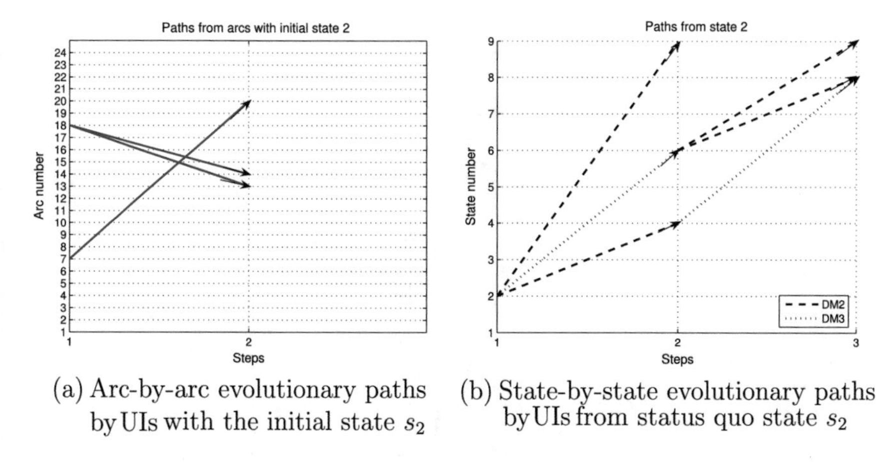

(a) Arc-by-arc evolutionary paths by UIs with the initial state s_2

(b) State-by-state evolutionary paths by UIs from status quo state s_2

Fig. 9.6 Evolutionary paths by UIs with status quo state s_2

$$s_3 \succ_3 s_4 \succ_3 s_7 \succ_3 s_8 \succ_3 s_5 \succ_3 s_6 \succ_3 s_1 \succ_3 s_2.$$

Therefore, the diagonal weight matrix, the diagonal UM weight matrix, the diagonal UI weight matrix, and the diagonal UIUM weight matrix of the Gisborne conflict are constructed in Table 9.12.

Based on the extended preference structure with uncertainty, Li et al. (2004a) redefine Nash stability, general metarationality, symmetric metarationality, and sequential stability for graph models with preference uncertainty. According to whether uncertain preferences are deemed as sufficient incentives to motivate the focal DM leaving the current state and credible sanctions to deter the focal DM from doing so, the aforesaid four types of stability are redefined in four different manners and indexed a, b, c, and d. These four extensions are conceived to depict DMs with distinct risk profiles in face of uncertainty. Li et al. (2004a) identify states s_4, s_6 and s_8 as equilibria under extensions b and d for the Gisborne conflict using logical stability definitions and the same results are obtained by using matrix representation of stabilities in Sect. 5.4. Note that for the stability definitions under extensions b and d, the focal DM is conservative in deciding whether to move away from the current state, since it would only move to preferred states (UIs). For details, one can refer to Sect. 5.2.1. In parallel to extensions b and d that predict the three equilibria s_4, s_6, and s_8, one examines the evolution paths PA^+ (allowing UIs only) from a status quo to the three equilibria. Based on the UI weight matrix W^+ constructed in Table 9.12, let

$$F(B^{(W^+)}) = [(B_{in}^{(W^+)})^T \cdot (B_{out}^{(W^+)})] \circ (E_l - D)$$

denote a conversion function that transforms the labeled multidigraph in Fig. 9.7a to the reduced line digraph in Fig. 9.7b including UI arcs only that is a simple digraph with no color constraints. Therefore, finding colored UI paths in Fig. 9.7a is equivalent to searching paths in Fig. 9.7b without constraints. If the status quo is s_1, it is obvious

Table 9.12 UM, UI, and UIUM weight matrices for the Gisborne conflict

Arc number	a_1	a_2	a_3	a_4	a_5	a_6	a_7	a_8	a_9	a_{10}	a_{11}	a_{12}
W	P_w	N_w	P_w	N_w	P_w	N_w	P_w	N_w	U_w	U_w	U_w	U_w
$W^{(UM)}$	1	1	1	1	1	1	1	1	1	1	1	1
W^+	1	0	1	0	1	0	1	0	0	0	0	0
$W^{+,U}$	1	0	1	0	1	0	1	0	1	1	1	1

Arc number	a_{13}	a_{14}	a_{15}	a_{16}	a_{17}	a_{18}	a_{19}	a_{20}	a_{21}	a_{22}	a_{23}	a_{24}
W	U_w	U_w	U_w	U_w	P_w	P_w	N_w	N_w	N_w	N_w	P_w	P_w
$W^{(UM)}$	1	1	1	1	1	1	1	1	1	1	1	1
W^+	0	0	0	0	1	1	0	0	0	0	1	1
$W^{+,U}$	1	1	1	1	1	1	0	0	0	0	1	1

that the equilibria s_4 and s_8 cannot be reached by legal UIs and the equilibrium s_6 is the only equilibrium that is attainable from the status quo. Specifically, the evolutionary paths $PA^+(s_1, s_6)$ can be described below:

$$a_1 \longrightarrow a_{18} \Longleftrightarrow s_1 \longrightarrow s_2 \longrightarrow s_6,$$

$$a_{17} \longrightarrow a_5 \Longleftrightarrow s_1 \longrightarrow s_5 \longrightarrow s_6.$$

However, if UIUMs are allowed, equilibrium s_8 is attainable from the status quo s_1. The UIUM weight matrix $W^{+,U}$ is defined in Table 9.12. Using conversion matrix $B^{(W^{+,U})}$, the labeled graph in Fig. 9.7a is reduced to Fig. 9.8a that illustrates the evolution of the graph model for the Gisborne conflict with allowing UIUMs only. By the conversion function $F(\cdot)$, the colored multidigraph in Fig. 9.8a is transformed to the reduced line digraph in Fig. 9.8b. Searching colored paths $PA^{+,U}(s_1, s_8)$ in Fig. 9.8a is equivalent to finding paths $PA^{+,U}(a_1, a_{14})$, $PA^{+,U}(a_1, a_7)$, $PA^{+,U}(a_9, a_{14})$, $PA^{+,U}(a_9, a_7)$, $PA^{+,U}(a_{17}, a_{14})$, and $PA^{+,U}(a_{17}, a_7)$ in Fig. 9.8b. Therefore, the evolution of the Gisborne conflict by the legal UIUMs from status quo state s_1 to equilibrium s_8 is illustrated as follows:

$$a_1 \longrightarrow a_{18} \longrightarrow a_{14},$$

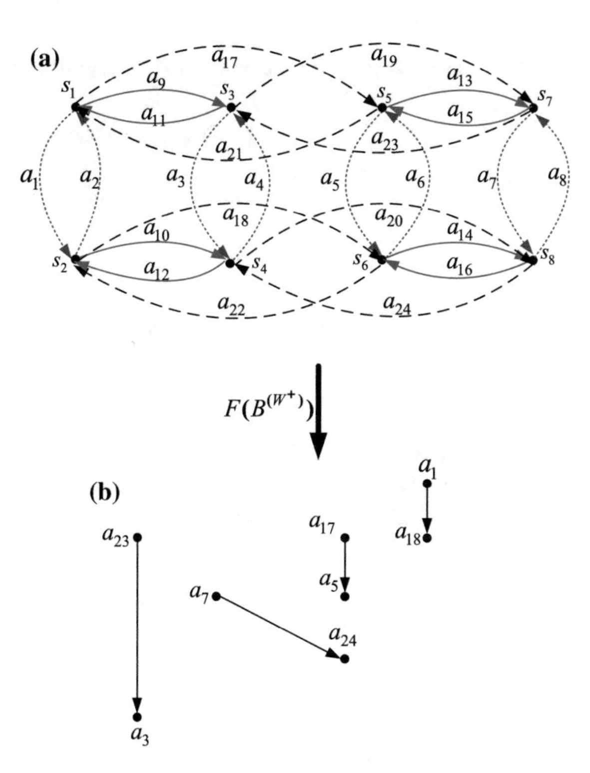

Fig. 9.7 Conversion graph for finding the evolutionary UI paths for the Gisborne conflict

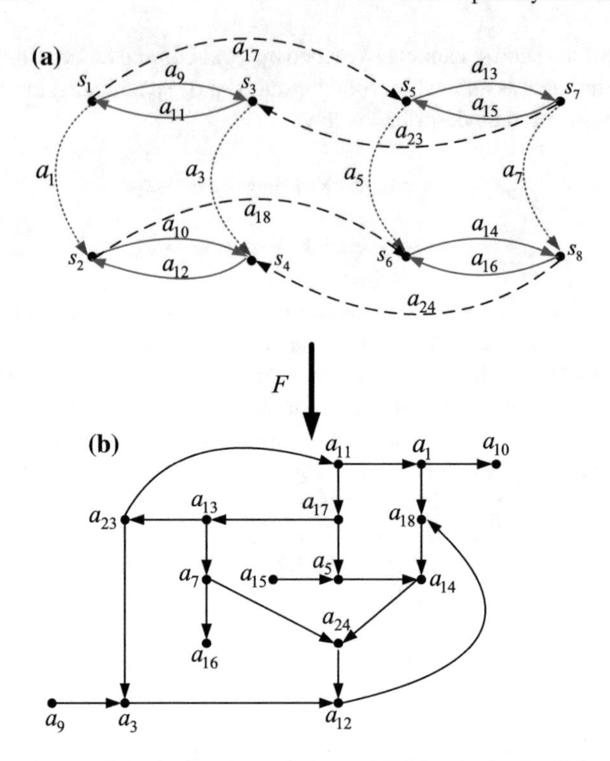

Fig. 9.8 Conversion graph for finding the evolutionary UIUM paths for the Gisborne conflict

$$a_9 \longrightarrow a_3 \longrightarrow a_{12} \longrightarrow a_{18} \longrightarrow a_{14},$$

$$a_{17} \longrightarrow a_5 \longrightarrow a_{14},$$

$$a_{17} \longrightarrow a_{13} \longrightarrow a_{23} \longrightarrow a_3 \longrightarrow a_{12} \longrightarrow a_{18} \longrightarrow a_{14},$$

$$a_{17} \longrightarrow a_{13} \longrightarrow a_{23} \longrightarrow a_{11} \longrightarrow a_1 \longrightarrow a_{18} \longrightarrow a_{14},$$

$$a_{17} \longrightarrow a_{13} \longrightarrow a_7.$$

After transforming a colored multidigraph to a simple digraph under conversion functions, existing algorithms such as those reported in Migliore et al. (1990) and Xia and Wang (2000) can be used to find all paths or search for the shortest path.

9.3.4.3 Analysis of GDU Conflict Evolution with Three Degrees of Preference

As post-stability analysis, the status quo analysis aims at assessing whether predicted equilibria are reachable from the status quo or any other initial state. The background

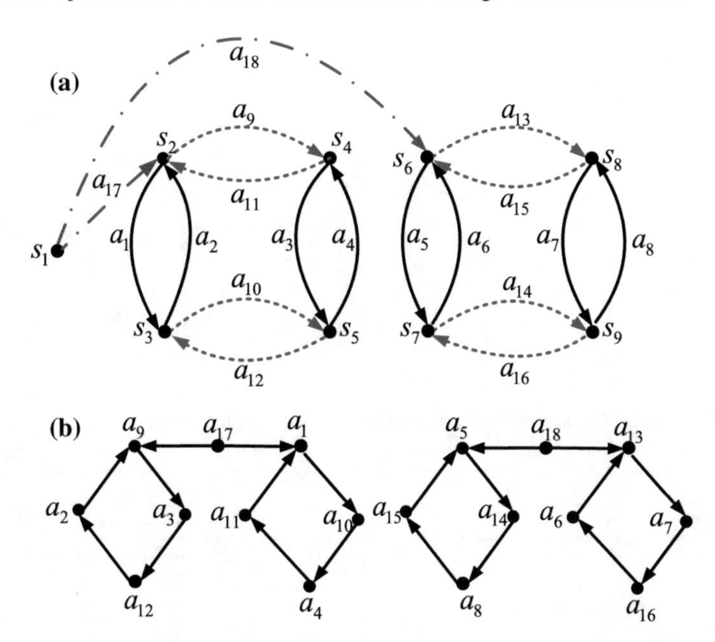

Fig. 9.9 Transformation of the graph model for the GDU conflict

of the GDU conflict is given in Sect. 6.6. The graph model for the GDU conflict displayed in Fig. 6.14 is equivalent to the labeled graph given in Fig. 9.9a. Based on preference information of the GDU conflict

$$s_2 >_1 s_4 >_1 s_3 >_1 s_5 >_1 s_1 >_1 s_6 >_1 s_9 >_1 s_7 \gg_1 s_8,$$

$$\{s_3 \sim_2 s_7\} >_2 \{s_5 \sim_2 s_9\} >_2 \{s_4 \sim_2 s_8\} \gg_2 \{s_1 \sim_2 s_2 \sim_2 s_6\},$$

$$\{s_2 \sim_3 s_3 \sim_3 s_4 \sim_3 s_5 \sim_3 s_6 \sim_3 s_7 \sim_3 s_8 \sim_3 s_9\} \gg_3 s_1,$$

the $l \times l$ diagonal weight matrix, the UM weight matrix, and the MSUI weight matrix are constructed in Table 9.13.

Table 9.13 Weight, UM weight, and MSUI weight matrices for the GDU conflict

Arc number	a_1	a_2	a_3	a_4	a_5	a_6	a_7	a_8	a_9	a_{10}	a_{11}	a_{12}	a_{13}	a_{14}	a_{15}	a_{16}	a_{17}	a_{18}
Weight matrix W	N_w	P_m	N_w	P_m	N_w	P_m	P_s	N_w	P_s	N_w	N_w	P_m	P_s	N_w	N_w	P_m	P_s	P_s
UM weight matrix $W^{(UM)}$	1	1	1	1	1	1	1	1	1	1	1	1	1	1	1	1	1	1
MSUI weight matrix $W^{+,++}$	0	1	0	1	0	1	1	0	1	0	0	1	1	0	0	1	1	1

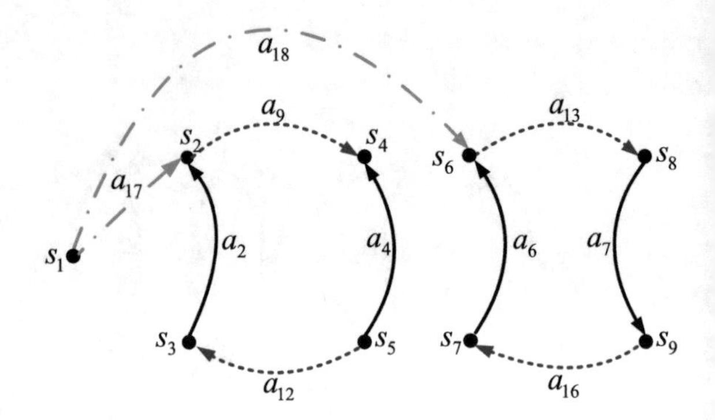

Fig. 9.10 The reduced graph allowing MSUIs only for the GDU conflict

Table 9.14 The GDU conflict evolution from the status quo s_1 to state s_9

DMs	Status quo		Transitional States				Equilibrium
USS							
1. Proceed	Y		Y		Y	⟶	N
2. Modify	N		N		N	⟶	Y
CDO							
3. Legal	N		N	⟶	Y		Y
IJC							
4. Completion	N		N		N		N
5. Modification	N	⟶	Y		Y		Y
State	s_1		s_6		s_8		s_9

By taking status quo analysis into account, additional insights are revealed about the attainability of any potential resolution. Given state s_4 is a strong equilibrium for Nash stability, GMR, SMR, and SEQ. When state s_1 is selected as a status quo, all possible UM evolutionary paths of the GDU conflict from s_1 to the equilibrium s_4 are obtained using the following steps:

- Using the UM weight matrix provided by Table 9.13, construct the conversion function
$$F(B^{(W^{(UM)})}) = [(B_{in}^{(W^{(UM)})})^T \cdot (B_{out}^{(W^{(UM)})})] \circ (E_l - D);$$
- This conversion function transforms the labeled multidigraph in Fig. 9.9a to the reduced line digraph in Fig. 9.9b including all UM arcs that is a simple digraph with no color constraints;
- Searching the colored paths $PA(s_1, s_4)$ between two vertices s_1 and s_4 in Fig. 9.9a is equivalent to finding all paths $PA(a, b)$ for $a \in A_S$, and $b \in A_E$ in Fig. 9.9b, where A_S and A_E are the two sets of arcs starting from vertex s_1 and arcs ending at vertex s_4;
- $A_S = \{a_{17}, a_{18}\}$ and $A_E = \{a_4, a_9\}$;

- Finding the legal UM paths $PA(a_{17}, a_4)$, $PA(a_{17}, a_9)$, $PA(a_{18}, a_4)$, and $PA(a_{18}, a_9)$ in the simple digraph shown in Fig. 9.9b;
- $PA(a_{17}, a_4) : a_{17} \rightarrow a_1 \rightarrow a_{10} \rightarrow a_4$; $PA(a_{17}, a_9) : a_{17} \rightarrow a_9$;
- Find paths between two vertices, s_1 and s_4, using the paths between corresponding two arcs:

$$a_{17} \rightarrow a_1 \rightarrow a_{10} \rightarrow a_4 \Leftrightarrow s_1 \rightarrow s_2 \rightarrow s_3 \rightarrow s_5 \rightarrow s_4,$$

$$a_{17} \rightarrow a_9 \Leftrightarrow s_1 \rightarrow s_2 \rightarrow s_4.$$

If a conversion function is designed by $F(B) = B \cdot W^{+,++}$, then the original graph in Fig. 9.9a is reduced to the graph shown in Fig. 9.10 including MSUIs only. Given state s_9 is a strong equilibrium for GMR and SEQ. The dynamics of the GDU conflict evolution from the status quo stateevolution from the status quo state s_1 to the desirable equilibrium state s_9 by the legal MSUIs is portrayed in Table 9.14. Specifically, the evolution path $PA^{+,++}(s_1, s_9)$ of the GDU conflict from state s_1 to state s_9 is

$$s_1 \rightarrow s_6 \rightarrow s_8 \rightarrow s_9.$$

9.4 Important Ideas

After carrying out individual and coalitional stability analyses, a valuable type of follow-up analysis to perform is to ascertain possible paths that could be followed from a specified initial or status quo state to a final desired state of interest. For instance, one may wish to determine if a beneficial climate change equilibrium can be reached between a government and its provinces within a nation starting from a status quo state at which there is no legally binding arrangement to an equilibrium having reasonable reductions in greenhouse gas emissions. Because the tracing of paths between two states can be carried out with respect to four types of preference situations (simple, unknown, degree and hybrid), the approach is indeed very flexible. Moreover, a logical interpretation of path tracing permits one to easily understand how it works. However, a matrix procedure effectively converts path finding from a logical structure to a flexible algebraic system of finding paths in a simple digraph with no color constrains. Consequently, the algebraic structure constitutes a versatile and encompassing framework for realistically tracing paths ranging from small graph models to very large and complicated ones.

Within Sect. 1.2.4 and in Fig. 1.1 near the start of the book, two important kinds of follow-up analysis are mentioned:

- evolution of a conflict (Sect. 1.2.4.1)
- Sensitivity analyses (Sect. 1.2.4.2)

In a sensitivity analysis, one wishes to determine how making meaningful small changes in a basic model input parameter, like the preferences of one or more DMs, influences the stability findings. As explained in Sect. 10.2.4 and depicted in Fig. 10.5, flexible sensitivity analysis procedures should be incorporated into the output subsystem of a decision support system (DSS) for GMCR based on having a matrix stability analysis engine. Besides having the capability to trace the possible evolutions of a dispute in the output subsystem of a DSS, another type of informative follow-up analysis is categorizing equilibria according to common features. For instance, in a climate change negotiation dispute over percentage by which carbon dioxide emissions should be lowered, one may be interested to categorize equilibria into one classification for which the overall percentage reduction in emission is over 80% within a 20-year period, and another category of equilibria for which reductions in greenhouse gas emissions are not sufficient to maintain a sustainable climate.

9.5 Problems

9.5.1 Qualitatively explain what is meant by a stability analysis and what happens when tracing the possible evolutions of a dispute. Mention the key differences between these two procedures.

9.5.2 The Elmira groundwater contamination dispute first mentioned in Sect. 1.2.2 is used in this chapter to explain how to calculate paths between two states using a matrix representation. Employing a logical interpretation and assuming simple preference, show by hand how the final cooperative equilibrium can be reached from the status quo state using option form. In your diagram of the evolution of the dispute, point out the assumptions underlying the moves from state to state.

9.5.3 The Lake Gisborne water export conflict is utilized in this chapter to show how to determine the paths between two states when employing a matrix representation. Assuming a logical interpretations of this path finding and simple preference, draw a diagram in option form to portray how the final equilibrium can be reached. Explain the assumptions underlying the moves that are allowed.

9.5.4 The Garrison Diversion Unit conflict is a large scale irrigation conflict used as an illustration for path tracing in option form. Under the assumption of simple preference and a logical representation, draw a graph based on option form showing the evolution of the dispute from the status quo to the final equilibrium. What assumptions are you making about the kinds of moves you permit from state to state.

9.5.5 In the path finding procedures of this chapter, four kinds of preferences are entertained: simple, unknown, degree, and hybrid. Beyond simple preference, what other kind of preference makes sense with respect to the Elmira conflict when tracing paths from the status quo to the final resolution? Show the matrix formulations and calculations for determining this path. Explain what is happening in the diagram that you draw using option form to depict these moves.

9.5.6 Select a conflict which is of high interest to you. After building a conflict model, carry out a stability analysis. Draw a diagram depicting the evolution of the dispute in option form from the status quo to the most desirable equilibrium that you found. Explain what is happening in the diagram and show your calculations using a matrix interpretation.

9.5.7 Allowing for different kinds of preferences can be interpreted as a kind of sensitivity analysis. For the case of the Elmira dispute, explain where you think preference may be unknown for one of the DMs. Determine how this affects your stability findings as well as the evolution of the dispute from the status quo to a final equilibrium.

9.5.8 For the case of the Lake Gisborne dispute, assume that part of the preferences is different for one of the DMs. Show how this difference in preferences affects the stability findings, if at all. Using a diagram, depict how the different preferences influence the evolution of the dispute, if at all.

9.5.9 Present a conflict of your choice for which you think three degrees of preference may be important to consider because of the basic nature of the dispute. Carry out a stability analysis with and without three degrees of preference and comment upon the differences in strategic results. Using diagrams, show how the three degrees of preference influence or not the evolution of the dispute from the status quo state to the final resolution.

9.5.10 Qualitatively explain what is meant by a sensitivity analysis. Make a list of the kinds of sensitivity analyses you think are most important to consider in a GMCR study.

9.5.11 In more detail than that given in Sect. 10.2.4 for the output subsystem for GMCR, explain how you would incorporate follow-up analyses into the theoretical design of a DSS for GMCR. Use examples of actual conflict situations to justify your design.

9.5.12 Provide a general discussion on follow-up analyses and why you think they are important. Beyond tracing paths, various kind of sensitivity analyses and classifying equilibria according to commonalties, describe other kinds of follow-up analyses that you think are important. How do you think designing creative solutions to better resolving conflict could be incorporated into follow-up analyses?

References

Dieste, R. (1997). *Graph theory*. New York: Springer. https://doi.org/10.1007/978-1-4419-1153-7_402.

Fang, L., Hipel, K. W., Kilgour, D. M., & Peng, X. (2003a). A decision support system for interactive decision making, part 1: Model formulation. *IEEE Transactions on Systems, Man and Cybernetics, Part C: Applications and Reviews, 33*(1), 42–55. https://doi.org/10.1109/tsmcc.2003.809361.

Fang, L., Hipel, K. W., Kilgour, D. M., and Peng, X. (2003b). A decision support system for interactive decision making, part 2: Analysis and output interpretation. *IEEE Transactions on Systems, Man and Cybernetics, Part C: Applications and Reviews, 33*(1), 56–66. https://doi.org/10.1109/tsmcc.2003.809360.

Godsil, C. & Royle, G. (2001). *Algebraic graph theory*. New York: Springer. https://doi.org/10.1007/978-1-4613-0163-9.

Gondran, M., & Minoux, M. (1979). *Graphs and algorithms*. New York: Wiley.

Hoffman, A. J. & Schiebe, B. (2001). The edge versus path incidence matrix of series-parallel graphs and greedy packing. *Discrete Applied Mathematics*, *113*, 275–284. https://doi.org/10.1016/s0166-218x(00)00294-8.

Kilgour, D. M. & Hipel, K. W. (2010). Conflict analysis methods: The graph model for conflict resolution. In D. M. Kilgour & C. Eden (Eds.), *Handbook of group decision and negotiation* (pp. 203–222). Dordrecht, The Netherlands: Springer. https://doi.org/10.1007/978-90-481-9097-3_13.

Li, K. W., Hipel, K. W., Kilgour, D. M., & Fang, L. (2004a). Preference uncertainty in the graph model for conflict resolution. *IEEE Transactions on Systems, Man, and Cybernetics, Part A, Systems and Humans*, *34*(4), 507–520. https://doi.org/10.1109/tsmca.2004.826282.

Li, K. W., Kilgour, D. M., & Hipel, K. W. (2004b). Status quo analysis of the Flathead river conflict. *Water Resources Research*, *40*(5). https://doi.org/10.1029/2003wr002596.

Li, K. W., Hipel, K. W., Kilgour, D. M., & Noakes, D. J. (2005a). Integrating uncertain preferences into status quo analysis with application to an environmental conflict. *Group Decision and Negotiation*, *14*(6):461–479. https://doi.org/10.1007/s10726-005-9003-9.

Li, K. W., Kilgour, D. M., & Hipel, K. W. (2005b). Status quo analysis in the graph model for conflict resolution. *Journal of the Operational Research Society*, *56*, 699–707. https://doi.org/10.1057/palgrave.jors.2601870.

Migliore, M., Martorana, V., & Sciortino, F. (1990). An algorithm to find all paths between two nodes in a graph. *Journal of Computational Physics*, *87*, 231–236. https://doi.org/10.1016/0021-9991(90)90235-s.

Shiny, A. K. & Pujari, A. K. (1998). Computation of prime implicants using matrix and paths. *Journal of Logic and Computation*, *8*(2), 135–145. https://doi.org/10.1093/logcom/8.2.135.

Xia, Y., & Wang, J. (2000). A discrete-time recurrent neural network for shortest-path routing. *IEEE Transactions on Automatic Control*, *45*(11), 2129–2134. https://doi.org/10.1109/9.887639.

Xu, H., Li, K. W., Hipel, K. W., & Kilgour, D. M. (2009a). A matrix approach to status quo analysis in the graph model for conflict resolution. *Applied Mathematics and Computation*, *212*(2), 470–480. https://doi.org/10.1016/j.amc.2009.02.051.

Xu, H., Li, K. W., Kilgour, D. M., & Hipel, K. W. (2009b). A matrix-based approach to searching colored paths in a weighted colored multidigraph. *Applied Mathematics and Computation*, *215*, 353–366. https://doi.org/10.1016/j.amc.2009.04.086.

Xu, H., Hipel, K. W., Kilgour, D. M., & Chen, Y. (2010a). Combining strength and uncertainty for preferences in the graph model for conflict resolution with multiple decision makers. *Theory and Decision*, *69*(4), 497–521. https://doi.org/10.1007/s11238-009-9134-6.

Xu, H., Kilgour, D. M., Hipel, K. W., & Kemkes, G. (2010b). Using matrices to link conflict evolution and resolution in a graph model. *European Journal of Operational Research*, *207*, 318–329. https://doi.org/10.1016/j.ejor.2010.03.025.

Chapter 10
Design of a Decision Support System for Conflict Resolution

A rich range of basic concepts in conflict resolution are presented in Chaps. 3–9 in this book. These ideas were mathematically designed to reflect different types of conflict situations that can arise in the real-world. Moreover, by having a basic mathematical design and associated capabilities that mirror the key characteristics of actual disputes, meaningful strategic advice can be discovered for resolving conflicts in the best possible way. By suitably accounting for the value systems of the different decision makers (DMs) involved in a specific conflict, one can determine the potential resolutions for the dispute when DMs act according to their own individual interests by behaving in a competitive and noncooperative way, or by cooperating with one another via coalitions to see if they can do even better strategically. For instance, by forming alliances, it may be possible to reach a win/win resolution.

Table 10.1 provides a summary of the key ideas presented in earlier chapters in the book as categorized according to types of preferences. Throughout these chapters, the solution concepts consisting of Nash, general metarational, symmetric metarational and sequential stabilities are utilized for both noncooperative and cooperative behavior. The evolution of a conflict is also provided for the different kinds of preference situations. In order for researchers, practitioners, teachers and students to be able to immediately employ these and other future developments in conflict resolution, a flexible decision support system (DSS) is required to permit extensive analyses to be expeditiously executed. The objective of this chapter is to present a general or universal design of a flexible DSS for GMCR that will capture all of the progress made to date as well as permit easy expansion of the DSS as new theoretical and practical developments are achieved. As can be appreciated from Table 10.1 which summarizes many recent ideas presented in this book, a really powerful set of tools are available now for addressing conflicts ranging from simple to complex. For instance, the matrix design of the graph model permits the construction of a truly powerful analysis engine for efficiently producing strategic findings for all of the ideas given in Table 10.1 as well as for ongoing and future developments presented in Sects. 10.3.1 and 10.3.2, respectively. One can envision this design as being similar to using the Danish toy invention called "lego" in which lego building blocks are

H. Xu et al., *Conflict Resolution Using the Graph Model: Strategic Interactions in Competition and Cooperation*, Studies in Systems, Decision and Control 153,
https://doi.org/10.1007/978-3-319-77670-5_10

Table 10.1 Stability analysis and conflict evolution for different preference structures

Types of preferences	Kinds of analyses	Locations
Graph model with simple preference	Individual stabilities	Chap. 4
	Coalitional stabilities	Chap. 8
	Conflict evolution	Chap. 9
Graph model with unknown preference	Individual stabilities	Chap. 5
	Coalitional stabilities	Chap. 8
	Conflict evolution	Chap. 9
Graph model with degrees of preference	Individual stabilities	Chap. 6
	Coalitional stabilities	Chap. 8
	Conflict evolution	Chap. 9
Graph model with hybrid preference	Individual stabilities	Chap. 7
	Coalitional stabilities	Chap. 8
	Conflict evolution	Chap. 9

snapped together to build a structure such as a boat or a fortress in which one can easily add extra features in the future as attachments are designed and made available to customers. Likewise, a well-designed DSS will not have to be completely reprogrammed but rather easily extended as new features, such as attitudes, emotions and misperceptions, as well as other expansions mentioned in Sect. 10.3 are added to the DSS.

The overall philosophy of having a well-conceived mathematical design for the graph model and its theoretical expansions, coupled with a continuously evolving DSS provides users with an exceptional paradigm for resolving a wide variety of complex conflicts. In the next section, an introduction to DSSs is provided along with an overview of existing DSSs for implementing the graph model. Subsequently, a universal design of a DSS consisting of the input, engine, and output systems is unveiled which encompasses the concepts put forward in the earlier chapters in this book, as well as other extensions of the graph model furnished in Sect. 10.3. Finally, ongoing and potential expansions of the graph model are discussed in Sect. 10.3 in conjunction with how they could be incorporated into the universal design of a DSS for GMCR.

10.1 Decision Support Systems

10.1.1 Introduction

As pointed out by Sage (1991) in his classic book on decisions support systems, "in very general terms, a decision support system (DSS) is a system that supports technological and managerial decision-making by assisting in the organization of knowledge about ill-structured, semi-structured issues." More specifically, a general DSS consists of four main components as shown in Fig. 10.1: the user inter-

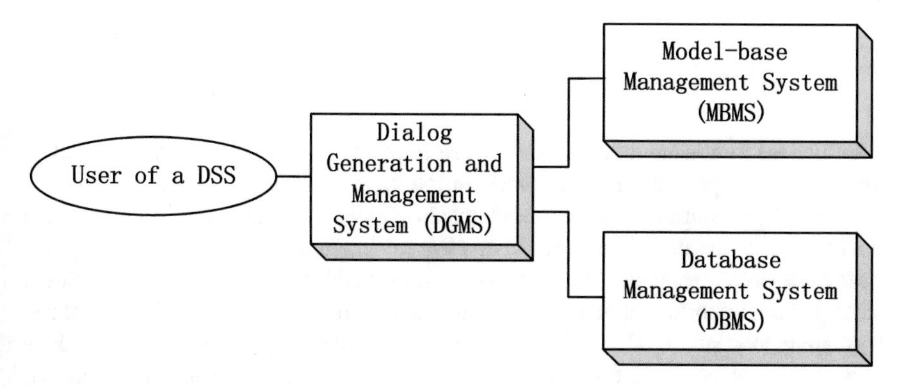

Fig. 10.1 The framework of a general DSS

face system, dialog generation and management system (DGMS), the model-base management system (MBMS), and database management system (DBMS) (Sage 1991). While a variety of DSSs exists, the above four components can be found in many DSS architectures and play an important role in their structure. In Fig. 10.1, the user interacts with the DSS through the DGMS. As can be seen, the DGMS communicates with the MBMS and DBMS in order to determine the appropriate sets of models from the model-base and required data from the database to utilize for investigating the problem under study.

As mentioned in Sect. 2.3.3, in most areas of study a rich range of models have been developed over the years for addressing challenging problems. The important field of water resources constitutes a key domain in which both physical and societal systems models have been designed for tackling tough situations based on a systems perspective in an integrative and adaptive fashion in which stakeholder values are purposefully taken into account as explained in Sect. 2.4 and by authors such as Hipel et al. (2008). For example, in the pollution of an underground aquifer, such as the one underlying the town of Elmira, Ontario, Canada (see Sect. 1.2), physical systems models based on stochastic partial differential equations are required to model the pollution plume as it spreads underground from the source. Societal systems models, such as various economic and conflict resolution models, are needed to model the societal aspects of the problem. In fact, the Elmira conflict is analyzed using GMCR at various locations in this book (see Sects. 4.5 and 9.2). Hipel et al. (2008) traced the evolution of systems models and associated DSSs in water resources management.

When a DSS is specifically designed for employment in negotiation processes, it is often referred to as a negotiation support system. Existing negotiation support systems have been reviewed and compared by various authors such as Jelassi and Foroughi (1989), Thiessen and Loucks (1992), and Kilgour et al. (1995) to explain and compare their capabilities and effectiveness. Software packages based on metagame analysis (Howard 1989), conflict analysis (Meister and Fraser 1994, DecisionMaker 1996), and decision systems analysis (Langlois 1994) were designed to assist in modeling and analyzing interactive decision situations.

10.1.2 Existing Decision Support Systems for the Graph Model

Conflict resolution methodologies require implementation algorithms to facilitate their practical application to real-world problems. To permit convenient and expeditious use by practitioners, a methodology should be computerized as a DSS, which includes all associated algorithms. In this way, the methodology is transformed into a realizable decision technology. Because decision-making and negotiations constitute common but important human activity, there is a great need for flexible DSSs that can systematically investigate a wide range of real-world strategic conflicts. To achieve this, a number of DSSs for implementing GMCR and various expansions thereof have been developed over the years. Four particular systems are briefly described with respect to their basic capabilities. Subsequently, in Sect. 10.2 an encompassing universal design is put forward for constructing new DSSs or expanding current ones for GMCR.

In combination with the publication of their 1993 book, Fang et al. (1993) provided a basic analysis engine called GMCR I for calculating stability. Specifically, the GMCR I engine determines stability for a conflict having two or more DMs for the solution concepts consisting of Nash stability, general metarationality (GMR), symmetric metarationality (SMR), sequential stability (SEQ) and limited-move stability (for horizons $h \geq 2$) (Zagare 1984, Kilgour 1985, Kilgour et al. 1987) and non-myopic stability (Brams and Wittman 1981, Kilgour 1984, 1985, Kilgour et al. 1987). In addition, for the case of two DMs, Stackelberg equilibria (von Stackelberg 1934) can be ascertained. A user's manual for GMCR I is provided in Appendix B of their book along with a disk containing the program for GMCR I (Fang et al. 1993).

GMCR II is the next generation DSS for implementing the graph model which was developed within a Windows environment. Some features of this system are described by Hipel et al. (1997, 2001) while the detailed design and implementation algorithms are presented by Fang et al. (2003a, b). In addition to having an engine adapted from GMCR I, the DSS GMCR II contains user-friendly input and output subsystems. Via a flexible user interface, a user can conveniently interact with the input and output subsystems. More specifically, by employing the input subsystem a user can construct a conflict model by utilizing option form defined in Sect. 3.1.2 and used with applications such as the Elmira groundwater dispute in Sects. 1.2 and 4.5, as well as elsewhere in this book. Given the DMs and options, GMCR II can automatically generate the mathematically possible states that could occur. Moreover, GMCR II prompts the user to specify infeasible situations that could not occur and are therefore removed from the overall conflict model. Circumstances or states that are essentially the same can be coalesced by GMCR II into a single state. Finally, a technique called option prioritization (see Sect. 1.2.2) can be utilized to obtain a ranking of states for each DM under the assumption of having transitive preference for the case of simple preference.

The GMCR II engine calculates stability for every state and each DM according to the aforementioned solution concepts encoded within the GMCR I and II engines.

These calculations are carried out using the logical definitions for the solution concepts. The output subsystem displays the individual and equilibrium findings in a number of informative fashions. For example, the equilibrium states can be displayed according to preference for any selected DM. Additionally, the output points out situations, if they exist, in which DMs can cooperate to jointly reach a more preferred equilibrium from another resolution under investigation.

Kinsara et al. (2015b, 2018) developed a DSS called GMCR+ which can calculate both individual and coalitional stability based on the solution concepts consisting of Nash, GMR, SMR, and SEQ for two or more DMs. The stability calculations can be executed using both the logical and matrix definitions for stability. GMCR+ also contains an exhaustive search algorithm for determining the preferences required by the DMs in a conflict to make a particular state an equilibrium with respect to the solution concepts encoded in its engine. This procedure is referred to as the inverse GMCR perspective as described in Sect. 10.3.2 and depicted in Fig. 10.6. The input and output subsystems possess similar capabilities to GMCR II along with graphical procedures to display movements among states, including an integrated graph model like the ones displayed in Figs. 3.2 and 3.5 in this book for the sustainable development and Elmira groundwater contamination disputes, respectively.

Jiang et al. (2015) provided a design for a DSS called the matrix representation for conflict resolution (MRCR). This DSS is called MRCR because the engine subsystem employs the matrix representation for individual and coalition definitions of stability to determine the stability findings for Nash, GMR, SMR and SEQ. A special feature of this DSS is its capability to handle not only simple preference, but also other preference structures such as three degrees of preference.

10.2 Universal Design of a Decision Support System for the Graph Model

10.2.1 Overall Design

In Sect. 10.1.2, the capabilities of four DSSs are outlined for permitting the GMCR methodology to be applied to actual disputes. Probably, other GMCR DSSs will be constructed in the future by researchers working at different locations around the globe. Accordingly, the objective of Sect. 10.2 is to present a universal design for a DSS for GMCR. By having a clever and effective design, a GMCR DSS can be readily utilized for systematically investigating conflict in situations like those described in Sect. 1.2.5. Moreover, the DSS can be easily expanded by simply adding new developments to its basic structure for tackling a broader variety of conflict problems. As can be seen, the framework displayed in Fig. 10.2 consists of an Input Subsystem, Analysis Engine and Output Subsystem, which are discussed in more detail in Sects. 10.2.2–10.2.4, respectively. Moreover, the User Interface permits a user to

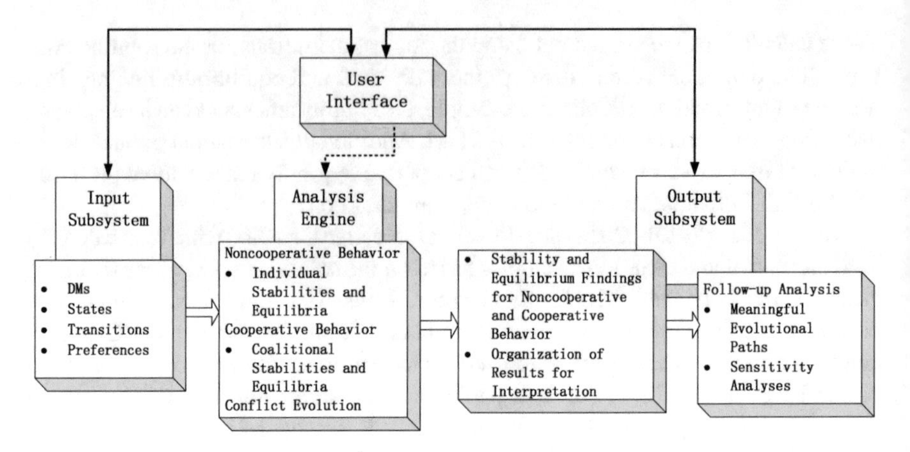

Fig. 10.2 The overall structure of a graph model based DSS

utilize the DSS in a highly interactive and creative fashion when systematically and intuitively investigating conflict ranging from simple to highly complex situations.

10.2.2 Input Subsystem

As indicated on the left in Fig. 10.2, the key inputs to a conflict model are the DMs, states, movements among states in one step controlled by each DM, and the relative preferences of the DMs among the feasible states. By far the most common and useful way for a user to interactively develop a conflict model using the GMCR DSS is to employ the option form defined in Sect. 3.1.2 and utilized in the groundwater contamination dispute in Sects. 1.2 and 4.5 and elsewhere throughout the book with this and other illustrative applications.

The left central part of Fig. 10.3 for the Input Subsystem portrays the entry of the model in option form while the right branch indicates model entry using another form such as graphical. The main steps in modeling and analyzing a conflict using GMCR are given in Fig. 1.1 in Chap. 1. Notice in the third enclosure from the top in Fig. 10.3, that for option form, the DSS requests the user to enter the DMs participating in the dispute under study as well as the options or courses of actions available to each DM. An explanation for this, as well as other steps in the modeling stage are provided in Sect. 1.2.2 for the case of the Elmira groundwater contamination dispute.

Because some states cannot occur in the real-world, procedures have been developed for removing states that could not possibly occur in the real-world. The Input Subsystem can provide suggestions to the user for eliciting situations that cannot occur in terms of combinations of option selections or circumstances that must take place. For instance, the most recent three existing DSSs mentioned in Sect. 10.1.2 prompt the user for the following information:

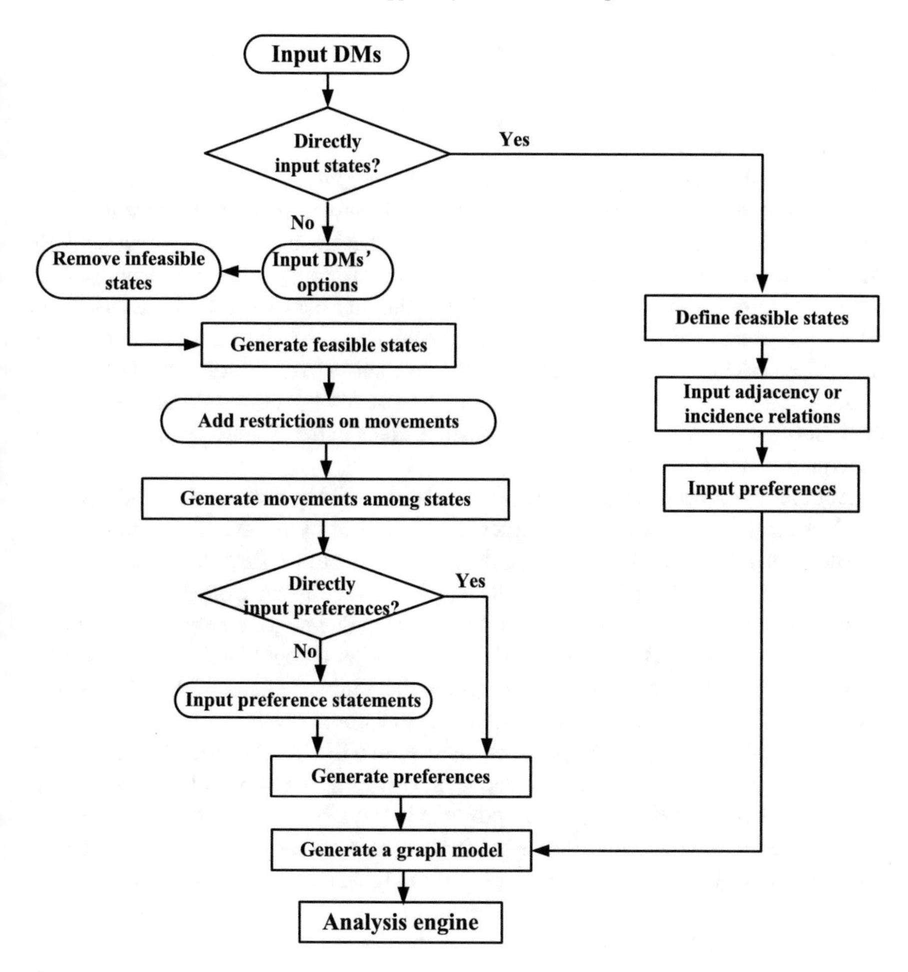

Fig. 10.3 Input subsystem

(i) mutually exclusive options (In the Elmira conflict, for instance, in Sect. 1.2.2, the company Uniroyal (UR) can only select at most one of its three options that it controls.),
(ii) option selections that must take place (In the Elmira conflict, UR is expected to do something, so it will choose one of its three mutually exclusive options.),
(iii) dependent options (A certain situation can only occur if something else is not done first. For instance, an aircraft company can only sell its airplanes if it has at least one firm order.), and
(iv) any impossible combination of option choices that could occur.

Through numerous applications of GMCR to actual conflicts, the authors have found that the first two situations are most common. Whatever the case, subsequent to considering the aforesaid methods, a final simplification technique is to coalesce states that are essentially the same. (In Table 1.2 for the Elmira conflict state s_9 in which UR closes down its factory contains 16 states which are essential the same since it does not matter what other options in the conflict model are selected if the plant is abandoned.)

Based on the aforesaid procedures, as well as other methods that may be developed in the future, the Input Subsystem generates the feasible states for the conflict being investigated. In practice, the authors have found that the number of mathematically possible states in the conflict are greatly reduced especially for larger disputes. (For the case of the Elmira conflict in Sect. 1.2.2, the number of states is reduced from 32 to only nine as displayed in Tables 1.2 and 1.3). Keep in mind, that the Input Subsystem asks for a minimum amount of information from the user: the DMs, their options and infeasible, necessary and common states (given as a coalesced single column of essentially one state). The feasible states in the dispute are automatically generated by the Input Subsystem.

Another component of the conflict that is automatically produced by the DSS are the unilateral moves in one step under the control of each DM. These moves can be shown using option form (see Table 1.7 for the Elmira dispute) or graphically (see Fig. 3.5 for the Elmira conflict). The user can require the system to display the directed graph of movement for each DM or as a single integrated graph as displayed in Fig. 3.5. In fact both the Input and Output Subsystems should be flexibly designed to take full advantage of the underlying graph theoretical design of GMCR to permit the user to visualize a rich range of situations that could occur.

With respect to the Input Subsystem, the system will initially assume that possible movements can take place in both directions. However, the user will be prompted to see if he or she would like to specify possible irreversible moves expressed in terms of option selections. (In the Elmira conflict an example of an irreversible move is where UR closes down its plant as displayed in Fig. 3.5). A user should also consider the type of situation that is being modeled. In the October 1962 Cuban Missile Crisis (Fraser and Hipel 1984, Hipel 2011), the US can threaten to bomb the Soviet missile sites that were being installed in Cuba. Of course, this threat can be removed. However, once the bombing takes place, the move from not bombing to bombing is irreversible. Users can interactively experiment with different combinations of irreversible moves to ascertain how that would affect the overall strategic findings.

The final step in the modeling stage is determining the relative preferences of each DM with respect to the feasible states in the conflict as indicated at the top of Fig. 1.1, left part of Fig. 10.2, and central section of Fig. 10.3. Three available techniques, in addition to others that may be designed in the future, for employment with option form are

(i) option prioritization,
(ii) option weighting, and
(iii) direct ranking on the computer screen.

The above methods assume that the preferences are transitive, which is usually the case for conflicts that are underway and fairly well understood. However, as mentioned in Sects. 1.2.2, 2.2.2, and 3.2.4, GMCR can handle both transitive and intransitive preferences. Accordingly, the DSS for GMCR should be designed to work for both types of preferences, even though transitive preferences are by far the most common. In the aforesaid three preference elicitation methods, option prioritization is an extremely user-friendly approach to capturing preference and was referred to by one consultant "as the best thing since sliced bread". Option prioritization was developed by Hipel et al. (1997) and Fang et al. (2003a) as a refinement of the preference tree approach first put forward by Fraser and Hipel (1988). In option prioritization, preference statements, expressed in terms of option selections for a given DM, are provided in hierarchical order from most to least important (see Table 1.4 for the Ontario Minister of Environment (MoE) for the Elmira groundwater pollution conflict). Assuming transitivity, a simple algorithm takes those preference statements and ranks the states from most to least preferred, where ties are allowed, for the DM being considered (see Tables 1.3 and 1.5 for the MoE in the Elmira controversy).

A "quick and dirty" way to order states for a specific DM is to provide weights for each option where a higher number for weight means more preferred when the option in a state is selected. After summing the weights for each state, the states can be ordered from most to least preferred for which a higher sum means more preferred. In a survey reported by Hipel et al. (2008), users of the GMCR methodology preferred option prioritization over option weighting. However, some users were attracted by the simplicity of option weighting even though it possesses no theoretical basis like option prioritization which was found to satisfy all the rules of first order logic.

A third way to order the feasible states according to preference for each state is to drag the states on the screen to place the states from most to least preferred for the DM under consideration where equal preference for a set of states can be indicated by a common color. For a small conflict, this approach can work well such as for the Sustainable Development Conflict in Sects. 3.1 and 3.2 which possesses only four feasible states. Additionally, after sorting states using option prioritization or option weighting for a particular DM, one can further adjust the ranking by simply dragging the states and perhaps also ordering a set of states that was previously considered to contain equally preferred states.

As indicated by the right branch in Fig. 10.3, one may employ another approach other than option form to define feasible states. For instance, in a brain storming session among officials for a company trying to figure out how to interact with its competitors to gain more market share, a facilitator may simply use circles to represent states with writing inside each circle to describe what this state means. Arrows could be drawn to indicate movement controlled by each DM among states and the circles could also be drawn from most to least preferred for each DM to reflect preference. Direct graphical input to a DSS, among other potential approaches, can be investigated in detail in future research.

A way to assist the option-form input for a graph model is to employ what is called a case-based reasoning expert system, which is a type of knowledge-based expert system developed within the information technology field of artificial or machine intelligence. In particular, conflicts arising within a particular application domain tend to have similar types of DMs, options, feasible states and relative preferences. For example, in a conflict over the pollution of groundwater by effluents from an industrial area, typical DMs include the government agency that is responsible for pollution control, polluter, local government which represents the interests of the local residents and environmental groups. Notice that the model for the Elmira groundwater contamination dispute developed in Sect. 1.2.2 contains the first three kinds of DMs along with options for each DM, some of which may reflect those which are modeled in a general situation. In addition, the general kinds of relative preferences for each DM may roughly correspond to what happens in this kind of water pollution controversy. The Input Subsystem could be equipped with a case-based reasoning component which utilizes stored previous conflict studies to suggest an initial generic model for a new user who wishes to investigate this kind of situation. Subsequently, this suggested model could be refined by a user so it fits the conflict under study before a reasonable model is sent to the Analysis Engine to calculate stability results that can be examined in the Output Subsystem. Ross et al. (2002) put forward the design of a system based upon studies complied over many years using GMCR.

10.2.3 Analysis Engine

Now that a conflict model is fully specified in terms of DMs, feasible states, movement and relative preference, one can have the DSS immediately carry out an exhaustive stability analysis as indicated in the lower portion of Fig. 1.1, central part of Fig. 10.2 and also in Fig. 10.4. A major benefit of having the GMCR methodology and an associated well-constructed DSS is that one can expeditiously determine the

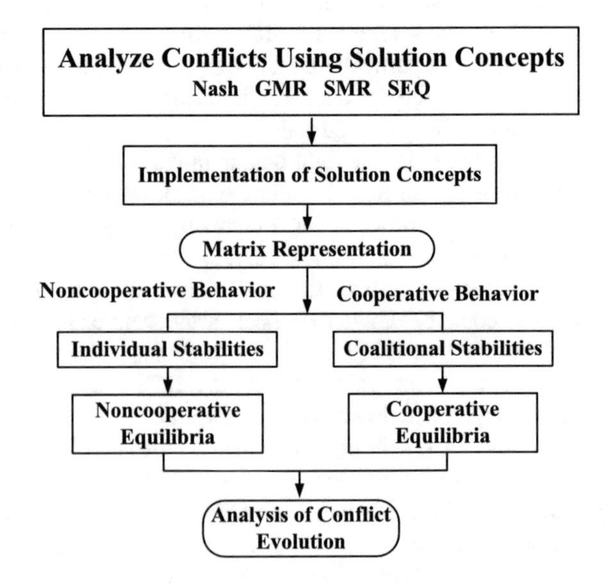

Fig. 10.4 Analysis engine of a graph model based DSS

strategic consequences and implications of the specific model being considered. If one is examining an ongoing dispute, for instance, one can determine in advance the strategic impact of what could occur if DMs behave according to the model. If the findings are not attractive one could examine other models and carry out additional stability analyses to see if better results can be achieved in advance of decisions actually being made. This means that one can readily experiment with a range of conflict models and seek answers to "what-if" questions in order to provide sage advice on how to enhance the decision-making process. This reflects what happens in engineering and the physical sciences where, for example, one will use a basketful of appropriate laws from physics, thermodynamics, fluid mechanics, and areas of science and engineering to design an aircraft, followed by testing a model in a wind tunnel prior to constructing a prototype and flying it. This kind of sensible approach to modeling and analysis is what GMCR permits users and DMs to do prior to actually making decisions which could end up being devastating to one or more parties if the strategic consequences of decisions are not fully tested beforehand. Moreover, this process should be carried out in an integrative and adaptive manner from a System of Systems viewpoint which reflects the value systems of stakeholders as explained in Sect. 2.4.

When using the metaphor of a chess player, as is done in Sect. 1.2.3, one wishes to think in terms of moves, and countermoves in order to determine the strategic result of selecting a certain strategy or combination of options. If one takes advantage of one or more unilateral moves, but could end up in a less preferred position with respect to the state being examined for stability, one is better off not to move. In combination with a rich range of solution concepts, the stability analysis stage produces the strategic findings of what different kinds of defined moves and countermoves can create. Additionally, one can use the DSS to precisely explain why a certain state is stable for a given DM in terms of moves and countermoves.

As stressed in Sects. 1.2.5 and 2.4 and elsewhere in the book, in a conflict study one wishes to first determine how well a DM can do on his or her own. Secondly, one wishes to ascertain if the DM can do even better via cooperating with others, which is often the case (In the Elmira groundwater contamination dispute the MoE and UR cooperated to reach the more desirable equilibrium for them as shown in Table 1.7 and explained in Sect. 1.2.3.). Notice that in the central part of Fig. 1.1, under the Analysis Engine in the central part of Fig. 10.2 and just below the middle of Fig. 10.4, both individual stability and coalitional stability analyses should always be executed in any conflict study.

Because people may behave differently in conflict situations, a range of solution concepts have been proposed. Four very useful concepts are Nash, GMR, SMR, and SEQ which are qualitatively explained in Sect. 1.2.3 as to how they work according to the four characteristics listed in the third to sixth columns from the left in Table 1.6. Precise mathematical definitions for different types of preference situations for individual stability and coalitional stability are given in Chaps. 4 to 7 and Chap. 8, respectively. At the top of Fig. 10.4, those four key solution concepts are listed but it is noted that other solution concepts such as limited-move stability (Zagare 1984, Kilgour 1985, Kilgour et al. 1987), Stackelberg equilibrium (von Stackelberg 1934), and metarational tree stability definitions (Zeng et al. 2006, 2007) could also be incorporated into an analysis engine. One should keep in mind that because Nash, GMR, SMR, and SEQ are defined in a way that only pairwise comparisons of preference of states are considered within the definitions, those solution concepts are valid for employment with both transitive and intransitive preferences. However, because of the backward induction process used with limited-move stability for logically calculating stability, limited-move stability is only valid for use with transitive preferences.

When determining stability for a specific state, the universal DSS exhaustively calculates stability for each DM for each solution concept embedded in the engine and for both individual and coalitional stability. If a state is stable for all DMs according to a specific solution concept then it forms an overall equilibrium or resolution which can be determined for both individual and coalitional stabilities.

A key breakthrough in the advancement of the GMCR methodology was the development of the matrix representation of the graph model by Xu et al. (2007a, b). As defined in depth in each chapter in this book from Chap. 3 onwards, key information such as state transitions and preferences are contained within matrices for each DM and matrix or algebraic stability calculations are defined for the four main solution concepts for both individual and coalitional stabilities. Within the Input Subsystem, the movement and preference matrices for each DM can be determined as input to the Analysis Engine as indicated by the larger arrow connecting the Input and Analysis stages in the bottom two boxes in the left of Fig. 10.2. The central part of Fig. 10.4 shows that a matrix representation is utilized to carry out all stability calculations. This matrix representation is central to the design of the Analysis Engine, as well as its connections to the Input and Output Subsystems. Furthermore, it permits this overall design of a universal DSS to be conveniently expanded in a "lego-like fashion" as new advances are added.

Based on the calculation of various kinds of individual and coalitional stability which can be utilized to explain the strategic impacts of decisions based on the conflict model being considered, one may wish to trace the evolution of a conflict from a specified state to another state of interest. For example, one may wish to determine whether or not a desirable state can be reached from the status quo state by following the research of Li et al. (2005a, b) (explained in Chap. 9). One can program the Analysis Engine to determine various situations for examining the evolution of a dispute such as for only allowing DMs to invoke unilateral improvements. Moreover, one could permit joint unilateral improvements which are entertained in coalition analysis.

The matrix representation design of the powerful Analysis Engine permits it to be readily expanded to handle preference uncertainty in combination with appropriate information received from the Input Subsystem. In this book, theoretical matrix representation results are provided for simple preferences (Chap. 4), unknown preferences (Chap. 5), degrees or strengths of preferences (Chap. 6) and hybrid preferences (unknown combined with degrees of preferences) (Chap. 7). Other kinds of uncertain preferences that can be entertained include fuzzy (Hipel et al. 2011, Bashar et al. 2012; 2016), grey (Kuang et al. 2015a, Zhao and Xu 2017) and probabilistic (Rego and dos Santos 2015). One could also program a DSS containing any meaningful

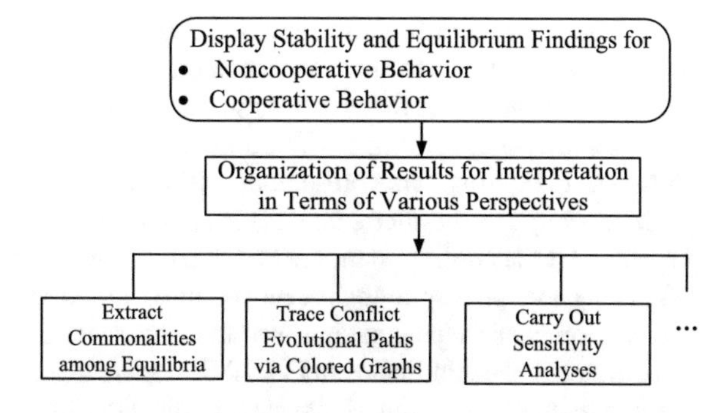

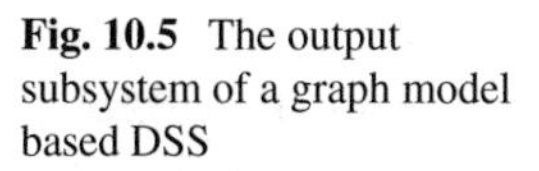

Fig. 10.5 The output subsystem of a graph model based DSS

combination of the aforementioned approaches to modeling uncertainty. Demand for a specific approach or combination of methods to model specific uncertain situations that arise in the real-world provide motivation as to where the DSS should be expanded. Whatever the case, the matrix-based solver is more efficient than logical-based definitions for determining stability because carrying out matrix calculations is highly developed and more efficient than following long logical loops for ascertaining stability, as is done in the logical approach. For an explanation of what the findings mean in actuality, it is more informative to use the logical interpretation, as is done for many of the findings in the Output Subsystem discussed next. Finally, all of the current and new approaches to conflict analysis discussed in Sect. 10.3 can be also incorporated into the universal GMCR DSS when utilizing a matrix based engine.

The overarching objective of the Output Subsystem is to furnish useful strategic advice in a user-friendly way for enhancing the decision-making process. The Output Subsystem takes findings produced by the Analysis Engine and translates them into tables, graphs, and typed explanation following an insightful presentation format that permits a user to fully understand the strategic consequences of the current model of reality which is under consideration. As indicated in Fig. 10.5, this can be done for both noncooperative and cooperative behavior so the user will be fully aware when cooperation can be advantageous and perhaps this can be achieved via clearer communication among participants.

10.2.4 Output Subsystem

The Analysis Engine calculates a large array of stability findings for both individual and coalition behavior. The Output Subsystem can prompt the user as to how the stability findings can be displayed both in writing and graphically. For example, a list could be available from which the user can decide how to display individual and equilibrium findings. As indicated in the bottom left box in Fig. 10.5, the user may wish to see the equilibria listed for which specific options are taken or not. For larger conflicts, many possible equilibria may be found and hence the user may want to categorize them according to commonalities. When modeling potential war between two groups of nations, for instances, one may wish to categorize equilibria according to which ones are peaceful and those which involve going to war. One may also wish to list equilibria according to the preferences of a specific DM from most to least preferred. Flexibility in the presentation of results is crucial in the Output Subsystem.

In the Output Subsystem, one might also want to find out why a state is stable for a specific DM according to moves and countermoves. This could be explained by the DSS using a graph displaying the moves and countermoves. This type of "chess-like" explanation constitutes a natural way to discuss stability in strategic situations.

As noted in the middle box in Fig. 10.5 and also in the bottom right box in Fig. 10.2, one can request the Output Subsystem to display a graph which traces the evolution of the conflict from say a status quo state to a desirable equilibrium to see if the equilibrium can be reached and how this can be achieved. Colors in the graphs can be used to indicate which DM, or set of DMs, controls the movement between states.

Based on his or her knowledge of the conflict, a user can request the DSS to execute appropriate kinds of sensitivity analyses as pointed out in the bottom right box in Figs. 10.2 and 10.5. As explained in Sect. 1.2.4, in a sensitivity analysis, one can determine how changes, usually small but not necessarily so, in the conflict model alter the strategic findings. For instance, a user may want to know if a meaningful change in a DM's preferences can cause a better resolution to take place that may be win/win for all parties in a conflict or only those who are members of a specific coalition. If the key equilibria remain unchanged then one can conclude that these equilibria are "robust" with respect to this preference change. The demand for a particular sensitivity analysis within the Output Subsystem will automatically be calculated by the Analysis Engine and the findings quickly displayed to the user.

10.3 Ongoing and Future Developments in the Graph Model Methodology

As mentioned at the start of this chapter and summarized in Table 10.1, significant progress has been archived in advancing graph model methodologies which can be implemented in practice using a DSS. In the next subsection, ongoing expansions to the graph model beyond those covered in the book are outlined and put into perspective. All of these advancements can be added like "lego" blocks to an existing DSS when it is designed according to the principles of a universal DSS described in Sect. 10.2. In Sect. 10.3.2, future developments of the graph model are explained within the context of a systems investigation. The encompassing systems view of the future of the graph model paradigm confirms that much remains to be accomplished for realistically tackling a truly broad and rich range of systems problems.

10.3.1 Ongoing Expansions of the Graph Model

Inspired by real-world situations in which complex conflicts must somehow be realistically resolved, a number of advances in the graph model approaches are ongoing and, in some cases, substantial progress has already been made. Four moving frontiers in extending the graph model are discussed in this subsection: different ways to handling uncertain preferences, psychological factors, power, and human behavior. One unique technique to account for unknown preferences is defined and explained in Chaps. 5, 8 and 9, as shown in Table 10.1. For the unknown preference, one assumes a minimum amount of information: one simply acknowledges that certain preference information is unknown which is then cleverly incorporated into individual (Chap. 5) and coalitional (Chap. 8) stability definitions. In Chap. 9, it is even embedded into the evolution of a conflict. Within Chap. 6, the degree of preference is formally defined. An example of three degrees of preference is when one allows for strength of preference in which, for instance, an environmental agency may state that it greatly prefers situations in which industry properly treats its wastes over states where it does not. For other pairwise comparison of states, one state may be simply more preferred, equally preferred or less preferred with respect to another, when the environmental agency compares them.

Table 10.2 lists three specific ways in which uncertainty in preferences can be captured. From the least to the most amount of information that is needed to calibrate

Table 10.2 Further extensions of uncertain preference in the graph model

Type	Explanation
Unknown preferences	See Chap. 5
Fuzzy preferences	Allowing preferences to be "fuzzy" is one way to model uncertain preference in which, for example, a DM may more or less prefer one state over another. Fuzzy preference information can be utilized to determine fuzzy stability for both noncooperative and cooperative behavior (Al-Mutairi et al. 2008a, b, Hipel et al. 2011, Bashar et al. 2012, 2014, 2015, 2016, 2018).
Grey preferences	Permitting preferences to be "grey" is a means to capture uncertainty in preferences consisting of either discrete real numbers, intervals of real numbers, or combinations of them. Grey preference information can be utilized to determine grey stability (Kuang et al. 2015a, b, c, Zhao and Xu 2017).
Probabilistic preferences	Allowing preferences to include probability is another way to entertain uncertainty which may involve risk in GMCR. Solution concepts containing probability concerns are available (Rego and dos Santos 2015, Silva et al. 2017).
Combination of preferences	Have any combination of the above preference structures, and other, such as the hybrid preferences (unknown plus degree of preferences) presented in Chap. 7.

Table 10.3 Psychological factors in the graph model

Type	Explanation
Attitudes	In a conflict, a given DM can have a positive, neutral or negative attitude towards him or herself and others. One would expect that when all DMs in a dispute possess positive attitudes towards themselves and others a better resolution for everyone concerned can be reached. In practice, a win/win resolution is a common occurrence (Inohara et al. 2007, Yousefi et al. 2010a, b, c, Bernath Walker and Hipel 2012, Bernath Walker et al. 2012, 2013).
Emotions	Emotions often arise in conflict situations and can propel a given dispute towards a range of final outcomes depending upon the mix of emotions that are present (Obeidi and Hipel 2005, Obeidi et al. 2009a, b). Some emotions can be expressed via strength of preference, which can be formerly modeled within the Graph Model methodology.
Hypergames	Hypergame analysis is a way for formally modeling misunderstandings and determining their strategic consequences. Because a conflict may contain misperceptions, based upon faulty interpretations of reality, hypergames possess multiple levels of perception (Takahashi et al. 1984, Hipel et al. 1988, Wang et al. 1988, 1989, Wang and Hipel 2009, Aljefri et al. 2014, 2016, 2018). The concept of a perceptual graph model can also capture misperceptions (Obeidi et al. 2009a, b).

them, the ordering is grey, fuzzy, and probabilistic preferences. In fact, unknown preferences require even less information than these three.

As can be seen in the bottom row of Table 10.2, one can also consider any combination of the aforementioned approach to preference uncertainty. For example, as indicated in the lower row of Table 10.1, one can consider hybrid preference in which unknown preference is combined with degree or strength of preference (three degrees) for employment in individual (Chap. 7) and coalition (Chap. 8) stability calculations. In reality, one could have any sensible combination of unknown, grey, fuzzy and probabilistic preference with degree of preference. Moreover, both logical and matrix formulations of these combinations could be studied. Table 10.1 provides a summary of the key ideas presented in earlier chapters in the book, as well as other concepts, such as attitude-base preferences (Inohara et al. 2007, Bernath Walker et al. 2012) and misperceived preferences (Wang et al. 1989), which are not addressed in this book. Significant gains have been made with continuing research being carried out with respect to attitudes, emotions and mispreceptions (hypergames) as explained in Table 10.3. Can positive, neutral or negative attitudes affect the evolution of a dispute? The answer is a resounding yes and this idea has been operationalized within the graph model. When combined with appropriate definitions for stability, one can calculate what equilibria could occur according to the different attitudes of the DMs involved in a dispute. In a given conflict, for instance, one may have DMs who are positive, neutral or negative for which possible resolutions can be determined. When all DMs have a positive attitude, as would be expected, often win/win resolution can be achieved. One way to capture emotions is to use strength of preference as

Table 10.4 Power in the graph model

Type	Explanation
Hierarchical structures	In a hierarchical conflict a common decision maker may be involved in two or more conflicts, such as when a federal government interacts separately with each province over an environmental issue (He et al. 2013, 2014, 2017a, b, 2018).
Power asymmetry	A DM in a conflict can influence the preferences of other DMs by taking advantage of additional options reflecting the particular DM's more powerful position (Yu et al. 2015).

explained in the right column of Table 10.3. However, other means are being pursued for operationalizing the concept of emotions. For instance, when one is extremely angry one may not think creatively under this self-induced mentality and thereby not envision obvious ways to resolve a dispute. This blocking of potential insightful solutions can be captured by what is called a perceptual graph (Obeidi et al. 2009a, b).

Within a hypergame situation mentioned in the bottom row of Table 10.3, one or more DMs in a dispute have a misperception about what is taking place. For example, in a military situation, one may falsely imagine that an opponent is stronger than expected when this is not the case. In fact, a DM can have a misunderstanding about preferences, options, states and even which DMs are participating in a conflict or any combination of these misperceptions. Hypergame modeling and analysis have been defined within the conflict analysis approach of Fraser and Hipel (1979, 1984) and applied to a range of different kinds of conflicts. However, work is ongoing to formalize the hypergame idea within the graph model paradigm (Aljefri et al. 2018) and provide matrix definitions so it can be implemented as a part of a DSS.

One may naturally think that a more powerful DM should tend to steer a conflict in a direction that will be more desirable for itself. Power, in reality, can be taken into account in the existing version of the graph model. For instance, a more powerful DM may have stronger options to invoke under conflict. In the Vietnam war, the United States had the capability to support the South Vietnamese regime and to attack its enemies whereas North Vietnam did not have the option of directly attacking the US mainland. Scenarios in which the US directly invaded North Vietnam or bombed military and industrial targets in the North would be greatly less preferred to the North Vietnamese to those situations in which the US did not directly attack the North.

Table 10.4 describes two advances that have been made in directly expanding the graph model to formally handle power: hierarchical conflicts and power asymmetry. Countries like France, China and Russia which hold significant amounts of power within their central governments can deal more effectively with different competing regions in their countries as well as with external disputes such as negotiations involving reductions in greenhouse gas releases to avert catastrophic climate change from taking place. As indicated in the lower row of Table 10.4, another way to formally handle power in the graph model is via power asymmetry.

Table 10.5 Further extensions of solution concepts in the graph model

Type	Explanation
Limited-move stability (LS)	Under LS at horizon h, a given DM can think h moves and countermoves into the future when assessing stability. If the DM is better off to stay at the initial state, this state is stable at horizon h (Zagare 1984, Kilgour 1985, Kilgour et al. 1987).
Non-myopic stability (NM)	Non-myopic stability is the limiting case of limited-move stability as the horizon h increases without bound (Brams and Wittman 1981, Kilgour 1984, Kilgour 1985, Kilgour et al. 1987).
Stackelberg equilibrium (ST)	For Stackelberg equilibrium, the DM who holds the more powerful position is called the leader, and the other, who reacts to the leader, is called the follower (von Stackelberg 1934).
Policy stability	A policy is defined as a plan of actions for a DM that specifies the DM's intended action starting at every possible state in a graph model of a conflict (Zeng et al. 2005, 2007).
Generalized metarational stability	A metarational tree is defined within GMCR, providing a general framework within which rational behavior among DMs can be described for any number of moves (Zeng et al. 2006, 2007).

As mentioned at the start of this chapter and indicated in Table 10.1, the four types of solution concepts for explaining human behavior under conflict for both competitive and more collaborative human behavior and discussed in this book are Nash, general metarational, symmetric metarational, and sequential stabilities. Table 10.5 furnishes a summary of additional solution concepts which have been defined and are being expanded for modeling potential human interactions in a conflict situation. As more kinds of human behavior in competitive situations are observed in the future, one can mathematically define them along with devising appropriate logical and matrix based algorithms to permit them to be incorporated into a flexible DSS.

Limited-move stability (LS) (Zagare 1984, Kilgour 1985, Kilgour et al. 1987), non-myopic stability (NM) (Brams and Wittman 1981, Kilgour 1984, 1985, Kilgour et al. 1987), and Stackelberg equilibrium (ST) (von Stackelberg 1934) all assume that the preferences of DMs are ordinal or transitive. On the other hand, Nash stability, GMR, SMR, and SEQ, as well as the two stabilities given at the bottom of Table 10.5 only require relative pairwise preference information, which could be intransitive or transitive.

10.3.2 Expansions of Systems Investigations in Conflict Resolution

A system of systems (SoS) engineering approach to sensibly addressing complex problems is described in Sect. 2.4.1. In general, a paradigm for categorizing approaches for solving complex SoS situations is according to forward investiga-

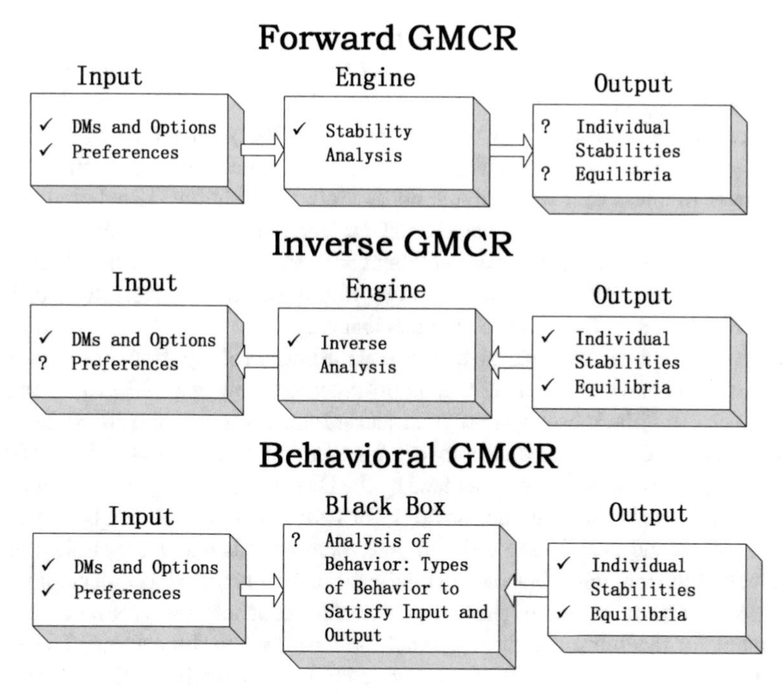

Fig. 10.6 The perspective of future development in GMCR

tions, inverse engineering and "black box system" determination. Figure 10.6 displays these three basic circumstances with respect to problem solving using the GMCR methodology.

As can be seen, the top diagram displays how a forward investigation is carried out. Given the input to a graph model in terms of DMs, options, preferences and other related information, a GMCR engine executes stability calculations which produce the output consisting of individual stability findings for each DM with respect to a range of solution concepts, equilibria, and insightful findings. This forward investigation, which is the approach utilized in the earlier chapters in this book, is incorporated into the three existing DSS, mentioned in Sect. 10.1.2. However, one may also wish to employ the other two kinds of approaches shown as the middle and lower graphs in Fig. 10.6.

In the inverse investigation, an analyst may wish to know what preferences are required by DMs in order to reach a desired equilibrium such as a win/win resolution. In a conflict having three DMs, for instance, one may wish to ascertain what preference structures by all three DMs may produce a desirable equilibrium based on specified types of solution concepts. In practice, an analyst may wish to not change the preferences of two out of three "reasonable" DMs but want to know the preferences needed by an "unreasonable" third DM to create a better final equilibrium. Moreover, for this DM, an analyst may "fix" part of the DM's preference and then want to determine how the remaining preferences over certain states need to change to create a final desirable resolution. The existing system called GMCR+ in Sect. 10.1.2 can handle certain aspects of inverse GMCR using an "exhaustive enumeration" approach to carry out an inverse GMCR study. This type of decision technology could be useful in what is called Third Party Intervention, in which a facilitator tries to influence disputants participating in a negotiation to change their preferences in a way that will bring about a win/win resolution (Hipel et al. 2016). Research is needed to develop an analytical approach to the inverse problem in order to design an efficient engine for employment in inverse GMCR. In related research, Garcia and Hipel (2017) developed an algorithm to determine preferences of DMs in a conflict based on the observed actions that they take.

In the third type of investigation depicted at the bottom of Fig. 10.6, one is addressing the problem in which the input and output are known and one wishes to determine the behavior or "black box system" that caused this. Wang et al. (2018) designed an analytical procedure based on a matrix formulation of GMCR (see Sect. 3.3) for determining the behavior being practiced by the DMs in a conflict given the input and output. A physical system model analogy for behavioral GMCR can be explained in terms of ground penetrating radar to determine what is under the surface of the land, such as the remains of an ancient civilization. A technician has radar equipment for shooting known beams of ground penetrating radar into the ground as well as equipment for receiving the signals which bounce back to the surface. Therefore, given the known radar input and the measured radar output what is the "black box" or system under the surface of the earth, as depicted in Fig. 10.7?

As can be appreciated from the discussion surrounding Fig. 10.6, a rich range of research remains to be done in all three types of investigations. Section 10.3.1 mentions ongoing and future studies for enhancing the forward investigation displayed at the top of Fig. 10.6. In fact, very little research has been carried out in the inverse investigation except for initial research by Sakakibara et al. (2002) and Kinsara et al. (2015a). Accordingly, all of the developments made so far within a forward investigation have to be "re-engineered" and expanded to handle the inverse investigation

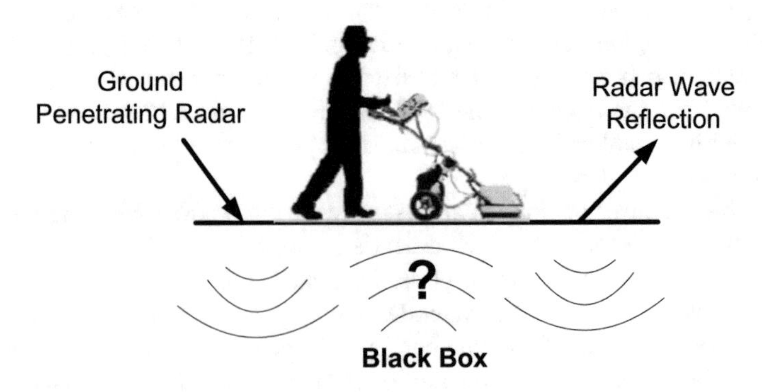

Fig. 10.7 The "black box system" of ground penetrating radar

and "black box" study displayed as the middle and lower graphs in Fig. 10.6, respectively. Some of the new challenging methodological developments will certainly be inspired by complex real-world problems that must be more realistically addressed.

10.4 Problems

10.4.1 Select three key references which define what constitutes a decision support system. Summarize these definitions and explain which particular definition you prefer and why.

10.4.2 Find a recent paper dealing with decision support systems which appeared in a water resources journal such as Water Resources Research, Journal of Water Resources Planning and Management, Journal of Hydrology, and Journal of the American Water Resources Association. In not more than two pages, outline the capabilities of the particular DSS developed in the paper with an emphasis on the features that you particularly like. Describe some drawbacks of the DSS and explain how you think they could be overcome.

10.4.3 The decision support system called GMCR II for implementing various capabilities of the Graph Model for Conflict Resolution (GMCR) methodology is mentioned in Sect. 10.1.2. By referring to the papers by Fang et al. (2003a, b) and Hipel et al. (1997), explain in more detail than the overview given in Sect. 10.1.2 the main capabilities of GMCR II. In which directions do you think it would be most worthwhile to expand GMCR II?

10.4.4 Table 1.8 in Sect. 1.2.5 in the book lists application areas to which the GMCR methodology has been applied. Some of these applications involved the employment of the decision support system GMCR II referred to in Sect. 10.1.2. Select a specific application in which GMCR II has been utilized which is of direct interest to you. Describe in not more than one page the strategic insights that the authors of the paper which you selected discovered when they used GMCR II to formally study their conflict.

10.4.5 GMCR+, which is mentioned in Sect. 10.1.2, is a decision support system for applying the GMCR methodology to actual disputes. In more detail than provided in Sect. 10.1.2, describe the main capabilities of GMCR+ by reading the papers of Kinsara et al. (2015b, 2018). How would you like to see GMCR+ expanded and enhanced?

10.4.6 A universal design of a decision support system for implementing the GMCR methodology in practice is put forward in Sect. 10.2. What are the particular characteristics of this basic design that you especially like and can you provide recommendations on how it can be improved? What particular programming language or set of programming languages would you recommend for coding this universal design?

10.4.7 The components of the Output Subsystem for a universal design of a GMCR decision support system are put forward in Sect. 10.2.4. In your opinion, what specific features are the most important to include in the Output Subsystem in order to enhance the strategic insights that you can gain in a formal conflict investigation?

10.4.8 In Sect. 10.3.1, ongoing expansions of the GMCR methodology are outlined. If you were to prioritize enhancements that should be incorporated into a decision support system, in what order would you rank them? Explain why you selected your particular ordering.

10.4.9 Table 10.2 and the first part of Sect. 10.3.2 provide an explanation as to how the GMCR methodology could be expanded from a systems thinking perspective. By referring to the published paper by Wang et al. (2018), explain how the behavioral GMCR procedure works. Why is this important in practical applications?

10.4.10 Using a physical systems explanation, explain how the forward, inverse and behavioral (or engine) procedure in Fig. 10.6 would work. You may wish to do this in terms of a specific physical systems problem.

10.4.11 Table 10.5 furnishes a list of further extensions of solution concepts within the GMCR paradigm and the associated implementation with a decision support system. Which particular expansion would be of particular importance to you? Explain why.

10.4.12 Beyond opportunities mentioned in Sect. 10.3, how would you like to see the GMCR methodology enhanced and expanded? Explain the reasons for your opinion.

References

Al-Mutairi, M. S., Hipel, K. W., & Kamel, M. S. (2008a). Fuzzy preferences in conflicts. *Journal of Systems Science and Systems Engineering, 17*(3), 257–276. https://doi.org/10.1007/s11518-008-5088-4.

Al-Mutairi, M. S., Hipel, K. W., & Kamel, M. S. (2008b). Trust and cooperation from a fuzzy perspective. *Mathematics and Computers in Simulation, 6*(2007), 430–446. https://doi.org/10.1016/j.matcom.2007.04.006.

Aljefri, Y. M., Fang, L., & Hipel, K. W. (2014). Modeling misperception of options and preferences in the graph model for conflict resolution. *Proceedings of the 2014 IEEE International Conference on Systems, Man and Cybernetics* (pp. 1573–1578). San Diego, CA. https://doi.org/10.1109/SMC.2014.6974140.

Aljefri, Y. M., Hipel, K. W., Fang, L., & Bashar, M. A. (2016). Misperception in nationalization of the Suez Canal. *Proceedings of the 2016 IEEE International Conference on Systems, Man and Cybernetics* (pp. 355–360). Budapest, Hungary. https://doi.org/10.1109/SMC.2016.7844266.

Aljefri, Y. M., Bashar, M. A., Fang, L., & Hipel, K. W. (2018). First-level hypergame for investigating misperception in conflicts. *IEEE Transactions on Systems, Man, and Cybernetics: Systems, PP*(99), 1–18. https://doi.org/10.1109/TSMC.2017.2690619.

Bashar, M., Kilgour, D. M., & Hipel, K. W. (2012). Fuzzy preferences in the graph model for conflict resolution. *IEEE Transactions on Fuzzy Systems, 20*(4), 760–770. https://doi.org/10.1109/tfuzz.2012.2183603.

Bashar, M. A., Kilgour, D. M., & Hipel, K. W. (2014). Fuzzy option prioritization for the graph model for conflict resolution. *Fuzzy Sets and Systems, 26*, 34–48. https://doi.org/10.1016/j.fss.2014.02.011.

Bashar, M. A., Hipel, K. W., Kilgour, D. M., & Obeidi, A. (2015). Coalition fuzzy stability analysis in the graph model for conflict resolution. *Journal of Intelligent and Fuzzy Systems, 29*(2), 593–607. https://doi.org/10.3233/ifs-141336.

Bashar, M. A., Obeidi, A., Kilgour, D. M., & Hipel, K. W. (2016). Modeling fuzzy and interval fuzzy preferences within a graph model framework. *IEEE Transactions on Fuzzy Systems, 24*(4), 765–778. https://doi.org/10.1109/tfuzz.2015.2446536.

Bashar, M. A., Hipel, K. W., Kilgour, D. M., & Obeidi, A. (2018). Interval fuzzy preferences in the graph model for conflict resolution. *Fuzzy Optimization and Decision Making*. https://doi.org/10.1007/s10700-017-9279-7.

Bernath Walker, S. & Hipel, K. W. (2012). Dominating attitudes in the graph model for conflict resolution. *Journal of Systems Science and Systems Engineering, 21*(3), 316–336. https://doi.org/10.1007/s11518-012-5198-x.

Bernath Walker, S., Hipel, K. W., & Inohara, T. (2012). Attitudes and preferences: Approaches to representing decision maker desires. *Applied Mathematics and Computation, 218*(12), 6637–6647. https://doi.org/10.1016/j.amc.2011.11.102.

Bernath Walker, S., Hipel, K. W., & Xu, H. (2013). A matrix representation of attitudes in conflicts. *IEEE Transactions on Systems, Man, and Cybernetics: Systems, 43*(6), 1328–1342. https://doi.org/10.1109/tsmc.2013.2260536.

Brams, S. J. & Wittman, D. (1981). Nonmyopic equilibria in 2×2 games. *Conflict Management and Peace Science, 6*(1), 39–62. https://doi.org/10.1177/073889428100600103.

DecisionMaker. (1996). *DecisionMaker: The Conflict Analysis Program*. Waterloo, Ontario, Canada: Open Options.

Fang, L., Hipel, K. W., & Kilgour, D. M. (1993). *Interactive decision making: The graph model for conflict resolution*. New York: Wiley. https://doi.org/10.2307/2583940.

Fang, L., Hipel, K. W., Kilgour, D. M., & Peng, X. (2003a). A decision support system for interactive decision making, part 1: Model formulation. *IEEE Transactions on Systems, Man and Cybernetics, Part C: Applications and Reviews, 33*(1), 42–55. https://doi.org/10.1109/tsmcc.2003.809361.

Fang, L., Hipel, K. W., Kilgour, D. M., & Peng, X. (2003b). A decision support system for interactive decision making, part 2: Analysis and output interpretation. *IEEE Transactions on Systems, Man and Cybernetics, Part C: Applications and Reviews, 33*(1), 56–66. https://doi.org/10.1109/tsmcc.2003.809360.

Fraser, N. M. & Hipel, K. W. (1979). Solving complex conflicts. *IEEE Transactions on Systems, Man, and Cybernetics, 9*(12), 805–816. https://doi.org/10.1109/tsmc.1979.4310131.

Fraser, N. M. & Hipel, K. W. (1984). *Conflict analysis: Models and resolutions*. North-Holland, New York. https://doi.org/10.2307/2582031.

Fraser, N. M. & Hipel, K. W. (1988). Decision support systems for conflict analysis. In M. Singh, D. Salassa & K. Hindi (Eds.), *Proceedings of the IMACS/IFOR First International Colloquium on Managerial Decision Support Systems and Knowledge-Based Systems* (pp. 13–21). Manchester, United Kingdom. Amsterdam, North-Holland.

Garcia, A. & Hipel, K. W. (2017). Inverse engineering preferences in simple games. *Applied Mathematics and Computation, 311*, 184–194. https://doi.org/10.1016/j.amc.2017.05.016.

He, S., Kilgour, D. M., Hipel, K. W., & Bashar, M. A. (2013). A basic hierarchical graph model for conflict resolution with application to water diversion conflicts in China. *INFOR, 51*(3), 103–119. https://doi.org/10.3138/infor.51.3.103.

He, S., Hipel, K. W., & Kilgour, D. M. (2014). Water diversion conflicts in China: A hierarchical perspective. *Water Resources Management, 28*(7), 1823–1837. https://doi.org/10.1007/s11269-014-0550-1.

He, S., Hipel, K. W., & Kilgour, D. M. (2017a). Analyzing market competition between Airbus and Boeing using a duo hierarchical graph model for conflict resolution. *Journal of Systems Science and Systems Engineering, 26*(6), 683–710. https://doi.org/10.1007/s11518-017-5351-7.

He, S., Kilgour, D. M., & Hipel, K. W. (2017b). A general hierarchical graph model for conflict resolution with application to greenhouse gas emission disputes between USA and China. *European Journal of Operational Research, 257*(3), 919–932. https://doi.org/10.1016/j.ejor.2016.08.014.

He, S., Kilgour, D. M., & Hipel, K. W. (2018). A basic hierarchical graph model for conflict resolution with weighted preference. *Journal of Environmental Informatics,31*(1), 15–29. https://doi.org/10.3808/jei.201700382.

Hipel, K. W. (2011). A systems engineering approach to conflict resolution in command and control. *The International C2 Journal, 5*(1), 1–56.

Hipel, K. W., Dagnino, A., & Fraser, N. M. (1988). A hypergame algorithm for modelling misperceptions in bargaining. *Journal of Environmental Management*, *27*, 131–152.

Hipel, K. W., Kilgour, D. M., Fang, L., & Peng, X. (1997). The decision support system GMCR in environmental conflict management. *Applied Mathematics and Computation, 83*(2-3), 117–152. https://doi.org/10.1016/s0096-3003(96)00170-1.

Hipel, K. W., Kilgour, D. M., Fang, L., & Peng, X. (2001). Strategic decision support for the services industry. *IEEE Transactions on Engineering Management, 48*(3), 358–369. https://doi.org/10.1109/17.946535.

Hipel, K. W., Fang, L., & Kilgour, D. M. (2008). Decision support systems in water resources and environmental management. *Journal of Hydrologic Engineering, 13*(9), 761–770. https://doi.org/10.1061/(asce)1084-0699(2008)13:9(761).

Hipel, K. W., Kilgour, D. M., & Bashar, M. A. (2011). Fuzzy preferences in multiple participant decision making. *Scientia Iranica, Transactions D: Computer Science and Engineering and Electrical Engineering, 18*(3(D1)), 627–638. https://doi.org/10.1016/j.scient.2011.04.016.

Hipel, K. W., Sakamoto, M., & Hagihara, Y. (2016). Third party intevention in conflict resolution: Dispute between Bangladesh and India over control of the Ganges River. In M. Hagihara & C. Asahi (Eds.), *Coping with regional vulberability: Preventing and mitigating damages from environmental disasters* (pp. 329–355). Tokyo, Japan: Springer. https://doi.org/10.1007/978-4-431-55169-0_17.

Howard, N. (1989). *CONAN 3.0, Metagame Analysis Program*. Nigel Howard Systems, Birmingham, U.K.

Inohara, T., Hipel, K. W., & Walker, S. (2007). Conflict analysis approaches for investigating attitudes and misperceptions in the War of 1812. *Journal of Systems Science and Systems Engineering, 16*(2), 181–201. https://doi.org/10.1007/s11518-007-5042-x.

Jelassi, M. T. & Foroughi, A. (1989). Negotiation support systems: An overview of design issues and existing software. *Decision Support System, 5*, 167–181. https://doi.org/10.1016/0167-9236(89)90005-5.

Jiang, J., Xu, H., & Jiang, Y. (2015). A decision support system for solving the conflict between human and environment. In K. W. Hipel, L. Fang, J. Cullmann, & M. Bristow (Eds.), *Conflict resolution in water resources and environmental management* (pp. 213–225). Cham, Switzerland: Springer. https://doi.org/10.1007/978-3-319-14215-9_12.

Kilgour, D. M. (1984). Equilibria for far-sighted players. *Theory and Decision, 16*(2), 135–157. https://doi.org/10.1007/bf00125875.

Kilgour, D. M. (1985). Anticipation and stability in two-person noncooperative games. In M. D. Ward & U. Luterbacher (Eds.), *Dynamic model of international conflict* (pp. 26–51). Boulder CO: Lynne Rienner Press.

Kilgour, D. M., Hipel, K. W., & Fang, L. (1987). The graph model for conflicts. *Automatica, 23*, 41–55. https://doi.org/10.1016/0005-1098(87)90117-8.

Kilgour, D. M., Fang, L., & Hipel, K. W. (1995). GMCR in negotiations. *Negotiation Journal, 11*(2), 151–156. https://doi.org/10.1111/j.1571-9979.1995.tb00056.x.

Kinsara, R. A., Kilgour, D. M., & Hipel, K. W. (2015a). Inverse approach to the graph model for conflict resolution. *IEEE Transactions on Systems, Man, and Cybernetics: Systems*, *45*(5), 734–742.

Kinsara, R. A., Petersons, O., Hipel, K. W., & Kilgour, D. M. (2015b). Advanced decision support for the graph model for conflict resolution. *Journal of Decision Systems, 24*(2), 117–145. https://doi.org/10.1080/12460125.2015.1046682.

Kinsara, R. A., Kilgour, D. M., & Hipel, K. W. (2018). Communication features in a DSS for conflict resolution based on the graph model. *International Journal of Information and Decision Sciences, 10*(1), 39–56. https://doi.org/10.1504/IJIDS.2018.090668.

Kuang, H., Bashar, M. A., Hipel, K. W., & Kilgour, D. M. (2015a). Grey-based preference in a graph model for conflict resolution with multiple decision makers. *IEEE Transactions on Systems, Man and Cybernetics: Systems, 45*(9), 1254–1267. https://doi.org/10.1109/tsmc.2014.2387096.

Kuang, H., Bashar, M. A., Kilgour, D. M., & Hipel, K. W. (2015b). Strategic analysis of a brownfield revitalization conflict using the grey-based graph model for conflict resolution. *EURO Journal on Decision Processes, 3*(3), 219–248. https://doi.org/10.1007/s40070-015-0042-4.

Kuang, H., Kilgour, D. M., & Hipel, K. W. (2015c). Grey-based PROMETHEE II with application to evaluation of source water protection strategies. *Information Sciences, 294*, 376–389. https://doi.org/10.1016/j.ins.2014.09.035.

Langlois, J. P. (1994). *Decision systems analysis version 3.3*. San Francisco, CA: San Fransisco State University.

Li, K. W., Hipel, K. W., Kilgour, D. M., & Noakes, D. J. (2005a). Integrating uncertain preferences into status quo analysis with application to an environmental conflict. *Group Decision and Negotiation, 14*(6), 461–479. https://doi.org/10.1007/s10726-005-9003-9.

Li, K. W., Kilgour, D. M., & Hipel, K. W. (2005b). Status quo analysis in the graph model for conflict resolution. *Journal of the Operational Research Society, 56*, 699–707. https://doi.org/10.1057/palgrave.jors.2601870.

Meister, D. B. & Fraser, N. M. (1994). Conflict analysis technologies for negotiation support. *Group Decision and Negotiation, 3*(3), 333–345. https://doi.org/10.1007/bf01384333.

Obeidi, A. & Hipel, K. W. (2005). Strategic and dilemma analyses of a water export conflict. *INFOR, 43*(3), 247–270. https://doi.org/10.1080/03155986.2005.11732727.

Obeidi, A., Kilgour, D. M., & Hipel, K. W. (2009a). Perceptual graph model systems. *Group Decision and Negotiation, 18*(3), 261–277. https://doi.org/10.1007/s10726-008-9154-6.

Obeidi, A., Kilgour, D. M., & Hipel, K. W. (2009b). Perceptual stability analysis of a graph model system. *IEEE Transactions on Systems, Man, and Cybernetics, Part A, Systems and Humans, 39*(5), 993–1006. https://doi.org/10.1109/tsmca.2009.2020686.

Rego, L. C. & dos Santos, A. M. (2015). Probabilistic preferences in the graph model for conflict resolution. *IEEE Transactions on Systems, Man, and Cybernetics: Systems, 45*(4), 595–608. https://doi.org/10.1109/tsmc.2014.2379626.

Ross, S., Fang, L., & Hipel, K. W. (2002). Case-based reasoning system for conflict resolution: Design and implementation. *Engineering Applications of Artificial Intelligence, 15*(3, 4), 369–383. https://doi.org/10.1016/s0952-1976(02)00065-9.

Sage, A. P. (1991). *Decision support systems engineering*. New York: Wiley.

Sakakibara, H., Okada, N., & Nakase, D. (2002). The application of robustness analysis to the conflict with incomplete information. *IEEE Transactions on Systems, Man, and Cybernetics, Part C: Applications and Reviews, 32*(1), 14–23. https://doi.org/10.1109/tsmcc.2002.1009122.

Silva, M. M., Kilgour, D. M., Hipel, K. W., & Costa, A. P. C. S. (2017). Probabilistic composition of preferences in the graph model with application to the New Recife project. *Journal of Legal Affairs and Dispute Resolution in Engineering and Construction, 9*(3), 05017004. https://doi.org/10.1061/(asce)la.1943-4170.0000235.

Takahashi, M. A., Fraser, N. M., & Hipel, K. W. (1984). A procedure for analyzing hypergames. *European Journal of Operational Research, 18*(1), 111–122. https://doi.org/10.1016/0377-2217(84)90268-6.

Thiessen, E. M. & Loucks, D. P. (1992). Computer-assisted negotiation of multiobjective water resources conflicts. *Water Resources Bulletin, 28*(1), 163–177. https://doi.org/10.1111/j.1752-1688.1992.tb03162.x.

Von Stackelberg, H. (1934). *Marktform und Gleichgewicht.* Vienna: Springer. https://doi.org/10.2307/2549070.

Wang, J., Hipel, K. W., Fang, L., Xu, H., & Kilgour, D. M. (2018). Behavioral analysis in the graph model for conflict resolution. *IEEE Transactions on Systems, Man, and Cybernetics: Systems, PP*(99), 1–13. https://doi.org/10.1109/tsmc.2017.2689004.

Wang, M., & Hipel, K. W. (2009). Misperceptions and hypergame models of conflict. In K. W. Hipel (Ed.), *Conflict resolution* (Vol. 2, pp. 167–188). Oxford, United Kingdom: Eolss Publishers.

Wang, M., Hipel, K. W., & Fraser, N. M. (1988). Modelling misperceptions in games. *Behavioural Science, 33*(3), 207–223. https://doi.org/10.1002/bs.3830330305.

Wang, M., Hipel, K. W., & Fraser, N. M. (1989). Solution concepts in hypergames. *Applied Mathematics and Computation, 34*(3), 147–171. https://doi.org/10.1016/0096-3003(89)90102-1.

Xu, H., Hipel, K. W., & Kilgour, D. M. (2007a). Matrix representation of conflicts with two decision makers. *Proceedings of the 2007 IEEE International Conference on Systems, Man, and Cybernetics* (pp. 1764–1769). Montreal, Canada. https://doi.org/10.1109/icsmc.2007.4413988.

Xu, H., Kilgour, D. M., & Hipel, K. W. (2007b). Matrix representation of solution concepts in graph models for two decision-makers with preference uncertainty. *Dynamics of Continuous, Discrete and Impulsive Systems, 14*(S1), 703–707.

Yousefi, S., Hipel, K. W., & Hegazy, T. (2010a). Attitude-based negotiation methodology for the management of construction disputes. *Journal of Management in Engineering, 26*, 114–122. https://doi.org/10.1061/(asce)me.1943-5479.0000013.

Yousefi, S., Hipel, K. W., & Hegazy, T. (2010b). Attitude-based strategic negotiation for conflict management in construction projects. *Project Management Journal, 41*(4), 99–107. https://doi.org/10.1002/pmj.20193.

Yousefi, S., Hipel, K. W., & Hegazy, T. (2010c). Considering attitudes in strategic negotiation over brownfield disputes. *ASCE Journal of Legal Affairs and Dispute Resolution in Engineering and Construction, 2*(4), 1–10. https://doi.org/10.1061/(asce)la.1943-4170.0000034.

Yu, J., Kilgour, D. M., Hipel, K. W., & Zhao, M. (2015). Power asymmetry in conflict resolution with application to a water pollution dispute in China. *Water Resources Research, 51*(10), 8627–8645. https://doi.org/10.1002/2014wr016257.

Zagare, F. C. (1984). Limited-move equilibria in games 2 × 2 games. *Theory and Decision, 16*, 1–19. https://doi.org/10.1007/BF00141672.

Zeng, D. -Z., Fang, L., Hipel, K. W., & Kilgour, D. M. (2005). Policy stable states in the graph model for conflict resolution. *Theory and Decision, 57*, 345–365. https://doi.org/10.1007/s11238-005-2459-x.

Zeng, D.-Z., Fang, L., Hipel, K. W., & Kilgour, D. M. (2006). Generalized metarationalities in the graph model for conflict resolution. *Discrete Applied Mathematics, 154*(16), 2430–2443. https://doi.org/10.1016/j.dam.2006.04.021.

Zeng, D.-Z., Fang, L., Hipel, K. W., & Kilgour, D. M. (2007). Policy equilibrium and generalized metarationalities for multiple decision-maker conflicts. *IEEE Transactions on Systems, Man, and Cybernetics, Part A, Systems and Humans, 37*(4), 456–463. https://doi.org/10.1109/tsmca.2007.897704.

Zhao, S., & Xu, H. (2017). Grey option prioritization for the graph model for conflict resolution. *Journal of Grey System, 29*(3), 14–25.

Index

H. Xu et al., *Conflict Resolution Using the Graph Model: Strategic Interactions in Competition and Cooperation*, Studies in Systems, Decision and Control 153,
https://doi.org/10.1007/978-3-319-77670-5

Printed in the United States
By Bookmasters